WITHDRA
ST

LASER VELOCIMETRY AND PARTICLE SIZING

Edited by

H. Doyle Thompson

and

Warren H. Stevenson

both of
School of Mechanical Engineering
Purdue University
West Lafayette, Indiana

⬤HEMISPHERE PUBLISHING CORPORATION

Washington   New York   London

UNIVERSITY COLLEGE
LIBRARY
CARDIFF

Proceedings of the Third International Workshop on Laser Velocimetry (LV-III) held at Purdue University, July 11–13, 1978, under the joint sponsorship of the Air Force Office of Scientific Research, the U.S. Army Air Mobility Research and the Development Laboratory, the Department of Energy, and the Office of Naval Research.

LASER VELOCIMETRY AND PARTICLE SIZING

Copyright © 1979 by Hemisphere Publishing Corporation. All rights reserved.
Printed in the United States of America. No part of this pubication may
be reproduced, stored in a retrieval system, or transmitted, in any form
or by any means, electronic, mechanical, photocopying, recording, or
otherwise, without the prior written permission of the publisher.

1 2 3 4 5 6 7 8 9 0   B R B R   7 8 3 2 1 0 9

Library of Congress Cataloging in Publication Data

International Workshop on Laser Velocimetry,
   3d, Purdue University, 1978.
   Laser velocimetry and particle sizing.

   Proceedings of the third International Workshop on
Laser Velocimetry, held at Purdue University, July
11–13, 1978 under the sponsorship of the Air Force
Office of Scientific Research, and others.
   Bibliography: p.
   Includes index.
   1. Laser Doppler velocimeter—Congresses.
I. Thompson, Howard Doyle, date   II. Stevenson,
Warren H.   III. United States. Air Force. Office
of Scientific Research.   IV. Title
TA357.I595   1978   620'.43   79-59
ISBN 0-89116-150-3

Contents

SESSION IXA **GENERAL APPLICATIONS II**
Chairman: H. D. Thompson

SESSION IXB **PARTICLE DIAGNOSTICS II**
Chairman: W. H. Stevenson

SESSION X **WILD CARD SESSION**
Chairman: F. Durst

SESSION XI **PANEL DISCUSSION**
Moderator: H. D. Thompson   **539**

Preface

The use of the laser Doppler velocimeter (LDV) for flow measurement was first demonstrated in 1964. Since that time, it has evolved from a novel laboratory instrument into a practical tool for research and industrial use. The LDV's obvious advantage is its ability to make measurements without perturbing the flow under conditions where other instruments provide questionable results or cannot be used at all.

Although the basic physical principles upon which the LDV operates are easy to grasp, it becomes a practical instrument only when numerous challenging problems posed by requirements on the optical system and the electronic signal processing are understood and overcome. In addition proper interpretation of the data is sometimes difficult. For these reasons much of the early reseach in the LDV field has been directed toward the instrument itself and its behavior in various types of flows. Many of the major advances in the field have been reported at meetings held in the United States and Europe. The first of these was the First International Workshop on Laser Velocimetry (LV-I) held at Purdue University, March 9-10, 1972.* That meeting brought together active workers in this relatively new field from the United States and Western Europe for two days of very stimulating discussion. The success of LV-I led to the Second International Workshop on Laser Velocimetry (LV-II) held at Purdue University, March 27-29, 1974.† Both LV-I and LV-II were sponsored by the Purdue

*Stevenson, W. H., and Thompson, H. D., The Use of the Laser Doppler Velocimeter for Flow Measurements, *Proceedings of a Workshop Co-sponsored by Project SQUID (ONR) and the U.S. Army Missile Command*, March 9-10, 1972. (Supplies of these proceedings from the original printing are exhausted. However, copies may be obtained as AD-753243.)

†Thompson, H. D., and Stevenson, W. H., *Proceedings of the Second International Workshop on Laser Velocimetry*, vols. I and II, March 27-29, 1974. (Copies of these two volumes are available for $25 per set, postpaid; make checks payable to Purdue University, while the supply lasts. For overseas orders, other than Canada, add $5 for additional air postage and handling. Address request to: Professor H. D. Thompson, School of Mechanical Engineering, Purdue University, West Lafayette, IN 47907.)

University School of Mechanical Engineering, the U.S. Army Missile Command, and Project SQUID (Office of Naval Research).

Several meetings devoted to laser velocimetry were held subsequent to LV-II at various locations, attesting to the continued rapid development of the field. In all of these, attention was primarily focused on instrumentation advances rather than flow measurement. Thus, the need was felt for a meeting emphasizing measurements, particularly in hostile environments such as highly turbulent and reacting flows. Therefore, the Third International Workshop on Laser Velocimetry (LV-III) was held at Purdue University, July 11–13, 1978, under the joint sponsorship of the Air Force Office of Scientific Research, the U.S. Army Air Mobility Research and Development Laboratory, the Department of Energy, and the Office of Naval Research. This volume represents the proceedings of that meeting.

Although flow measurements were emphasized at LV-III, it is obvious that instrument development and data analysis are still receiving attention. Particle sizing with LDV or related laser instruments is a very active area that was the subject of many of the presentations. The total program included over forty formal presentations and a number of informal presentations on work in progress. Technical sessions included:

- Developments in Instrumentation
- Data Analysis (*2 sessions*)
- Combustion Measurements
- Measurements in Turbulent Flows
- Measurements in Internal Combustion Engines
- General Applications (*2 sessions*)
- Particle Diagnostics (*2 sessions*)
- Wild Card
- Summary Panel Discussion

Major conclusions of the workshop were:

- The laser velocimeter is finding wide acceptance as a measuring instrument in difficult environments.
- In difficult measuring situations the laser velocimeter needs to be carefully designed for the specific application.
- Additional work remains to be done on data collection and analysis in highly turbulent and reacting flows. Significant questions of data biasing in turbulent flows remain.
- The problem of in situ particle sizing using the fringe (laser velocimeter) method and other optical techniques is being studied by a number of groups. A great deal of progress has been made, but many questions remain.
- The two-spot or dual-focus velocimeter is receiving increasing attention and is advantageous in some applications. Questions remain about its performance in highly turbulent flows.

A few comments on the editing of the proceedings are in order. Our previous experience with trying to record and publish the discussion following each presentation led us to avoid this desirable but very difficult task. Therefore, except for the wild card session and the summary panel discussion, informal comments were not recorded and are not included. We did record these two sessions and have edited the discussion in a way that we hope conveys as accurately as possible the important points made by the speakers. In most cases the technical presentations are reproduced exactly as they were submitted by the authors except for minor typing corrections and some editing needed to produce a relatively uniform format. We did find it necessary to edit some papers rather heavily and hope this was done without introducing errors, since time did not allow us to return them to the authors for review. One paper, "Laser Doppler Measurements in Complex Flow Situations" by Dr. Franz Durst, could not be completed in time for publication due to unexpected commitments. Dr. Durst asked us to convey his regrets to the readers.

We believe that this collection of information on laser velocimetry and particle sizing will be of considerable value to those already working in these fields as well as those contemplating the application of these powerful measuring techniques to new problems. We thank the authors for their contributions and cooperation. Sincere thanks are also due the sponsors of LV-III for their support and encouragement which made all of this possible.

Warren H. Stevenson
H. Doyle Thompson

A Historical Review of Laser Velocimetry

WARREN H. STEVENSON
Applied Optics Laboratory
School of Mechanical Engineering
Purdue University
West Lafayette, IN 47907

ABSTRACT

The laser Doppler velocimeter has been the subject of intense research for over a decade. This instrument has now reached a rather mature level of development and is widely used as a research tool in fluid mechanics. It is therefore appropriate to review the development of the field and draw attention to those areas where further research is needed. In this paper major advances which have occurred are discussed to put past accomplishments in perspective and serve as an introduction to the succeeding papers in this volume.

INTRODUCTION

The Laser Doppler Velocimeter (LDV) made its debut in 1964 with the appearance of the classic paper by Yeh and Cummins [1]. They presented the basic theory for a reference beam LDV and included excellent data obtained for the laminar velocity profile in a circular tube. The potential of such a non-interfering technique was immediately recognized by those involved in supersonic flow research and a significant share of the early activity was in this area. Other elements of the fluid mechanics community entered the game as instrument capabilities increased and commercial devices became available. In early investigations it was often a major accomplishment to obtain a reasonably good Doppler signal from a simple flow and oscilloscope traces of the signal were prominently displayed in publications. Now the problem is often what to do with the massive amounts of data which can be generated in a short time.

It is evident that the level of development of laser velocimetry has followed the familiar S-shaped curve depicted in Fig. 1. We are now somewhere near the point at which the instrument is "fully developed" and, barring major conceptual advances, only minor improvements in capability can be expected. As the papers at this meeting clearly indicate, the focus is now on applications - many of them in hostile environments. Each application poses unique problems, but techniques used in one investigation can often be employed to advantage in new situations. Therefore, a collection of papers such as this serves a valuable function.

In order to put all of this in perspective, this paper will review the history of laser velocimetry with emphasis on specific advances in optical design, signal processing, and data analysis which have led to the current state-of-the-art. It is impossible to adequately cover the field in this short article and an exhaustive review is not intended. Literature references cited were chosen to illustrate particular points, but in most cases other works of equal merit

1

could have been selected. Those interested in a more detailed survey should
consult a recent review article [2] and the books which are now available [3,4].
The proceedings of previous meetings also contain valuable information [5,6,7,8,
9,10].

OPTICAL SYSTEM DEVELOPMENT

The heart of an LDV is the optical system and, in retrospect, developments
in this area were extremely slow. Yeh and Cummins did their pioneering work
with a reference beam system which was basically a variant of the familiar Mach-
Zehnder interferometer (Fig. 2). Such a configuration is unsuitable for general
use due to the difficulty of maintaining the critical alignment of the many
independent optical elements. Goldstein and Kreid [11] designed an improved
reference beam system. However, it was not until about 1970 that the seemingly
obvious differential Doppler arrangement consisting of a beam divider to generate
two parallel beams and a single lens to focus them at a point was proposed by
several investigators. This design, based on the fringe model of Rudd [12], is
now in almost universal use. A single prism version due to Durst is shown in
Fig. 3.

The next major advance in optical system design came with recognition that
the Gaussian nature of the laser beam must be taken into consideration. When
such a Gaussian beam is focused by a lens, the minimum beam diameter (the "waist")
is found at the back focal plane of the lens only if a waist also exists at the
front focal plane. Hanson [13] showed that if this condition is not met the
wavefront curvature outside the waist region leads to a variation in fringe
spacing, and thus Doppler frequency, along the probe volume as illustrated in
Fig. 4. Durst and Stevenson carried out a general analysis of this problem and
verified their results experimentally [14,15]. Correction lenses placed between
the laser and beam divider can be used to obtain the proper waist position. It
has been found that this not only eliminates the undesirable variation in fringe
spacing, but also considerably improves the signal-to-noise ratio, since the
minimum beam diameter and maximum intensity are now centered in the probe volume.
Unfortunately some of the simplicity afforded by a single prism beam divider is
necessarily lost when Gaussian beam correction is employed.

Often the number of fringes in the probe volume is much greater than that
needed by current signal processors to obtain an accurate measurement from
individual burst signals. For this reason it is becoming more common to reduce
the beam diameter in the probe volume with a corresponding increase in the peak
intensity. This can be done by increasing the beam diameter before the beam
divider or by inserting additional lens elements between the beam divider and
focusing lens. The latter approach is shown in Fig. 5. With proper adjustment
these lenses can also take care of the Gaussian beam correction. Since reduction
by a factor of two in the focused beam diameter leads to a factor of four increase
in peak intensity, the signal level from a given size laser can be improved
significantly.

It is probably unnecessary to emphasize the importance of using aberration
free lenses in both the transmitting and receiving optics. The use of precision
pinholes of the proper size can also yield substantial improvements in signal
quality. Suitable anti-reflection coatings on optical surfaces must obviously
be employed and polarization effects in such coatings accounted for. In the
case of backscatter systems proper optical design is especially important,
since the signal is inherently weak. Complete separation of the transmitting
and receiving paths as illustrated in Fig. 6 is desirable in backscatter systems
to minimize noise arising from reflections in the optics.

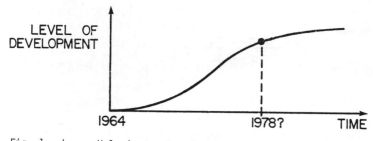

Fig. 1. Laser Velocimeter Development

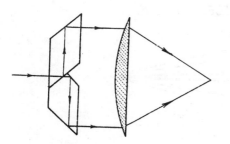

Fig. 2. Yeh and Cummins System (1964)

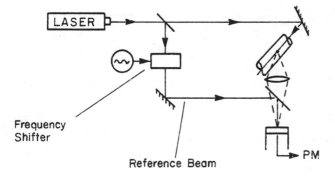

Fig. 3. Single Prism Beam Divider

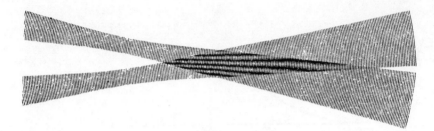

Fig. 4. Finge Spacing Variation in an Improperly Aligned System

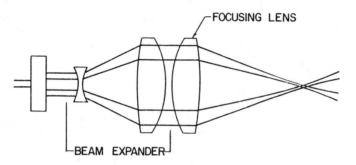

Fig. 5.   Probe Volume Size Reduction by Means of
          a Beam Expander

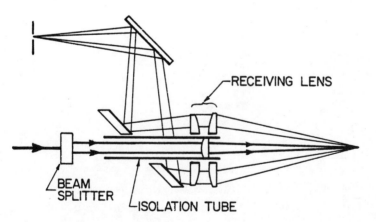

Fig. 6.   Backscatter Optics

LDV optical system design is based on only a few rather straightforward principles. Nevertheless it is only after more than a decade that the factors discussed here have all been generally recognized and incorporated in new designs. This slow evolution probably occurred because it is rather easy to get what appears to be a good Doppler signal even with poor optics. However, proper attention to optical design can markedly reduce laser power requirements and improve overall LDV system performance. Therefore it is important to be aware of the basic design principles whether one is designing his own instrument or purchasing a commercial device.

SIGNAL PROCESSOR DEVELOPMENT

The nature of the signal obtained at the photodetector output depends strongly on particle seeding density, particle size, laser power, and background noise due to stray reflections. Figure 7 illustrates the four general types of signal output commonly encountered. A variety of processing devices have been developed to meet the requirements of different applications.

Early LDV signal processing was done with available spectrum analyzers. These are still useful in some cases, but most work is done with other devices specifically developed for handling the signals encountered in LDV measurements. Three primary instruments have evolved for general application: the frequency tracker, the burst processor (counter), and the photon correlator. The history of these devices is well known in the field and only a few comments will be made.

Frequency trackers provide a means of generating an output voltage proportional to the instantaneous velocity. They work best with a nearly continuous signal as provided by fairly heavy particle seeding. Contemporary LDV trackers employ dropout controls to prevent instabilities during periods when the signal is absent. However, if the Doppler frequency changes too rapidly from one burst to the next it is possible to lose track. Therefore trackers tend to be best suited to heavily seeded flows of modest turbulence intensity. Their upper frequency limit of about 50 MHz is below that of currently available counters.

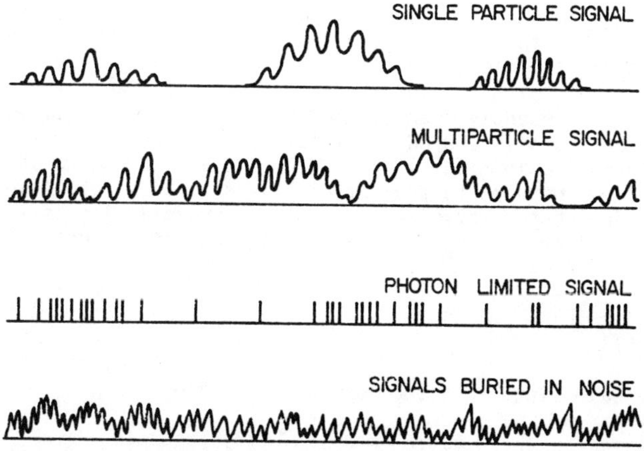

SINGLE PARTICLE SIGNAL

MULTIPARTICLE SIGNAL

PHOTON LIMITED SIGNAL

SIGNALS BURIED IN NOISE

Fig. 7.   Doppler Signals Observed Under
          Various Conditions

Counter type (burst) processors were fathered by laser velocimetry and were first developed for studies of sparsely seeded high velocity gas flows [16]. They operate by producing a square pulse initiated by the first zero crossing in the Doppler signal after a preset amplitude threshold has been crossed and terminated after a selected number of zero crossings has occurred. Thus the pulse length is equal in time to a certain number of Doppler cycles - eight is a typical number. This time is often compared to that for a different number of zero crossings as a means of validating "good" signals. Timing is normally done by means of a high frequency digital clock, although time to amplitude converters have also been used [17].

Early burst processers used a 4/8 comparison. At a later stage it was believed that there was some statistical advantage to comparing an even and odd number of cycles such as 4/9 or 5/8. However, this complicates the circuitry and has been found to be unnecessary. Newer units allow a selectable number of Doppler cycles to be counted and also provide a total burst mode which makes use of the entire signal and employs end of burst detection rather than a ratio comparison for signal validation. Currently available commercial counters have a 200 MHz Doppler frequency limit due to inherent limitations in available circuit components. While this may be pushed somewhat higher with later generation components it is unlikely that such capability would be needed except in rare instances. A more desirable result of component advances would be a reduction in processor cost.

Photon correlation was first introduced as a means of obtaining LDV measurements at extremely low scattered light levels in 1972 [18]. The low Doppler frequency capability of correlators and the long integration time needed to build up the correlation have generally restricted their use to special applications where processing methods fail due to the weak signals. It would seem that photon correlation could also be used in cases where very poor signal to noise conditions exist, but this does not seem to have been demonstrated in engineering applications. Improvements in the frequency capabilities and processing time of photon correlators are on the horizon and their field of application may widen somewhat as a result. They also find application in dual-focus velocimetry where correlation is a fundamental requirement of the data analysis.

A new processing approach which offers some attractive features is direct recording of the Doppler signal with a high speed transient recorder followed by digital computer processing of the stored signal [19]. This allows a variety of efficient digital processing schemes to be employed and signals of very poor quality ($S/N \approx 0.01$) can be dealt with. A comprehensive analytical and experimental study of this scheme has recently been carried out [20]. The digitizing rate of current transient recorders is 100 MHz and therefore the Doppler frequency must be much less than this in general to obtain adequate sampling. Also the data rate is extremely low compared to that attainable with burst processors and, as in the case of current photon correlation, real time information is lost. Nevertheless the flexibility offered by this approach will undoubtedly motivate future development.

DEVELOPMENTS IN DATA ANALYSIS

LDV data analysis may be divided into two general categories:

1. Correction for biases in the measured mean velocity resulting from particle seeding effects, probe volume effects, and processor effects.

2. Correction for inherent errors in measured turbulence parameters and the
    turbulence power spectrum due to processor characteristics and the
    nature of the Doppler signal.

It has always been recognized that large particles will not follow the flow
as well as small particles and therefore a bias error in the measured mean velocity
can exist due to particle lag. If a broad particle size distribution is present
there will also be a spread in the measured velocity distribution at a point.
Improvements in particle seeders and rejection of the high amplitude signals
from large particles can eliminate the particle lag problem in most applications.

A second bias due to particles is the individual realization or velocity
bias first treated by McLaughlin and Tiederman in 1973 [21]. They noted that
more high velocity than low velocity particles per unit time would pass through
the measuring volume, leading to an erroneously high mean velocity if a simple
average of the data was taken. Proper correction for this effect was a source
of controversy for some time, but it is now accepted that time averaging rather
than particle averaging of the data will eliminate the error. This requires that
time as well as velocity data be taken.

A third bias of some importance is the incomplete signal bias which may
arise in highly turbulent flows when particle paths through the probe volume
vary significantly from the direction perpendicular to the fringes. This leads
to a reduction in the number of fringes crossed and can cause rejection of these
signals by a processor requiring a fixed number of zero crossings for signal
validation (see Fig. 8). Thus the probe volume has a variable "polar response"
[22] and the measured mean velocity would be high for that velocity component.
Various methods for minimizing this bias error exist including the use of fewer
fringes or the total burst mode of processing.

· A fourth bias which is also due to both particle and processor effects
occurs when particle acceleration in the probe volume changes the Doppler period
significantly during a single burst. Processors using a ratio comparison will
then reject a certain fraction of these signals. A number of other biases have
been identified [23]. However, the ones mentioned are the most significant.
Fortunately these can usually be corrected for easily, at least with burst
processor measurements. When trackers are used the problem is not as clear cut,
but the general considerations are similar.

The analysis of turbulence measurements, particularly the power spectrum,
presents a somewhat more difficult task and the treatment of tracker data is much
different from that used for burst processor data. In the case of trackers a
high seeding density is normally employed and phase fluctuations occur in the
Doppler signal due to population changes and both mean and turbulent velocity
gradients in the probe volume. This leads to errors in the high frequency end
of the turbulence spectrum. These spectrum broadening effects were an important
and often confusing issue in early studies. However, they are now well understood. George has outlined this problem in detail and described correction
procedures [24].

Burst processors provide a discontinuous data output and the problem in this
case is one of constructing a spectrum from a randomly sampled velocity signal.
(Most burst processors can give a "continuous" output by holding the last
frequency measurement until a new particle signal is validated. However, this
is not a true picture of the velocity-time behavior.) If the particle seeding
density was relatively high (but still less than one particle in the probe
volume) one could in principle sample at a rate high enough to permit a direct
spectral analysis. However, this generally proves to be impossible in practice

[25]. Thus the power spectrum must be obtained by correlation methods from data samples taken at a rate well below the highest turbulence frequency of interest. A number of papers presented at the 1974 Purdue meeting address this problem and suitable analytical approaches have been developed, although a rather large investment in computer time is required.

SYSTEM COMPONENT ADVANCES

Here we are concerned primarily with components which are independent of the LDV field itself, namely lasers, detectors, and frequency shifting devices. Obviously advances in these important system elements might have significant impact. However, it can be generally observed that only modest improvements have occurred over the past ten years which can be utilized to advantage. Helium-neon and argon lasers continue to be the normal sources used. Aside from some improvement in argon laser mode structure and stability no major improvements in this area have taken place and none can be foreseen. The situation is much the same for detectors. Some advances in photomultiplier sensitivity and response have been made. Avalanche photodiodes have come on the scene and are sometimes used, but their lower sensitivity and higher noise limit applicability. Photo-diodes are less subject to overloading and can be employed in cases where signals from occasional large particles saturate photomultipliers, however.

Frequency shifting of one or both beams is a requirement in many LDV systems and two approaches have been employed. One is the use of acousto-optic cells and the other is the rotating diffraction grating. The major improvements in

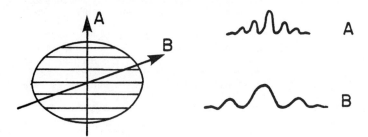

Fig. 8.   Effect of Particle Path on Number of
          Doppler Cycles

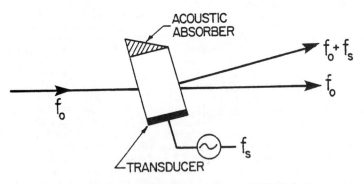

Fig. 9.   Angular Deviation of Frequency Shifted Beam
          Produced by a Bragg Cell

acousto-optic cells have been a reduction in cost and an increase in efficiency. These cells are normally used in the Bragg mode and an angular deviation of the beam dependent on drive frequency is always present (see Fig. 9). This can be corrected for by various means, but all introduce complexities in the optical system, especially if a variable frequency shift is desired. A shift range of about 10 MHz centered near 40 MHz is the typical limit for efficient operation, which means that two units must often be employed to give the desired performance. Obviously a single sideband modulator which provided a wide frequency shift range without beam deflection would be desirable. However, prospects for the development of such a device appear remote.

Frequency shifting by means of a rotating diffraction grating was first proposed in 1970 [26]. Gratings with extremely good diffraction efficiencies and frequency shift capability up to 10 MHz have been developed [27]. A typical system is shown in Fig. 10. Such gratings have been used successfully in a number of studies, but for some reason their adoption by commercial firms has been very limited. It would seem that this is a fruitful area for further development.

ALTERNATIVE APPROACHES

The fringe mode laser Doppler system has dominated the field of laser velocimetry, but there are other means of measurement. The most important of these is the dual-focus velocimeter which, although proposed many years ago, has seen extensive development only recently. Much of this work has been at the DFVLR in Germany [28,29]. This instrument is based on measuring the time required for particles to traverse the space between two sharply focused beams. Some data obtained in turbomachinery studies has been reported [30], but a clear picture of the capabilities and limitations of the method is not yet available. Further investigations are required, particularly in highly turbulent flows.

A less well known approach relies on the passage of particles through a single focused Gaussian beam. This was first proposed by Rudd [31] and has been recently expanded by Hirleman [32]. The basic concept is that the time between the $1/e^2$ points of the detected signal is dependent only on velocity and independent of particle path or particle size.

Hirleman has also shown that with two beams it is possible to get two velocity components and the particle size. This is a surprising result. There are some obvious difficulties with signal processing, but this technique certainly merits further study.

Another method which deserves mention is one based on correlation of the arrival rate of particles passing through a single beam [33]. This does not provide explicit velocity information, but can yield the velocity correlation function in a turbulent flow.

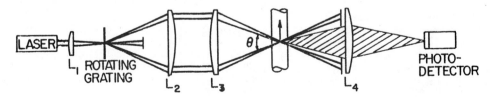

Fig. 10.   Velocimeter Employing a Rotating Grating to Obtain a Frequency Shift

FUTURE DIRECTIONS

It is obvious that the LDV technique has reached a high level of maturity
and further advances in the instrument itself will probably be minor. One
possible exception to this conclusion is in the area of true three-dimensional
measurements. There is a definite need for instrument development here. The
problems of obtaining all three velocity components simultaneously from a
single particle are formidable, however. Some attempts have been made, but no
optical design suitable for general purpose use has been developed and tested.
The rotating diffraction grating may offer some hope as the basis for a relatively simple 3-D instrument [27].

Some improvements in signal processing and data analysis may take place, but
these problems seem to be well understood and current methods are adequate. In
most cases, therefore, instrument performance is not a significant factor in LDV
studies and attention has now shifted toward new applications, as clearly demonstrated by the papers presented at this meeting. Each application presents
unique problems to be overcome. Questions which have not been answered are
therefore primarily in relation to specific areas of application.

Combustion is an important current area of LDV research. Here the problems
of optical access and refractive index fluctuations are often severe. Measurements have been made successfully in an 0.12 meter diameter research gas turbine
combustor [34], but the technique essentially failed when applied to a five meter
industrial furnace [35]. Therefore the actual limit for making successful
measurements in combustion systems is still in question. Basic research needs
to be done in order to quantify the problem in general terms so that one can
predict the probability of success in a given situation.

Another active area of study is particle sizing as evidenced by the many
papers on this topic presented at this meeting. Advancement has been rapid over
the past few years, but there is still work to be done. Signal processing and
data analysis in particular need considerable attention.

Two-phase flows present unique problems, particularly when large natural
particles are present. Only a limited amount of research has been done in this
field, but there seems to be considerable potential for LDV use since other
methods often fail completely. The use of fluorescent particles may offer some
hope here [36,37].

Undoubtedly many other points of this nature could be raised. Laser
velocimetry has proven to be a much more complex field than originally expected
and we now have answers to questions which had not even been asked a few years
ago. The papers presented at this meeting provide a good definition of the
current state-of-the-art and a foundation for future advances.

REFERENCES

1.   Yeh, Y. and Cummins, H. Z., "Localized Fluid Flow Measurements with an
     He-Ne Laser Spectrometer," Applied Physics Letters, 4, 176 (1964).

2.   Stevenson, W. H., "Principles of Laser Velocimetry," pp. 307-336,
     Experimental Diagnostics in Gas Phase Combustion Systems, Vol. 53 of
     Progress in Astronautics and Aeronautics,(B. Zinn, ed.) American Institute
     of Aeronautics and Astronautics, New York (1977).

3.   Durst, F., Melling, A. and Whitelaw, J. H., Principles and Practice of
     Laser Anemometry, Academic Press, New York (1976).

4. Durrani, T. S. and Greated, C. A., Laser Systems in Flow Measurement, Plenum, New York (1976).

5. Stevenson, W. H. and Thompson, H. D. (eds.), The Use of the Laser Doppler Velocimeter for Flow Measurements, Proceedings of the First International Workshop on Laser Velocimetry, Purdue University (1972) AD-753243.

6. Thompson, H. D. and Stevenson, W. H. (eds.), Proceedings of the Second International Workshop on Laser Velocimetry, Purdue University (1974).

7. Proceedings of the LDA-Symposium, Technical University of Denmark (1975).

8. Eckert, E. R. G. (ed.), Proceedings of the Minnesota Symposium on Laser Anemometry, University of Minnesota (1975).

9. Proceedings of the NATO/AGARD Symposium on Applications of Non-Intrusive Instrumentation in Fluid Flow Research, St.-Louis, France, AGARD Conference Proceedings No. 193 (1976).

10. Pfeifer, H. J. and Haertig, J. (eds.) Proceedings of the ISL/AGARD Workshop on Laser Anemometry, German-French Research Institute, St.-Louis, France (1976).

11. Goldstein, R. J. and Kreid, D. K., "Measurement of Laminar Flow Development in a Square Duct using a Laser Doppler Flowmeter," Journal of Applied Mechanics, 34, 813 (1967).

12. Rudd, M. J., "A New Theoretical Model for the Laser Dopplermeter." Journal of Physics E, 2 55 (1969).

13. Hanson, S., "Broadening of the Measured Frequency Spectrum in a Differential Laser Anemometer due to Interference Plane Gradients," Journal of Physics D, 6, 164 (1973).

14. Durst, F. and Stevenson, W. H., "Properties of Focused Laser Beams and the Influence on Optical Anemometer Signals," p. 371 in Ref. 8 (1975).

15. Durst, F. and Stevenson, W. H., "Influence of Gaussian Beam Properties on Laser Doppler Signals," Applied Optics (To be published in 1979).

16. Brayton, D. B., Kalb, H. T., and Crosswg, F. L., "A Two-Component, Dual-Scatter Laser Doppler Velocimeter with Frequency Burst Signal Readout," p. 52 of Ref. 5 (1972).

17. Zammit, R. E., Pedigo, M. K., Stevenson, W. H., and Owen, A. K., "LDV Processor for High Velocity Flows," p. 204 of Ref. 6, Vol. 1 (1974).

18. Pike, E. R., "The Application of Photon Correlation Spectroscopy to Laser Doppler Measurements, Journal of Physics D, 5, L23 (1972).

19. Peterson, J. C. and Maurer, F., "A Method for the Analysis of Laser Doppler Signals using a Computer in Connection with a Fast A/D Converter," p. 312 in Ref. 7 (1975).

20. Durst, F. and Tropea, C., "Processing of Laser Doppler Signals by Means of a Transient Recorder and a Digital Computer," SFB Report 80/E/118, University of Karlsruhe (1977).

21. McLaughlin, D. K. and Tiederman, W. G., "Biasing Correction for Individual Realization of Laser Anemometer Measurements in Turbulent Flows," Physics of Fluids, 16, 2082 (1973).

22. Whiffen, M. C., "Polar Response of an LV Measurement Volume," p. 589 in Ref. 8 (1975).

23. Thompson, H. D. and Flack, R. D., "An Application of Laser Velocimetry to the Interpretation of Turbulent Structure," p. 189 in Ref. 10 (1976).

24. George, W. K., Jr., "Limitations to Measuring Accuracy Inherent in the Laser Doppler Signal," p. 20 in Ref. 7 (1975).

25. Whiffen, M. C. and Meadows, D. M., "Two-Axis Single Particle Laser Velocimeter System for Turbulence Spectral Analysis," p. 1 in Ref. 6, Vol. 1 (1974).

26. Stevenson, W. H., "Optical Frequency Shifting by Means of a Rotating Diffraction Grating," Applied Optics, 9, 649 (1970).

27. Oldengarm, J., "The Use of Rotating Radial Diffraction Gratings in Laser Doppler Velocimetry," Paper No. 22 in Ref. 9 (1976).

28. Schodl, R., "On the Development of a New Optical Method for Flow Measurements in Turbomachines," ASME Gas Turbine Conference, Zurich, Switzerland (1974)

29. Schodl, R., "Laser-Two-Focus Velocimetry for Use in Aero Engines," Paper No. 4, AGARD Lecture Series No. 90, Laser Optical Measurement Methods for Aero Engine Research and Development (1977).

30. Echardt, D., "Detailed Flow Investigations Within a High Speed Centrifugal Compressor, ASME Journal of Fluids Engineering, 98, 390 (1976).

31. Rudd, M. J., "Non-Doppler Methods of Laser Velocimetry," p. 390 in Ref. 6, Vol. II (1974).

32. Hirleman, E.D., "Laser Technique for Simultaneous Particle Size and Velocity Measurements," Optics Letters, 3, 19 (1978).

33. Erdmann, J. C. and Gellert, R. I., "Particle Arrival Statistics in Laser Anemometry of Turbulent Flow," Applied Physics Letters, 29, 408 (1976).

34. Owen, F. K., "Laser Velocimeter Measurements of a Confined Turbulent Diffusion Flame Burner," pp. 373-394, Experimental Diagnostics in Gas Phase Combustion Systems, Vol. 53 of Progress in Astronautics and Aeronautics, (B. Zinn, ed.) AIAA, New York (1977).

35. Whitelaw, J. H., Comments in Round Table Discussion, Ref. 9 (1976).

36. Stevenson, W. H., Santos, R., and Mettler, S. C., "A Laser Velocimeter Utilizing Laser-Induced Fluorescence," Applied Physics Letters, 27, 395 (1975).

37. Stevenson, W. H., Santos, R. and Mettler, S. C., "Fringe Mode Fluorescence Velocimetry," Paper No. 20 in Ref. 9 (1976).

DEVELOPMENTS IN INSTRUMENTATION

Chairman: **R. L. SIMPSON**
Southern Methodist University

Operational Two-dimensional Laser Velocimeter for Various Wind-Tunnel Measurements

ALAIN BOUTIER, JEAN LEFEVRE,
CLAUDE PEROUZE, and OSTAPE PAPIRNIK
Office National d'Etudes et de Recherches Aerospatiales (ONERA)
92320 Chatillon, France

SUMMARY

An operational two-dimensional, two-color, fringe laser velocimeter, built for various flow studies in wind tunnels, is described. This strong, rigid and modular instrument can be operated in a noisy and highly vibrating environment.

Backscattering or forward scattering mode can be used. The probe volume can be translated along three orthogonal axes, over 200 mm. The velocity sign is determined on each component thanks to acousto-optic modulators.

The signals are processed by two counting systems, on line with a mini-computer.

Three seeding boxes are available to seed the flows with incense smoke or DOP, or refractory powders.

Results obtained in an air-inlet are presented.

1. INTRODUCTION

The need for velocity measurements has always been high in all aerodynamic flow studies. Many flow configurations necessitate non-intrusive measurements, such as transonic flows, recirculation and separation zones, flames, etc.

For all these needs, laser velocimetry has been developed at ONERA for eight years ; in a first stage, four one-dimensional laser velocimeters (i.e. able to measure successively different components of the velocity) have been put into operation [1]. The many flow investigations performed with these instruments revealed the necessity to transform these laser velocimeters into two-dimensional ones, in order to obtain more precise results more rapidly. So, in a second stage, one of the one-dimensional velocimeters was transformed into a two-color two-dimensional laser velocimeter (i.e. able to measure simultaneously two velocity components).

The purpose of this paper is to describe in detail this instrument, built for rise in the ONERA facilities and now fully operational.

2. REQUIREMENTS FOR VELOCIMETERS USED IN ONERA FACILITIES

Laser velocimetry is applied in different kinds of flows such as : free jets (cold and hot), subsonic, transonic and supersonic wind tunnels, hot flames, compressors, highly turbulent flows.This great variety of applications implies that the velocimeter can be easily transformed for its adaptation to each particular problem.

As in any application, the purpose is to establish a velocity and turbulence map of the flow ; so the probe volume has to be moved along three axes ; the fringe direction must be adjustable so that the interesting velocity components can be measured.

In the environment of transonic or supersonic wind tunnels (or free jets) the noise level is often very high, and acoustic vibrations are generated ; in the vicinity of hot flows (flames, hot free jets) the temperature

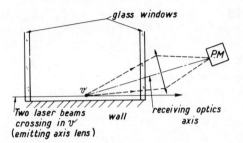

Fig. 1 — Setting of the receiving optics when the probe volume is near a wall.

may become high. In such environmental conditions, the optical device must remain adjusted ; for this reason, a very strong and stable mechanism must be built. This leads to the concept of an apparatus composed of modular optics and mechanics, undisturbed by vibration.

In boundary layer explorations, the probe volume has to be approached very close to the wall (a few tenths of mm.) To keep a good signal to noise ratio near this wall, the solid angle of the scattered light collection optics must not be truncated by the wall itself. This is achieved when the receiving optics axis is not coaxial with the emitting lens axis, as shown on figure 1.

As the receiving optics position can vary from one experiment to another (forward or backscattering mode), it must be modular and easily set at various positions relative to the emitting optics. For instance, figure 1 shows laser beams tangential above an horizontal floor ; but in another flow study they may be tangential under a ceiling and the collection optics axis is then under the emitting axis. This leads to the necessity to be able of assembling the optical and mechanical elements in various ways.

A means for measuring the velocity sign is absolutely necessary to study aerodynamic flows, because when the turbulence level increases (above 30 %), correct measurements are impossible without such a means, particularly in separated or recirculation zones, or free jet boundaries.

It is desirable to know the order of magnitude of the velocities to be measured, because it allows the definition of the optical components ; the higher the velocities, the more desirable is the forward scattered light configuration. The optics are calculated as a function of the maximum frequency the electronics can process (with a good precision) and the maximum velocity to be measured. As velocity increases, the measurements are more difficult : high power lasers (more than 2 watts on each wavelength used) are required so that the photodetectors could be sensitive to signals due to very small particles.

The path length of the laser beams through flows presenting high gradients of the refraction index (such as flames, shock-waves) must be limited to 100 mm in order not to disturb too much the probe volume position.

Usual counting techniques allow high data acquisition rates : but the flows must be seeded with convenient submicron particles, such as DOP vapour or incense smoke in cold flows, and refractory powders (Z_rO_2) in high temperature flames.

For turbulence studies in two-dimensional flows, with a one-dimensional laser velocimeter it is necessary in practice to measure successively three coplanar velocity components to obtain the following quantities :

$\overline{u}$: mean velocity vector value
$\overline{\alpha}$: direction of the mean velocity vector
$\overline{u'^2}$: axial fluctuation
$\overline{v'^2}$: transverse fluctuation
$\overline{u'v'}$: Reynolds shear stress.

But for more detailed and precise turbulence studies, a two-dimensional laser velocimeter provides all the interesting informations within a shorter time, and its measurements do not call upon any hypothesis concerning the flow stationarity, as it provides at once the instantaneous velocity vector.

A computer on line with the laser velocimeter signal processor is needed to record the velocity histogram at each mesured spot, and then calculate the mean velocity and the various turbulence parameters.

It can also be mentioned that the application of laser velocimetry in wind tunnels requires for the facility :

- plane glass windows, large enough to allow the passage of the laser beams and of the collected scattered light solid angle ; perspex windows must be avoided because much light of the incident laser beams is scattered within it, thus increasing the stray light ; in a circular test chamber, it is better to have a smaller plane window than a broader cylindrical window : in fact with a cylindrical window, when the probe volume is moved in a plane normal to the cylinder axis, the adjustment between the probe volume and the photomultiplier pinhole is not maintained ; moreover when the plane defined by the two laser beams outside the wind tunnel do not contain the cylinder-axis, the laser beams do not cross in the probe volume ;

- the possibility to seed the flow either locally or in a plenum chamber upstream, without disturbing the flow in the measurement section ;

- electric power supply for the argon laser ranging from 10 to 40 kW, according to the argon laser model ;

- a water source to cool the laser.

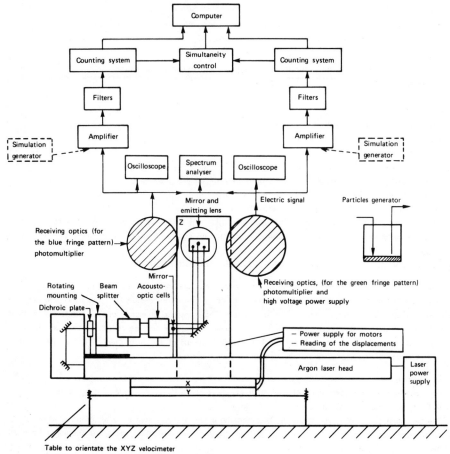

Fig. 2 — General scheme of the ONERA two-dimensional laser velocimeter.

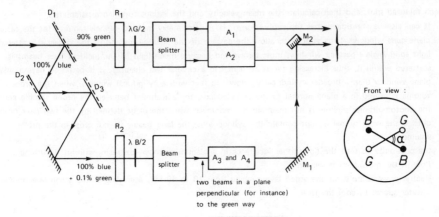

Fig. 3 — Top view of the beam splitting part.

D_1, D_2, D_3 : three identical dichroïc plates ; R_1, R_2 : two independent rotating mounting ;
A_1, A_2, A_3, A_4 : acousto-optic cells ; M_1, M_2 : mirrors ; M_2 : will be replaced by a dichroïc plate.

Suming up all these needs, we can outline two lists of requirements, one concerning the facility, the other the velocimeter :

facility

- plane glass window,
- possibility of particle injection,
- electric power and water of the laser ;

laser velocimeter

- modular elements, strong and rigidly assembled,
- easy displacement of the probe volume and rotation of the fringe pattern,
- possibility to fix the receiving part in various positions,
- velocity sign determination very often needed,
- simultaneous measurement of two velocity components : the successive measurements of three coplanar velocity components provides mean informations which do not seem sufficient for many experiments.

3. DESCRIPTION OF THE TWO-DIMENSIONAL LASER VELOCIMETER

The basic principles of the two-dimensional laser velocimeter are the following : a green and a blue fringe patterns are created in the same probe volume ; they can be orientated independently ; the fringe motions (for velocity sign determination) can be adjusted independently. There are two collecting optics, one for each fringe pattern. Two counting systems process the simultaneous signals in parallel and are connected to a computer.

Figures 2 and 3 show the arrangement of all the main elements of this instrument. The photograph of figure 4 represents all these components gathered and ready to work.

3,1. The emitting part

The light source is an argon laser with an all lines power of 15 watts, i.e. about 6 watts in the green line $\lambda = 514,5$ nm and the same on the blue line $\lambda = 488$ nm.

The two most powerful wavelengths, the blue and the green, are separated by a set of three identical dichroïc plates, as represented on figure 3. Each dichroïc plate transmits 90 % of the green line and reflects all the other wavelengths, therefore 100 % of the blue line as well as 10 % of green line. After the passage through three identical dichroïc plates, the green power present in the blue path can be neglected (about 0.1 %), so that no useful signal can be obtained from the green fringe pattern parallel to the blue fringe pattern. The other wavelengths of the argon laser are not taken into account, because in the receiving part one photomultiplier is

Fig. 4 — General view of ONERA two-dimensional laser velocimeter.

sensitive only to the green wavelength and the other one only to the blue wavelength, thanks to interferential filters.

On each laser beam (one green, and the other blue) a separating system is placed, that includes :

— A beam splitter which delivers two beams of equal intensity, symmetric and parallel relative to the incident laser beam, with a fixed distance of 30 mm ; just before the beam splitter, a half-wavelength plate is placed in order to conveniently orientate the polarisation direction of the laser beam, relative to the beam-splitter incidence plane.

— Acousto-optic modulators (Bragg cells) : on one beam the frequency shift is F_1 and on the other one it is $F_1 + \nu$:

  • for the green way F_1 is 200 MHz, ν being 2 ; 4 ; 6 ; 8 MHz ;
  • for the blue way F_1 is 36 MHz, ν being 2.5 ; 5 ; 7.5 MHz.

The use of two Bragg cells for each way is due to the fact that the signal processors used accept frequencies up to 100 MHz, but must not be used beyond 38 MHz if a good precision has to be kept.

There are no basic differences between the green and blue ways, but only different technologic means are used.

In both systems, at the output, the beam deviations are compensated by prisms.

— Beam separation reduction to 20 or 10 mm, achieved by a couple of plane parallel glasses.

Each separating system is fixed on a rotating mounting ; the two systems can be rotated independently.

The four beams are recombined thanks to the mirror M_2 situated in the middle of the two green laser beams, so that the beam configuration of figure 3 (front view) is obtained on the emitting lens.

The photograph of figure 5 shows this optical arrangement ; the photograph of figure 6 is the same, but with the Bragg cells in place.

The focal length of the emitting objectives may be 300, 500, 800, 1000, 1500, 2000 mm : the longitudinal chromatism is perfectly corrected for the two interesting wavelengths. The objectives having focal lengths of 800, 1000, 1500, 2000 mm can be replaced by tele-objectives having the same focal lengths but a constant front length of 500 mm.

The optical axis of the emitting lens is generally horizontal, but its direction can be adjusted in some particular applications.

3,2. The receiving part

The receiving optics are such that they can collect light at a distance varying between 300 mm and 2 m, thanks to a set of lenses and Cassegrain telescopes ; the aperture varies between f/3 and f/10. In fact there are

Fig. 5 — Beam splitting system without Bragg cells.

Fig. 6 — Beam dividing system equipped with Bragg cells.

two receiving optics, placed on both sides of the emitting optical axis ; an interferential filter of 10 nm bandwidth is placed in front of each photomultiplier : one is centered at 488 nm and the other at 514.5 nm.

The photomultipliers have a S20 sensitive photocathode coating and 14 amplification stages ; they necessitate a—1800 volt supply.

The receiving part may be directly linked to the emitting part as on photograph 4 in a backscatter configuration. It is easy to transform this set-up in pure forward configuration (or in forward configuration with a mirror) : a rigid mechanical beam may be fixed on the Z vertical translation movement and pass over the wind tunnel test section : at the extremity of this mechanical beam the receiving part (or a mirror) can then be placed.

The optical axes of the receiving optics may be inclined by a $\pm 10°$ angle from the horizontal axis of the emitting optics in order to make the optical configuration of figure 1 possible when exploring boundary layers.

3,3. Probe volume displacement

The whole optical and mechanical apparatus is fixed on a support which can move the probe volume along three rectangular axes with a 200 mm translation amplitude ; these axes can be set parallel to the reference axes of the flow. These translations are achieved thanks to stepping motors ; their amplitudes are read by optical encoders with a 0.01 mm precision. The motors command and the translation readings can be installed at a distance of 15 m from the velocimeter, which allows the experimentators to be far from the noisy environment of the flow.

The probe volume altitude from the ground is adjusted by a set of mechanical wedges which can be introduced between the X, Y horizontal movements and the Z vertical movement.

3,4. Signal processing

The signals are processed thanks to counting techniques. In a first stage of development two systems as those defined at ISL (German − French Institute of Saint-Louis) in 1975 [2] are used in parallel ; each of them is composed of one precise counter, a validation unit, an amplifier, filters, a miniprogrammer and a print-recorder. An auxiliary unit has been built at ONERA to make sure that the valid signals which are taken into account are simultaneous. This system has a low data rate acquisition (one hundred measurement couples per second), a good precision (0.2 %) and provides at the output of the print-recorder the mean velocity value of the two measured components and three quantities allowing one to further calculate the three usual mean turbulence parameters $(\overline{u'^2}, \overline{v'^2}, \overline{u'v})$. But instead of the print-recorder a mini-computer can be used which calculates in real time the same parameters, and moreover provides histograms of the velocity distribution, the modulus and angle of the mean velocity vector and the higher orders of the turbulence.

In a second stage of development we intend to replace these two counting systems by two fast commercially available counters on line with a computer, in order to increase the data acquisition rate. The counters possibilities are about 10^5 measurements per second ; but the limitation will come from the particles flow through the probe volume, which does not exceed 10^4 per second.

A spectrum analyser is always associated to the counting system to control the signal to noise ratio and to know the measured frequencies : many errors can thus be avoided concerning the choice of high- and low- pass filters.

3,5. Particles injection

Three seeding boxes have been built (cf. fig. 7) to emit incense smoke (A), or DOP (dioctyl-phtalate) vapour (B) in aerodynamic flows, solid particles (C) (powders of ZrO_2, TiO_2, etc...) in hot flows.

3,6. General comments

Measurement precision is limited by three factors :
− the particle bahaviour (submicron particles must be used) [3],
− the signal processor (less than 1 %),
− the measurement of the fringe spacing (or the angle between the crossing beams by a theodolite) : 0.1 to 0.3 %.

Fig. 7 − Three seeding boxes

A. incense smoke ; B. DOP (dioctyl-phtalate) vapour ; C. solid particles (ZrO_2 powder).

The transformation of the one-dimensional laser velocimeter into a two-dimensional one consisted in adding the following components, the laser emitting all lines instead of only one :
- the dichroïc plates,
- the blue beam splitting system,
- the two mirrors M_1 and M_2 (cf. fig. 3),
- a second receiving optics,
- a second signal processor, with the simultaneity control of the measurements.

It must be mentioned that the use of a powerful argon laser (more than 15 watts) necessitates very clean optical surfaces, in order to avoid local heating of the materials, which would distort the laser beam.

It is planned to replace in the near future the mirror M_2 by a dichroic plate allowing the two fringe patterns to be set parallel in the probe volume ; then, when rotating the mirror M_1, the blue fringe pattern will move relative to the green fringe pattern, which allows one to easily perform spatial correlations on short distances (a few tens of millimiters).

4. TYPICAL RESULTS OBTAINED IN AN AIR INLET AT THE S3 CHALAIS-MEUDON WIND TUNNEL

The two-dimensional laser velocimeter has been used in a transonic wind tunnel, in order to explore the velocity field inside an air inlet situated in a high incidence configuration ; the laser velocimeter eliminates all the interaction problems linked to classical probes.

The general characteristics of this transonic wind tunnel are :
- Mach number ranging from $M_0 = 0.1$ to $M_0 \simeq 1$;
- installed power : 2 MW ;
- as the closed circuit is not air-tight, the stagnation temperature increases as a function of the Mach number up to $340°K$, the stagnation pressure remaining equal to atmospheric ;
- the test chamber is 1.75 m long and has a 0.66 m^2 octogonal section, inscribed in a 1-m-dia. circle ; the lateral walls, separated by 900 mm, are equipped with visualization windows.

The measurements performed enter the framework of a general study of flows in air inlets with high incidence angles ; a "two-dimensional" model was used, the characteristics of which are given in figure 8 :
- rectangular inlet section, 35 mm high, 140 mm wide ; incidence angle : 40°,
- parallel lateral walls with two windows,
- air duct 280 mm in length, i.e. eight times the inlet section height, it is termined by a geometrical sonic throat.

The aspiration flow rate through the air intake is ensured by a system creating a depression, situated downstream of the model ; a critical flow rate is imposed at the throat level ($M_c = 1$) ; then upstream a mean Mach number $M_2 = 0.2$ is obtained, which is similar to the conditions existing in a compressor inlet plane.

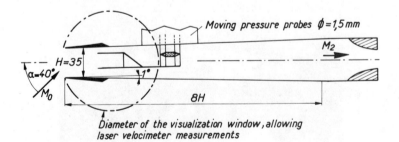

Fig. 8 — General scheme of the air inlet in 40° incidence.

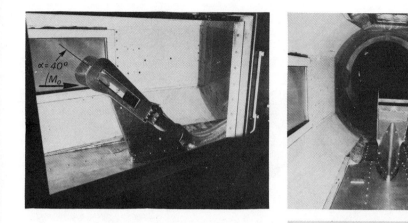

Fig. 9 — Photographs of the air-inlet model installed in the
test chamber of the S3 transonic wind tunnel at
Chalais-Meudon.

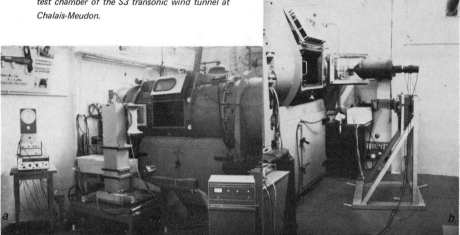

Fig. 10 — The two-dimensional laser velocimeter installed at S3Ch wind tunnel.
a. Emitting optics in one side of the wind tunnel ;
b. Receiving optics in the other side of the wind tunnel.

The photographs of figure 9 show the air inlet model installed in the test chamber of the wind-tunnel.

The two-dimensional laser velocimeter was used in the forward scattered light mode ; all the emitting optics were situated on one side of the wind tunnel (cf. figure 10a) and all the receiving optics were fixed on a second XYZ translation system, on the other side of the wind tunnel (cf. figure 10b).

In this experiment, the laser beams, and the forward scattered light, pass through two windows ; these conditions are very uncomfortable to obtain a good signal to noise ratio, because a great amount of stray-light is created.

The measurements were performed in the vertical median plane of the model, along horizontal and vertical axes. The flow was seeded by incense smoke, introduced in the plenum chamber.

The external flow velocity has been adjusted successively at $M_0 = 0.3$ and $M_0 = 0.65$; in these conditions, the captured flow coefficients ε were respectively about 0.75 and 0.35.

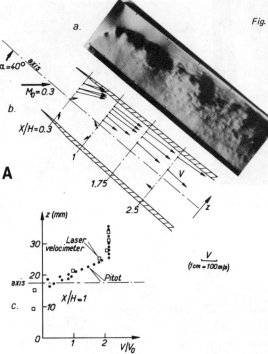

Fig. 11 — Direct results obtained in the air inlet.

A : $M_0 = 0.3$; $M_2 = 0.2$
B : $M_0 = 0.65$; $M_2 = 0.2$

a. schlieren picture of the flow,
b. representation of the velocity
   vectors in four sections,
c. evolution of the velocity
   modulus at $X/H = 1$, compared
   with pitot measurements.

The direct results are presented on figures 11 A and B for the two Mach numbers. Schlieren views, with a short time exposure (about 1 μs), show a highly turbulent flow inside the air inlet ; so all velocimeter measurements were performed with Bragg cells on the green and blue ways. On these pictures three different types of flows can be recognized from the top to the bottom of the air inlet : a quasi laminar flow, eddies, a separated flow with recirculation zone.

Four sections of the flow were explored ; figures 11 b represent the velocity vector evolution : in the recirculation zone this vector varies rapidly in modulus and direction. The results are compared to those deduced from a pitot probe in the section $X/H = 1$ (figures 11 c). A good agreement is obtained in figure A for $M_0 = 0.3$, but important

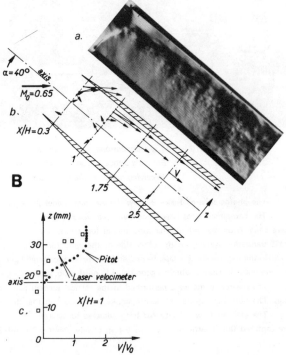

differences appear in figure 11 B for M_0 = 0.65. These differences may be due on one hand to the fact that a probe, present in a high pressure gradient flow, disturbs this flow. On the other hand, as the turbulence is very high, laser velocimeter measurements are biased.

5. CONCLUSION

A two-dimensional laser velocimeter has been built for various aerodynamic applications. It can be considered as operational when it works in the backscatter mode as regards mechanical and optical considerations ; but the measured velocities must not exceed 50 m/s if a good signal to noise ratio is to be obtained in this backscatter mode. For high velocity flow studies, forward scattered light must be used and it is then necessary to foresee a mechanical adaptation for each facility which requires the forward scattered light mode.

The results obtained in an air inlet show that it is possible with such an instrument to perform measurements in very complicated flows.

These experiments were a first approach of the problem and the detailed exploitation must now be done. Further experiments will require a computer on line with the velocimeter, necessary to be able one to make statistics on histograms containing a higher number of individual velocity informations.

The greatest effort will now concern the software of the computer associated to the laser velocimeter, so that the data acquisition and the probe volume displacement become automatic. This process does not exclude the necessity to obtain at each measured point the velocity histogram in order to control the validity of the calculated results.

REFERENCES

1. Boutier, A., Fertin, G., and Lefèvre, J. Laser velocimeter for wind tunnel measurements. To be published in IEEE/AES.

2. Pfeifer, H.J., Schafer, H.J. A single counter technique for data processing in laser anemometry. ISL Report No. 30 (74/1974).

3. Haertig, J. Particle behaviour in ISL Report R 117/76 (1976). Proceedings of the ISL/AGARD workshop on laser anemometry May 5-7, 1976, p. 1-39.

The Time-of-Flight Laser Anemometer versus the Laser Doppler Anemometer

LARS LADING
Risø National Laboratory
DK-4000 Roskilde, Denmark

ABSTRACT

A theoretical comparison is made between the laser Doppler anemometer (LDA) and the time-of-flight laser anemometer (TFLA). The TFLA is found to be superior to the LDA in many cases.
Three applications are described where the LDA has been inferior to the time-of-flight anemometer; they are: measurements in a two-phase flow around a simulated fuel rod, remote measurement of wind velocity and boundary layer measurements close to a wall.

NOMENCLATURE

ω_o	true Doppler frequency
τ_o	true time-of-flight
$\mathrm{var}\{\hat{\omega}_o\}$	variance on the estimated Doppler frequency
ε	relative uncertainty (e.g. $\varepsilon_d^2 = \mathrm{var}\{\hat{\omega}_o\}/\omega_o^2$)
a	peak signal (count rate)
b	background signal (count rate)
A	mean square of the signal
N	spectral noise density
Δ	half bandwidth
κ	relative broadening (e.g. $\kappa_d = 2\Delta_d/\omega_o$)
T	averaging time
d_u	size of the focal volume in the u-direction (u=x,y or z)
V	focal volume
l	beam or fringe spacing

Subscripts

d   Doppler system

t   time-of-flight system

x,y or z   x, y or z component

INTRODUCTION

     The laser Doppler technique is well established for measure-
ments in fluids. However, the Doppler technique is only one of
many laser techniques. The laser Doppler anemometer (LDA) is
characterized by a "wave packet" intensity distribution - as seen
from the detector. Other intensity distributions are certainly
possible and some of them may in fact prove superior to the
Doppler method. We shall here compare the Doppler system with
the time-of-flight system. The time-of-flight anemometer (TFLA)
is characterized by an intensity distribution in the form of two
displaced peaks. What appears as a frequency in the LDA will in
the TFLA appear as a time lag.

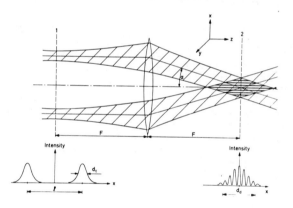

Fig. 1. Fourier transforming set-up. The field
distribution in plane 2 is given by the Fourier
transform of the distribution in plane 1.

THE LDA AND THE TFLA - A THEORETICAL COMPARISON

     Consider the set-up in Fig. 1.   The lens generates the Fourier
transform in one focal plane of the scalar field distribution in
the other focal plane (assuming that the para-axial approximation
is valid). If the field distribution in the first focal plane is
that of the beam waists of two beams parallel with the optical

axis, then we get a "wave packet" in the other plane, which squared gives the well known intensity distribution of the measuring plane (volume) in a LDA. In this system information about the velocity is derived by determining the frequency of the temporal wave packets, which are generated by the scattering from particles passing the measuring volume.

As an alternative to this system, we may detect the light scattered by particles passing in or near the first focal plane, and derive the velocity by a determination of the time of flight between the two spots. A velocimeter (anemometer) based on this principle will be called a time-of-flight laser anemometer (TFLA) [1], [2].

We shall here compare the two systems under three different assumptions. The comparison is made by computing the variances of maximum likelihood estimators [3] for the Doppler frequency ω_o and the time of flight τ_o, respectively; or, where this technique does not work, by computing the variances of the first moments of the power spectrum (LDA) and the correlation function (TFLA), respectively.

The three cases to be considered are:
1) Single particle detection with a dominating background.
2) Many particles in the measuring volume, no electronic and/or photon noise.
3) Many particles in the measuring volume and a dominating background.

A dominating background means that the photocurrent contains a DC-component that is larger than the peak signal-current. This background may be caused by e.g. scattering from a near-by wall or contaminated windows and/or light from the sun; it may also be given by the large DC-component of the scattered light power, as it is found in an incoherent set-up with many particles in the measuring volume.

The three cases have been chosen because they are computationally simple, but also because they may represent experimentally difficult situations.

The relative uncertainties of the three cases are computed in appendix I.

For case 1) we get

$$\varepsilon_t^2 = \frac{2}{\sqrt{\pi}} \frac{b_t}{a_t^2} \kappa_t^2 \Delta_t \; ; \; \varepsilon_d^2 = \frac{1}{\sqrt{\pi}} \frac{b_d}{a_d^2} \kappa_d^2 \Delta_d \tag{1}$$

for case 2)

$$\varepsilon_t^2 \simeq \frac{1}{4} \frac{\kappa^2}{\Delta_t T} \; ; \; \varepsilon_d^2 \simeq \frac{1}{4} \frac{\kappa^2}{\Delta_d T} \tag{2}$$

and for case 3)

$$\varepsilon_t^2 = \frac{1}{2\sqrt{\pi}} \kappa^2 \frac{\Delta_t}{T} \frac{N_t^2}{A_t^2} \qquad \varepsilon_d^2 = \frac{1}{2\sqrt{\pi}} \kappa^2 \frac{\Delta_d}{T} \frac{N_d^2}{A_d^2} \tag{3}$$

In order to compare the two systems we have to impose some assumptions and constraints. They are:

(a) Same given spatial resolution in the x and y directions: the focal beam diameter of the LDA is equal to the beam spacing of the TFLA, i.e. $d_{xd} = l_t$; in the y-direction, $d_{yd} = d_{yt}$.

(b) Same given spatial bandwidth. The spatial bandwidth is given by the solid angle through which light can be focused into the measuring volume. Assumptions (a) and (b) imply that the relative broadenings of the two systems are identical, $\kappa_t = \kappa_d$ and $d_{xt} = \kappa d_{xd}$ which for a given velocity implies $\kappa\Delta_t = \Delta_d$ [12]. It is also assumed that the light collecting solid angles are identical.

(c) Same uniform background illumination; the photocurrent of this background is proportional to the area of the pinhole(s) of the receiver. The pinhole is matched to the image of the measuring volume. For the extension of the focal volumes along the optical axis (z-axis) we get $d_{zt} = \kappa d_{zd}$.

(d) Same laser power, i.e. (by assumption a) the peak intensity of the LDA is 2κ x the peak intensity of the TFLA.

Using these assumptions in Equations 1-3 yields the following results concerning the performance of the two systems:

1) Single particle detection; dominating background.
    Assumption (d) yields $\kappa a_t = a_d$ and assumption (c) $2b_t = \kappa b_d$ which together with assumption d and Equations 1 yields

$$\frac{\varepsilon_t^2}{\varepsilon_d^2} = 4 \kappa^2 \qquad\qquad\qquad (4)$$

For $\kappa = 1/2$ the two systems are of equal performance, which is consistent with the fact, that for this value of κ the two systems essentially have the same intensity distribution in the measuring volumes. In general, that is for $\kappa < 1/2$, the TFLA is superior to the LDA.

2) Many particles in the measuring volume, no electronic and/or photon noise.
    Assumption (b) and Equation 2 yields

$$\frac{\varepsilon_t^2}{\varepsilon_d^2} \sim \kappa \qquad\qquad\qquad (5)$$

    - again the TFLA is superior to the LDA. (On the basis of the same argument as used for case 1) $\varepsilon_t^2 = \varepsilon_d^2$ for $\kappa = 1/2$ and the proportionality constant of Equation 5 should be 2).

3) Many particles in the measuring volume; a dominating background given by particle scattering (incoherent detection). The correlation functions are given by Equations A-12 and A-13. The DC-signal is proportional to the number of illuminated particles; this number is proportional to the size(s) of the illuminated volume(s). Assumption (b) and (c) yields $V_t = 2 \kappa^2 V_d$. The spectral noise density (shot noise) N is here proportional to the mean photocurrent, i.e. $N_t = 2 \kappa^2 N_d$. A is the mean square of the photocurrent and proportional

to the number of illuminated particles and thus to the illuminated volume, i.e. $A_d = A_t$ (for rigorous calculations
see e.g. Refs. 4 and 2).
Substituting these results into Equation 3 yields

$$\frac{\varepsilon_t^2}{\varepsilon_d^2} \cong \kappa \qquad\qquad\qquad (6)$$

MEASUREMENTS IN A TWO-PHASE FLOW AROUND A SIMULATED FUEL ROD

Measurements have been performed in a two-phase flow configuration as shown in Figure 2.

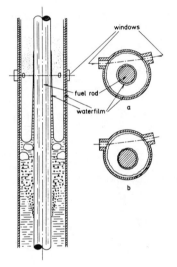

Fig. 2. Simulated fuel rod in annulus geometry. The diameter of the
rod and the inner diameter of the
tube are 17 mm and 27 mm, respectively.

Fig. 3. (below) Set-up used for
two-phase flow measurements. The
beam spacing is 2.6 mm. The spots
are elliptic with a size (outside
the flow) of 0.1 mm x 2 mm.

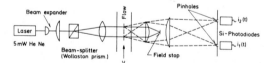

The basic configuration of the optical configuration is shown
in Fig. 3. As beam splitter a Wollaston prism is used which con-
verts the incident beam into two angularly - but not necessarily
spatially - separated beams of orthogonal polarization. In a
"tough" environment there may be a considerable cross-coupling between the two channels, despite the spatial filtering ability of
the receiver. This cross-coupling can be eliminated by the orthogonal polarizations of the transmitted beams in conjunction
with polarization filters in front of the detectors. The beams
are refracted and focused by the transmitter lens. With the beam

expander in front of the Wollaston prism it is possible to control the size of the focal spots.

The receiver incorporates a light collecting lens that generates a real enlarged image of the focal spots at the pinholes. A field stop blocks the directly transmitted beams - they carry no information.

The outputs of the photodiodes are correlated in a multi-channel correlator. A "delay lock-loop" [15] was also used in order to get a signal proportional to the velocity averaged over a time shorter than the dominating turbulence scales. However, "instantaneous" measurements were not successful and most measurements were performed with the multichannel correlator and a "long" averaging time.

In our set-up the windows had to withstand quite severe impact from the flow. We tried several types of glass and quartz, but did not succeed in finding a material which could keep its optical quality for more than 8 hours when exposed to the high-pressure, high-temperature two-phase flow. The windows were shaped as cylindric rods, 15 mm long and 6 mm in diameter. One end was cylindric in order to follow the surface of the tube. The rods were so mounted that they could be pushed a maximum of 3 mm into the flow.

Comments on the measurements and the performance of the TFLA

We succeeded in measuring the average in-core velocity in a large variety of flow conditions (temperature/pressure and mass flow). A typical correlation curve is shown in Fig. 4a. The velocities determined on the basis of the peak [2] of the correlation curves are in quite good agreement with the calculated values (a numerical error of less than 10%).

Of special interest were the measurements in a set-up with an eccentrically placed rod. The velocity was measured as a function of the angular position around the rod; an example is shown in Fig. 6.

The transit time broadening of the set-up was supposed to be of the order of 5%. This figure is based both on calculations and measurements on a rotating Perspex disc. As it can be seen (Fig. 4) the actual broadenings are much larger. This can be due either to velocity gradients or velocity fluctuations within the averaging time (turbulence) or both [2].

Another cause could be particles (drops) larger than the focal spots - in such a case the particle diameter is included in the expressions for the broadenings instead of the focal diameter. It is actually possible to select a specific particle size (range) by adequate prefiltering of the detector signals - as described in Ref. [5]. One would expect that the larger particle, under the influence of gravity, would move more slowly than the smaller ones. Only in a few cases did we obtain measurements indicating that this could be the case.

However, in addition to these effects it seems very likely that the actual focal diameters were much larger than predicted, and the spatial filtering was less efficient. The pinhole diameters could be increased almost by an order of magnitude without any influence on the signal other than that the signal strength increased. Considering the character of the flow, it does not seem likely that anything close to diffraction-limited focusing could be achieved under the given circumstances. The strength of this "defocusing" effect has still to be investigated. As far

as the spatial resolution is concerned, this effect does especially
decrease the resolution along the optical axis (it is calculated
to ∿ 2 mm).

Another phenomenon, which is very probably also caused by this
effect, is optical cross-coupling between the two channels. In
the first series of experiments we performed, the two beams were
of equal polarization, and the recorded cross-correlation looked
like a superposition of an auto and a cross-correlation curve
(Fig. 5b). The introduction of orthogonal polarization eliminated
this problem.

The fact that instantaneous measurements gave no more than
time averaged measurements (rather less) indicates velocity gradients, large particles and de-focusing effects of the water film
(Fig. 2) to be the primary causes of broadening rather than temporal fluctuations (turbulence).

Experiments with the LDA

The Doppler technique has also been tried out in a dual-beam
mode with both coherent and incoherent detection. We failed to
get any sensible results by using spectral analysis and certainly
not by using a "tracker".

It is believed that the fringe pattern is "washed out" by
the random fluctuations in the optical path lengths; or (alternatively) the beams to be heterodyned on the detector contain
random phase fluctuations which "average" the beat-note out.

It is noted that experiments performed under easier
conditions does not indicate that the LDA should be more sensitive
to weak phase disturbances (i.e. $\Delta\phi << \pi$) than the TFLA. The
measurements were performed by injecting a hot gas-stream in front
of the transmitter lens on set-ups which fulfilled the assumptions
a-d of the previous section [6].

The test loop and some (non-laser) measurements are described
in Reference 7.

REMOTE MEASUREMENT OF WIND VELOCITY

Measurements in the atomsphere have been performed at ranges
up to 70 m without artificial seeding. The initial measurements
are described in Reference 8. Figure 6 shows a slightly modified
version of the original optical configuration. The first part of
the transmitter is essentially built as the transmitter in Fig.. 3.
The large lens system generates an image of the first focal plane
at the desired measuring distance. At 10 m the two spots are 1 mm
apart and 0.1 mm in diameter. These dimensions scales with the
range. The receiver is from the Wollaston prism and out a dupli-
cate of the transmitter. The Wollaston prism W2 in conjunction
with lens L5   generates one image of the two spots, which is
filtered by a matched pinhole.

This configuration makes it possible to rotate the measuring
direction by rotating the two Wollaston prisms. In order to avoid
rotating the photomultiplier configuration a $\lambda/2$-plate is rotated
half the angle of W1 and W2, and thus restores the original polarization of the two beams received with respect to the polarizing
beam-splitter.

The first processing used consisted of two discriminators and
a multichannel analyzer which gave a histogram of the times-of-
flight (Fig. 7).

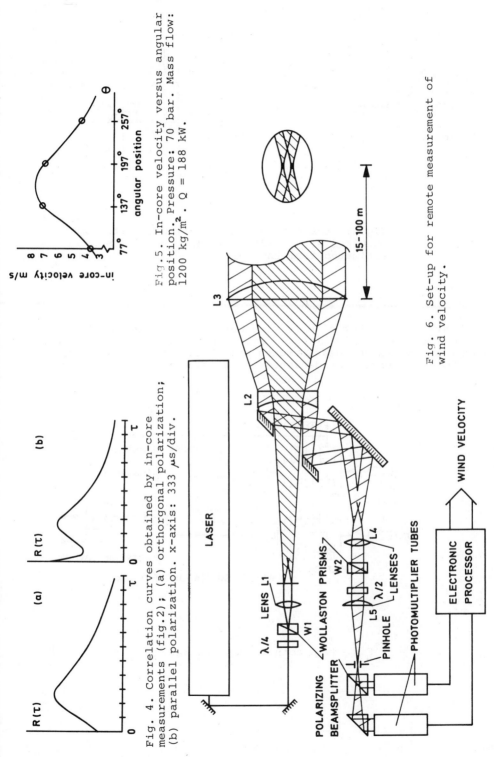

Fig.5. In-core velocity versus angular position. Pressure: 70 bar. Mass flow: 1200 kg/m^2. Q = 188 kW.

Fig. 4. Correlation curves obtained by in-core measurements (fig.2); (a) orthorgonal polarization; (b) parallel polarization. x-axis: 333 μs/div.

Fig. 6. Set-up for remote measurement of wind velocity.

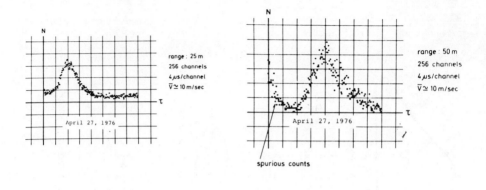

Fig. 7. Examples of time-of-flight histograms
obtained with a MCA. The mean velocities are
estimated from the peaks.

The mean number of particles detected was of the order of
10-50 per sec. at 10 m/sec. This sampling rate is high enough to
make instantaneous measurements possible. For this purpose a
special processor was designed: a coincidence tracker (Fig. 8).

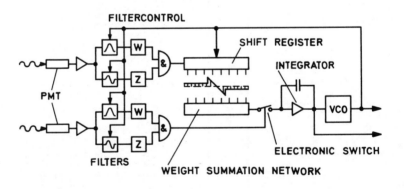

Fig. 8. A coincidence tracker.

The filter /discriminator performs a (near) maximum likelihood
estimation of the temporal position of the detected pulses. The
discriminator pulses are fed to a shift register, which is taped
with the shown weight function. Each time a time-of-flight
measurement is performed the clock frequency is changed so the
"new" $\hat{\tau}_0$ corresponds to the zero-crossing of the taped shift
register. However, this will only occur if the difference between two consecutive measurements is smaller than the "lock-in"
range; this in order to take advantage of the fact, that the
probability for a large difference between two consecutive
measurements is much smaller than the probability for a small difference; that is when the sampling rate is higher than the spectral
width of the turbulence.

A coincidence tracker has been tested on an analog computer,
and it performed as expected. An actual instrument has just been
built and is presently being tested.

The limitations on the performance of the present system seem
to be given by the noise associated with the background light.
This light originates partly from scattering/diffraction in the
instrument itself and partly from scattering by the many small
particles in the measuring volumes and in some cases also from
sunlight. (It is noted that with the incoherent receiver configuration many small particles will give a signal with low modulation depth unsuited for the signal processing used). The
"spurious counts" (Fig. 7) are caused by uncorrelated triggerings
of the discriminators; these "triggerings" are here caused by the
noise of the background.

Comparable measurements have been made with a differential
Doppler set-up with a "single-burst" photon-count correlator/
spectrum analyzer [9]. Even though the Doppler set-up suffered
from problems both with the optics and the processor, we feel
it is safe to conclude that, under the present conditions, the
TFLA with single-particle timing is superior to the differential
Doppler system. This conclusion is in agreement with the results
published in Reference 10.

BOUNDARY LAYER MEASUREMENTS

A back-scattering set-up has been built, which can be easily
converted from a Doppler system to a time-of-flight system. The
set-up is designed according to the assumptions and constraints
discussed in the theoretical part of this paper, except for circular spots in the time-of-flight mode. Only one photomultiplier
is used. As processor for the time-of-flight system is used the
multichannel analyzer set-up mentioned in the previous section
and described in Reference 8. For the Doppler mode is used a
bandpass filter (taped delay line) with a relative width of 1/5.
This filter is followed by a zero-crossing counter gated by the
envelope of the filtered signal. The temporal distance between
the first and the 16th count gives a time interval which is
processed like the times-of-flight.

Measurements were performed in a free jet into which a thin
black plate could be positioned. The plate would be perpendicular
to the optical axis.

The thresholds of the discriminator and envelope detector were
adjusted so that the parasitic particle count-rate was negligible
with the plate far from the measuring volume.

For a set-up with a focal beam diameter (LDA) or a beam spacing (TFLA) of 160 μm the following results were obtained: With the plate 22 mm from the centre of the measuring volume the parasitic count-rate was half the total count-rate (number of measurements) for the Doppler mode; for the time-of-flight mode this distance was 3 mm. By adjusting the thresholds sensible measurements could be performed with the plate 3 mm from the measuring volume in the Doppler case and 0.2 mm in the time-of-flight case.

It is noted that elliptic spots would give a poorer performance for the TFLA, but still better than the LDA.

CONCLUSION

If the total solid angle through which light can be focused is the limiting factor, the TFLA will be superior to the LDA. If the f-number of available lenses is the limiting factor, but these can be duplicated, the LDA may be equally good.

The theoretical analysis given here does not include all possible cases; especially not the single particle case, where the photon noise of the signal itself dominates, and the many particle case with coherent detection. However, although it may be impossible to devise optimum estimators, the general structure of the problems does not indicate, that the results for these cases would be basically different from those given here.

Concerning the implementation of the measuring set-ups: the optics of the TFLA may be more demanding than for the LDA whereas the electronic processor may be simpler for the TFLA than for the LDA.

ACKNOWLEDGEMENT

I wish to thank A. Skov Jensen, R.V. Edwards and C. Fog for useful discussions on various matters related to the subject of this paper. E. Rasmussen has been most helpful in implementing the set-up for remote wind measurements. C. Fog was responsible for the implementation of the coincidence tracker and the Doppler processor.

APPENDIX

The variances of the estimated time-of-flight, $\hat{\tau}_o$, and the estimated Doppler frequency, $\hat{\omega}_o$.

Single particle detection; mean square uncertainty for a maximum likelihood estimator.

Let the number of photon-counts in a time interval Δt at time t be n(t)

$$n(t) = [s(t) + b(t)]\Delta t \qquad (A-1)$$

where s(t) is the signal photon count-rate and b(t) the background count-rate.

The probability distribution for n is a Poisson distribution

$$p(n) = \frac{<n>^n}{n!} e^{-<n>} \qquad (A-2)$$

where $<n> = E\{n\}$, the expected value of n at a given time.

Let
$$<n(t)> = <n(t,\alpha)>$$

where α is the parameter to be estimated. The maximum likelihood estimator (MLE) for α is the value $\hat{\alpha}$ that maximizes the conditional probability distribution [3] $p(\{n\}|\alpha)$ where $\{n\} = (...n(-\Delta t), n(o), n(\Delta t)...)$.

The count numbers at different times are statistically independent, therefore

$$p(\{n\}|\alpha) = \Pi p(n(t_i)|\alpha) \qquad (A-3)$$

$p(\{n\}|\alpha)$ will obtain its maximum for the same value of α as ln $p(\{n\}|\alpha)$. ln $p(\{n\}|\alpha)$ is easier to handle.

From A-1 and A-2 we get that a MLE of α is the value $\hat{\alpha}$ that maximizes (the correlation function)

$$\Sigma_i n_i \ln<n_i> \qquad (A-4)$$

If the background dominates, the detected signal + background has to be filtered with the expected signal: a matched filter.

A lower bound on the variance on α is given by [3] (the Cramer-Rao bound)

$$var\{\alpha\} \geq - [E\{\partial^2 \ln p(\{n\}|\alpha)/\partial\alpha^2\}]^{-1} \qquad (A-5)$$

Substituting A-1, A-2 and A-3 into A-5 letting $\Delta t \rightarrow o$ and assuming $<b(t)> >> <s(t)>$ for any t yields

$$E\{\frac{\partial^2}{\partial\alpha^2}\{\ln p(\{n\}|\alpha)\}\} = -\int \frac{1}{<b(t)>}(\frac{\partial}{\partial\alpha} <s(t,\alpha)>)^2 dt \qquad (A-6)$$

If $var\{b(t)\} <s(t)^2$ then A-5 is in fact the variance on $\hat{\alpha}$ [11].

The LDA case:

Let the expected signal be

$$<s(t) + b(t)> = a_d \exp\{-\tfrac{1}{2}(t+t')^2 \Delta_d^2\}(1+\cos\omega_o(t+t')) +b_d \qquad (A-7)$$

which substituted into A-6 and A-6 into A-5 yields

$$\text{var}\{\omega_o\} = \frac{4}{\sqrt{\pi}} \; \frac{b_d}{a_d^2} \; \Delta_d^3$$

The relative variance is

$$\varepsilon_d^2 = \frac{1}{\sqrt{\pi}} \; \frac{b_d}{a_d^2} \; \kappa^2 \; \Delta_d \tag{A-8}$$

where
$$\kappa = 2\Delta_d/\omega_d .$$

The TFLA case:

Let the expected signal be

$$<s(t)+b(t)> = a_t(\exp\{-\tfrac{1}{2}(t+t' - \frac{\tau_o}{2})^2 \Delta_t^2\}+$$

$$+\exp\{-\tfrac{1}{2}(t+t' + \frac{\tau_o}{2})^2 \; \Delta_t^2\} + b_t$$

which gives

$$\text{var}\{\tau_o\} = \frac{1}{\sqrt{\pi}} \; \frac{b_t}{a_t^2} \; \frac{1}{\Delta_t}$$

and

$$\varepsilon_t^2 = \frac{2}{\sqrt{\pi}} \; \frac{b_t}{a_t^2} \; \kappa^2 \; \Delta_t \tag{A-9}$$

where
$$\kappa = \frac{2}{\Delta_t \tau_o}$$

Many particles in the measuring volume, no photon/electronic noise

The Cramer-Rao bound is not applicable in this case because the statistics of the signals does not make it possible to have A-5 fulfilled as an equation. We shall therefore use a heuristic approach; this can be justified by rather lengthy rigid mathematical calculations. Space limitation does not allow them to be included here.

We will assume that the power spectrum (LDA) and the correlation function (TFLA) are obtained with an averaging time T. The power spectrum and the correlation function are considered as distribution functions for the frequency and time delay, respectively. The estimated Doppler frequency and time of flight are then given by the mean of the estimated curves; this estimate is a random variable. The variance on the estimate of the mean is reduced by the number of trials with which we obtain a distribution function, i.e.

$$\text{var}\{\hat{\omega}_o\} \sim \frac{1}{N_d} \quad \text{and} \quad \text{var}\{\hat{\tau}_o\} \sim \frac{1}{N_t},$$

where the $\hat{}$ stands for estimate. N must be equal to the time-bandwidth product, i.e. $N = \Delta T$, where T is the averaging time. The variance on the estimate of the mean for $N = 1$ must be equal to the broadening, i.e.

$$\text{var}\{\hat{\omega}_o\} \simeq \frac{\Delta_d^2}{\Delta_d T}$$

$$\text{var}\{\hat{\tau}_o\} \simeq \frac{t^2}{\Delta_t T} \quad . \tag{A-10}$$

The normalized errors are then

$$\varepsilon_d^2 = \frac{\text{var}\{\hat{\omega}_o\}}{\omega_o^2} \simeq \frac{1}{4} \frac{\kappa_d^2}{\Delta_d T}$$

$$\varepsilon_t^2 = \frac{\text{var}\{\hat{\tau}_o\}}{\tau_o^2} \simeq \frac{1}{4} \frac{\kappa_c^2}{\Delta_t T} \tag{A-11}$$

It should be noted that equation A-10 is in agreement with the results that can be derived on the basis of Rice's analysis of the fluctuations in the frequency (defined as the derivative of the phase) for a narrow band process [12].

Many particles in the measuring volume; a dominating background

Our starting point is the correlation functions

$$R_d(\tau) = A_d e^{-\frac{1}{2}(\tau\Delta_d)^2}(e^{j\omega_o\tau}+1)+N_d\delta(\tau) \tag{A-12}$$

$$R_t(\tau) = A_t e^{-\frac{1}{2}(\tau-\tau_o)^2\Delta_t^2} + N_t\delta(\tau) \tag{A-13}$$

Let N >> A.

The distribution function for $\hat{R}(\tau)$ is rather complicated [13], however, for a large T it will approach a Gaussian (this can be shown using the central limit theorem) with a variance [14]

$$\text{var}\{R(\tau)\} \simeq \frac{1}{T}\int R^2(\tau)d \simeq \frac{1}{T}N^2(\delta\tau)^{-1}$$

where $\delta\tau = 1/\omega_a$ and ω_a is the pre-amplifier bandwidth.

Using the same procedure as before we get

$$\text{var}\{\hat{\omega}_o\} \simeq [\int_N^T \frac{1}{N^2} \frac{\partial R_d(\tau,\omega_o)}{\partial\omega_o}^2 d\tau]^{-1} \tag{A-14}$$

and a similar expression for $\text{var}\{\hat{\tau}_o\}$ assuming $\omega_o > \Delta_d$, $R_d(\tau,\omega_o)$ is the ω_o dependent part of A-12. By use of A-12 and A-13 we get

$$\varepsilon_d^2 = \frac{1}{2\sqrt{\pi}} \frac{N_d^2}{A_d^2} \kappa_d^2 \frac{\Delta_d}{T}$$

and

$$\varepsilon_t^2 = \frac{1}{2\sqrt{\pi}} \frac{N_t^2}{A_d^2} \kappa_t^2 \frac{\Delta_t}{T} \tag{A-15}$$

REFERENCES

1. R. Schodl, "The Laser-Dual-Focus Flow Velocimeter", AGARD-
    CP-193, paper 21.

2. L. Lading, "The Time-of-Flight Laser Anemometer", AGARD-CP-
    193, paper 23.

3. A.D. Wahlen, Detection of Signals in Noise, New York,
    Academic Press, 1971, chapter 10.

4. L. Lading, "Analysis of Signal-to-Noise ratio of the Laser
    Doppler Velocimeter", Opto-electronics 5, 175-187 (1973).

5. L. Lading, "Analysis of a Laser Correlation Anemometer",
    Proceedings of the Third Symposium on Turbulence in Liquids,
    1973, University of Missouri, Rolla, pp. 205-219.

6. Further experiments and analysis on the effect of phase
    disturbances are in progress.

7. A. Jensen, "Description of Loop and Test Section for SDS
    Annulus Experiments". Risø SDS-64 (Danish Atomic Energy
    Commission, Research Establishment Risø, DK-4000 Roskilde,
    Denmark).

8. L. Lading et al., "Time-of-Flight Laser Anemometer for
    Velocity Measurements in the Atmosphere", Appl. Opt. 17
    pp. 1486, (1978)

9. C. Fog, "A Photon-Statistical Correlator for LDA Application",
    Proceedings of the LDA-Symposium, Copenhagen, 1975, pp 336-349.

10. K.G. Bartlett and C.Y. She, "Single-Particle Correlated Time-
    of-Flight Velocimeter for Remote Wind-Speed Measurement",
    Optics Lett. 1, 175 (1977).

11. H.L. van Trees, Detection, Estimation and Modulation Theory,
    New York, Wiley, 1968, chapter 4.

12. L. Lading, "Comparing a Laser Doppler Anemometer with a Laser
    Correlation Anemometer", in The Engineering Uses of Coherent
    Optics, Ed. E.R. Robertson, Cambridge University Press,
    Cambridge, 1976, pp 493-510.

13. B. Epstein, "Some Applications of the Mellin Transforms in
    Statistics", Ann. Math. Statist. 19, 370-379 (1948).

14. J.S. Bendat and A.G. Piersol, Random Data: Analysis and
    Measurement Procedures, New York, Wiley, 1971, p. 184.

15. See e.g. H. Meyer, "Delay-Lock Tracking of Stochastic Signals",
    IEEE Trans. Com., COM-24, 331-339 (1976).

Burst Correlation Processing in LDV

E. R. PIKE
Royal Signals and Radar Establishment
Malvern, Worcestershire, England

ABSTRACT

A 10 ns burst correlator will be described which comprises a 10 ns photon correlator followed by 50 µs Fourier processor. This new instrument gives real-time velocity tracking up to 10 kHz with the highest accuracy and sensitivity.

INTRODUCTION

In the two previous Purdue meetings and elsewhere we have described the principles of photon correlation LDV. In short, the method exploits to the full the quantal nature of the detection of light by the use of detectors and circuitry of the highest speed and sensitivity to preserve the necessarily digital form of the input signal arising from the individual photons scattered into the detector. The statistical information provided by the random photon detections, random particle arrivals in the scattering volume and random fluctuations of velocity is extracted by ever more sophisticated treatments of the photon train autocorrelation function.

Up to the present, however, the widest use of the method has been in situations where the turbulence has been statistically stationary. Emphasis has been placed on obtaining the velocity probability distribution or its lower moments rather than frequency spectra or the temporal behaviour of the velocity fluctuations.

Where such velocity fluctuations have needed to be correlated with other variables such as pressure, concentration or simply the phase of a rotating machine, means have been developed successfully to "gate" the photon-correlator to obtain the required information.

The ultimate need for frequency spectra can be met less wastefully either by using the recently developed recurrence-rate correlation methods of Dr Erdmann [1] or by actually attempting to follow the real-time behaviour of velocity within the sample volume. This latter entails forming a velocity estimate on a time scale short compared with the maximum fluctuation frequency expected. Analogue methods of LDV such as frequency tracking or burst counting have perforce to operate in this way and therefore, as is well known, demand signals of high information content to work successfully.

Historically, photon correlators were developed at Malvern for measuring the Brownian motions of macromolecules, and this, and related chemical and biological uses have continued to employ photon correlation to the point where it is fair to say that analogue methods are almost no longer used. In LDV this dominance is not so prominent since the signal has much higher intrinsic information capacity, as we will discuss in another paper at this meeting. A greater inhibiting factor, however, has been the lack of a photon correlator designed

41

specifically to follow high-frequency velocity fluctuations directly which would, therefore, include the modes of performance of present-day analogue processors.

To remedy this situation we have spent the last three years designing and building such an LDV system. There are five major changes over previous photon-correlation LDV equipments. These are:

(i)   ECL 10 ns circuitry

(ii)   Custom LSI store

(iii)   50 µs real-time Fourier output processor

(iv)   Microprocessor control of all facilities

(v)   New "combined optics" package for fringe and two-spot operation

The principles upon which photon correlation LDV are based do not change for short sampling and processing times save for the a priori assumption that a single velocity value only is to be output for each consecutive sample period (of 50 µs or longer as selected). The advantages of statistical processing are retained, all the information arriving in each experiment, whatever the number of particles and regardless of how many photons they contribute, is accumulated and utilised to build up the correlation function, down to single "burst" signals from the transit of individual particles. The difference between this "burst correlator" and the conventional "burst" counter processor is again one of sensitivity and accuracy allowing lower laser power, smaller seeding particles and backscatter optics to be used. Such a correlator is also admirably suited to process signals from a two-spot anemometer either by photon correlation or by multiscaling of the output of a peak-detecting discriminator.

CORRELATOR CIRCUITRY

The new burst correlator processor is of the same design as the previous "Malvern" correlator but with a faster front end. ECL circuitry has been employed to reduce the minimum photon sampling time to 10 ns and the facilities of scaling, multiscaling, cross-correlation and probability analysis are retained. At this speed the familiar problems of photomultiplier performance become accentuated and we are working closely with manufacturers to improve the range of suitable tubes available. Pulse derandomising circuitry is employed as in the earlier instruments. Sample time and all other experimental variables are set using the microprocessor system from the keyboard as will be discussed later.

CUSTOM LSI STORE

Each correlation coefficient is accumulated in an independent 32 bit 10 ns scaler. There are 128 such channels in the full store and a specially designed LSI chip has been commissioned for use in this correlator. The first byte is accessed by a fast pre-scaler and readout is byte organised for compatibility with a standard computor input bus. The store from outside thus looks like part of the microprocessor's own memory. The chip represents a considerable invest-ment but the cost, size and power consumption of an instrument without it would make a much less attractive proposition.

In the "burst tracking" mode only 64 coefficients are utilised and these are assembled into the Fourier transform input autoranger buffer store, the whole set being transferred at rates up to a maximum of 20 KHz.

FOURIER PROCESSOR

The hypothesis of a single-frequency input allows a theoretically infinite accuracy to be achieved in Fourier inversion of the correlogram by an interpolation algorithm based on a known beam radius and experiment sampling time [3]. The relation is rather complex and, in fact, ad-hoc algorithms that are easily generated in hardware and give acceptable accuracy are used instead.

The velocity accuracy will thus be limited by the finite photon number and, perhaps, by any systematic departure from the physical conditions hypothesised.

The Fourier cosine transform (FCT) of the autocorrelation function yields a spectral estimator identical to the periodogram of the orirginal photon event train except for a scale error and an additional 'DC' contribution to each coefficient. The absence of a zero point in the correlator gives a further 'DC' shift but all these differences are made irrelevant by using relative magnitudes only in the calculations.

A typical photon train, its correlation function and the subsequent FCT are shown in Fig 1. Each point of the FCT alternates about an axis of symmetry with a phase change at the coefficients which bracket the input frequency. These differences are the two largest and are simply located; interpolation follows using their relative magnitude.

Some preprocessing is necessary to scale the input data from the correlator to fit an acceptable input dynamic range; in this case eight bits are used. An autoranging circuit notes the maximum and minimum values of the correlation function and subtracts the minimum value from all coefficients. A scale expansion factor is calculated to fill the dynamic range of the FCT and the data set is scaled to meet the optimum expansion.

This data set is transferred to an output buffer store at the end of each input data sample, the previous output buffer store is then used as the next internal buffer. The autoranging circuit is a useful design for other signal processing applications. It requires some 60 MSI packages and consumes 30 watts when operating continuously at 50 μs rates. It can be housed as an independent unit.

The FCT exploits recently available fast serial-parallel multipliers in LS TTL and reverts to the simple N^2 operation method using 64 "coefficient arithmetic units" (CAU's) in parallel, each executing 64 operations for each spectral estimate. The serial input data, available at all CAU's simultaneously, is multiplied by the relevant cosine term from a look-up table and generates a 16 bit product of which the 10 most significant bits are summed for the N products. Overlapped serial timing allows this 24 bit arithmetic to be performed in 17 clock cycles which, with a clock of 25 MHz, takes 45 μs for each transform. The CAU's are then unloaded into the output buffer in 5 μs thus meeting the 50 μs target for the complete cycle. Interpolation logic follows, one 50 μs period behind. The FCT is constructed in eight 6" x 8½" PCB's, each containing 8 CAU's, together with two further boards for timing logic for computation and readout. At 25 MHz the total power dissipation is 100 W. A block schematic diagram of the complete system is shown in Fig 2.

MICROPROCESSOR CONTROL SYSTEM

The microprocessor has indeed brought a revolution into experimental science and in the case of photon correlators has changed the scene completely to one of full operator-oriented operation. Anything suggested as a possibility invariably becomes an absolute necessity at the earliest possible date! The system design concepts adopted and their implementation cannot be treated in detail here, many types of command interpreter were considered with the aim of retaining as much flexibility for future developments as possible. A list of specifications for the system was drawn up as follows.

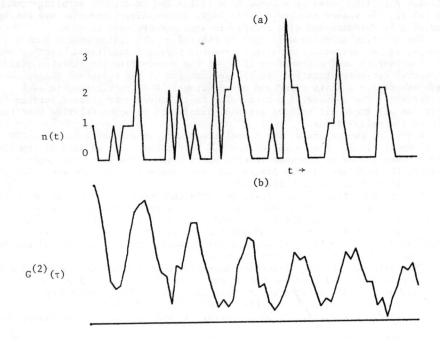

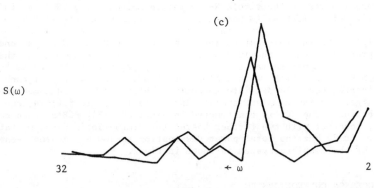

Fig 1

(a) A typical photon burst, n(t).

(b) Its correlation function clipped at 1, $G^{(2)}(\tau)$.

(c) The subsequent Fourier cosine transform, $S(\omega)$.

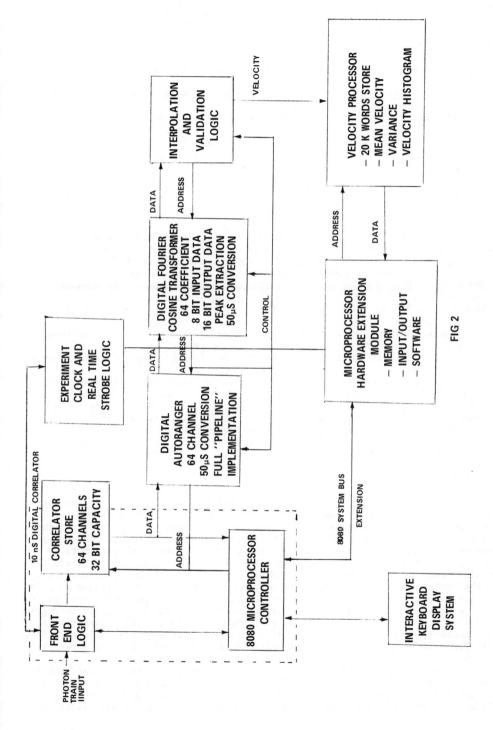

FIG 2

1.   All facilities available on previous correlators including
     cross correlation, signal averaging and probability analysis.

2.   External computor control of whole equipment.

3.   Simultaneous display of analogue and alpha-numeric data.

4.   Multiple "pages" to be available for output of experimental data.

5.   Simultaneous data output to five or more peripherals.

6.   Automatic experiment sequencing and control.

7.   String "macros" as desired by the operator.

8.   Recognized add-on PROM routines.

9.   Post-experimental analysis and display.

Requirement three was judged essential on past experience and dictated a   TV
display console rather than a VDU, this is coupled to an ASC II keyboard.
"Limited-display" devices such as TTY or VDU may also be attached.
     A multiple-level interrupt system has been implemented using the INTEL
8080 system. Commands are entered by the operator in three letter mnemonic
form with full tell back if error free. Command strings can be entered and
given a "macro" name, the subsequent use of which will cause execution of the
complete string.
     Some 30 commands occupying 12 Kbytes are available at present with room
for 98 more at the disposal of individual operators.
     Under 9 the microprocessor will be capable of performing a Fourier analysis
of velocity sequences for display on the TV console.
     The system is capable of recognising, in the "switch-on" procedure, any
add-on PROM routines. In writing such routines each piece of code needs to be
indexed with its command name, address and storage requirements. The software
enhancements may be written in any language compatible with relocatable and
linkable 8080 assembly language. The high-level CORAL 66 is particularly
convenient for this purpose. The software structure is indicated in Fig 3.

"COMBINED OPTICS" PACKAGE

     Time and space does not permit a full description of the optical anemometer
which has been designed to operate with the new burst correlator system. As
mentioned above, the anemometer is a synthesis of two-spot and fringe instruments
in one package with rapid change possible between operating modes. The design
has been developed over a considerable period with regard to many factors.
Perhaps the most important feature of the design was the decision to rotate the
whole system about its main optical axis. The two possible main alternatives of
either rotating output and input components in synchronism or utilising shearing
optics common to both paths were rejected after detailed comparative studies.
     Extreme care has been taken to reduce reflection and glass losses through-
out and, again, great importance has been placed on flexibility. Essentially
all known modes of operation are possible, ranging from a simple one-dimensional
low-power He-Ne laser system to two-colour backscatter Ar^+ ion laser arrangements. In all, some fifty options are available.
     The basic fringe anemometer is outlined in Fig 4. The input laser beam is
passed through a Brewster-angled offset prism and then through a recently per-
fected Brewster-angled low-loss beamsplitter. The two equal-power emerging
beams are incident upon a short focal length lens which crosses and waists them

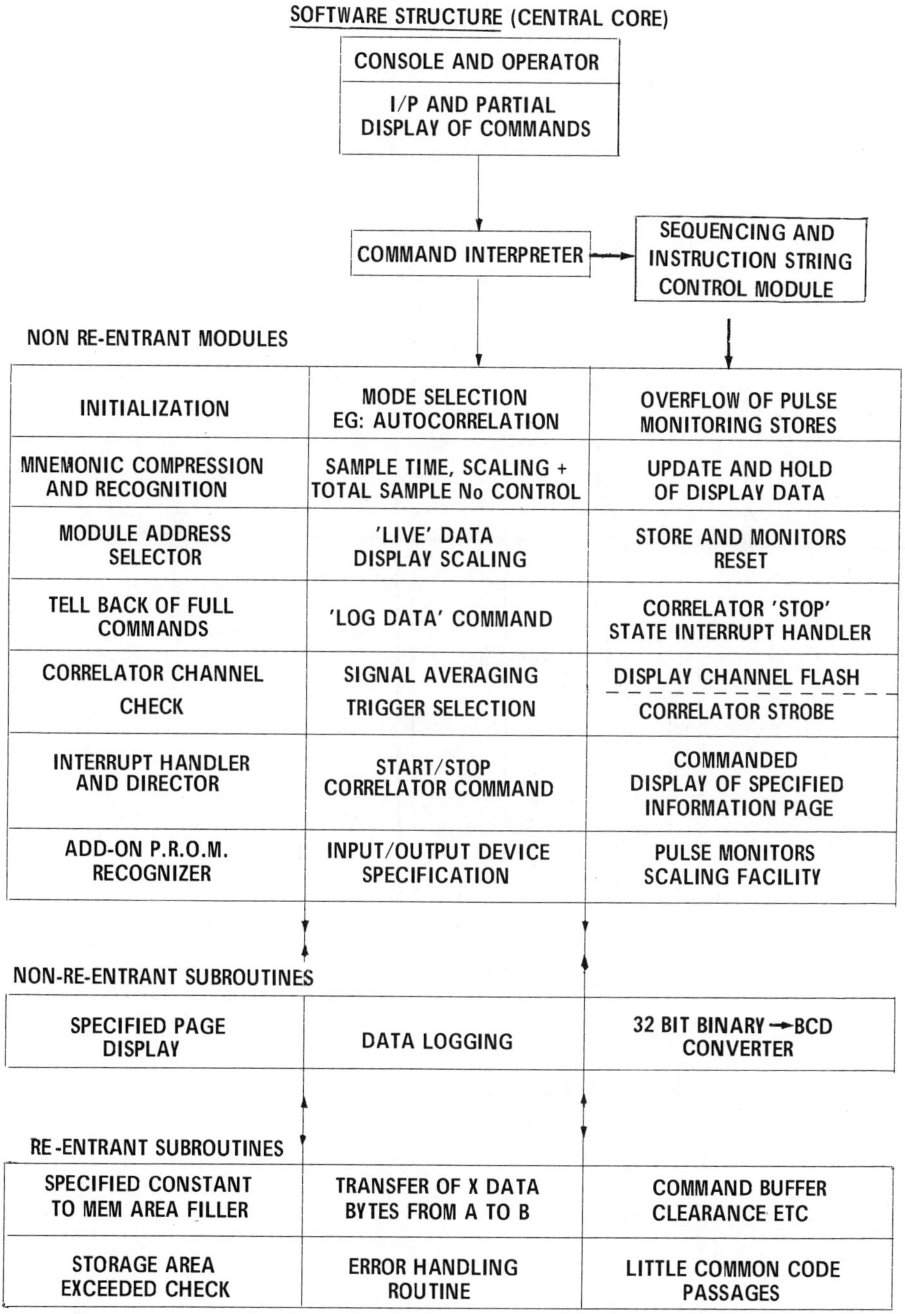

SOFTWARE STRUCTURE (CENTRAL CORE)

CONSOLE AND OPERATOR

I/P AND PARTIAL
DISPLAY OF COMMANDS

COMMAND INTERPRETER

SEQUENCING AND
INSTRUCTION STRING
CONTROL MODULE

NON RE-ENTRANT MODULES

INITIALIZATION	MODE SELECTION EG: AUTOCORRELATION	OVERFLOW OF PULSE MONITORING STORES
MNEMONIC COMPRESSION AND RECOGNITION	SAMPLE TIME, SCALING + TOTAL SAMPLE No CONTROL	UPDATE AND HOLD OF DISPLAY DATA
MODULE ADDRESS SELECTOR	'LIVE' DATA DISPLAY SCALING	STORE AND MONITORS RESET
TELL BACK OF FULL COMMANDS	'LOG DATA' COMMAND	CORRELATOR 'STOP' STATE INTERRUPT HANDLER
CORRELATOR CHANNEL CHECK	SIGNAL AVERAGING TRIGGER SELECTION	DISPLAY CHANNEL FLASH CORRELATOR STROBE
INTERRUPT HANDLER AND DIRECTOR	START/STOP CORRELATOR COMMAND	COMMANDED DISPLAY OF SPECIFIED INFORMATION PAGE
ADD-ON P.R.O.M. RECOGNIZER	INPUT/OUTPUT DEVICE SPECIFICATION	PULSE MONITORS SCALING FACILITY

NON-RE-ENTRANT SUBROUTINES

SPECIFIED PAGE DISPLAY	DATA LOGGING	32 BIT BINARY → BCD CONVERTER

RE-ENTRANT SUBROUTINES

SPECIFIED CONSTANT TO MEM AREA FILLER	TRANSFER OF X DATA BYTES FROM A TO B	COMMAND BUFFER CLEARANCE ETC
STORAGE AREA EXCEEDED CHECK	ERROR HANDLING ROUTINE	LITTLE COMMON CODE PASSAGES

FIG 3

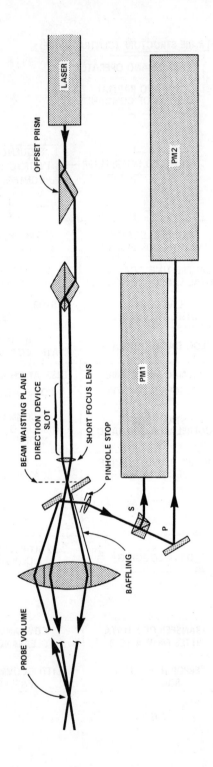

FIG 4 Basic Fringe Anemometer Optics

at the image plane of the main relay lens from which they are projected to cross
again at the probe volume. The relay lens acts as the first element of the
receiver optics. Light collected over an outer annulus is imaged via a mirror
to a pinhole stop and then, via a short focal length lens and polarizing cube,
to a photomultiplier detector. Directional and high turbulence capability in
the Doppler-difference mode may be achieved by inserting an appropriate device,
for example a phase modulator [2] , in the transmitter. Directional information
may alternatively be obtained simply by inserting a $\lambda/4$ plate in the transmission path and cross correlating the outputs of the two detectors[3] .

     Conversion of the basic Doppler-difference system to the two-spot con-
figuration is straightforward. The short focal length lens of the transmitter
is exchanged for one of shorter focal length. The offset prism is rotated
through a small fixed angle to achieve the desired beam separation at the probe
volume. A $\lambda/2$ plate is introduced into one beam and path compensation in the
other. The receiver modifications amount simply to replacing the single pinhole with a double one. Light from each spot is directed, via the polarizing
cube, to a separate photomultiplier, the outputs of which are cross-correlated.

     The basic "fringe" and "two-spot" modes may be used in a number of ways.
The fringe mode may be extended to measure two components simultaneously by
using either two orthogonal polarizations or two colours. When using two colours
a dichroic beam splitter replaces the polarizing cube. In both these cases the
four beams are created by an arrangement of three standard Brewster-angled beam-
splitters. By using dichroic beam-splitters in both transmit and receive paths
a powerful two-colour transit anemometer may also be formed.

     A central feature of all these arrangements is the relay lens. The
requirements on this lens are very stringent. Essentially diffraction-limited
performance must be achieved, coupled with high aperture. Internal flare and
scatter as well as loss must be at low levels. To meet these requirements a
special aspheric lens has been designed having the required performance at two
pairs of fixed conjugates,giving working distances of 50 and 100 cm respectively.
The closer spaced pair are chosen by adding a supplementary component. Between
other conjugates the performance will drop. Desirable as a "zoom" lens might
be no design could satisfy the above conditions.

     This optical anemometer, when coupled to the 10 ns correlator described
above, will form an LDV system of very great flexibility, sensitivity and
accuracy, suitable for a wide variety of stringent applications.

ACKNOWLEDGEMENTS

     During the development of this instrument we have benefitted from dis-
cussions with Mr J Abbiss of RAE Farnborough, Dr R Elder of the Cranfield
Institute of Technology and Dr P Richards of the Whittle Laboratory, Cambridge
University.

REFERENCES

1.   Erdmann, J.C., and Gellert, R.I. 1976.
     Particle Arrival Statistics in Laser Anemometry of Turbulent Flow.
     Applied Physics Letters, 29 408-411.

2.   Foord, R., Harvey, A.F., Jones, R., Pike, E.R. and Vaughan, J.M. 1974.
     A Solid-State Electro-Optic Phase Modulator for Laser Doppler Anemometry.
     Journal of Physics D, 7. L36-L39.

3.   Pike, E.R. 1976. Photon Correlation Velocimetry in Photon Correlation
     Spectroscopy and Velocimetry. Eds H.Z. Cummins and E.R. Pike. Plenum Press
     pp 246-343.

Evaluation of Various Particles for Their Suitability as Seeds in Laser Velocimetry

JUGAL K. AGARWAL and LEROY M. FINGERSON
TSI Inc.
P.O. Box 43394
St. Paul, MN 55164

ABSTRACT

Smaller aerodynamic diameter and good light scattering properties are the two basic requirements of LV seed particles. To evaluate seed particles a test facility was built to simultaneously measure the aerodynamic diameter and the light scattering properties of particles. It consists of an accelerating nozzle and a laser velocimeter.

The facility was designed so that the test particles go through a small central position of the measuring volume. Since the particle lags behind the gas according to its size, the measured particle velocity is a direct indication of the aerodynamic particle diameter. The peak height of the LV signal represents the light scattering properties of the particle. The facility was calibrated by using DOP particles generated by a vibrating orifice monodisperse aerosol generator.

The facility was used to evaluate alumina particles and plastic microballoons. Results showed that the alumina powder contained many particles which scattered 4 to 8 times more light than DOP particles of equivalent aerodynamic diameter. The microballoon contained some particles which scattered as much as 20 times more light than DOP particles of equivalent aerodynamic diameter.

INTRODUCTION

The laser velocimeter (LV) signal originates from the light scattered by the tracer particles (seeds) present in the fluid and the LV actually measures the velocity of these particles. Hence, it is important that these tracer particles follow the fluid with high fidelity. Also, it is necessary that these tracer particles scatter sufficient light so that the signal produced by the photodetector can be measured with the signal processor used.

The ability of a particle to follow the flow can be described by a single particle parameter called aerodynamic diameter. It is defined as the "diameter of a unity density sphere with same settling velocity as the particle in question." The aerodynamic diameter of a particle depends on the shape, size and density of the particle. For LV seeding, particles with smaller aerodynamic diameter are desirable as they follow the flow more readily.

The strength of the LV signal depends on the light scattering properties of the particle, which depends primarily on the cross-sectional area and the refractive index of the particle [1]. For LV seeding, particles with better light scattering properties are desirable as they will produce larger LV signal.

Unfortunately, the general requirement for better light scattering properties and smaller aerodynamic diameter are contradictory. A particle with larger cross-sectional area generally implies a larger diameter and a particle

with higher refractive index usually has higher density. In spite of this general rule, it is still possible that some particles scatter more light than
other particles of equivalent aerodynamic size. Especially, it is possible
that hollow seed materials like glass bubbles or plastic microballoons scatter
more light compared to other solid particles of equivalent aerodynamic diameter.

We at TSI are interested in developing LV seeders which produce or disperse
particles of known aerodynamic size and known light scattering properties. We
have developed a test facility in which we can simultaneously measure the aerodynamic diameter and the light scattering properties of any aerosol particle.
The purpose of this paper is to describe this facility and present some of our
results regarding alumina powder and plastic microballoons.

DESCRIPTION

Yanta [2] first showed that the aerodynamic diameter of aerosol particles
can be determined by measuring the particle velocity in a rapidly accelerating
(or decelerating) flow. Dahneke [3] and Liu et al [4] have used the sheath air
approach to confine the aerosol particles to the center streamline of the flow
through a converging nozzle. Wilson [5] constructed and calibrated an aerodynamic particle sizer in which he made the aerosol particles flow through the
center of a converging nozzle by using the sheath air approach. He measured
the velocity of the particles at the exit of the nozzle by a LV, and showed
that his device has adequate resolution down to 0.5 μm particles.

Our test facility is schematically shown in Figure 1. The accelerating
nozzle is a 1 mm diameternozzle attached to the end of a 12.5 mm diameter tube.
The aerosol particles are introduced at the center of this nozzle through a
second, 0.5 mm diameter, nozzle attached to the end of a 3.1 mm tube. Filtered
air is introduced through the annular space between the two concentric tubes.

The laser velocimeter is set up in a dual beam forward scattering mode
with a 15 mW He-Ne laser and 600 mm focal length lens. The spacing between the
two laser beams entering the lens is 10 mm. The receiving optics is a 120 mm/
200 mm combination lens. The diameter of the aperture on the photomultiplier
tube is 0.89 mm. For this optical arrangement, the calculated parameters of
the LV measuring volume are given below:

Distance between fringes, d_f = 37.97 μm,
Diameter of the measuring volume, d_m = 439.47 μm,
Length of the measuring volume, ℓ_m = 52.74 mm,
Number of fringes = d_m/d_f = 11.57.

The space between the focusing lens and the receiving lens is enclosed with
plexiglas tubing to form a vacuum chamber (see Figure 1). This enclosed chamber
was maintained at -254 cm of water by a vacuum pump. Aerosol as well as filtered air is sucked into the vacuum chamber through the aerosol inlet and the
filtered air inlet respectively. The air velocity at the exit of the outer
nozzle is about 200 m/sec.

The aerosol particles exit through the center of the 1 mm accelerating
nozzle as a narrow focused aerosol beam. The diameter of this aerosol beam is
smaller than the diameter of the LV viewing volume. Thus, when properly
positioned, all the aerosol particles pass through the center of the viewing
volume. Since the intensity of the laser light is nearly constant at the center
of the viewing volume, the pedestal height of the LV Doppler signal is a direct
indication of the light scattering properties of the particle.

Near the exit of the accelerating nozzle, the particles lag behind the
flow proportional to their aerodynamic size. The larger particles move slowly,
while the smaller particles move at a velocity closer to the flow velocity.
The velocity, V, of an individual particle is given by the following equation:

$V = d_f\ f = 37.97\ f$

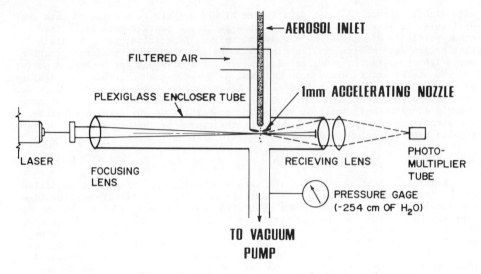

Figure 1. Schematic diagram of the test facility

where f is the Doppler frequency of the LV signal. In our facility, the aerodynamic diameter of a particle is measured by measuring the Doppler frequency
of the LV signal.

CALIBRATION

    The test facility was first calibrated with aerosol particles of known
size. Di-octyl phthalate (DOP) particles in the size range of 1.4 to 10 µm
were generated by using a vibrating orifice monodisperse aerosol generator
(TSI Model 3050). This is a unique instrument to generate uniform particles
from any alcohol or water soluble material and is widely used by aerosol researchers for instrument calibration [4] and for other aerosol studies [6]. A
brief description of this instrument is given in Appendix A of this paper.
Polystyrene latex (PSL) uniform spheres were used to calibrate the facility down
to 0.79 µm. The method of generating PSL aerosol is described in Appendix B of
this paper. The aerodynamic diameter, d_a, of these calibration aerosol particles
was calculated by the equation:

$$d_a = \sqrt{\rho} \cdot d$$

where ρ is the density and d is the diameter of the particle.
    The calibration aerosols were sampled by our facility through the aerosol
inlet. Since these particles are highly uniform, as they pass through the
center of the viewing volume they produce identical signals. Figure 2 is a
photograph of a typical Doppler signal displayed on an oscilloscope. It should
be emphasized that the signal shown in Figure 2 is not a single trace, but
represents a number of superimposed traces. From Figure 2 it is evident that
(a) the particles are highly uniform and (b) all the particles go through the
center of the measuring volume.
    For each particle size, the test facility was calibrated for (a) signal
strength and (b) particle velocity. Throughout this investigation, the gain
of the photomultiplier tube (TSI Model 965) was set at maximum. When calibrating the facility with larger particles, we used a neutral density filter to
attenuate the laser beam (until the PM tube did not saturate), then measured

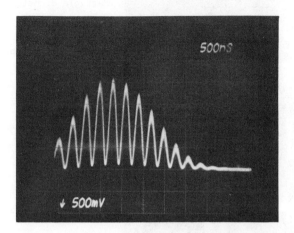

Figure 2. Typical laser Doppler signal (continuous display)
produced by monodisperse DOP aerosol. (Particle
diameter 2.7 μm, laser power 18 mW)

the intensity of this attenuated laser beam with a power meter (Spectra Physics
Model No. 404). The signal strength for that particle size was calculated by
dividing the peak signal height by the laser power. For example, Figure 2 shows
the signal for 2.7 μm particles obtained with a laser power (for 2 beams) of 18
mW. The peak signal height for this signal is 2 volts. Hence, the signal
strength for 2.7 mm DOP particles is 0.11 V/mW. Figure 3 shows the Doppler
signal for 3 different particle sizes. The laser power, peak signal height and
the calculated signal strength are shown on the right hand side of the photographs. The signal strengths for various sizes of DOP and PSL particles are
listed in Table 1. Figure 4 shows the relationship between the signal strength
and the aerodynamic particle diameter. A calibration curve was established by
drawing a line through the maximum signal strength points.

The Doppler frequency, hence the particle velocity, could also be measured
from the oscilloscope display. However, for better accuracy we used a counter
type signal processor (TSI Model 1990) and a probability analyzer (TSI Model
1077) to measure the mean Doppler frequency. For each LV signal, the counter
produces a digital signal which is inversely proportional to the Doppler frequency. The probability analyzer takes the digital signal and classifies it
into one of 256 channels. Figure 5 shows the probability analyzer output for
4.8 μm monodisperse DOP particles. Approximately 99.4% of the frequencies were
sampled between channel 97 to 102. The peak was in channel 99. The Doppler
frequency and the particle velocity corresponding to channel 99 are 2.53 MHz and
95.8 m/sec respectively. Channel 112 contained 0.6% of the frequencies. These
frequencies represent larger particles (doublets) produced by the coalenscence
of two particles.

The measured Doppler frequencies and the particle velocities corresponding to the various particle sizes are given in Table I. Figure 6 shows the
relationship between the particle velocity and aerodynamic particle diameter.

EVALUATION OF ALUMINA POWDER

One of the seeding materials we have tested is 1 - 3 μm alumina powder
provided by Micro-Abrasive Corporation (720 Southampton Road, Westfield,

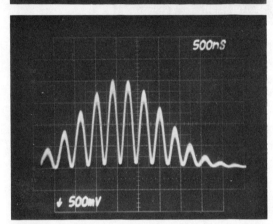

Particle diameter   = 3.2 μm

Laser power   = 8.0 mW

Peak signal height = 2.1 V

Signal strength   = 0.26 V/mW

Doppler frequency   = 2.98 MHz

Particle velocity   = 112.9 m/sec

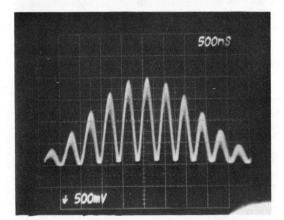

Particle diameter   = 4.8 μm

Laser power   = 3.0 mW

Peak signal height = 2.0 V

Signal strength   = 0.67 V/mW

Doppler frequency   = 2.53 MHz

Particle velocity   = 95.8 m/sec

Particle diameter   = 6.9 μm

Laser power   = 1.7 mW

Peak signal height = 2.0 V

Signal strength   = 1.14 V/mW

Doppler frequency   = 2.16 MHz

Particle velocity   = 81.7 m/sec

Figure 3. Typical laser Doppler signals for 3
sizes of monodisperse DOP aerosols.

TABLE 1.

Aerodynamic Diameter (μm)	Doppler Frequency (MHz)	Particle Velocity (m/sec)	Peak Signal (V)	Laser Power (mW)	Signal Strength (V/mW)	Remarks
0.8	4.50	170.7	0.5	17.5	0.029	PSL
1.1	4.10	155.4	0.4	17.0	0.023	$\rho = 1.05$ gm/cm^3
2.0	3.50	132.6	2.4	17.0	0.14	
1.4	3.97	150.5	1.1	34.0	0.030	DOP
1.7	3.82	144.7	2.0	26.0	0.077	$\rho = 0.98$ gm/cm^3
2.4	3.40	129.0	2.1	16.5	0.13	
2.7	3.22	122.3	2.0	18.0	0.11	
3.2	2.98	112.9	2.1	8.0	0.26	
4.8	2.53	95.8	2.0	3.0	0.67	
6.9	2.16	81.7	2.0	1.7	1.14	
9.6	1.80	68.2	2.0	1.2	1.67	

Calibration data of the test facility with
monodisperse DOP and PSL aerosol.

Mass. 01085). We dispersed these particles in a constant-feed fluidized bed
[7] aerosol generator (TSI Model 3400) shown in Figure 7. In this generator,
the powder to be dispersed is stored in a chamber adjacent to the main fluidized bed chamber and is continuously transported to the fluidized bed by a
bead chain drive.

When this alumina aerosol was sampled by our facility, the Doppler frequencies varied over a wide range. However, by visual observation on an
oscilloscope, we found that the Doppler frequency of the majority of these
signals was about 2.6 MHz which corresponds to an aerodynamic diameter of 4.5 μm.
The geometric mean diameter of this powder was measured by the supplier with a
Coulter counter (Model T, Coulter Electronics Inc., 590 W. 20th St., Hieleah,
Fla. 33010) and found to be 2.0 μm. Assuming the density of the alumina powder
to be 3.96 gm/cm^3, the mean aerodynamic diameter of these particles should be
about 4.0 μm. This is in good agreement with our observation.

However, we observed that the amplitude of these signals (signals which
had approximately 2.6 MHz Doppler frequency) varied over a wide range. Figure 8
shows single traces of 3 signals which were obtained while sampling the
alumina powder. All of these signals have a Doppler frequency of nearly 2.6 MHz,
but the signal strength varied from 8.1 V/mW for 8(a) to 0.4 V/mW for 8(c).
Figure 9 shows the signal from a 4.4 μm DOP particle. The strength of this
signal is 0.86 V/mW. Thus, even though some signals were of low amplitude,
most of the alumina powder particles gave signals larger than those obtained
from DOP particles. For example, the signal shown in Figure 8(a) represents an
alumina particle which gave nearly 8 times larger signal than the DOP particle
of equivalent aerodynamic diameter.

It is not obvious why the signal strength for alumina powder varied over
such a wide range. We did not observe any similar variation in amplitude while
sampling coal dust. With coal dust, all the signals were either equal to or
lower than the signals produced by DOP particles of equivalent aerodynamic
diameter. Hence, we do not suspect that the observed variation in signal
amplitude is caused by any spurious effect in our experimental technique.

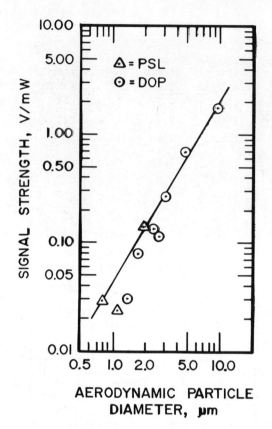

Figure 4. Relationship between aerodynamic diameter and
         signal strength for DOP and PSL particles.
         (Curve is drawn through maximum signal strength
         points)

EVALUATION OF PLASTIC MICROBALLOONS

    We also have tested a sample of plastic hallow spheres, normally known
as plastic microballoons (Bakelite phenolic, BJO-09-30, Chemical and Plastic
Division, Union Carbide). These microballoons were also dispersed by the constant feed fluidized bed aerosol generator (TSI Model 3400). However, while
dispersing these microballoons, no brass beads were used in the fluidized bed
chamber of the aerosol generator. The bead chain transported the microballoons
to the empty fluidized bed chamber. The air flowing from the bottom of the
fluidized bed chamber entrained these microballoons from the bead chain and
carried them with it.
    Figure 10 shows the Doppler signal produced by the microballoons. The
mean Doppler frequency of this signal is 4.3 MHz which corresponds to a mean
aerodynamic particle diameter of 0.95 μm. The signal strength for this signal
is 0.028 volts/mW which is nearly the same as that obtained from 0.95 μm DOP
particles.
    However, with the plastic microballoons, some large amplitude, high fre-
quency signals were observed. Figure 11 shows a photograph of one such signal.
This is a 4 MHz signal corresponding to a particle of 1.25 μm but it represents
a particle which scattered 20 times more light than a DOP particle of the same
aerodynamic diameter.

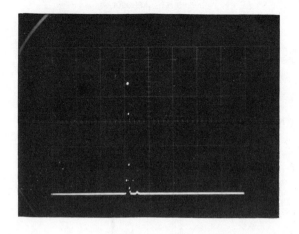

Figure 5. Probability analyzer output for
4.8 μm monodisperse DOP aerosol.

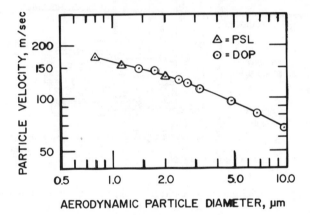

AERODYNAMIC PARTICLE DIAMETER, μm

Figure 6. Relationship between aerodynamic diameter
and particle velocity in the test facility
(Chamber pressure –254 cm of H_2O)

CONCLUSIONS

We have developed a test facility to simultaneously measure the aerodynamic diameter and the light scattering properties of any particle within the
range of 0.5 – 10 μm. By using this test facility, we have observed that
alumina powder contains many particles which scatter more light than DOP particles of equivalent aerodynamic diameter. We also found that, even though
plastic microballoons contain some particles which scatter as much as 20 times
more light than DOP particles of equivalent aerodynamic diameter, most of them
scatter about the same light as DOP particles. However, the mean aerodynamic
diameter of these plastic balloons is about 1 μm and they are easy to disperse.

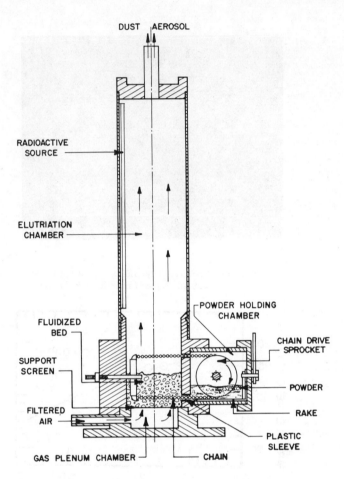

Figure 7. Schematic diagram of continuous feed
fluidized bed aerosol generator.

The major component of our facility is the LV. By adding an accelerating
nozzle to the LV system, one can measure the aerodynamic size of the seed
material. Such facility can also be used for many other studies. For example,
by using monodisperse DOP particles, the signals obtained by two photomultiplier
tubes can be compared, the quality of lenses can be checked, and the effect of
the size and shape of the opening through the collecting lens can be studied.

APPENDIX A

The vibrating orifice monodisperse aerosol generator (TSI Model 3050) used
in the present study was developed by Berglund and Liu [8] at the University of
Minnesota. This instrument is based upon the principle of instability and uniform breakup of a cylindrical liquid jet. A cylindrical liquid jet shown in
Figure 12 is naturally unstable and tends to breakup into droplets. When left
uncontrolled, the breakup process produces non-uniform droplets. However, by
applying a periodic disturbance of appropriate frequency to the liquid jet, the
breakup process can be controlled to produce exceedingly uniform droplets.

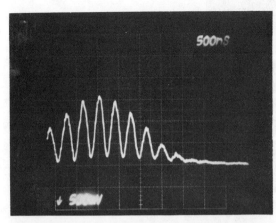

Doppler frequency   = 2.56 MHz

Particle velocity   = 97.2 m/sec

Aerodynamic particle = 4.7 μm
  diameter

Laser power   = 0.185 mW

Peak signal height   = 1.5 V

Signal strength   = 8.1 V/mW

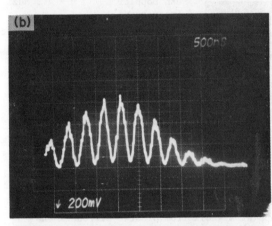

Doppler frequency   = 2.40 MHz

Particle velocity   = 91.0 m/sec

Aerodynamic particle = 5.5 μm
  diameter

Laser power   = 0.185 mW

Peak signal height   = 0.675 V

Signal strength   = 3.6 V/mW

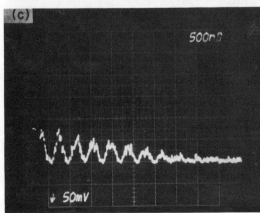

Doppler frequency   = 2.45 MHz

Particle velocity   = 92.9 m/sec

Aerodynamic particle = 5.2 μm
  diameter

Laser power   = 0.185 mW

Peak signal height   = 0.075 V

Signal strength   = 0.40 V/mW

Figure 8. Variation in the peak signal height for 5 μm
aerodynamic diameter alumina particles.

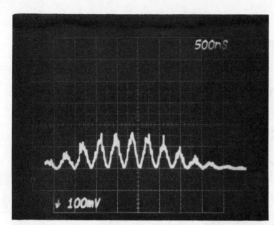

Doppler frequency	= 2.60 MHz
Particle velocity	= 98.9 m/sec
Aerodynamic particle diameter	= 4.4 μm
Laser power	= 0.185 mW
Peak signal height	= 0.16 V
Signal strength	= 0.86 V/mW

Figure 9. Signal from a 4.4 μm DOP particle obtained under the same conditions as those in Figure 8.

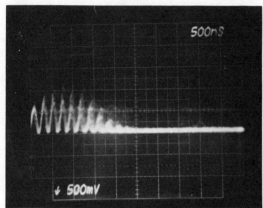

Mean Doppler frequency	= 4.3 MHz
Mean particle velocity	= 163 m/sec
Mean aerodynamic particle diameter	= 0.95 μm
Laser power	= 35 mW
Mean peak signal height	= 1 V
Mean signal strength	= 0.028 V/mW

Figure 10. Multiple-trace laser Doppler signals from plastic microballoons

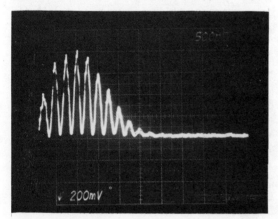

Doppler frequency	= 4 MHz
Particle velocity	= 152 m/sec
Aerodynamic particle diameter	= 1.25 μm
Laser power	= 0.6 mW
Peak signal height	= 0.8 V
Signal strength	= 1.3 V/mW

Figure 11. Signal obtained from a microballoon particle which scattered 20 times more light than a DOP particle of equivalent aerodynamic diameter.

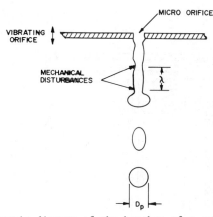

Figure 12. Schematic diagram of the breakup of a cylindrical liquid jet

Furthermore, since the generator produces one droplet per cycle of disturbance, the volume of a droplet can be precisely calculated from the liquid feed rate and the frequency of the distrubance.

The size of the aerosol particles can be varied over a wide range by using the technique of solvent evaporation. The liquid sprayed through the orifice is a solution of a non-volatile solute in a volatile solvent. When the solvent evaporates from the droplets, smaller solute particles remain. The size of the final solute particle can be varied by simply changing the concentration of the solute in the solvent. The size of the solute particle d_p is calculated from the following equation:

$$d_p = \left(\frac{6Q}{\pi f}\, C\right)^{1/3}$$

where Q is the liquid feed rate, f is the disturbance frequency, C is the concentration of the non-volatile solute in the volatile solvent.

The droplet generator of the TSI Model 3050 is shown in Figure 13. A syringe pump feeds the liquid through the orifice at a constant rate. The mechanical disturbance of the orifice is created by driving a piezoelectric ceramic with an a.c. signal. The function of the dispersion air is to provide a turbulent air jet around the droplet stream so that they do not coalesce significantly.

The complete vibrating orifice monodisperse aerosol generator is schematically shown in Figure 14. The dispersed aerosol stream merges with a flow of dilution air. This diluted aerosol passes through a vertical drying and neutralizing chamber containing a radioactive Kr-85 source. The Kr-85 source ionizes the dilution air which in turn neutralizes the electrostatic charge induced on the aerosol particles during the generation process. The residence time in the vertical chamber allows the volatile solvent to evaporate from the liquid droplets. Monodisperse aerosol exits at the top of the instrument and is transported to the sampling inlet through a plastic vacuum cleaner hose.

APPENDIX B

Polystyrene latex aerosol was generated by using a system shown in Figure 15. Uniform polystyrene latex spheres in suspension were purchased from Dow Chemical Company (Dow Diagnostics, P.O. Box 8511, Indianapolis, Indiana 46268). A very dilute suspension of these spheres was prepared by mixing one drop of the original suspension in about 400 ml of distilled water. This suspension was atomized in a Rotec Aerosol Generator (Model RAG, Cavitron, 7922 Haskell Avenue, Van Nuys, Ca. 91406).

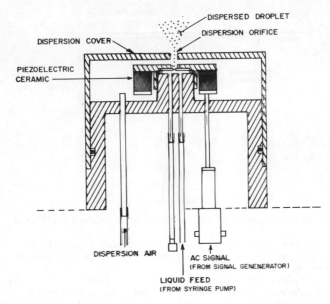

Figure 13. Schematic diagram of the droplet generator of the vibrating
liquid jet.

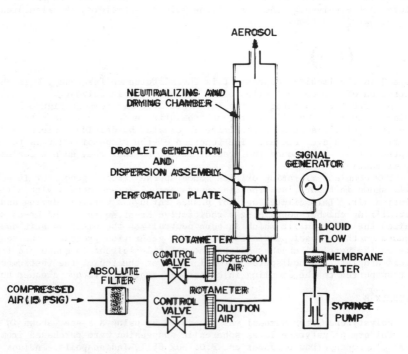

Figure 14. Schematic diagram of the vibrating orifice monodisperse aerosol
generator.

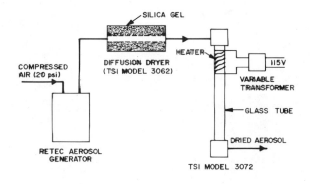

Figure 15. Schematic of the monodisperse PSL aerosol generation system.

The resulting aerosol was passed through a diffusion drier (TSI Model 3063). The diffusion drier consists of two coaxial cylinders, the outer one of plastic and the inner one of wire mesh. The space between the cylinders contains silica gel. Aerosol passes down the axis of the unobstructed inner cylinder. Moisture diffuses to the silica gel. The aerosol is then passed through a heated glass tube (TSI Model 3072) to ensure complete drying of the PSL particles.

REFERENCES

1. Adrian, R.J., 1978. Estimation of LDA Signal Strength and Signal-To-Noise Ratio. TSI Quarterly, TSI Incorporated, P.O. Box 43394, St. Paul, Mn. 55164. Vol. IV, Issue 1.

2. Yanta, W., 1973. Measurements of aerosol size distribution with a laser Doppler velocimeter. Presented at AIAA 6th Fluid and Plasma Dynamics Meeting, Palm Springs, Ca.

3. Dahneke, B. and Flachsbart, H. 1972. An Aerosol Beam Spectrometer. J. Aerosol Sci. 3:345-349.

4. Liu, B.Y.H., Berglund, R.N., and Agarwal, J.K. 1974. Experimental Studies of Optical Particle Counters. Atmospheric Environment, Vol. 8, 717-732.

5. Wilson, J.C. 1977. Aerodynamic Particle Size Measurement by Laser-Doppler Velocimetry. Ph.D. Thesis, University of Minnesota, Minneapolis, Mn. 55455.

6. Liu, B.Y.H, and Agarwal, J.K. 1974. Experimental Studies of Aerosol Deposition in Turbulent Flow. Aerosol Science, Vol. 5, 145-155.

7. Marple, V.A., Liu, B.Y.H., and Rubow, K.L. 1978. A Dust Generator for Laboratory Use. Am. Ind. Hyg. Assoc. J. 39:26-32.

8. Berglund, R.N., and Liu, B.Y.H. 1973. Generation of Monodisperse Aerosol Standards. Environ. Sci. Technol. 7:147-153.

SESSION II
DATA ANALYSIS I

Chairman: **E. R. PIKE**
Royal Signals and Radar Establishment

Advances in Digital Data Processing

INVITED PAPER

H. J. PFEIFER
German-French Research Institute
68301 Saint-Louis, France

ABSTRACT

Digital processing of data from counter type (burst) processors proves to be the most accurate and reliable approach in the majority of LDA applications. Its use is of particular interest if the data rate is limited because of poor mean signal-to-noise ratio or because of low particle number density. The latter is true in nearly all gaseous flows and especially in high speed phenomena including high speed combustion flows.

Data acquisition and processing systems based on zero crossing detection in individual bursts are, therefore, widely used in LDA. Many different systems existing today allow one to determine mean velocities and turbulence intensities. This paper describes a new type of zero crossing system based on a single counter which is simpler than current dual counter devices. The performance of this instrument is examined with particular regard to the effect of signal-to-noise ratio on accuracy.

The present paper also describes how additional information can be obtained including power spectra and auto- and cross-correlation functions. Power spectra can be found either by a direct FOURIER transform of the individual realizations or by FOURIER-transforming the auto-correlation function. As will be shown, the choice between the two approaches strongly depends on the maximum data rate which can be handled by the instrumentation and the available number of signal bursts with a given signal-to-noise ratio.

INTRODUCTION

In many LDA applications there is a poor signal-to-noise ratio of the scattered light signals. This may be true for combustion studies where artificial particles are difficult to seed into the flow or for backscatter experiments where light scattering by particles is much less effective than in forward scattering, but it may also be true for low gas density flows as they occur in many wind tunnels [1].

In counter type data acquisition systems poor signal-to-noise ratio generally leads either to high measuring error or to low signal rate. Since the background noise is constant, measuring error and signal rate are not independent of each other, because both are controlled by the trigger level which initiates the individual realizations. The choice of a high trigger level results in a high signal-to-noise ratio and hence in a low measuring error, but the data rate drops down and vice versa.

In many cases it is desirable to obtain a high data rate for auto- and cross-correlation measurements and power spectra at the expense of accuracy. This can be done in flow situations where the turbulence level is high. In laminar flows, however, even quite a small amount of noise on the signals can completely prevent calculating correlation functions or spectra.

In this paper counter based data acquisition systems for LDA applications are described. Hereby emphasis is directed upon accuracy, data rate, and reliability of the data to be obtained under difficult circumstances. In the first part dual counter systems will be described followed by single counter systems in the second part. Finally, methods for the determination of auto- and cross-correlations as well as algorithms for power spectra calculations will be outlined.

PERFORMANCE OF DUAL COUNTER SYSTEMS

Since the beginning of Laser Doppler Anemometry and especially since the fringe type LDA has been described for the first time many different instruments have been constructed for data acquisition purposes [2]. They work either in the frequency domain, as is done by frequency analyzers and frequency trackers, or they work in the time domain like photon correlators and zero-crossing counters. Although counter based systems have been invented in an early stage of LDA development [3], they became widely used in the last three years only. Nowadays, they are probably more often applied than any other data acquisition device. The main reason for this is supposed to be the fact that signals stem from individual particles and only counter based systems allow to obtain an individual velocity value attributed to an individual particle. Among various types of counter methods the dual counter system which was described in 1972 for the first time [4,5] has found the widest spread use so far in research and test institutions. In order to discuss the performance of this type of instrument its mode of operation will briefly be outlined by means of Figure 1:

The data acquisition process is started as soon as the high-pass filtered signal ① of Figure 1 exceeds the trigger level TL. The zero crossing detector ② delivers a pulse for each positive going zero crossing of the signal. Counter ③ then measures the

time interval t_1 cor-
responding to a number
n_1 of oscillations,
whereas, for the same
signal, counter ④
measures the time in-
terval t_2 which cor-
responds to a number
n_2 of oscillations. A
signal is only proc-
essed if the follow-
ing condition is met:

$$\left| \frac{n_1}{t_1} \cdot \frac{t_2}{n_2} - 1 \right| < \varepsilon \quad (1)$$

where ε is much less
than unity.

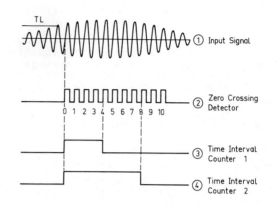

Fig. 1: Principle of dual counter system

In various approaches calculations according to Equation (1) have been made possible. The ratio n_1/n_2 was in most cases 5/3 or 8/4 or both together [6,7,8,9]. The comparison between n_1/t_1 and n_2/t_2 has been made either in an analog or in a digital way or after digitizing each individual signal by means of a minicomputer.

One must mention in this context that nearly all of these in-struments require a minicomputer for data reduction. Only very few of them are capable of calculating mean velocity values out of a given number of individual realizations and even less are able to deduce turbulence intensities as stand alone systems.

In dual counter systems the use of the second counter was o-riginally found necessary for two reasons. The first one was that because of the simultaneous presence of two or more light scatter-ing particles in the probe volume, destructive interferences may occur leading to erroneous measurements. At medium or low signal rates, however, this risk will scarcely occur as will be shown in the following estimation.

The probability of particle occurrence in the probe volume is covered by the POISSON probability distribution. If $\bar{n}$ denotes the mean particle rate as defined by the quotient from the mean time of presence in the probe volume divided by the mean time interval between succeeding particles, then the following statement applies: The probability P_n to find more than one particle in the probe volume at one instant related to the probability P_1 to detect one particle is given by the following relationship:

$$\frac{P_n}{P_1} = \frac{e^{\bar{n}} - 1}{\bar{n}} - 1 \approx \frac{\bar{n}}{2} . \quad (2)$$

Experience has shown that in many cases the mean particle rate is less than a few percent or less than one percent. Even in the case of simultaneous presence of two particles, no errors will gen-erally occur. As a matter of fact the signal amplitude decreases significantly if destructive interference is observed so that such signals are not accepted by most of the data acquisition systems.

Even in the worst case the maximum error is half an oscillation if
no precautions are provided for strong amplitude variations during
the measuring time.

        The second reason for which the use of two counters seemed to
be necessary was the possibility that the data acquisition element
becomes ready for a new measurement at the instant when a particle
nearly ended its passage through the probe volume. This would lead
to an insufficient number of oscillations in this signal and the
counter would continue to count until the next signal arrives. The
same event may occur if a particle flies nearly parallel to the
fringes and no frequency shifting device is used.

        So, by means of the second counter and applying Equation (1)
it is possible to remove erroneous velocity values from further
processing. One must emphasize, however, that the use of a second
counter and application of Equation (1) does not improve the result at a poor signal-to-noise ratio of the scattered light signals [10]. This can be
explained by means of
Table 1. Here, noise
was added to a 0.5 MHz
signal frequency. The
bandwidth of the pink
noise ranged from 0.1
MHz to 1 MHz. The first
column of this Table
contains various sig-
nal-to-noise rms values.
The ε values, according
to Eq.(1), were chosen
to be between 10 % and

S/N ε	10 %	2 %	1 %	0.5 %	100 %
1.00	9.22	8.87	7.41	7.46	12.67
1.25	7.66	7.22	5.94	6.19	10.67
2.00	3.84	3.53	3.17	2.81	6.21
5.00	0.71	0.74	0.70	0.70	0.73
∞	0.06	0.06	0.06	0.06	0.06

Table 1: Standard deviation σ in percent as a func-
         tion of the signal-to-noise ratio and ε.

0.5 %, and in the last column ε was set to 100 %. The turbulence
levels in percent, as they are pretended by the noise, are listed
in this Table and it can clearly be seen that the choice of ε
has nearly no influence on the result at a poor signal-to-noise
ratio and no influence at all for signal-to-noise ratios above 5.
Since only the latter case allows reliable turbulence values to be
obtained, it may be concluded that the choice of ε serves no purpose
and that the dual counter system is much more elaborate than necessary. In spite of this, the dual counter system has been demonstrated
in the past to be very useful for data acquisition under various and
especially under difficult flow conditions. Nevertheless it is supposed that the future trend is going to single counter systems which
present at least the same accuracy and the same reliability as the
dual counter systems, but which consist of fewer components.

SINGLE COUNTER SYSTEM

        For the above mentioned reasons it is possible to substanti-
ally simplify the data acquisition systems in laser anemometry. In
fact, it suffices to replace the second counter by a logical circuit. It is the purpose of this circuit to verify if, at the end
of the measuring time of the counter, the signal amplitude is still
lying above a previously chosen level. This process is depicted in
Figure 2. If the input signal ① exceeds a prechosen trigger level
TL, the data acquisition system is put into operation and the zero
crossing detector ② delivers a pulse for each zero crossing of
the input signal as long as the trigger level is exceeded. A select-

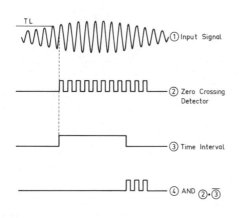

Fig. 2: Principle of single counter system

able time interval is given in which the number of oscillations occurring in the signal is determined. The signal emanating from the zero crossing detector ② as well as the inverted signal from the time interval ③ are AND-gated. The resultant signal ④ is the flag pulse for further signal processing.

A further logical circuit consisting of a retriggerable mono-flop may be used. This circuit is not included in Figure 2. It checks whether the signal amplitude falls significantly under the trigger level during the measuring time or not. Such an event may occur by destructive interference between two particles in the probe volume.

Because of its electronic arrangement, the single counter method is more straightforward in use than the dual counter method. In effect a somewhat complex counter is here replaced by a logical circuit. Moreover it offers the advantage of not necessitating computational operations for Equation (1). A comparison made between the single counter and the dual counter methods shows both systems to be of equal reliability.

Several LDA data acquisition systems have been built according to this principle [11,12], one of them will be explained in some more detail [13]. The basic philosophy for construction of this system was that the overall measuring error in determining the mean velocity by LDA methods is given by the error in determining the fringe spacing. Errors in the order of 0.5 % are supposed to be realistic if enough care is taken in the measuring procedure. From this point of view no effort is worthwhile to lower the absolute error in mean frequency measurement to values below 0.1 %. Relative errors for individual signals come from jitter in zero crossings due to noise and furthermore in most data acquisition systems from the ± 1 count

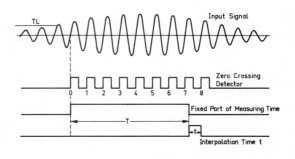

Fig. 3: Single counter system with interpolation

ambiguity. Since noise cannot be removed completely, errors of individual realizations in the order of 0.1 % generally have to be
tolerated even under favorable conditions.

For these reasons, in the system described here, no clock is
used for frequency measurement. As the filtered input signal exceeds the trigger level TL, Figure 3, the data acquisition process
is put into operation. The measuring cycle starts at the first following zero crossing by triggering a monoflop which defines the
fixed part of the measuring time. The negative going edge of this
monoflop sets the interpolation pulse t. The interpolation pulse
is reset by the next positive going zero crossing of the input signal. At the same time, the gate of the counter is closed. By this,
the duration of the interpolation pulse determines the interval
between the end of the fixed part of the measuring time T and the
next zero crossing of the input signal. The interpolation pulse is
applied to the input of a high-speed integrator whose output voltage
is proportional to pulse duration t. Using a high-speed analog-to-
digital converter, this output voltage is converted into a digital
value after a settling time of 1 μs. The total measuring time T + t
for an integer number N of cycles starts and stops exactly with zero
crossings of the input signal. Hence, the signal frequency is given
by

$$f = \frac{N}{T + \tau} \; .$$

Tests of the system with artificial signals showed that absolute errors of the mean frequency are in the order of 0.1 % over
the total frequency range from 2.6
MHz to 20 MHz. With perfect artificial signals the relative standard deviation was always less than
0.07 %.

Again the influence of noise
on the measured relative standard deviation was determined.
Figure 4 gives an idea of the
results. The ordinate shows the
relative standard deviation as
it is pretended by noise and
the abscissa the signal-to-noise
ratio. Two different signal
frequencies and two different
fixed parts T of the measuring
time were used. It is worth
mentioning that doubling the
measuring time results in
cutting the measuring error
in half as already has been
found for dual counter systems [14]. From this point
of view single counter systems
with variable measuring time
according to the individual signal duration are preferable to
fixed time systems [15]. The
variable measuring time, how-
ever, requires a much more elaborate technique which unfortunately
is also more time consuming.

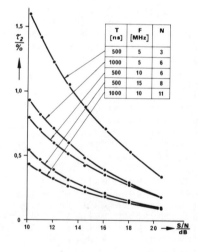

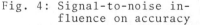

Fig. 4: Signal-to-noise influence on accuracy

In spite of the high accuracy attained, the data acquisition system described here has a high data rate. Minimum time interval between succeeding individual realizations is only 5 μs, a value which is adapted to the maximum possible transfer rate into minicomputer memories via direct memory access. Figure 5 shows some results of the turbulence intensity which have been obtained from measurements on the axis of a high speed free jet at various mean velocities.

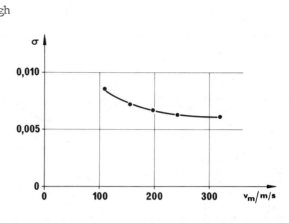

Fig. 5: Measured standard deviations

For the turbulence intensity σ the following formula holds

$$\sigma = \lim_{Z \to \infty} \left(\frac{1}{Z} \frac{\sum (v_i - v_m)^2}{v_m^2} \right)^{1/2}$$

where v_i is the instantaneous velocity and v_m the mean velocity. No bias correction was needed for these examples. Figure 6 shows a histogram based on 5000 individual realizations. Figure 5 and Figure 6 give an idea of the accuracy of the system and of its reliability with respect to false data recognition.

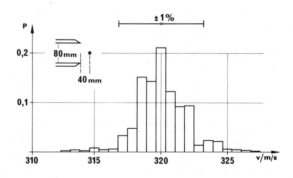

Fig. 6: Velocity histogram in laminar flow

4. CORRELATION MEASUREMENTS AND POWER SPECTRA

During the last years, auto- and cross-correlation measurements and determination of power spectra with LDA techniques attracted increasing interest [16,17]. The reason for this is the fact that LDA is a nonintrusive method. In contrast to mechanical probes it avoids all perturbations brought into the flow by the measuring procedure. Such perturbations produced, for example, by hot wires, make the results of correlation measurements often very uncertain [18]. For high speed gaseous flows it is certainly true that LDA is the only method which allows cross-correlation measurements of a physical quantity, i.e. the flow velocity, in absolute units.

In Figure 7 two random processes are supposed to be a function of time. These processes are the velocity fluctuations v_A' at point A and v_B' at point B, respectively. Time intervals are denoted by τ. The cross-correlation R is defined by Equation (3), whereas the cross-correlation coefficient is given by Equation (5).

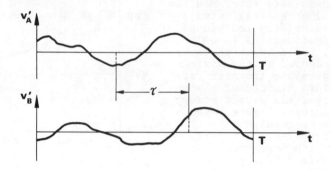

Fig. 7: Principle of cross-correlation measurements

$$(3) \quad R_{v_A' v_B'}(\tau) = \lim_{T \to \infty} \frac{1}{T} \int_0^T v_A'(t) v_B'(t + \tau) dt$$

$$(4) \quad R_{v_A' v_B'}(-\tau) = R_{v_B' v_A'}(\tau)$$

$$(5) \quad \rho_{v_A' v_B'}(\tau) = \frac{R_{v_A' v_B'}(\tau)}{\sqrt{R_{v_A'}(0) \ R_{v_B'}(0)}}$$

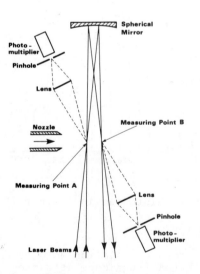

For calculation of the auto-correlation of velocity fluctuations at one point in the flow, the same three Equations from (3) to (5) hold and only the indices A and B must be cancelled. In this case Equation (4) indicates that auto-correlation is an even function.

The optical system for cross-correlation measurements is shown in Figure 8. Two measuring points are formed by a single laser and a spherical mirror. From both probe volumes the forward scattered light is observed so that no mutual disturbance occurs between them. The spacing between the two points can be changed by adjusting the spherical mirror and the observation system for probe volume B.

Fig. 8: Optical setup for cross-correlation measurements

For data processing purposes it must be taken into account that in individual realization systems the particles arrive at ran-

dom instants and that no information on velocity is available in
the time intervals between. For any correlation measurement, therefore, the indi-
vidual time in-
tervals between
succeeding sam-
ples must be re-
corded and stored.
For the compu-
tation of corre-
lation estimates
a time grid is
formed, after
this, according
to which corre-
lation elements
are calculated
for different τ
values. The prin-
ciple of a data
processing system

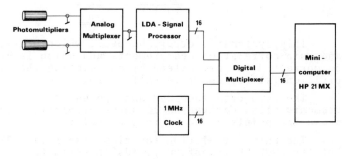

Fig. 9: Data acquisition system for cross-corre-
        lation measurements

for cross-correlation measurements is depicted in Figure 9.

Only one LDA signal processor is needed to process signals
from the two probe volumes. This is made possible by an analog multiplexer between the photodetectors on the one side and the input of
the LDA processor on the other. At the moment when a signal is validated by the processor, the momentary value of a 1 MHz clock is fed
into the input of a digital multiplexer which connects the processor
and the 1 MHz clock over one single 16 bit wide data line with a
minicomputer. So the direct memory access of minicomputers can easily be used. This facilitates the software and accelerates the data
transfer. Figure 10 shows a registration example of cross-correlation coefficients which were obtained in the course of free jet
studies. The data characterizing this experiment are included in
the Figure. The total recording time was 138 milliseconds, only.
An amount of 1500 individual velocity values from each of the in-

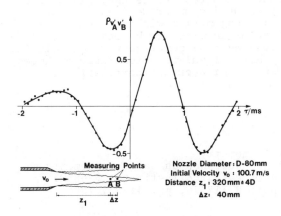

put channels along with
the corresponding clock
values were recorded.
This means that the
average signal rate was
approximately 11000 sig-
nals per second. For
Figure 10 the mean val-
ues of the cross-corre-
lation coefficients
over time intervals of
64 μs were calculated
and plotted.

Fig. 10: Example for cross-correlation
         coefficients

Again a few words
on noise in correlation
measurements: If $\overline{v'^2}$
denotes the mean square
of true velocity vari-
ations and $\overline{n^2}$ denotes
the mean square veloci-
ty variations pretended

by noise, then it is easy to demonstrate [19] that the measured correlation coefficient E_m is lower than the true correlation coefficient E_t:

(6)
$$\frac{E_m\{\rho(\tau)\}}{E_t\{\rho(\tau)\}} = \frac{1}{1 + \dfrac{n^2}{\overline{v'^2}}}$$

This systematic error, however, which occurs for both cross-
and auto-correlation measurements and which is independent of τ can
be corrected to some extend by calculating the auto-correlation co-
efficient(s) for the time shift zero.

The same data recorded and stored for cross-correlation measurements allow derivation of auto-correlation functions at the
measuring points A and B, respectively.

The auto-correlation function itself is only of minor interest
but its FOURIER transform yields an estimate of the power spectrum
which in nearly all cases gives very useful information on the
flow characteristics [20]. Several papers have been published on
this subject and for a long time the auto-correlation function was
supposed to be the only way to get power spectra in individual realization LDA systems. More recently, however [21], a theoretical comparison between FOURIER transform of randomly sampled values and
FOURIER transforming the auto-correlation function was carried out.
Assuming the same conditions, the variabilities of the spectral
estimates in the first case were found to be higher than in the
second one. Moreover, due to the characteristics of the POISSON
distribution even extremely low data rates theoretically let one
calculate auto-correlation values for very short time shifts, which
results in high frequencies in the spectrum. An advantage of the
direct FOURIER transform is the smaller amount of computational
effort.

CONCLUSIONS

The trend in counter-based LDA signal processors will
probably be toward single counter devices, since they are of the
same accuracy and reliability as dual counter devices, but contain fewer components. Used in conjunction with minicomputers
they allow all statistics of major interest to be computed including auto- and cross-correlations and the power spectra of velocity
fluctuations. A sufficiently high signal-to-noise ratio of known
magnitude is a required condition for accurate correlation
measurements.

REFERENCES

1. A. Boutier
    Data Processing in Laser Anemometry
    Proceedings of the ISL/AGARD Workshop on Laser Anemometry,
    May 5-7, 1976, German-French Research Institute, F-68301
    St-Louis, France, ISL-Report R 117/76, pp. 41-53, 1976

2. H.J. Pfeifer
    The Development and Use of Laser Anemometers in Transonic
    and Supersonic Gaseous Flows (Review Article)
    J. Phys. E: Scientific Instr., Vol. 8, No. 4, pp. 245-252, 1975

3. A.E. Lennert, F.H. Smith, Jr., H.T. Kalb
   Application of Dual Scatter, Laser, Doppler Velocimeters for
   Wind Tunnel Measurements
   ICIASF'71 Record, IEEE Publication 73-C-33
   AES, pp. 157-167, 1971

4. H.J. Pfeifer, H.D. vom Stein
   An Automatic Data Processing System for Laser Anemometers
   IEEE Transactions on Aerospace and Electronic Systems,
   Vol. AES-8, No. 3, pp. 345-349, May 1972

5. Asher, J.A.
   Laser Doppler Velocimeter System Development and Testing
   General Electric Technical Information Series, P.O. Box 43,
   Bldg. 5, Schenectady N.Y. 12301, Oct. 1972

6. R.E. Zammit, M.K. Pedigo, W.H. Stevenson, A.K. Owen
   LDV Processor for High Velocity Flows
   Proceedings of the Second Intern. Workshop on Laser Velocimetry,
   Edited by H.D. Thompson and W.H. Stevenson, Engineering Ex-
   periment Station, Purdue University, W. Lafayette, Indiana
   47907, Vol. 1, pp. 204-223, 1974

7. H.T. Kalb, V.A. Cline
   New Technique in the Data Processing and Handling of Laser
   Velocimeter Burst Data
   Rev.Sci.Instrum., Vol. 47, No. 6, pp. 708-711, June 1976

8. J.C.F. Wang
   Measurement Accuracy of Flow Velocity via a Digital-Frequency-
   Counter Laser Velocimeter Processor
   Proceedings of the LDA-Symposium Copenhagen, P.O. Box 70,
   DK-2740 Skovlunde, Denmark, pp. 150-175, 1976

9. A.E. Lennert
   Fundamentals of Laser Doppler Velocimetry
   AGARD Lecture Series No. 90, paper 3, AGARD, 7, rue Ancelle,
   92200 Neuilly sur Seine, France, 1977

10. H.J. Pfeifer, H.J. Schäfer
    A Single-Counter Technique for Data Processing in Laser Ane-
    mometry
    German French Research Institute, St.-Louis (ISL), F-68301
    St.-Louis, France, ISL Report 30/74, 1974

11. M. König
    LDA-Datenerfassungsgerät zur Bewertung und schnellen Auf-
    bereitung von LDA-Signalen für einen Minicomputer
    German French Research Institute, St.-Louis (ISL), F-68301
    St.-Louis, France, ISL Report RT 509/77, 1977

12. W. Hösel, W. Rodi
    Eine genaue digitale Auswertemethode für Laser-Doppler-
    Velocimeter-Signale
    Technisches Messen atm, Heft 2, pp. 43-48, and Heft 3, pp. 105-
    110, 1978

13. M. König, H.J. Pfeifer
    LDA Data Acquisition Systems with High Accuracy and High Data
    Rate, ICIASF'77 Record, Int. Congress on Instrumentation in
    Aerospace Simulation Facilities, Shrivenham, England
    Sept. 6-8, 1977, IEEE Publication 77 Ch 1251-8 AES, pp. 13-21,
    1977

14. F.V. Steenstrup
     Counting Techniques Applied to Laser Doppler Anemometry
     DISA-Information, No. 18, pp. 21 - 25, September 1975

15. W. Hoesel, W. Rodi
     A Highly Accurate Method for Digital Processing of Laser
     Doppler Velocimeter Signals
     SFB 80, University of Karlsruhe, D-7500 Karlsruhe, Germany
     Report SFB 80/E/67, 1975

16. J.C.F. Wang
     Laser Velocimeter Turbulence Spectra Measurements
     Proceedings of the Minnesota Symposium on Laser Anemometry,
     Oct. 22-24, 1975, Edited by E.R.G. Eckert, University of
     Minnesota, pp. 538-567

17. B. Koch, H.J. Pfeifer
     Detection of Large Scale Coherent Structures in Free Jets
     by Laser Anemometry and Crossed Beam Schlieren Techniques
     Preprints for the Fifth Biennial Symposium on Turbulence,
     Oct. 3-5, 1977, Dept. of Chem. Engineering, University
     of Missouri, Rolla, Paper V. 8-1, 1977

18. N.W.M. Ko, P.O.A.L. Davies
     Interference Effect of Hot Wires
     IEEE Transactions on Instrumentation and Measurement IM 20
     pp. 76-78, 1971

19. X. Bouis
     Applications récentes, à l'ISL, de la vélocimetrie Laser aux
     mesures dans des écoulements turbulents
     AGARD-CP-193 on Applications of Non-Intrusive Instrumentation
     in Fluid Flow Research, paper 7, May 1976

20. D.M. Smith, D.M. Meadows
     Power Spectra from Random-Time Samples for Turbulence
     Measurements with a Laser Velocimeter
     Proceedings of the Second Intern. Workshop on Laser Velo-
     cimetry, Edited by H.D. Thompson and W.H. Stevenson,
     Engineering Experiment Station, Purdue University,
     W. Lafayette, Indiana 47907, Vol. 1, pp. 27-46, 1974

21. M. Gastner, J.B. Roberts
     The Spectral Analysis of Randomly Sampled Records by a
     Direct Fourier Transform
     Proc. R. Soc. London A.354, 27-58, 1977

How Real Are Particle Bias Errors?

ROBERT V. EDWARDS
Chemical Engineering Department
Case Western Reserve University
Cleveland, OH 44106

ABSTRACT

A review of the current literature will reveal that there is a disagreement among different investigators as to the form or even the existence of particle bias errors. The effect of the interaction of velocity fluctuations, mean gradients, particle density and detection method will be explored for three different types of detectors:

1) Spectrum Analyzers
2) Trackers
3) Counters.

Estimates of mean velocities, gradients and velocity fluctuation intensities will be calculated. By taking into account the time scales of the flow and the detectors, some of the contradictions in the literature will be resolved. Experimental strategies for avoiding problems with particle biases will be suggested.

INTRODUCTION

Particle bias errors will be defined as those effects due to the sampling statistics of a particular velocity being dependent on the velocity. This can be caused by either temporal or spatial changes in the velocity as seen by the detector. How real are particle bias errors? It depends on the characteristics of the signal processor. The original work [1] on particle bias errors was predicated on the idea of a so-called "individual realization" detector. Such a detector would measure the velocity of each particle that passed through the sample volume and give equal weight to each measurement. The weighting factor for each velocity is clearly proportional to the volume of fluid swept through the sample volume by velocity $\vec{v}$, $d\vec{v}$ during the measurement period T where $d\vec{v} = dv_1\ dv_2\ dv_3$ is a volume element in velocity space.

Let $V(\vec{v})$ be the volume of fluid swept through the sample volume in time T by velocity $\vec{v}$, $d\vec{v}$; $P(\vec{v})d\vec{v}$ be the probability of observing a velocity at $\vec{v}, d\vec{v}$; and $d_v t$ be the amount of time the velocity $\vec{v}$, $d\vec{v}$ is observed during the measurement period T. Then let $A(\vec{v}/|v|)$ be the cross section of the sample volume as seen by the detector for a particle coming from the direction defined by $\vec{v}/|v|$. Then

$$V(\vec{v}) = \vec{v} \cdot \vec{A}\ d_v t \tag{1}$$

79

But $P(\vec{v})d\vec{v} = \dfrac{d_v t}{T}$, so

$$V(\vec{v}) = T\,\vec{v}\cdot\vec{A}\;P(\vec{v})d\vec{v} \tag{2}$$

    The scattering particles are assumed to be Poisson distributed throughout the system, so that the expected number of measurements of velocity $\vec{v}$, $d\vec{v}$ is

$$\rho V(\vec{v}) = \rho T\vec{v}\cdot\vec{A}\;P(\vec{v})d\vec{v} \tag{3}$$

where ρ is the particle number density.

    The total number of measurements of velocity during the time T is

$$N = \rho T \int_{-\infty}^{\infty} \vec{v}\;\;\vec{A}\;P(\vec{v})d\vec{v} \tag{4}$$

The average velocity measured, $\overline{\vec{v}}_M$, is

$$\overline{\vec{v}}_M = T\int_{-\infty}^{\infty} \vec{v}\;(\vec{v}\cdot\vec{A})\;P(\vec{v})\;d\vec{v}/N \tag{5}$$

    Assuming the system is temporally   fluctuating but statistically stationary

$$\overline{\vec{v}}_M = \frac{\displaystyle\int_{0}^{T} \vec{v}(t)(\vec{v}\cdot\vec{A})dt}{\displaystyle\int_{0}^{T} (\vec{v}\cdot\vec{A})dt} \tag{6}$$

    Buchhave [2] has carefully tabulated values of these formulas for an ellipsoidal shaped sample volume in a turbulent flow with a Gaussian velocity distribution.

    For purposes of illustration, let us assume a one dimensional oscillating flow of the form

$$v = (1 + B\sin\omega t), \text{ where } |B| < 1 \tag{7}$$

For this flow $\overline{v}_M = 1 + \dfrac{B^2}{2}$ but the true Eurlerian mean velocity $\overline{v}$ is 1. A bias is obtained.

    In general [1,2],

$$\bar{v}_M/\bar{v} = 1 + \overline{\Delta v^2}/\bar{v}^2 \tag{8}$$

where $\overline{\Delta v^2}$ is the variance of the velocity distribution.

Several things should be noted about the above derivation:

1) The result is independent of particle density. This is in contrast to the result of Barnett and Bentley [3] who claim the bias should vanish in the limit of low particle density. This derivation simply does not support that result.

2) The detector is assumed to be ideal in that it cannot be saturated by too many samples arriving per unit time.

3) The result does not depend on the correlation time of the flow.

The above result does not mean, however, that all methods of signal detection and processing will have this particle bias. The oldest method of signal processing-spectral analysis, does not have the particle bias error.

The spectrum $S(\omega)$ for the signal from a single particle of velocity $\vec{v}$, $d\vec{v}$ passing through the sample volume is

$$S(\omega) = \left| \frac{\hat{R}}{|v|^2} \left[(\vec{K} \cdot \frac{\vec{v}}{|v|}) - \frac{\omega}{|v|} \right] \right|^2 \tag{9}$$

where $\hat{R}$ is the spatial Fourier transform of the envelope of the sample volume, $\vec{K}$ is the scattering vector and ω is the radial frequency. The number of samples of velocity $\vec{v}$, $d\vec{v}$ is still

$$dN = \rho T \vec{A} \cdot \vec{v} \, P(\vec{v}) d\vec{v} \tag{10}$$

The expected spectrum is thus

$$S(\omega) = \rho T \int_{-\infty}^{\infty} \frac{|\hat{R}|^2}{|v|} (\frac{\vec{v}}{|v|} \cdot \vec{A}) P(\vec{v}) d\vec{v} \tag{11}$$

Note that the term describing the effect of the sample volume cross section, $(\vec{v}/|v|) \cdot \vec{A}$, only depends on the direction of $\vec{v}$ not its magnitude. Now compute the average of the spectrum, $\bar{\omega}$.

$$\bar{\omega} = \frac{\int_{-\infty}^{\infty} \omega \, S(\omega) d\omega}{\int_{-\infty}^{\infty} S(\omega) d\omega} \tag{12}$$

$$\bar{\omega} = \frac{\int\limits_{-\infty}^{\infty} \omega \int\limits_{-\infty}^{\infty} \left|\frac{\hat{R}}{|v|}\left[(\vec{K} \cdot \frac{\vec{v}}{|v|}) - \frac{\omega}{|v|}\right]\right|^2 (\frac{\vec{v}}{v} \cdot \vec{A})P(\vec{v})d\vec{v}\ d\omega}{\int\limits_{-\infty}^{\infty}\int\limits_{-\infty}^{\infty} \frac{|\hat{R}|^2}{|v|}(\frac{\vec{v}}{|v|} \cdot \vec{A})\ P(\vec{v})d\vec{v}\ d\omega} \tag{12a}$$

If we switch the order of integration and assume $|\hat{R}|^2$ is symmetric about zero, we get

$$\bar{\omega} = \frac{\int\limits_{-\infty}^{\infty} (\vec{K}\cdot\vec{v})(\frac{\vec{v}}{|v|} \cdot \vec{A})\ P(\vec{v})d\vec{v}}{\int\limits_{-\infty}^{\infty} (\frac{\vec{v}}{|v|} \cdot \vec{A})\ P(\vec{v})d\vec{v}} \tag{13}$$

Now perform the integrations in two directions perpendicular to $\vec{K}$. The term $\vec{A} \cdot (\vec{v}/|v|)$ integrates to a constant which is the same in the numerator and denominator and thus drops out. Thus,

$$\bar{\omega} = \frac{\int\limits_{-\infty}^{\infty} \vec{K} \cdot \vec{v}\ P(v_K)\ dv_K}{\int\limits_{-\infty}^{\infty} P(v_K)\ dv_K} \tag{14}$$

$\bar{\omega}$ is proportional to the true Eulerian average velocity in the $\vec{K}$ direction.

The author realizes that this derivation is short and thus may hide details of interest to some readers. They are referred to References 4 and 5 which contain an expanded derivation of the same result.

It can be seen that the existence of a particle bias error is a function of the manner in which the signal is processed.

GRADIENTS

In the above discussion, the terminology has implicitly assumed that a temporally fluctuating flow was being examined. However, the same considerations apply to a steady flow with a gradient in velocity. The product of the cross-section and the velocity probability can be interpreted as the spatial weighting factor of the sample volume. Because of the intensity distribution across the sample volume, there are positions in the volume where measurements are less possible than others.

Example:

Consider a counter processing the signal from a steady flow with a mean gradient A. Let the direction of A be perpendicular to the mean flow direction, the x-direction. Let the counter be set so that it will only measure particles that enter between L, -L on the x-axis. Let the width of the volume be W.

The volume of fluid swept through the sample volume at x, dx is vTWdx.

The mean number of measurements of that velocity is thus $\rho vTWdx$.

But $v = v_0 + Ax$, where v_0 is the velocity at the center of the volume. Therefore, the mean number of measurements of v, dv is $\rho vTWdv/A$ when $(v_0 - AL) \leq v \leq (v_0 + AL)$.

This is the same form as the above equations, therefore the results are true for mean gradients as well as fluctuating velocities.

SATURATION EFFECTS

An individual realization detector is well approximated by a counter. However, no real counter can process signals (bursts) infinitely fast, it will have a maximum rate at which it can process the "realizations." One should therefore suspect that for high particle densities or for high velocities, the counter will cease to behave like an individual realization detector and the form of the particle bias should change. In order to investigate this effect, consider the following approximation to a saturated detector:

1) A counter or its output device is taking samples at a constant rate of $1/\tau$ secs.

2) This rate is much shorter than the correlation time of the flow, so that even though the counter may be measuring each particle, the velocity undergoes negligible change during each cycle. Thus, the output of each cycle is the same as long as one or more particles were detected.

3. Each cycle that records a measurement is given the same weight, 1. A cycle with no measurement is given the weight zero.

The expected number of times the velocity in the sample volume will be $\vec{v}$, $d\vec{v}$ during a measurement time T is given approximately by

$$n_v \doteq \frac{T}{\tau} P(\vec{v}) \qquad (15)$$

The probability of no measurement of velocity during one of those periods is $e^{-\rho(\vec{A} \cdot \vec{v})\tau}$ since the particles have Poisson statistics. The probability of at least one measurement during each occurrence of velocity $\vec{v}$ is thus $1 - e^{-\rho(\vec{A} \cdot \vec{v})\tau}$. The overall weight of velocity $\vec{v}$ occurring times the probability of a measurement during those periods is

$$W_v = \frac{T}{\tau} P(\vec{v}) (1 - e^{-\rho(\vec{A} \cdot \vec{v})\tau}) \qquad (16)$$

For small densities ρ,

$$W_v = \frac{T}{\tau} P(\vec{v}) (1 - e^{-\rho(\vec{A} \cdot \vec{v})\tau} \doteq \rho(\vec{A} \cdot \vec{v}) T P(\vec{v}) \qquad (17)$$

Thus the weighting factor takes on the form of the individual realization detector. However, as ρ gets large,

$$W_v \doteq \frac{T}{\tau} P(\vec{v}) \tag{18}$$

$$\overline{\vec{v}}_{MG} = \frac{\int_{-\infty}^{\infty} \vec{v} \, P(\vec{v})(1-e^{-\rho(\vec{A}\cdot\vec{v})\tau})d\vec{v}}{\int_{-\infty}^{\infty} (1 - e^{-\rho(\vec{A}\cdot\vec{v})\tau}) \, P(\vec{v})d\vec{v}} = \frac{\int_{-\infty}^{\infty} \vec{v}P(\vec{v})d\vec{v}}{\int_{-\infty}^{\infty} P(\vec{v})d\vec{v}} \tag{19}$$

In the limit of high particle density, the particle bias will vanish.

TRACKERS

George and Lumley [6] showed that the laser anemometer signal from a heavily seeded flow (more than 5 particles per sample volume) can be well approximated by white noise passed through a Gaussian shaped band pass filter. Rice [7] showed that an ideal frequency detector processing this signal would measure the same mean as that obtained from spectral analysis.

Lading and I have shown [8,9] that for certain ranges of the detector parameters a second order phaselock loop is an excellent approximation to an ideal detector. Therefore, a properly designed tracker in a high particle density flow should have no particle bias error.

A necessary condition that a tracker approximate an ideal detector is that its bandwidth be at least as wide as the bandwidth of the anemometer signal minus the turbulence contribution.

We have recently performed some measurements with a properly adjusted tracker in laminar flow of water in a tube, looking for particle bias errors. The system has a lot of dust in it so it behaves as if heavily seeded. Simultaneous spectral measurements and tracker measurements were made. Recall from Equation 8 that a particle bias error will appear as a measured mean velocity larger than the true velocity. The fractional error is $\Delta f^2/f^2$, where Δf is the bandwidth of the anemometer signal. So far our results indicate no particle bias error. The errors observed are of the wrong sign and do not depend on the term $\Delta f^2/f^2$.

Some work by Owen and Rodgers [10] at first blush seems to indicate just the opposite. They measured apparent particle bias errors using a tracker. The tracker they used has a wide bandwidth, so that incorrect adjustment of the tracker was apparently not the problem. The particle density was not reported, but in their paper they mention another experiment where they counted the number of particles that passed through the volume by counting the envelopes of the bursts. This indicates that the particle density was such that there was less than one particle in the sample volume at a time. Under these circumstances, the signal is not well approximated by filtered white noise. Indeed the tracker is measuring each particle as it passes through the volume. The tracker they used has a circuit that averages the signal only when the tracker can see a particle, thus the behavior of the tracker approximated an "individual realization detector."

The results to date indicate that a tracker used on a system with a high particle density has no particle bias error, but a tracker used on a low particle density system does have a bias.

CONCLUSIONS

The existence and form of particle bias errors depends on the type of signal processing and the seeding of the system. The error should be most apparent, except for spectrum analyzers, for low seeding densities. At high particle densities, the errors should tend to vanish for trackers and "saturated" counters.

ACKNOWLEDGMENTS

I wish to than L. Lading for many inspirational conversations to keep me going. The author also acknowledges the financial support of NSF ENG75-19185 and ENG76-22965A01).

REFERENCES

1. D. K. McLaughlin and W. G. Tiederman, "Biasing Corrections for Individual Realization of Laser Anemometer Measurements in Turbulent Flow," Phys. of Fluids, 16, 2082 (1973).

2. P. Buchhave, "Biasing Errors in Individual Particle Measurements with the LDA-Counter Signal Processor," Proceedings of the LDA Symposium 75 (Copenhagen) p. 258,(1975).

3. D. O. Barnett and H. T. Bentley III, "Statistical Bias of Individual Realization Laser Velocimeters," Proceedings of the Second International Workshop on Laser Velocimetry, p. 428, (1974)

4. R. V. Edwards, J. C. Angus, M. J. French, and J. W. Dunning, Jr., "Spectral Analysis of the Signal from the Laser Doppler Flowmeter: Time Independent Systems," J. Appl. Phys., 42(), 837-850, (1971).

5. R. V. Edwards, J. C. Angus, J. W. Dunning, Jr., "Spectral Analysis of the Signal from the Laser Doppler Velocimeter: Turbulent Flows," J. Appl. Phys. 44(4), 1694-1698 (1973).

6. W. K. George and J. L. Lumley, "The Laser Doppler Velocimeter and its Application to the measurement of Turbulence," J. Fluid Mech., 60, p. 321, (1973).

7. S. O. Rice, "Statistical Properties of a Sine Wave Plus Random Noise," Bell Sys. Tech. Journal, 27, p. 109-157 (1948).

8. L. Lading and R. V. Edwards, "The Effect of Measurement Volume on Laser Doppler Anemometer Measurements as Measured on Simulated Signals," Proceedings LDA-Symposium, Copenhagen (1975), p. 64.

9. R. V. Edwards, L. Lading and F. Coffield, "Design of a Frequency Tracker for Laser Anemometer Measurements," Fifth Biennial Symposium on Turbulence, Rolla, MO. (1977).

10. J. M. Owens and R. H. Rogers, "Velocity Biasing in Laser Doppler Anemometers," Proceedings LDA-Symposium, Copenhagen (1975), p. 89.

Analytical and Experimental Study of Statistical Bias in Laser Velocimetry[1]

T. V. GIEL[2] and D. O. BARNETT[3]
Sverdrup/ARO, Inc., AEDC Division
Arnold Air Force Station, TN

ABSTRACT

It is widely recognized that arithmetic averages of individual realization laser velocimeter data may be statistically biased in time varying flows because the velocity sampling process may be dependent on the velocities being sampled. Accordingly, several data reduction algorithms have been proposed to negate this bias. In order to assess the validity of these algorithms an experiment was conducted which attempted to identify statistical bias in actual laser velocimeter data. Individual realization laser velocimeter measurement taken in a highly turbulent, uniformly seeded jet flow were used to model the conditions that theoretically result in statistical bias. In an attempt to reduce particle dynamic effects the jet was seeded with monodisperse micron-sized seed particles. Vector velocimeter measurements of the jet axial velocity and the times between successive samples were recorded at different data rates and averaged. The resulting averages were compared with hot wire measurements in the same jet, but no clear evidence of statistical bias was found. An analysis is presented to show the bias reduction that can occur when the velocities of some particles crossing the probe volume are not recorded and when the seed distribution is not perfectly uniform. In addition, the effect of seeding rate and flow structure on bias is discussed.

INTRODUCTION

In studying the sampling characteristics of laser velocimeters operated in the individual realization mode, Mayo [1] observed that the interval between particle arrivals was governed by a Poisson probability distribution having a rate parameter which was proportional to the flux of particles passing through the probe volume. He concluded that since the sampling rate and measured velocity were inter-related, averages of individual realization laser velocimeter (IRLV) data would overestimate the temporal mean velocity. A similar conclusion

---

[1]The research reported herein was performed by the Arnold Engineering Development Center (AEDC), Air Force Systems Command. Work and analysis for this research was done by personnel of ARO, Inc., a Sverdrup Corporation Company, operating contractor of AEDC. Further reproduction is authorized to satisfy needs of the U. S. Government.

[2]Research Engineer, Fluid Mechanics Section, Technology Applications Branch, Engine Test Facility, ARO, Inc. (This paper is based on author's Ph.D. dissertation, University of Notre Dame, Indiana).

[3]Senior Research Engineer, presently Assoc. Professor, School of Engineering, University of Alabama in Birmingham, Birmingham, Alabama.

was independently reached by McLaughlin and Tiederman [2] who termed the phenomenon statistical bias. Barnett and Bentley [3] showed that the mean velocity
was biased in direct proportion to the turbulence intensity and consequently,
was a potentially serious source of error in turbulent flows.

Since the publication of these pioneering studies, considerable effort has
been devoted to determining the conditions under which bias occurs and the development of data acquisition or processing techniques to eliminate bias [4-10].
Considerably less work has been devoted to the experimental verification of the
magnitude of bias [11]. IRLV data obtained at the AEDC [12] do not show con-
clusive evidence that the theoretically predicted bias exists. Despite the lack
of experimental verification, many IRLV users routinely correct for bias and
some commercial IRLV data processors have been designed to use algorithms developed to eliminate bias. Unfortunately, if bias does not always occur, such routine bias corrections can lead to significant errors. Therefore, the present
study was initiated to experimentally determine when statistical bias might significantly contribute to measurement uncertainty in turbulent flow to warrant
applying correction techniques.

EXPERIMENTAL SIMULATION

The analysis of Ref. 13 implies that an experiment to detect bias should
have (1) a constant fluid density, (2) spatially and temporally invariant particle number density, (3) negligible particle lag, (4) single counting of each
particle so that a rapid data acquisition system and a negligible fluid recirculation are necessary, and (5) particles entering the probe volume frequently
enough that the average velocity between particle detections, $\hat{V}_i$, is approximately the measured particle velocity, V_i. To improve the probability of detecting bias, an effort was made to attain these conditions in the experimental
study.

An air jet was selected for the experiments to provide a wide range of
turbulent intensities in all component directions so that the arithmetic means
of IRLV one-component data would be sufficiently biased for experimental identification in at least part of the flow. The jet issues from a 25 to 1 area
ratio converging nozzle which was contoured to match calculated potential flow
streamlines to smoothly accelerate the air and particles to a uniform nozzle
exit velocity. The jet exit Mach number was kept below 0.1 to ensure that the
fluid density could be considered constant. A fairly constant particle number
density was obtained by filtering the supply air to remove all detectable particles and then seeding with a fluidized-bed seeder. Additionally, the jet was
enclosed so that recirculation external to the jet tended to spatially equalize
particle concentration. The recirculation chamber is 18 nozzle diameters long
and 5 diameters square.

By varying the supply pressure to the fluidized-bed seeder, the seed concentration could be controlled and a range of data rates could be obtained.
Uniform one-micron-diameter particles were used in the seeder because particle
dynamic calculations indicated that for the range of conditions to be encountered in the jet particles of that size would accurately respond to fluid
transients.

LV INSTRUMENTATION

In order to measure statistical bias experimentally, the precision uncertainty in the IRLV measurements must be less than the bias. The IRLV and traverse systems were designed to provide the necessary precision. A one-component,
Bragg-diffracted LV with off-axis backscatter collector optics was used. References 12 and 13 describe in detail the LV system and associated electronics.
Precision uncertainties for velocities measured by this system were calculated
with the procedures given in Refs. 12, 14, and 15 and are summarized in Table 1.

Because a finite number of measurements is used to determine IRLV averages, the statistical confidence limits of the averages also needs to be considered in evaluating potential error. To establish a 1.5-percent confidence interval for the mean velocity, all averages were calculated from at least 1,500 samples. The confidence interval for the turbulence intensity is about 3 percent for this sample size.

Table 1. LV precision uncertainty.

U_x		ΔU_x	
m/sec	ft/sec	m/sec	ft/sec
0	0	±0.20	±0.68
±15.6	±50	±0.76	±2.50
30.5	±100	±1.33	±4.35

The uncertainty in positioning the IRLV probe volume will also contribute to uncertainty in the IRLV measurements. The probe volume could be moved along any of three mutually orthogonal axes and positioned to within ±0.08 cm by properly accounting for hysteresis. However, there is an additional uncertainty in locating the LV measurements due to the finite size of the probe volume. The diameter of the probe volume is taken as the diffraction limit of the focusing lens. For the 381-mm focal length lens and 1.5-mm beam diameters at the lens a probe volume diameter of 0.17 mm results. The apparent probe volume length was reduced to approximately 1.7 mm by viewing the probe volume through a 500-μm-diam pinhole at 20-deg off-axis in backscatter.

Data were processed using an AEDC-developed processor. Details of the operation of this processor are given in Refs. 16 and 17. The specific processor used could determine fringe crossing periods to ±0.005 nsec [18]. This processor was modified by the addition of interval counting circuitry to measure the time between successive validated signals, Δt_i. The interval counting resolution was ±0.5 μsec. This modification allowed determination of approximate time-integrated velocity moments [3 and 10] and data rates, in addition to the usual arithmetically-averaged velocity moments.

To attain near maximum IRLV data rates, the data were recorded using direct memory access to a minicomputer. This system could typically record over 13,000 fringe periods and time intervals per second. However, the flow was never seeded heavily enough to obtain data rates above about 4,500/sec to reduce the probability that LV signals would result from multiple particle traversals of the probe volume.

SUPPORTING INSTRUMENTATION

Statistical biasing will not occur in averages of velocities measured by devices continuously monitoring the jet flow. Such continuous measurements were made using both pressure and hot-wire instrumentation in order to identify statistical bias by comparing the average LV measurements with the average pressure and hot-wire measurements. However, such identification will be possible only if the uncertainty in the pressure and hot-wire measurements is less than the bias. Furthermore, the pressure and temperature stilling chamber measurements used to monitor the jet flow must be sufficiently accurate for a comparison of the independent measurements to be meaningful.

To attain the necessary accuracy in continuous pressure measurements, Barocel® pressure transducers were used. These transducers maintained an accuracy of ±0.15 percent of the measured pressure throughout their ±10 mm of Hg pressure range as determined by calibration against a laboratory standard Hook gauge which is traceable to the National Bureau of Standards (NBS). The ambient

pressure, P_a, was measured with a bourdon-tube pressure gauge, and the air temperature, T, was monitored using a shielded, self-aspirated, copper constantan
thermocouple. Both instruments were calibrated against NBS laboratory standards.
The precision of the resulting velocity measurements, determined as in Refs. 12
and 15, was ±1.5 percent of the velocity for the nominal jet exit velocity of
30.5 m/sec.

The hot-wire measurements were made using Thermo-Systems, Inc., 1050
series constant temperature anemometry equipment and a 3.8-μm-diam tungsten
sensor. The frequency response of the sensor is estimated to be 24 KHz when
eddies are convected past the sensor at the nominal jet exit velocity. Therefore, signal frequency components which were greater than 30 KHz were considered
noise and filtered out. The hot-wire bridge output was recorded on FM modulated
analog tape maintaining a 30-KHz frequency response. To attain the needed accuracy in the hot-wire measurements, the recorded bridge output was digitized at
a 50-KHz rate and each sample was converted to a velocity using King's law [19].

$$e_i^2 = A + B \ U_i^{1/n} - CU_i \tag{1}$$

The constants A, B, C, and n were determined by a nonlinear least-squares fit of
Eq. (1) to the hot-wire calibrations [20]. Calibrations were performed every
30 min in the test apparatus using velocities measured with the pressure instrumentation. Because Eq. (1) fit all the calibration data extremely well, the
uncertainties in the hot-wire velocity measurements are only slightly larger
than the uncertainties in the pressure determined calibration velocities. Furthermore, by using approximately 200,000 digital hot-wire samples when averaging the hot-wire data, the 95 percent statistical confidence intervals were 0.1
percent of the mean velocities and approximately 0.2 percent of the turbulence
intensities and should not add significantly to the hot-wire measurement uncertainty. The hot-wire and pitot probe position uncertainty was ±0.5 mm.

RESULTS

The experimental examination of statistical bias was conducted in three
phases. First, the jet flow was evaluated to determine how well the flow conditions which theoretically result in bias were met; second, the flow was
mapped with pitot and hot-wire probes; and third, the flow was mapped with the
IRLV. The second and third phase results were then compared to determine the
magnitude of any bias present.

In the first phase the operation of the seeder was examined to determine
if it provided a steady, regular, controllable output. Successive LV detections of particles in the jet core were recorded and the time between each detection, Δt_i, was measured by the LV data processor. Histograms of the detection times show that nearly equal times separate most particle detections [13].
In the smooth core flow equal separation times indicate equal fluid mass separating each particle, and thus, a uniform particle concentration. It should be
noted that no particles were detected with the seeder off verifying that only
particles from the seeder comprise the LV data sets obtained in the tests.

The distribution of the seed concentration in the test chamber was also
examined by timing successive LV detections and evaluating the average detec-
tion time, $<\Delta t>$. If the ratio $<\Delta t_{C_L}>/<\Delta t(r)>$ is distributed as $\overline{U}(r)/\overline{U}_{C_L}$ throughout the jet, then it can be concluded that the particle concentration is on the
average uniform throughout the jet. Figure 1 shows that radial distributions
of $<\Delta t_{C_L}>/<\Delta t(r)>$ correspond to distributions of $\overline{U}(r)/\overline{U}_{C_L}$ determined from pitot
probe measurements illustrating that the particle concentration is essentially
uniform, on the average, throughout the jet.

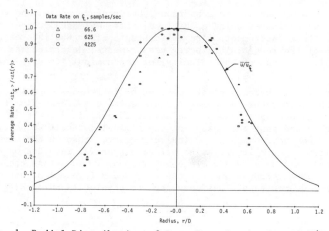

Fig. 1   Radial Distribution of Data Rate in the Jet at X/D = 4

To examine the size of particles in the test flow, particle samples were collected near the test chamber exit. Photomicrographs taken of the samples using a scanning electron microscope show that some coagulation of the 1-μm seed occurs and that most particles have 1 to 3-μm diameters [13]. A few larger particles were visible in the photomicrographs but they are chain-shaped coagulations that probably were created during the particle collection. Particle dynamic calculations indicate that 3-μm particles would not appreciably lag the flow in the present jet so the departure from the desired 1-μm size should have little effect on the results.

Once the seeding characteristics were determined, mapping of the flow field by conventional instrumentation was initiated. Radial surveys were made at axial stations of 2, 4, 6, and 8 nozzle diameters with both pitot and hot-wire probes. The mean velocities, shown in Fig. 2, have been fit with the semi-empirical expression,

$$\frac{\overline{U}(x,r)}{\overline{U}_{C_L}(x)} = \exp\left(-0.693\left(\frac{r - r_o(x)}{b(x)}\right)^2\right) \qquad (2)$$

as an aid for comparison to the LV results. The centerline velocity, $\overline{U}_{C_L}(x)$, inner radius of the jet shear layer, $r_o(x)$, and the shear layer width to the half-velocity point, $b(x)$, were determined using the curve fitting method of Ref. 20. While only hot-wire data are indicated in Fig. 2, both hot-wire and pitot measurements were used in deriving the curve fits. Although measurements were made across the entire jet, only the data from the inner jet region, where the local turbulence intensity, $\sigma_{\overline{u}}$, is less than 0.33, is used for bias identification because data aliasing due to negative velocities can be significant in pitot and hot-wire measurements at the higher values of turbulent intensity.

The hot-wire measurements of the turbulence intensities inside the $\sigma_{\overline{u}} = 0.33$ limit were least-squares-fitted to the second order polynomial:

$$\sigma_{\overline{u}}(x,r) = \begin{cases} A(x) + B(x)r_o + C(x) r_o^2, & \text{for } r \le r_o \\ A(x) + B(x)r + C(x) r^2, & \text{for } r > r_o \end{cases} \qquad (3)$$

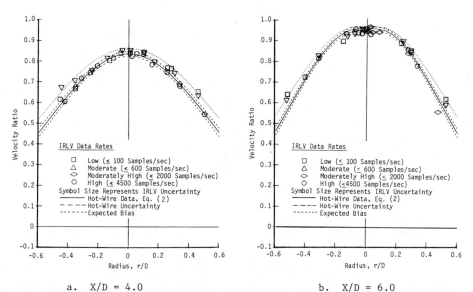

a. X/D = 4.0   b. X/D = 6.0
Fig. 2   Comparison of Radial Distributions of Mean Axial Velocity

The resulting expressions were used to predict the maximum mean velocity bias
that could be anticipated in the LV measurements [3, 4, and 13]. The predicted
bias and the hot-wire measurement uncertainty, predicted assuming that unidenti-
fied hot-wire uncertainties could double the hot-wire calibration uncertainty,
are also shown in Fig. 2 to indicate that bias should be identifiable in these
tests.

Having established the "unbiased" flow field characteristics, IRLV radial
surveys were made at axial stations of 2, 4, 6, and 8 nozzle diameters. The
IRLV radial surveys were repeated at several seeder settings whereby any bias
variation with seed density might be defined. Typical mean velocity surveys are
compared in Fig. 2 with the "unbiased" velocity curves. It can be seen that the
IRLV data more closely approximate the "unbiased" velocity curves than the theoretical bias curves regardless of the seeding rate. Typical IRLV measurements
of turbulence intensity, presented in Fig. 3, also agree well with the "unbiased"
velocity curves, indicating that statistical biasing is not nearly as significant as predicted for these tests.

To date the only AEDC data that show differences in IRLV and hot-wire
measurements which might be attributed to bias were reported in Ref. 12. The
difference in the two measurements was most obvious in the centerline axial velocity decay. Accordingly, the axial velocity decay and turbulence intensity
distribution, as determined by pitot, hot-wire, and IRLV data averages in the
present experiment, are shown in Fig. 4. These data also show no evidence of
statistical bias.

Bias was to be identified in this study by comparing averages of IRLV
data with unbiased measurements. It is stated in Refs. 3 and 13 that if the
particle interarrival times are random, no bias will result. Reference 9 further suggests that random selection of samples from an IRLV data set will lead
to unbiased averages of the velocity moments. Therefore, in a final attempt to
identify statistical bias the IRLV averaged mean velocities and turbulence intensities were compared with averages of randomly selected data. The selection
was done several times, first selecting one sample at random from each 200 recorded samples, then one sample from each 600 and finally one per 2,000 recorded

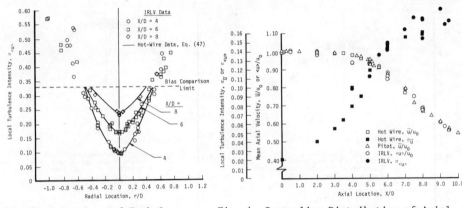

Fig. 3   Comparison of Turbulence
          Intensities Obtained from
          the LV and Hot-Wire Probe

Fig. 4   Centerline Distribution of Axial
          Velocity and Turbulence Intensity

samples. The random selections were taken from data sets containing at least
2,000 samples and the random selection process was completed when 500 samples
were selected. Figure 5 shows the ratio of the original IRLV data average to
the randomly selected data average for various radial locations at X/D of 4
and 6. There is no evidence of statistical bias in either data set.

INFLUENTIAL FACTORS

The preceding results indicate that the magnitude of IRLV statistical bias
is much less than predicted by the theory outlined in Refs. 3, 4, and 13.
Several factors that can reduce the bias magnitude include (1) the relationship
between sampling rate and turbulence structure, (2) processing rates possible
with a given data acquisition system, and (3) fluctuations in the local particle
concentration.

Sampling Rate

In Ref. 3 Barnett and Bentley concluded that the bias between the arithmetic and time-integrated averages of IRLV data will decrease as the sampling
rate per primary flow frequency goes down. To examine the possible dependence
of statistical bias on data rate, IRLV signals were simulated in a numerical
experiment. In this simulation a velocity-time function was generated to represent the flow velocity in the probe volume. Velocities were recorded as particle velocities every time a specified amount of fluid crossed the probe
volume. For consistency with the bias theory, negative velocities were not

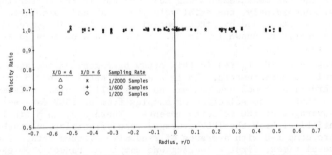

Fig. 5   Ratio of IRLV Average Mean to Randomized Data Mean

allowed to recirculate particles through the probe volume. Initially, a sinu-soidal velocity $V = V_o + V_1 \sin(\omega t + \Phi)$ was assumed to simplify the evaluation of bias as a function of the sampling rate per flow frequency, $\frac{1}{\langle \Delta t \rangle}/\omega$. The quantity $\frac{1}{\langle \Delta t \rangle}/\omega$ was varied by changing either ω or the specified fluid mass separating particles. Both the arithmetic average velocity and the approximate-time-integrated average velocity are compared with the true time-averaged ve-locity in Fig. 6. Since all theorized statistical bias conditions are met per-fectly in this simulation, the quantity $(\langle V \rangle / \hat{V} - 1)/\sigma_v^2$ should vary between unity for completely biased cases and zero for unbiased cases. Excursions out-side these limits are caused by data aliasing which occurs when the sampling rate is equal to the fluctuation rate. Figure 6 indicates that bias in arith-metic averages does not decrease as the sampling rate per flow frequency de-creases in opposition to the conclusions of Ref. 3. Figure 6, on the other hand, shows that at low sampling rates the approximate time-integrated velocity becomes biased which verifies Barnett and Bentley's hypothesis that the arith-metic mean velocity equals the time-integrated mean velocity as the particle interarrival time becomes long, but also indicates that time-integration is not a good approximation of the true mean velocity at large $\langle \Delta t \rangle$ as assumed in Ref. 3. The different symbols in Fig. 6 represent simulations for different ω, ϕ, and V_1/V_o values to illustrate that this conclusion is not limited to just one velocity-time history. Reference 13 shows that at low enough data rates the bias in arithmetic averages is reduced, but the required data rates are much too low (20 samples per hour) for practical use of an IRLV.

To examine the arithmetic and approximate time-integrated mean velocity for turbulent flow, the simulation was repeated using turbulent jet hot-wire measurements as the velocity-time function. The results, [13], again show no bias reduction for arithmetic averages of theoretical IRLV data and show a bias-ing of the approximate time-integrated means at large interarrival times, $\langle \Delta t \rangle$, per the integral time scale of the flow.

The predicted biasing in approximate time-integrated averages of IRLV data is particularly important if approximate time-integration is being used to avoid statistical biasing as suggested in Refs. 3 and 10. As the simulation, [13], illustrates, approximate time-integration will produce averages that can be

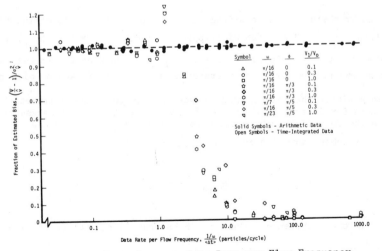

Fig. 6 Bias Dependence on Data Rate per Flow Frequency

guaranteed to be unbiased only if the data rates are high enough that at least
two samples are recorded per integral time scale. This would require data rates
of 4,000 to 10,000 samples/sec in the experiments conducted in this investigation. Most IRLV data systems are not able to process the data at such rates
and it is often not feasible to seed flows heavily enough to attain the necessary data rates without reducing the probability of individual realizations.

Processing Rate

It has been shown, [13], that statistical bias should result when the
time separating particle entrances into the probe volume, Δt_i, is inversely proportional to $\hat{V}_i$, the average fluid velocity evaluated over Δt_i. However, the
time between successive recorded IRLV realizations, Δt_j, may not be inversely
proportional to $\hat{V}_j$. This can happen if the time between some particles is less
than the time needed to evaluate and record LV realizations. For example, if
the IRLV instrumentation can only record 10 samples/sec but an average of 1,000
particles enter the probe volume each second, then a particle will probably be
in the probe whenever the LV is ready to process another sample. Thus, the interval between recorded realizations would be nearly constant at 0.1 sec and
arithmetic averages would be unbiased.

To analyze the effect that a processing time, k_t, can have on statistical
bias, the time between recorded samples can be expressed as

$$\Delta t_j = k_t + \Delta t_j^* \tag{4}$$

where Δt_j^* is the delay time after the completion of a data acquisition cycle until the next particle arrival. An approximate-time-integration for velocity
gives

$$\overline{V} = \frac{\sum\limits_{i=1}^{M} V_i(k_t + \Delta t_i^*)}{\sum\limits_{i=1}^{M} k_t + \Delta t_i^*} = <V>\left[1 + \frac{\frac{<\Delta t^*>}{k_t} R_{V_{\Delta t^*}}}{1 + \frac{<\Delta t^*>}{k_t}}\right] \tag{5}$$

where the brackets <> denote arithmetic averages. If the processing time is
short then $\frac{<\Delta t^*>}{k_t} >> 1$ and Δt^* will approach Δt so $\overline{V} \simeq <V> (1 + R_{V_{\Delta t}})$ indicating
that bias may result. However, if $<\Delta t^*>/k_t << 1$, as when the processing time is
long compared to the particle interarrival times, $\overline{V} \simeq <V>$ and no bias results.

To estimate the relationship between the temporal and arithmetic averages
of the data for intermediate values of processing time, it is necessary to
estimate the form of the correlation coefficient $R_{V_{\Delta t^*}}$. Reference 13 shows
that a reasonable approximation for $R_{V_{\Delta t^*}}$ is

$$R_{V_{\Delta t^*}} = \frac{\sigma_{\hat{v}}^2}{1 + \sigma_{\hat{v}}^2} \tag{6}$$

where σ_v is the turbulence intensity. Substituting Eq. (6) in (5) leads directly to

$$\overline{V} = <V> \frac{1 + \sigma\frac{2}{v} + \frac{<\Delta t^*>}{k_t}}{\left(1 + \sigma\frac{2}{v}\right)\left(1 + \frac{<\Delta t^*>}{k_t}\right)} \tag{7}$$

The bias parameter corresponding to this expression, shown in Fig. 7 as a function of the dimensionless processing delay $k_t/<\Delta t^*>$, is the upper limit for bias reduction due to processing delay [13]. It is apparent that as the delay becomes large the arithmetic and temporal averages converge so that the net effect of a relatively slow data acquisition system is to reduce bias.

Processing delay times are not absolutely constant as assumed in the preceding analysis. For example, the delay may be regarded as the sum of processor recycle time, signal processing time, and data recording time. The processor recycle time is approximately 0.04 μsec for the LV data processor used in this study. The data acquisition system could record data at rates over 13,000 samples/second so the maximum time associated with recording was 77 μsec. The signal frequency range in the experiment was 12.5 to 17 MHz so that, when the pulse stretching and required number of signal cycles associated with the processor are considered, signal processing times ranged from 60 to 80 μsec. Consequently, the total delay between successive samples was, at most, 160 μsec. The maximum average data rates encountered were approximately 4,000 samples/sec so that at most $k_t/<\Delta t^*> \simeq 0.7$. Therefore, Fig. 7 shows that the bias magnitudes associated with the present system should always be greater than 60 percent of the maximum possible in the experiments and, even when multidimensional effects and instrumentation uncertainties are also considered, the present experiments should have detected any consistent biasing of the IRLV results.

Particle Concentration Fluctuations

Statistical bias theories are generally initiated by assuming that the particle concentration at a point is invariant. This condition is extremely hard to obtain in practice as was noted in Ref. 21 and observed in the present experiments. As an aid to understanding the experimental results, it is instructive to consider those cases in which the particle concentration as well as fluid velocity is time dependent.

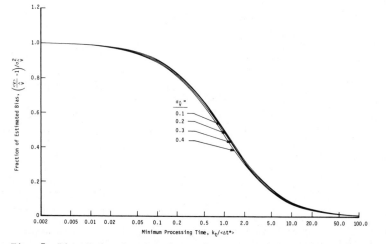

Fig. 7   Bias Reduction Resulting from Minimum Processing Time

For this purpose it is convenient to use a formal averaging procedure first reported by Cline and Bentley [22]. The fluid velocity at a point is, accordingly, defined by

$$\bar{f} = \sum_{k=1}^{m} P(f_k) f_k \tag{8}$$

where $P(f_k)$ is the probability that the variable f will have the particular value f_k. Likewise, the arithmetic average of the individual realizations of velocity is

$$<f> = \frac{1}{S} \sum_{k=1}^{m} S_k f_k \tag{9}$$

where S_k is the number of realizations that occur in some interval $f_k \pm \Delta f$ and S is the total number of samples. The ratio S_k/S is the probability that the particular value f_k occurs in the IRLV data. Now S_k is the product of the total sampling time, T, and λ_R, the sampling rate for the occurrence of the f_k events. Furthermore, the particle concentrations and velocities corresponding to the kth interval can be written as the sum of average and fluctuating terms, so that Eq. (9) for the mean velocity may be rewritten as

$$<V> = \bar{V} \frac{1 + \sigma_v^2 + 2R_{NV} + R_{NV}^2}{1 + R_{NV}} \tag{10}$$

in which the correlations between the concentration and velocity fluctuations are

$$R_{nv}^{n} = \frac{\overline{N'(v')^n}}{\bar{N} \bar{V}} \tag{11}$$

Similarly, the bias for the turbulence intensity can be evaluated for a fluctuating particle number density.

The predicted mean velocity and turbulence intensity bias for various number-density-velocity correlations are shown in Figs. 8 and 9 where the third

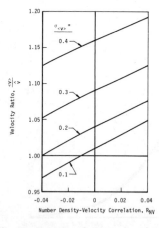

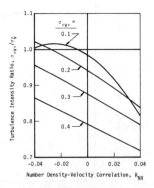

Fig. 8   Variation of Bias in the Mean
         Velocity for Fluctuating
         Number Density

Fig. 9   Variation of Bias in the
         Turbulent Intensity for
         Fluctuating Number Density

and higher order correlations are assumed to be zero. When $R_{NV} = 0$ the predicted bias is the same as predicted for uniform particle number density [3 and 4]. However, it is seen that even a small number density-velocity correlation can significantly alter the bias. No conclusions can be reached as to the sign of the R_{NV} correlations in the present experiment. However, the R_{NV} correlation can be expected to be similar to the species-velocity correlation in flows where several gases are mixing. These correlations are generally positive [23]. Such a positive correlation would increase the bias.

It is possible to hypothesize the existence of other mechanisms, such as particle lag effects, which may reduce the bias in IRLV experiments. However, on the basis of the preceding analyses, it appears that none of the postulated bias reducing mechanisms would have been adequate by themselves to eliminate bias from the present data. A possible explanation for the lack of bias in the present data may be simply that the cumulative effect of several mechanisms reduced the bias below measurable levels.

CONCLUDING REMARKS

An experimental investigation was conducted using a confined, subsonic air jet to examine statistical bias in velocity parameters determined by averaging individual realization laser velocimeter data. Efforts were made to conduct an experiment favorable to obtaining statistical bias. However, no consistent bias is evident in comparisons of either mean velocities or turbulence intensities determined from laser velocimeter data and pitot and hot-wire probes. The same lack of bias is evident when mean velocities and turbulence intensities determined from randomly sampled IRLV data are compared with the original LV results. Clearly, on the basis of these results, a significant statistical bias does not exist in all IRLV velocity measurements. Obviously, data should be corrected for bias only if a significant bias can be identified in that data.

The theory developed herein can be helpful for predicting possible bias in an experiment. The effects on bias of (1) time delays in IRLV data processing and recording, and (2) nonuniformities in seeding have been examined. Time delays introduced in acquiring the IRLV data will reduce bias, but seeding nonuniformities should increase bias if the particle number-density-velocity fluctuations are positively correlated, as anticipated. In this and possible other experiments it does not appear that any single mechanism was responsible for the lack of bias. However, the cumulative effects of several mechanisms might reduce bias below measurable levels. Some method of identifying statistical bias in experimental measurements is needed and random time sampling is suggested as a simple method.

Assuming that bias can be identified the selection of an appropriate bias correcting algorithm depends on experimental flow conditions, available equipment, and the required measurement accuracies. If appropriate timing equipment is available the approximate-time-integration bias correcting algorithm can be employed to attain correct velocity moments. However, approximate-time-integration also requires that high data rates are attained to allow several realizations to be made per the integral time scale of the flow. Thus, it should never be employed to reduce biased data recorded at low data rates. The residence-time algorithm may work adequately at low data rates if the flow is uniformly seeded, but again, appropriate data processors must be available. Random sampling, as suggested for bias identification, will reduce the bias for a high data rate; however, very large sampling times are needed for random sampling to be effective. If the flow is essentially one-dimensional and the velocity vector is normal to the fringes, the harmonic average can be employed successfully. Finally, sampling at low data rates should not be depended on to reduce bias, since only impractically low data rates will actually afford any significant reduction in bias.

REFERENCES

1. Mayo, W. T., Jr., "The Development of New Digital Data Processing Tech-
    niques," USAF Contract F40600-73-C-003, Third Monthly Progress Report,
    Appendix, March 1973.

2. McLaughlin, D. K. and Tiederman, W. G. "Biasing Correction for Individual
    Realization of Laser Anemometer Measurements in Turbulent Flows," The
    Physics of Fluids, Vol. 16, No. 12, December 1973.

3. Barnett, D. O. and Bentley, H. T., III. "Bias of Individual Realization
    Laser Velocimeters," Proceedings of the Second International Workshop on
    Laser Velocimetry, Purdue University, Vol. I, pp. 438-442, March 1974.

4. Kreid, D. K. "Laser Doppler Velocimeter Measurements in Non-Uniform Flow:
    Error Estimates," Applied Optics, Vol. 13, pp. 1872-1881, August 1974.

5. Karpuk, M. E. and Tiederman, W. G. "Effect of Finite-Size Probe Volume
    Upon Laser Doppler Anemometer Measurements," AIAA Journal, Vol. 14, pp.
    1099-1105, August 1976.

6. George, W. K., Jr. "Limitations of Measuring Accuracy Inherent in the
    Laser Doppler Signal," The Accuracy of Flow Measurements by Laser Doppler
    Methods, Proceedings of the LDA-Symposium Copenhagen, Technical University
    of Denmark, August 1975.

7. Owen, J. M. and Rogers, R. H. "Velocity Biasing in Laser Doppler Anemom-
    eters," The Accuracy of Flow Measurements by Laser Doppler Methods, Pro-
    ceedings of the LDA-Symposium Copenhagen, Technical University of Denmark,
    August 1975.

8. Buchhave, P. "Biasing Errors in Particle Measurements," The Accuracy of
    Flow Measurements by Laser Doppler Methods, Proceedings of the LDA-
    Symposium Copenhagen, Technical University of Denmark, August 1975.

9. Durao, D. F. G. and Whitelaw, J. H. "The Influence of Sampling Procedures
    on Velocity Bias in Turbulent Flows," The Accuracy of Flow Measurements
    by Laser Doppler Methods, Proceedings of the LDA-Symposium Copenhagen,
    Technical University of Denmark, August 1975.

10. Dimotakis, P. E. "Single Scattering Particle Laser Doppler Measurements of
    Turbulence," AGARD Symposium on Non-Intrusive Instrumentation in Fluid
    Flow Research, Saint Louis, France, May 1976.

11. Quigley, M. S. Experimental Evaluation of Sampling Bias in Individual
    Realization Laser Anemometry, Thesis, Oklahoma State University, December
    1975.

12. Barnett, D. O. and Giel, T. V., Jr. "Application of a Two-Component Bragg-
    Diffracted Laser Velocimeter to Turbulent Measurements in a Subsonic Jet,"
    AEDC-TR-76-36, May 1976.

13. Giel, T. V., Jr. Studies of Statistical Bias in Turbulent Flow Parameters
    Measured with a Laser Velocimeter, Dissertation, University of Notre Dame,
    August 1978.

14. Barnett, D. O. and Giel, T. V., Jr. "Application of a Two-Component,
    Bragg-Diffracted LV to Turbulence Parameter Determination in an Isothermal
    Subsonic Jet," Minnesota Symposium on Laser Anemometry, University of
    Minnesota, Minneapolis, pp. 146-183, October 1975.

15. Abernethy, R. B., et al., Pratt and Whitney Aircraft, and Thompson, J. W.,
    Jr., ARO, Inc. "Handbook of Uncertainty in Gas Turbine Measurements,"
    AEDC-TR-73-5 (AD755356), February 1973.

16. Brayton, D. B., Kalb, H. T. and Crosswy, F. L. "A Two-Component Dual Scatter Laser Doppler Velocimeter with Frequency Burst Signal Readout," Applied Optics, Vol. 12, No. 6, pp. 1145-1156, June 1973.

17. Crosswy, F. L. and Hornkohl, J. O. "Signal-Conditioning Electronics for a Vector Velocity Laser Velocimeter," AEDC-TR-71-192 (AD755842), February 1973.

18. Barnett, D. O. and Giel, T. V., Jr. "Laser Velocimeter Measurements in Moderately Heated Jet Flows." AEDC-TR-76-156, April 1977.

19. Bruun, H. H. "Linearization and Hot-Wire Anemometry," J. Sci. Instrum., Vol. 4, pp. 815-820, 1971.

20. Hartley, H. O. "The Modified Gauss-Newton Method for the Fitting of Non-Linear Regression Functions by Least Squares," Technometrics, 3, pp. 269-280, 1961.

21. Glass, M. and Kennedy, I. M. "On Improved Seeding Method for High Temperatures Laser Doppler Velocimeter," Combustion and Flame, Vol. 29, pp. 333-335, 1977.

22. Cline, V. A. and Bentley, H. T., III. "Application of a Dual Beam Laser Velocimeter to Turbulent Flow Measurements," AEDC-TR-74-56, September 1974.

23. Stanford, R. A. and Libby, P. "Further Applications of Hot-Wire Anemometry to Turbulence Measurements in Helium Air Mixtures," Project Squid Technical Report USCD 5PU, January 1974.

Experimental Evaluation of Sampling Bias in Naturally Seeded Flows

DAVID G. BOGARD and W. G. TIEDERMAN
Oklahoma State University
Stillwater, OK

ABSTRACT

Individual realization laser velocimeter measurements were made in the viscous sublayer of a turbulent channel flow. The Doppler signals were recorded on magnetic tape and then processed with systematic variations in the detection level in the counter processors. Naturally seeded flows were modelled by using 0-10 μm test dust for scattering centers while flows seeded with 5-10 μm test dust were used as a controlled comparison. Both the number density of the scattering particles and the detection level of the counter processors were independently varied to yield a wide range of particle arrival rates. The results show that sampling bias is not effected by the particle arrival rate. There was also no indication that sampling bias is either eliminated or decreased due to differences in the probability for detecting small particles that are moving slowly compared to small particles that are moving fast.

NOMENCLATURE

l_s average distance between particles yielding validated signals (ft)

N number of individual samples in a data ensemble

U_f frequency average estimate of the mean velocity (ft/sec)

U_i individual sample of the instantaneous velocity (ft/sec)

U_p period average estimate of the mean velocity (ft/sec)

T time for the ensemble of data (sec)

T_D average Doppler period for an ensemble of data (sec)

T_s average time between particles yielding validated signals (sec)

λ wave length of the laser light

θ angle of intersection between the laser beams

INTRODUCTION

Several of the fundamental problems involved in interpreting signals from a laser velocimeter system arise because the velocimeter measures the velocity of small particles entrained in the flow. Usually the investigator is attempting to acquire information about the fluid velocities and this information must be deduced from the particle velocities that are detected and measured by the velocimeter. This paper is concerned with the accurate interpretation of fluid velocities when the scattering particles are accurately following the fluid motion and the laser velocimeter is operating in the individual realization or counting mode (1).

The individual realization mode of operation is the natural result of a dilute concentration of particles in the fluid. This is typically the situation in large wind tunnel applications where the airflow is usually not seeded with scattering centers. In this mode of operation, an output signal occurs only when a particle that is sufficiently large to be detected passes through the measurement volume. Consequently the output signal is discontinuous and the occurrence of an output signal is dependent upon the arrival of a detectable particle in the measurement volume.

Because the signal is discontinuous, statistical methods must be used to estimate the desired quantities such as the time-average fluid velocity. If the sampling is random and unbiased, then an unweighted statistical analysis of the data will yield accurate estimates of the mean and root-mean-square velocities. If the sampling is biased because some velocities are detected more frequently than others, then suitable weighting factors must be used in the statistical analysis.

It has been postulated that biased sampling occurs when the scattering particles are homogeneously distributed in the fluid and when there is equal probability that each particle will yield a validated signal (2). Several analyses have been made and methods have been proposed to correct biased measurements (2,3,4,5,6,7). The corrections are significant when the velocity fluctuations are large.

However, it has been suggested that biased sampling is either totally or partially eliminated by a compensating effect. This compensating effect is based on the argument that signals from slower particles produce a higher signal to noise ratio than that from faster particles (8) and hence the probability for detecting slow particles is greater than the probability for detecting fast particles. One of the objectives of this study was to determine whether or not this effect could be detected. To model naturally seeded flows, in which the effect described above would be predominate due to the broad range of particle sizes, 0-10 μm test dust was used to seed the channel flow. Controlled seed measurements were simulated by using 5-10 μm test dust.

It has also been suggested (3) that biased sampling is influenced by the particle arrival rate. A second objective of this study was to vary the seed concentration while the flow conditions were maintained constant and thereby directly test this concept.

An attempt was also made to experimentally verify the existance of biased sampling in naturally seeded flows. This was done by making laser velocimeter measurements in the viscous sublayer of a fully developed, turbulent channel flow. In this near-wall region of the flow the velocity profile is linear and the slope was determined by making numerous velocity measurements as the laser velocimeter was traversed normal to the wall. Two data reduction techniques were applied to the laser velocimeter data and thus two different slopes were determined. One was an unweighted statistical analysis of the data and the second was a weighted statistical analysis that "corrects" for biased sampling. Since the turbulence intensity is high in the sublayer region, there is about a 10% difference between the unweighted and weighted estimates of the mean velocity.

The slope of the velocity profile at the wall was also estimated from a measurement of the pressure gradient along the wall. The gradient was deduced from a measurement of the pressure drop over an eighteen inch length of the channel. Under the assumption that the wall shear stress is constant around the perimeter and along the length of this eighteen inch long control volume which contains the velocimeter test section, the slope of the velocity profile at the wall can be determined from the pressure drop measurement. This slope was taken as the standard to which the slopes given by the laser velocimeter measurements were compared.

The unweighted estimates of the time-average fluid velocity were calculated from the expression

$$U_f = \sum_{i=1}^{N} U_i/N .$$ (1)

This estimate is called the frequency average velocity because each velocity realization U_i is proportional to the Doppler frequency of the laser velocimeter. Since the laser velocimeter was a single component device and since both counter processors were designed to analyze a fixed number of cycles, the one-dimensional weighting factor was used in the weighted estimates of the time-average velocity. Consequently the corrected velocity estimate was calculated from

$$U_p = N/ \sum_{i=1}^{N} (1/U_i).$$ (2)

This estimate is called the period average velocity because Equation 2 can be manipulated into the form

$$U_p = \frac{(\lambda/2) \sin (\theta/2)}{T_D}$$ (3)

where T_D is the average Doppler period for the ensemble of realizations.

EXPERIMENTAL APPARATUS

The basic apparatus, both water channel and laser velocimeter, used in this experiment has been described previously (9,10). Briefly, the measurements were done in a two dimensional channel with a sharp-edged Borda type entrance. The channel is 72 inches long, 12 inches high, and approximately one inch wide. The walls of the channel were bowed inwards by 0.060 inch at the mid-section along the entire length of the channel. This was done so that the velocimeter's probe volume could be traversed right up to the wall without the lower portion of the wall interfering with the incident laser beams. Measurements were made approximately 55 channel widths downstream of the entrance.

A single component, dual scatter, laser velocimeter was used with the sending optics modified so as to produce a 0.0098 inch by 0.0025 inch oval shaped probe volume. The shorter dimension was aligned normal to the wall to obtain the spatial resolution necessary for sublayer measurements. An optical technique for cancelling the pedestal frequency was used in the receiving optics (11). This technique was necessary since the common pedestal and Doppler frequencies that occur in this highly turbulent flow cannot be separated by an electronic filter.

Both receiving and sending optics were mounted on a single traversing mechanism which travels in a direction normal to the wall. This mechanism positioned the probe volume with an accuracy 0.0001 inch from one location to the next. It should be noted, however, that the position of the probe volume with respect to the wall was not known to the same accuracy.

Before all experiments, the 300 gallons of water typically used in the closed circuit water channel system was filtered to remove impurities above 0.5 μm. According to the requirements of the experiment seed particles of either 0-10 μm or 5-10 μm certified test dust were added to the system in quantities ranging from 1/8 gram to several grams.

The analog signal from the laser velocimeter was stored on magnetic tape using an Ampex FR1300 tape deck. The recorded signal was later processed using two different counter processors designed for individual realization laser velocimeter data. The Sequential Phase Comparator (SPC) operates on a Schmitt trigger pulse train derived from the Doppler signal. The device uses a system whereby the period of one cycle from a Doppler burst is sequentially compared with the time for the previous cycle from the same burst until a signal of ten cycles has been analyzed. If all ten time comparisons are within a tolerance level, then a digital counter outputs the average period for the ten cycles for subsequent data storage. A detailed description of the SPC is given by Salsman, Adcox, and McLaughlin (12).

The second processor used in this study was a DISA 55L90 LDA counter operated in the 5/8 comparison mode. This counter uses an internal Schmitt trigger to develop a pulse train which is subsequently analyzed. Since this Schmitt trigger is activated at a set level, the input Doppler signal was passed through a variable attenuator so that the Schmitt trigger level could be varied with respect to the signal. Compute accuracy settings of 1.5 and 3.0 percent were used on the counter during the data reduction.

Verified signal from both data processors was digitized and stored on punched paper tape using a Non-Linear Systems 2607 Serial Converter and a Tally P-120 Tape Perforator. This data was then analyzed on a Hewlett Packard 9820 computing calculator.

Pressure drop measurements were made at a vertical mid-section of the channel using a Gilmont G-1500 Micrometric Manometer. The two 0.0625 inch diameter pressure taps were 18 inches apart with the first tap 48 inches downstream of the entrance. Carbon tetrachloride with a specific gravity of 1.59 was used as the manometer fluid giving a sensitivity of approximately $\pm$ 10^{-5}psi. A static calibration was performed to verify the accuracy of the manometer.

RESULTS

One major advantage in having an analog tape of the laser velocimeter signal is that the output can be processed in different ways. Variation of the level at which signals are detected while holding all other processing variables constant yields results like those shown in Figure 1 for flows seeded with 5-10 μm particles. Notice that for detection levels greater than or equal to 0.3 volts both the frequency average and the period average velocities are constant. The bars on the frequency average velocities are estimates of the 95% confidence intervals for this data. As shown in part (a) of the Figure, this result is also independent of the counter used to process the signal. For these signals, the 0.3 volt level is slightly above a continuous signal that is probably a combination of noise and weak Doppler signals. Most operators of individual realization systems would choose to operate their velocimeter near this detection level in an attempt to maximize their data rate while also maintaining a high percentage of signal validations. As the detection level is raised only the larger amplitude Doppler signals are detected and the data rate declines. For example, the number of signals validated by the SPC processor varied from 2026 at the 0.3 volt level to 272 at the 1.8 volt level. At the 0.3 volt level the average time between validated signals was 0.35 seconds while the average

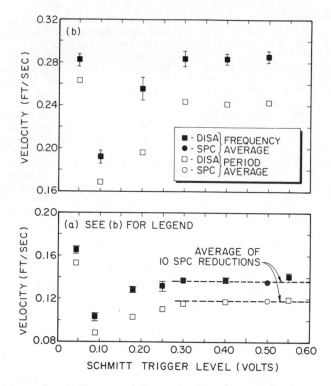

Figure 1.   Influence of Detection Levels on Velocity
            Estimates in a Flow Seeded with 5-10 μm Particles

time between validated signals was 2.62 seconds at the highest trigger level.
The important point is that over a considerable range of detection level the
average velocity estimated from the various ensembles of realizations was
constant.

The detection level was also set considerably below the intuitive optimal
level even though the validation percentage dropped well below the minimum
level recommended for reliable operation by DISA.   The reason for doing this
was an attempt to extract accurate velocity estimates from the continuous but
lower level signals and noise.   As the detection level was lowered the average
velocities first decreased as the number of validated signals in the very low
frequency portion of the ensemble increased.   Further decreases in the detection level yields an increase in the average velocity estimates to a level
markedly above the accurate estimate given by the higher detection levels.
From our limited experience this seems to be a characteristic of the processor.
We used this combination of low validation percentage and simultaneous increase
in the number of validated signals in the very low frequency range as indication
that the detection level was lower than that which yields accurate results.

Similar results were obtained when the flow was seeded with 0-10 μm particles as shown in Figure 2.   For these measurements the lowest accurate detection level was 0.4 volts.   Again the two processors yielded the same estimates
of both the frequency average and the period average velocity over a sutstantial
range of detection levels.   This is an important result because it indicates
that there was no evidence that the probability for detecting slow particles
had increased relative to the probability for detecting fast particles.   With

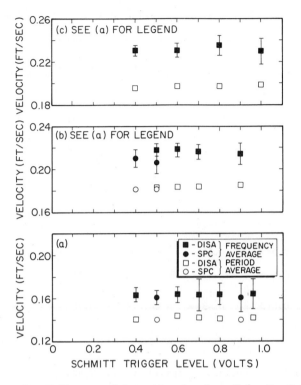

Figure 2.   Influence of Detection Level on Velocity Estimates
in a Flow Seeded with 0-10 μm Particles

this broad size distribution of seed particles it had been thought that the
average velocity estimates might decrease as the Schmitt trigger was lowered.
This would have occurred if particles with low velocities in the size range
of particles that yield marginal amplitude signals were detected and validated
while fast particles in this size range were not detected.   This did not occur
in the operational range of detection for the DISA and SPC processors in this
experiment.   The experiment is definitive because, for the 0-10 μm seed, the
number of particles in the marginally detectable size range must increase as
the detection level is lowered.

The arrival rate of particles that yield valid signals was varied over a
wide range in the water channel by simply varying the amount of seed added to
the filtered water.   The results for two experiments where the flow rate,
measurement location and detection level were held constant while the amount of
seed particles was varied in steps are shown in Table 1.   In reducing the tapes
from both experiments, the detection level was set well above the noise where
Figures 1 and 2 have shown that the velocity estimates are not effected by the
detection level.   There are two types of comparisons that can be made from this
data.   First, by considering only the results from runs with 5-10 μm seed, one
can see that an eight fold variation in arrival rate of the same type of particles does not change the velocity estimate.

Another interpretation that allows comparison of these results with those
from other experiments is to deduce an average distance between validated particles, l_s.   This was done by using the average time between validated signals,
T_s, and the period average estimate of the average velocity.   Specifically

Table 1. Effect of the Type and Concentration of Seed Particles on the
          Velocity Estimates

Run Number	U_f (ft/sec)	U_p (ft/sec)	T_s (sec)	1_s (ft)	Type and amount of seed in 300 gallons
7107	0.257±0.016	0.223	19.80	4.42	1/8 gm; 5-10 μm
7108	0.257±0.010	0.225	2.06	0.464	1 gm; 5-10 μm
7109	0.256±0.012	0.219	2.56	0.560	1 gm; 5-10 μm 1 gm; 0-10 μm
7110	0.263±0.013	0.227	2.05	0.464	1 gm; 5-10 μm 2 gm; 0-10 μm
7113	0.247±0.010	0.217	1.55	0.336	1 gm; 5-10 μm 3 gm; 0-10 μm
9002	0.493±0.013	0.444	6.00	2.66	1/4 gm; 5-10 μm
9003	0.490±0.009	0.434	0.784	0.340	1 gm; 5-10 μm
9005	0.480±0.010	0.434	0.824	0.360	1 gm; 5-10 μm 1 gm; 0-10 μm

$$T_s = T/N \tag{4}$$

where N is the number of validated signals in time T and

$$1_s = U_p T_s . \tag{5}$$

Using result from both of the 5-10 μm experimenta shown in Table 1 one can conclude that estimates of the mean velocity will not vary as the average distance
between validated particles varies from 0.34 to 4.42 feet. This range may be
expanded by using the results from Figure 1 where the velocity estimates were
constant over a wide range of detection levels. As mentioned earlier the number
of particles detected per unit time increases as the detection level is decreased. For the results in part (a) of Figure 1 the velocity estimates were
constant as the average distance between particles varied from 0.043 to 0.31
feet. Combining these results for 5-10 μm particles from Figure 1 and Table 1
we conclude that estimates of the mean velocity will not vary as the average
distance varies from 0.043 to 4.42 feet or from distances that range from ½ to
48 times the width of the channel.

The second comparison is between experiments where 1 gram of 0-10 μm seed
was added to fluid that already contained 1 gram of 5-10 μm seed. Notice that
there is a surprising increase in time between valid signals when 0-10 μm seed
is added (compare Run 7109 with 7108 and Run 9005 with 9003). The explanation
for this increase is simply that the S/N ratio is lower with the broader distribution of particle sizes and hence the percentage of signals validated by
the counter processor decreases.

Several experiments were performed in which the slope of the velocity profile was determined using both laser velocimeter measurements and simultaneous
measurement of the pressure drop in the water channel. Figure 3 is an example
of the laser velocimeter data showing the two different slopes which can be
deduced depending on whether the data is frequency averaged or period averaged.
Note that the two lines intersect the axis at a negative distance from the wall.

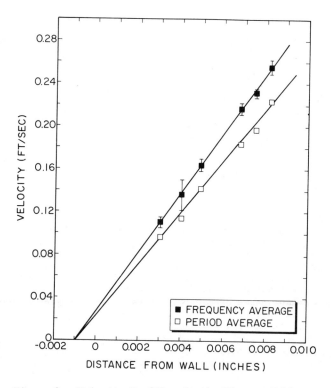

Figure 3. Velocity Profiles in the Viscous Sublayer

This is merely indicative of a slight error in the initial placement of the probe volume with respect to the wall. More significant is the linearity of both the frequency averaged and period averaged data which results in small uncertainty bands on the slopes estimated from this data.

However, when the above slopes are compared to the slopes obtained from the pressure drop measurements, it was found that they were both consistently lower than the pressure drop measurement (see Table 2). In most cases, the slope from the pressure drop measurement was much higher than both the frequency averaged and period averaged data. Consequently, this attempt to use pressure drop measurements to determine whether or not the velocity data was effected by sampling bias was not successful. It should be noted that the data from experiments 4000, 5000, and 6000 did not replicate earlier results that were obtained with essentially the same apparatus (10). The reason for this difference is not known.

CONCLUSIONS

Two types of flow seeding were examined during this study. Naturally seeded flows were modelled by using 0-10 μm test dust while flows with controlled seeding were modelled with 5-10 μm test dust. Two of the experiments listed in Table 1 show the effects of adding 0-10 μm seed. In effect, a controlled seed experiment was transformed to a natural seed experiment with all other variables being held constant. When examining the resulting data, one finds that the mean velocity, whether frequency averaged or period averaged,

Table 2. Comparison of the Velocity Profile Slopes with the Pressure Gradient

Run Number	Type and Amount of Seed	Slope of Velocity Profile at Wall(ft/sec/in)		
		From U_f Data	From U_p Data	From ΔP Data
3100	1 gm; 0–10μm	27.2	23.9	28.5
3300	1 gm; 0–10μm	27.7	24.0	34.2
4000	1 gm; 5–10μm	26.6	22.8	34.0
5000	1 gm; 5–10μm	22.0	20.1	28.3
6000	1 gm; 5–10μm	25.5	23.9	30.0

has not changed. Also, in Figure 1 (controlled seed) and in Figure 2 (natural seed), both seeding conditions are shown to yield consistent mean velocity values over a wide range of Schmitt trigger settings when the detection level is above the noise level. The combination of the above results leads to the conclusion that individual realization laser velocimeter measurements will yield the same result whether the flow is seeded with a broad range of particle sizes or with a narrow range of particle sizes. These results also suggest that there was little, if any, effect upon bias sampling due to differences in the light intensity scattered from slow and fast moving particles.

Several experiments were designed to investigate the effect of particle arrival rate upon sampling bias by varying the seed density in the flow over a wide range. The effect of particle arrival rate was further studied in flows with constant seed density by varying the number of validated signals through adjustments in the Schmitt trigger level. All the experimental data indicate that the particle arrival rate does not influence either the frequency average or the period average estimates of the mean velocity. Hence, the particle arrival rate does not influence bias sampling.

ACKNOWLEDGEMENTS

The assistance of Mr. Michael E. May in conducting several of these experiments is gratefully acknowledged. Mr. Jeff Moon assisted with the data reduction.

This research was sponsored by the Air Force Office of Scientific Research, Air Force Systems Command, USAF, under Grant No. AFOSR-77-3298. The United States Government is authorized to reproduce and distribute reprints for Governmental purposes not withstanding any copyright notation hereon.

REFERENCES

1.   Donohue, G.L., McLaughlin, D.K., and Tiederman, W.G. 1972. Turbulence Measurements with a Laser Anemometer Measuring Individual Realizations. Physics of Fluids. 15:1920-1926.

2.   McLaughlin, D.K., and Tiederman, W.G. 1973. Biasing Correction for Individual Realization Laser Anemometer Measurements in Turbulent Flows. Physics of Fluids. 16:2082-2088.

3.   Barnett, D.O., and Bentley, H.T. 1974. Statistical Bias of Individual Realization Laser Velocimeters. Proceedings of the Second International

Workshop on Laser Velocimetry, Purdue University, West Lafayette, Indiana 47907. 1:428–442.

4. Buchhave, P. 1976. Biasing Errors in Individual Particle Measurements with the LDA-Counter Signal Processor. Proceedings of the LDA-Symposium Copenhagen 1975. 235–278.

5. Dimotakis, P.E. 1976. Single Particle Laser Doppler Measurements of Turbulence, presented at the AGARD Symposium Non-intrusive Instrumentation in Fluid Flow Research. Saint-Louis, France. May 3–5.

6. Hoesel, W., and Rodi, W. 1977. New Biasing Elimination Method for Laser - Doppler Velocimeter Counter Processing. Rev. Sci. Instrum. 48:910–919.

7. Karpuk, M.E., and Tiederman, W.G. 1976. Effect of Finite-Size Probe Volume Upon Laser Doppler Anemometer Measurements. AIAA Journal. 14:1099–1105.

8. Durst, F., and Whitelaw, J.W. 1972. Theoretical Considerations of Significance to the Design of Optical Anemometers. Paper 72-HT-7, presented at the Heat Transfer Conference ASME, Denver, Colo., August 6–9. United Engineering Center, 345 East 47th Street, New York, N.Y. 10017. 9pp.

9. Karpuk, M.E., and Tiederman, W.G. 1974. A Laser Doppler Anemometer for Viscous Sublayer Measurements. Proceedings of the Second International Workshop on Laser Velocimetry, Purdue University, West Lafayette, Indiana 47907. 2:68–87.

10. Quigley, M.S., and Tiederman, W.G. 1977. Experimental Evaluation of Sampling Bias in Individual Realization Laser Anemometry. AIAA Journal. 15:266–268.

11. Bossel, H.H., Hiller, W.J., and Meier, G.E.A. 1972. Noise Canceling Signal Difference Method for Optical Velocity Measurements. J. Physics, E. Sci. Instr. 5:897–900.

12. Salsman, L.N., Adcox, W.R., and McLaughlin, D.K. 1974. Proceedings of the Second International Workshop on Laser Velocimetry, Purdue University, West Lafayette, Indiana 47907. 1:256–268.

Bias Corrections in Turbulence Measurements by the Laser Doppler Anemometer

PREBEN BUCHHAVE and WILLIAM K. GEORGE, JR.
Mechanical Engineering Department
State University of New York at Buffalo
Buffalo, NY 14214

INTRODUCTION

This paper describes work presently underway at the Turbulence Research Laboratory, Faculty of Engineering and Applied Sciences, State University of New York at Buffalo. The object of this program is partly to develop correct data analysis algorithms and software for bias-free estimation of statistical flow quantities based on randomly arriving LDA-counter data, partly to serve as a pilot program for the design of a larger experimental facility for jet noise studies, now in its initial phase.

The basic philosophy behind the development of data analysis programs is that the so-called individual realization LDA bias or statistical bias is not a mysterious and elusive property of some types of flow, but that the effect is always present in individual realization anemometry, and that the "bias" is only a result of incorrect data processing. Also, that the underlying sampling mechanism causing the bias (when data are treated incorrectly) is not something to try to avoid, but rather the "natural" way in which individual realization anemometers work, and that indeed the individual realization or single burst data collection is in many cases a desirable feature of such instruments, which may lead to new insight or alternative methods of measuring dynamic flow quantities.

To illustrate this we mention just a few of these possibilities: Seeding only one of the constituents in a turbulent mixing process, e.g. a jet issuing into an unseeded environment or vice-versa, presents a way of conditionally sampling fluid elements originally present in only one of the media. Also, the randomly sampled data provided by the individual realization anemometer allows alias-free spectral estimation and a time resolution beyond the usual Nyquist criterion.

The algorithms required for correct data processing of individual realization data are not more complicated than those used in normal arithmetic averaging and, apart from the requirement of being able to transfer more bits of information per measured data point, the hardware needed for bias-free data collection is not much more complicated than previously used LDA-counter circuitry.

The theoretical background for the individual realization data processing and the present status of LDA-measurements overall can be studied in Reference 1. A fuller account of the data processing and experimental evaluation of biased and unbiased LDA-data processing can be found in Reference 2.

110

EXPERIMENTAL SETUP AND DATA COLLECTION EQUIPMENT

The flow in which the experiments herein cited were carried out is a 3" diameter free jet in air. The air flow passes through a 12" diameter plenum chamber containing honey-comb and a number of screens. The flow is accelerated through a 16:1 contraction having a profile of matched cubics. The Laser ane- mometer is a DISA X-type optical system operating in forward scatter mode. The Laser is a Spectra-Physics 15 mW He-Ne laser Model 125. The optics defines a measuring volume approximately 0.1 mm in diameter and approximately 0.2 mm in length, located approximately 300 mm from the optical components. The whole optical system can be traversed in two directions in the horizontal plane. The whole environment around the jet and the optical system can be sealed off allow- ing uniform seeding of the whole flow field. Droplets (1-3μm dia.) of a 50:50 pct mixture of glycerin and water were used as seeding. The optical system in- cludes a Bragg cell frequency shift module operating at a constant frequency of 40 MHz. Electronic mixing with a local oscillator in the signal processing equipment (DISA 55N10 Dual-channel Frequency Shifter) phase-locked to the Bragg cell driver allows convenient selection of the effective frequency shift from 10 KHz to 10 MHz.

The signal processing equipment is composed of standard DISA LDA components including two 55L90a counters, a 55N10 Frequency Shifter, a 56G20 Interface and Buffer, a DEC LSI-11 mini-computer and a DEC DX-11 Dual Floppy Disk System. A block diagram is shown in Figure 1. The digital output from the counter is transferred to the Interface and Buffer, which handles up to eight parallel 16-bit input words and eight parallel, 16-bit output words. Two of the inter- face output lines are used to communicate with the LSI-11 minicomputer. The computer is controlled through a Beehive B-100 terminal and storage is provided by the DEC DX-11 Dual Floppy Disk System.

The Interface and Buffer contain a backplane with a number of print sockets and can be built up to the required complexity by inserting optional prints. The main ingredient is an input logic and buffer card, which handles incoming data, marks the data words so the source can later be identified, and stores the data in the 500 word SILO-type memory. The buffer can accept asynchronous data with a time spacing of only 1μs (or by replacement of memory elements with some- what more expensive but pin-for-pin compatible chips a spacing of only 70 ns). A high data reception rate is required by the interface to take advantage of the short time lags inherent in the random sampling of the LDA, especially if more channels are used. The function of the buffer is primarily to allow reception of asynchronous data and subsequent transmission to the computer. The buffer can be expanded if needed to 2000 words by insertion of additional memory cards with no change in logic circuitry. The function of the interface is controlled by the control logic card, and an output demultiplexer card allows communication through the remaining six 16-bit channels (in addition to the two occupied by the computer) for controlling external devices (flow parameters, traversing mechanisms, or other).

The LDA-counter mode of operation is an important factor in the data pro- cessing and is explained with reference to table 1. The DISA 55L90a counter may operate in four modes of which one (the fourth) is primarily of interest in time-of-flight measurements.

Mode-1 is the most common mode of operation of conventional LDA-counters. A measurement is initiated when the signal exceeds the fringe counter Schmidt trigger level. The first zero-crossing enables the counter and counting of clock oscillator pulses continues through the following eight zero-crossings if the counter is not reset by a detected error condition by the validation circuits (in which case the counter is immediately reset). The immediate reset allows

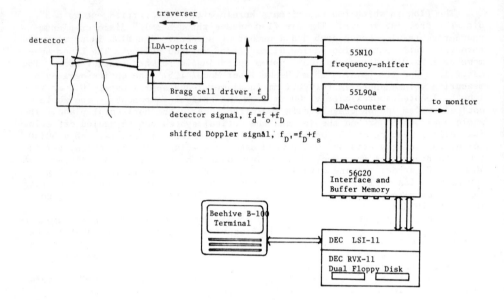

Figure 1. Block diagram of signal processing equipment

the highest possible data rate and ensures that a burst is measured even if a
fault should occur in the initial phase of the burst. In mode-1 new measurements
continue based on eight zero-crossings throughout the burst as long as Doppler
periods exceeding the Schmidt trigger level are available. The measurement is
checked by a 5:8 comparator circuit and three other validation schemes (burst
amplitude exceeding preset level, measured velocity below preset range minimum
and burst envelope dip detected by two-level sequency detector).

Mode-3 bases the measurement on the total number of periods in the burst
(from 2 to 256), only one measurement is output per burst, and in addition to
the velocity output the number of zero-crossings measured per burst, N_z, is also
available in binary form at the output. However, in this mode the 5:8 comparison
is not active and the counter is thus somewhat more sensitive to noise in this
mode. The "velocity" output is based on a variable number of zero crossings and
must therefore be "normalized" relative to N_z.

Mode-2 combines the positive features of mode-1 and mode-3. The measurement is still based on eight zero crossings, the 5:8 comparison is active, but
only one measurement is made per burst and the total number of zero crossings
is available at the output. Unless the S/N is high mode-2 works better than
mode-3, which can give many data points based on only 2,3 or a low number of
zero-crossings. Mode-3 may be somewhat less sensitive to the finite fringe
number angular effect described later if frequency shift is not used.

In the measurements described here only mode-2 was employed and variable
frequency shift was always available.

TABLE 1.   DISA   55L95 COUNTER MODES

MODE	N_z	VALIDATION	OPERATION	OUTPUT
1 CONTINUOUS	8	5:8 + 3	CONTINUOUS	V, T
2 COMBINED	8	5:8 + 3	ONE PER BURST	V, T, N_b
3 TOTAL BURST	N_b	- + 3	ONE PER BURST	"V", T, N_b
4 "STOP WATCH"	2	- + 3	CONTINUOUS	1/ T, T

N_z = NUMBER OF ZERO-CROSSINGS TIMED

N_f = NUMBER OF FRINGES IN MEASURING VOLUME

N_S = NUMBER OF PERIODS ADDED BY FREQUENCY SHIFT

N_b = NUMBER OF PERIODS IN MEASURED BURST = $N_f + N_s$

DATA PROCESSING

In Reference 1 and 2 it is shown that the digital samples measured in individual realization anemometers can be used to form the correct time-averaged
statistical quantities of the flow provided each realization enters the time
integral only during the time the particle is present in the measuring volume.
This forms the basis for the so-called residence-time weighted averaging in
which the measured residence time of each sample enters the formulas as a weighting factor to be applied to each individual data point.   The resulting formulas
for the time-averaged mean and mean-square velocity are given by:

Residence-time weighting

$$U = \frac{\sum_i u_i \Delta t_i}{\sum_i \Delta t_i} \qquad\qquad \overline{u'^2} = \frac{\sum_i (u_i - U)^2 \Delta t_i}{\sum_i \Delta t_i}$$

where Δt_i is the measured residence time of the ith realization.   Throughout the
paper the velocity vector is denoted by $\underline{u} = (u,v,w)$ and $\underline{U} = (U,V,W)$ and $\underline{u}' =$
(u',v',w') denote mean and fluctuating components respectively.
The corresponding formulas for the straight arithmetic averaging, which is
only correct for equidistantly sampled data, but which give bias-errors when
applied to individual realization anemometers, are:

Arithmetic averaging

$$U_{AR} = \frac{\sum_i u_i}{N} \qquad\qquad \overline{u'^2_{AR}} = \frac{\sum_i (u_i - U_{AR})^2}{N}$$

and the algorithms using the one-dimensional correction proposed originally by McLaughlin and Tiederman (Reference 3), which provide an approximate bias-correction at low turbulence intensities, are:

One-dimensional correction

$$U_{1D} = \frac{\sum_i u_i |u_i|^{-1}}{\sum_i |u_i|^{-1}} \qquad\qquad \overline{u'^2_{1D}} = \frac{\sum_i (u_i - U_{1D})^2 |u_i|^{-1}}{\sum_i |u_i|^{-1}}$$

In Reference 1 and 2 it is shown that the residence-time weighted data provides the correct answer assuming statistically uniform spatial particle distribution throughout the flow, and provided the measuring volume has the same effective size independent of flow direction. In other words, the finite fringe number effect due to the fact that the LDA-signal processor requires a minimum number of fringes for processing must be eliminated by the use of frequency shift or by other methods. Note that when the finite fringe number effect is negligible the measuring volume may have any shape (ellipsoidal or other) without causing any bias-effects.

A convenient, frequency independent method of measuring the residence time is provided by the binary output of the total number of zero-crossings N_z in a burst in conjunction with the measured frequency, f_m. The residence time is simply given by:

$$\Delta t_i = \frac{N_{z,i}}{f_{m,i}}$$

where $f_m = f_o + f_s$ is the sum of the Doppler frequency and the applied frequency shift. This formula is valid with or without frequency shift.

The angular effects due to the requirement of a finite number of zero-crossings were first studied in Reference 4 for arithmetic averaging (non-corrected data) and 1-D weighted data. The cross-sectional area from which measurements can result is a function of the direction of the particle and the ratio Q of the maximum number of fringes in the measuring volume N_f (particle trajectory along the x-axis) to the minimum number N_e required by the processor: $Q = N_e/N_f$. The limit Q→0, which corresponds to the case considered by McLaughlin and Tiederman, may be approached by requiring only a small N_e (such as two in mode-3) or by adding enough frequency shift so that a particle always produces more than N_e zero-crossings independent of velocity magnitude and direction. The biasing effects resulting from arithmetic and 1-D weighted averaging were considered in Reference 4 for various flow situations. Figure 2 shows the computed error in mean and mean-square velocity assuming a three-dimensional isotropic, Gaussian turbulence as a function of the turbulence intensity. Correctly averaged data would of course result in points on the x-axis.

It should be noted that the finite fringe number effect and the over-compensation introduced by the 1-D correction act in opposite directions and for a

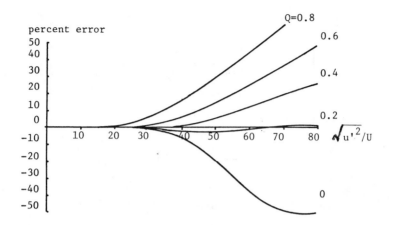

Figure 2.   Calculated bias error in mean velocity of 3-D Gaussian, isotropic
            turbulence including effect of measuring volume cross section

certain value of Q (Q≈0.2 for the particular flow considered in this example,
corresponding to e.g. 8 fringes required of a maximum of 40) nearly cancel even
for large turbulence intensity. This may explain why some authors have gotten
surprisingly good results with the approximative 1-D correction.

MEASUREMENTS

     Mean and mean-square velocities have been measured along the axis and
across the jet at various locations. The data have been computed using the
three sets of algorithms shown in the previous section for various amounts of
frequency shift. Some of the data are presented in the following figures.
     The measurements have been carried out all the way through the jet to
regions at the edge of the jet, where the turbulence intensity is very high.
The data were extremely stable as long as sufficiently long averaging time
was used in order to form averages over many integral time scales of the flow.
Ensembles of 2000 or more data points were used, and the variance of the measured data points was only a few percent.
     Figure 3 shows the measured mean values using the three different averaging
techniques in a scan along the axis of the jet, and the percentage difference
between the arithmetic and 1-D corrected data relative to the residence time
weighted results. Figure 4 shows similar results for a transverse scan at
X/D = 5.3. The difference in the two figures shows the influence of frequency
shift. The frequency shift used obtaining the data shown in Figure 4 was
sufficient to eliminate angular dependence due to the finite fringe number effect.
     Unfortunately it has not yet been possible to provide an absolute reference
by which to firmly establish the validity of the residence-time weighted results
in a three-dimensional, highly turbulent flow. Presently it is only possible
to establish the fact that large differences occur in using the three mentioned
averaging methods, and that the measured differences conform to the trend suggested by previous analysis. The influence of frequency shift also confirms
the trend expected from that analysis. From Figure 4B it is evident that the
opposite trends of the angular dependence and the 1-D compensation help reduce

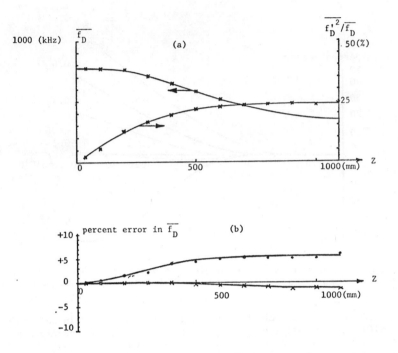

Figure 3. a) Measured axial mean and rms Doppler frequency along
             jet axis
          b) Measured percentage error in mean value relative to
             residence-time weighted data.

          frequency shift 300 kHz

          · arithmetic averaging
          x 1-D weighted data

the difference between the residence-time weighted data and the 1-D corrected
data whereas a larger difference exists in the case shown in Figure 4C where
the angular dependence has been minimized by frequency shift.

    More detailed measurements will be published in Reference 2 and it will
also be attempted to establish a reference to an unbiased measurement of sufficient accuracy to allow comparison with the residence-time weighted data.

AUTOCORRELATION FUNCTION

    The theory of residence-time weighting of statistical quantities involving
more than one realization is also developed in Reference 1 and 2. There it is
shown that the time-averaged autocorrelation function from individual realization anemometers can be computed by only integrating the realizations while
both are "on" at the same time. The resulting expression for the time-averaged
correlation function from individual realizations becomes:

(a)

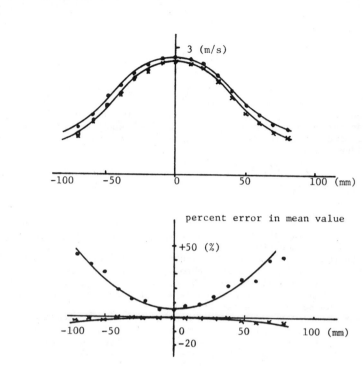

(b)

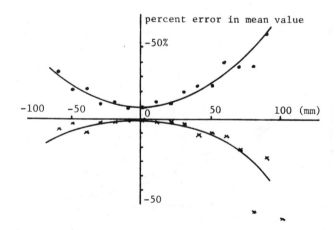

(c)

Figure 4. a) Measured axial mean velocity, transverse scan at X/D=5.3
          b) Errors in mean value relative to residence-time weighted
             data, frequency shift 0
          c) Same as (b) but frequency shift 200 kHz

          · Arithmetic averages
          x 1-D weighted data

$$R(\tau) = \frac{\sum\limits_{i<j} u_i u_j \Delta t_{ij}}{\sum\limits_{i<j} \Delta t_{ij}} \tag{1}$$

where $t_j - t_i$ are the actually occurring lags between pairs of realizations. Δt_{ij} is the "overlap-time", i.e. the time in which both the realization u_i and t_i and the delayed realization u_j and $t_j + \tau$ are present simultaneously.

It can further be shown that a time-slotted approximation to this estimator is equivalent to the following expression based directly on the measured residence times:

$$R(n_{\Delta\tau}) = \frac{T}{\Delta\tau} \frac{\sum\limits_{i<j} u_i u_j \Delta t_i \Delta t_j}{\sum\limits_{i<j} \Delta t_i \Delta t_j} \qquad (n-1/2)\,\Delta\tau \le t_j - t_i < (n+1/2)\Delta\tau \tag{2}$$

where T is the total time interval over which measurements are made. This expression provides an autocorrelation estimator based on the measured values of velocity and residence time, which is equivalent to the time-averaged auto-correlation function (i.e. bias-free) provided, as before, the particle distribution is statistically uniform throughout space and the finite fringe number effect has been eliminated. Note, that this estimator does not contain self-products and thus is in accordance with the estimators developed by Mayo (Ref. 5) and Gaster and Roberts (Ref. 6).

The estimator of Equation (2) and similar estimators based on arithmetic averaging and 1-D weighted data have been implemented with the data processing equipment described here, and some measurements have been made both in the centerline of the jet and in the highly turbulent region off-axis. Preliminary results indicate an appreciable difference in results obtained using the three different methods even at the jet centerline. Figure 5 shows typical autocorrelation coefficient results obtained at the jet centerline at X/D=5.3. The two curves show the result of arithmetic averaging compared to residence-time weighted data. Evidently the influence of the bias-errors is most marked for large time lags.

Further work on the influence of individual realization bias on autocorrelations and spectra is in progress and will be presented in Reference 2.

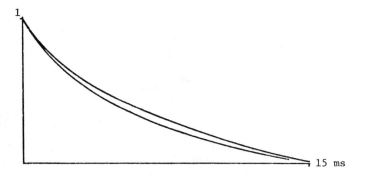

Figure 5   Autocorrelation function of axial velocity at X/D=5.3, 3.25 m/s, frequency shift 200 kHz

Upper curve: arithmetic averaging
Lower curve: residence-time weighted data.

References
1. P. Buchhave, W.K. George, Jr. and J.L. Lumley, "The Measurement of Turbulence with the Laser Doppler Anemometer", Prepared for publication in Annual Review of Fluid Mechanics, Annual Reviews, Inc., Palo Alto, California, 94306 (Van Dyke and Wehausen, Co-editors). Also issued as TRL-101 by Turbulence Research Laboratory, State University of New York at Buffalo.
2. P. Buchhave (1978) "Errors and Correction Methods in Turbulence Measurements with the LDA", Ph.D. Thesis, Department of Mechanical Engineering, State University of New York at Buffalo, Buffalo, New York.
3. D.K. McLaughlin and W.K. Tiederman (1973), "Biasing Correction for Individual Realization of Laser Anemometer Measurements in Turbulent Flows", Phys. of Fluids, 16, pp. 2082-2088.
4. P. Buchhave (1976), "Biasing Errors in Individual Particle Measurements with the LDA-Counter Signal Processor". In proceedings of the LDA-Symposium Copenhagen, 1975: pp. 258-278. Copenhagen: PO Box 70, 1740 Skovlunde, Denmark. 736 pp.
5. W.T. Mayo, M.T. Shay and S. Riter (1924) "The Development of New Digital Data Processing Techniques for Turbulence Measurements with a Laser Velocimeter". Final Report (AEDC-TR-74-53), USAF Contract Co. F40600-73-C-003, August 1974.
6. M. Gaster and T.B. Roberts (1975) "Spectral Analysis of Randomly Sampled Signals". J. Inst. Maths. Applic., 15, 195-216.

COMBUSTION MEASUREMENTS

Chairman: **H. D. THOMPSON**
Purdue University

Measurements in Combustion Systems

INVITED PAPER

F. K. OWEN
Consultant
P.O. Box 1697
Palo Alto, CA 94302

ABSTRACT

Laser velocimeter measurements made in the initial mixing regions of confined turbulent diffusion and spray flame burners will be described. In the
case of the diffusion flame burner, measurements of the axial and tangential
mean velocity profiles and the root mean square (rms) and probability density
distributions of the velocity fluctuations show that there are significant
variations of the mean and time-dependent flow fields with changes in ambient
combustor pressure and inlet air swirl.

The spray flame measurements suggest a picture of the fuel spray in which
a conical cone comprising larger fuel droplets, with velocities determined
by fuel nozzle characteristics, is surrounded by clouds of smaller fuel droplets
which respond relatively quickly to the ambient gas velocity. The inertia of
the larger droplets enable them to penetrate into the air stream with little
interaction. Droplet velocity fluctuations appear to occur primarily as a result of the interaction of the smaller droplets with the gas flow.

Measurements in both flow fields suggest that large scale motions play a
key role in the turbulent mixing processes and as such are an important influence on energy release and pollutant emissions.

INTRODUCTION

It is now generally accepted that changes in operating conditions,
which alter the flow patterns in combustion devices, can have significant
effects on combustion efficiency and pollutant emissions. In particular, the
size, shape, recirculated mass flow, and local turbulence levels associated
with recirculation zones are critical to flame stability, intensity, and overall
performance. Unfortunately, experimental examination of the interaction between
fluid-dynamic and chemical processes inside combustors is difficult, since the
mean flow fields and turbulence properties of combusting flows with recirculation
are difficult to determine with any degree of reliability using conventional instrumentation. For instance, flows with severe adverse-pressure gradients,
which normally give rise to separation and recirculation, are difficult to
document, as they are extremely sensitive to local geometry and probe interference. Problems associated with turbulent structure measurements are even more
acute, because linearized hot-wire data interpretations are not accurate in these
highly turbulent flows.

Fortunately, linear nonperturbing fluid mechanical measurements of complex three-dimensional flow fields are now possible with the laser velocimeter, provided light-scattering particles can be relied upon to follow the local fluid velocity. There are, to be sure, many sources of LDV errors (a grand total of 18 were reported at the Minnesota meeting in 1975) and a multitude of other reasons why accurate measurements are difficult to make in turbulent combustion systems. However, the measurements described in this paper illustrate that a great deal of useful quantitative information can be obtained with current instrumentation.

These studies of recirculating flows were made as part of a research program directed toward developing an understanding of the chemical and physical processes occurring in combustion devices and their effects on pollutant formation. Specifically, they were initiated to obtain reliable mean velocity and turbulence data in recirculating combusting flow fields and to determine the effects of swirl and ambient pressure on these parameters [1,2].

EXPERIMENTAL DETAILS

The laser velocimeter measurements were made in the combustor described in Ref. 1. This facility consisted of an axisymmetric combustor in which central gaseous or liquid droplet fuel streams mixed with coaxial annular airstreams. Details are shown schematically in Fig. 1. The 12.23cm diameter, 100cm long combustor was divided into three water-cooled zones of approximately equal length. A choke was installed downstream of the extender to raise the ambient combustor

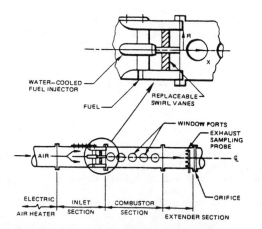

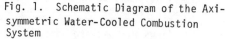

Fig. 1. Schematic Diagram of the Axi-symmetric Water-Cooled Combustion System

pressure when required. Quartz windows were installed in the instrument ports
to provide optical access for the laser measurements. Air from a 30-atm supply,
at flow rates up to 0.65 kg/sec, was preheated to provide inlet air temperatures
up to 1000°K. The air then entered the circular annulus of the replaceable in-
jectors installed in the 12.23-cm-dia inlet section. In the gaseous fuel tests
the fuel was introduced through three airfoil shaped struts into the central
delivery duct and brought into contact with the annular airstream at the exit of
the injector. Airflow swirl was produced by straight swirl vanes inserted into
the annular passage of the injector. In the liquid fuel tests air entered the
combustor flowing around a nominal 60° hollow cone fuel injector nozzle located
on the combustor center line. Swirl vanes were once again inserted into the
annular air passage.

The local air, gas and droplet velocity measurements were made with a dual-
beam velocimeter utilizing a crystal Bragg cell which acted as a beam splitter
and frequency-shifted the first deflected beam. A schematic of the optics and
signal processing instrumentation is shown in Fig. 2. The sensing volume determined by beam crossover volume, off-axis collection, and photomultiplier pinhole size was elliptic with principal axes of 0.2 and 2.0 mm, respectively.
Single-particle, time-domain signal processing was used to build up velocity
probability density distributions from which both the mean and rms velocities
were obtained. A minimum of 1000 instantaneous velocity determinations was
used to build up the probability densities. This resulted in a statistical
error of less than 5% in the computed values of both the mean and variance, with
a confidence level of 95%. On-line signal processing with a minicomputer was
used to determine the local mean velocity, turbulence intensity, and velocity
probability density function.

The optical sensitivity of the forward scatter system used in the present
study was such that naturally occurring submicron particles could be used for
the gas velocity determinations in regions far from the fuel spray. However,
to increase the signal-to-noise ratio and the data acquisition rate in these
regions, the air flow was seeded with micron-sized particles dispensed from a
fluidized bed. A number of materials which had previously been used to seed

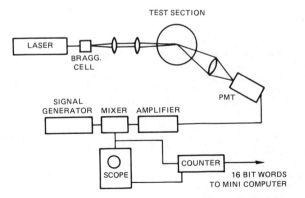

Fig. 2. Schematic Diagram of the Laser Velocimeter

small open flames were tested but none proved suitable for the present experimental arrangement. Both Al_2O_3 and TiO_2 deposited on the combustor windows degrading the Doppler signals to an unacceptable extent, and silicone oil droplets dispensed from a Laskin nozzle evaporated on or before reaching the combustion zone. However, nominal 5μm microballoons (hollow spheres) of bakelite phenolic resin were used successfully. Due to their low initial density (< 0.1 gm/cc) and to the fact that they charred to micron size in the combustion zone, these particles gave adequate turbulence response and excellent signal/noise ratio (> 10:1) without disturbing optical access.

Since for these flows more than 95% of the turbulence energy is likely to be in scales corresponding to Eulerian frequencies below 25 KHZ (turbulence scales less than 1/10 of the combustor diameter), errors due to particle response (in the Lagrangian frame) should be negligible. To determine gas velocity in regions of moderate fuel droplet concentration, the air flow was seeded selectively, as discussed later.

Directional ambiguity (which can result in data interpretation errors in highly turbulent and/or recirculating flows [3]) was avoided due to the zero velocity frequency offset provided by the Bragg cell. In the case of fuel droplet velocity studies, there is an additional compelling reason to use a Bragg cell (moving fringe) laser velocimeter system. Since most commercial counting devices require a minimum of eight cycles, this provides a lower fringe number limit. Most practical liquid fuel sprays have a droplet size distribution covering a wide range (up to 250 μm), with typical mean droplet sizes between 75 and 150 μm. The minimum fringe number requirement together with the broad particle size distribution raises a problem of spatial resolution with stationary fringe systems, since optimum signal/noise requirements dictate that the fringe spacing should be greater than or equal to the maximum fuel droplet diameter. Thus, for 250 μm droplets, the focal volume diameter should be approximately 2 mm to provide an adequate number of fringes. Even with off-axis light collection, the probe volume length is typically an order of magnitude greater than the focal volume diameter which would be much too large   ($\sim$ 2cm in this case). Fortunately, frequency biasing increases the number of effective fringes moving through the focal volume so that sufficient fringe crossings and reasonable spatial resolution can be obtained even with a large effective fringe spacing.

DISCUSSION OF RESULTS

Gaseous Fuel Tests
Axial Velocity Measurements

Representative mean axial velocity profiles are presented in Figs. 3 and 4. These profiles show the location and shape of the time-averaged recirculation zones and indicate their longitudinal and lateral extent. Both sets of data show a consistent trend toward uniform velocity profiles, which would be expected for highly turbulent flows downstream of the initial mixing region.

In the case of zero swirl at 3.5 atm (Fig. 3), a large spheriodal recirculation zone is present immediately downstream from the center (fuel) jet, with associated mean reverse velocities significantly larger than in any of the swirling-flow cases. This has a pronounced effect on the mean flow field, namely, flow curvature around the zone, which results in a high-shear layer in the outer annulus close to the wall. The associated radial flow angles indicated by the outward movement of the peak axial velocities are comparable to those induced by the 0.3-swirl vanes. In addition to this central zone, the velocity profile at X-1.8 cm shows evidence of a second recirculation zone behind the backward-facing step at the nozzle exit plane.

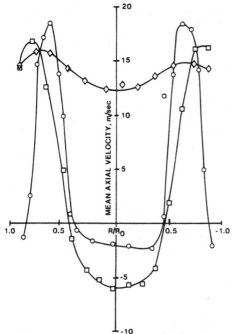

Fig. 3. Mean axial velocity profiles, zero swirl, 3.5 atm o-X = 1.8 cm, □- 2.3 cm, ◊ - 22 cm.

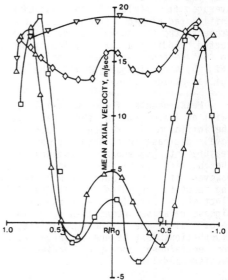

Fig. 4. Mean axial velocity profiles, 0.3 swirl, 3.5 atm, □- 3.6 cm, △- 6.7 cm, ◊- 18.1 cm, ▽- 36.8 cm.

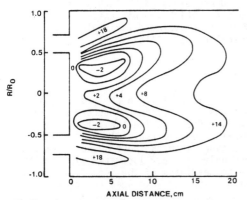

Fig. 5. Axial velocity contours for the case of 0.3 swirl at 3.5 atm.

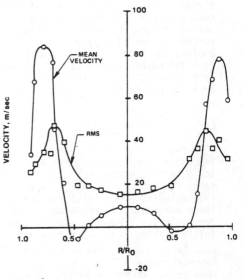

Fig. 6. Axial mean and rms velocity distributions at X = 2.0 cm for the case of 0.6 swirl at 1.0 atm.

The introduction of swirl brings about significant changes in the time-averaged flow field. In the case of 0.3 swirl at 3.5 atm, for instance, the initial radial flow is high, but a much smaller torroidal recirculation zone is present; the secondary recirculation zone is so reduced that it cannot be detected. Additional axial velocity profiles were obtained at this test condition so that contours of constant axial velocity could be determined. These contours (Fig. 5) show, in more detail, the structure of the axial flow field in the initial mixing region.

Measurements of the rms axial velocity fluctuations for the case of 0.6 swirl at 1 atm (Fig. 6) show that local turbulence levels are extremely high in the initial mixing region, although in general it was found that the fluctuation levels decreased with both increasing swirl and ambient pressure. Detailed measurements [4] show that the peak fluctuation levels essentially coincide with the locations of high-mean-shear, i.e., maximum local mean velocity gradient. However, it must be borne in mind that rms velocity fluctuation levels say nothing about the turbulent scales involved. For instance, there are significant fluctuations associated with the time-averaged recirculation zones with the highest relative intensities occurring close to the time-averaged axial velocity zeros. Since mean velocity gradients are relatively low in these regions, the fluctuations must be due to large-scale mixing associated with recirculation zone entrainment and/or unsteadiness about its mean location. This point will be discussed later.

Tangential Velocity Measurements

The principal feature of each set of tangential velocity profiles (Fig. 7, e.g.) is the progressive change from solid-body rotation (forced vortex) flow close to the injector toward a combined free/forced (Rankine vortex) flow downstream. Thus a region of nonrotational flow develops and progresses toward the center of the duct as the flow proceeds downstream. As a result, the point of maximum tangential velocity moves radially inward. This trend from forced to Rankine vortex flow becomes more pronounced with both increased swirl and ambient combustor pressure (increased residence time) [4]. Such a transition from forced to Rankine vortex requires a sink, which in this case is provided by the inward radial flow downstream of the time-averaged recirculation zones as evidenced by the increasing axial centerline velocities in Fig. 5.

Large-Scale Structure

High-speed cine pictures of the reacting flow field in the vicinity of the injection plane show significant fluctuations in the flame structure and large-scale motions associated with flow reversal. Quantitative insight into this large-scale turbulent motion associated with flow recirculation can be obtained from velocity probability density distributions such as those shown in Fig. 8. These measurements, which can be obtained only with a velocimeter system with zero velocity frequency offset, show the unsteadiness of the flow field in the initial mixing region. For instance, within the time-averaged recirculation zone ($r/r_0 = 0.35$), there are significant numbers of positive velocity occurrences (approximately 30%, which are the result of either instantaneous recirculation zone breakdown and/or extensive streamwise and lateral movement. This large-scale mixing structure results in significant deviations from Gaussian turbulence. In the corner region ($r/r_0 = 0.9$) approximately 25% of the instantaneous velocity occurrences are negative and the velocity probability density is again skewed.

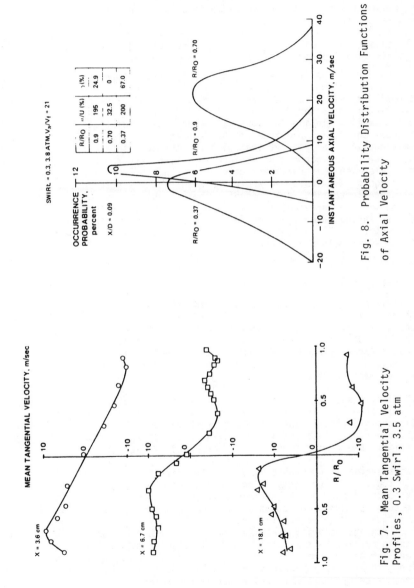

Fig. 8. Probability Distribution Functions of Axial Velocity

Fig. 7. Mean Tangential Velocity Profiles, 0.3 Swirl, 3.5 atm

Defining the directional intermittency (γ) as the fraction of the total observed velocity occurrences which are negative, we can construct contours of constant directional intermittency from the velocity probability densities. Such a plot for the case of 0.3 swirl at 3.5 atm is shown in Fig. 9. These data show that there are significant numbers of negative velocity occurrences over most of the initial mixing region, and that the probability of reverse flow into the fuel port and the region behind the backward-facing step is high. Since the negative intermittency never exceeded 80%, it also can be concluded that there are significant spatial and temporal motions in the region of the time-averaged recirculation zone.

To obtain additional insight into the relative magnitudes of the small and large-scale turbulent motions, we can consider the possible sources of the total velocity fluctuations (u'_T), namely, the small-scale turbulent fluctuations associated with forward and reverse flow (u' and u'_{REV}, respectively) and the additional large-scale source due to sign changes of mean velocity ($\bar{U} - \bar{U}_{REV}$) at the point in question. An indication of the relative rms intensities of the large and small-scale fluctuations may be obtained from Fig. 10. In this figure, which shows the radial profiles of the mean and rms velocities and directional intermittency, it is apparent that although the peak total rms velocity fluctuations occur in the regions of maximum mean velocity gradient, they certainly do not scale with this parameter generally. In the central and corner recirculation zones, for instance, where the mean velocity gradients are small, significant fluctuations are present. These fluctuations, approximately three times the local mean velocities, must be primarily the result of large-scale mixing, as indicated by the high values of directional intermittency. Thus it is apparent that the large-scale mixing represents a significant contribution to the total axial velocity fluctuations.

Spray Flame Tests

Axial velocity measurements have been taken across the entire flow field at a number of axial stations in the initial mixing region for central sprays of iso-octane at annular airstream swirl numbers of 0.3 and 0.6 and for a No. 2 distillate fuel spray at an annular airstream swirl number of 0.3 [5]. Possible sources of signal dropout and error, namely, fuel spray opaqueness, background luminosity, beam wander, phase distortion, and beam divergence, did not affect the measurements adversely.

Selective seeding of the air supply was used to distinguish between the local droplet and gas-phase velocities in regions of the flow where air seeding significantly increased the data rate or where the gas and fuel droplet velocities were substantially different. The characteristic bimodal velocity probability density functions were apparent as illustrated in Fig. 11.

With seeding, the probability density function at $r/r_0 = 0.75$ is weighted heavily toward the local gas velocity, since the seed particle number density dominates the number of fuel droplets in this region. The situation is reversed as the point of measurement approaches the mean droplet spray trajectory ($r/r_0 = 0.63$). Mode velocities are illustrated in Fig. 12. Without seeding, the fuel droplet velocity distribution is determined directly. This distribution then is normalized by the total number of velocity determinations used and subtracted from the bimodal distribution obtained at the same location with air seeding. The result represents the local gas velocity distribution from which the mean velocity and the variance can be determined.

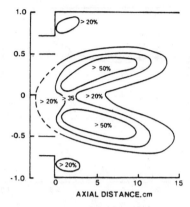

Fig. 9. Directional Intermittency Contours for the
Case of 0.3 Swirl at 3.5 atm

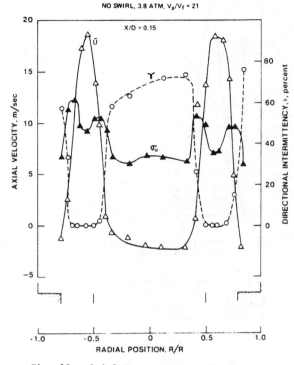

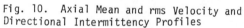

Fig. 10. Axial Mean and rms Velocity and
Directional Intermittency Profiles

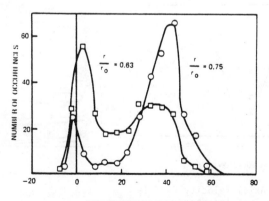

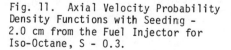

Fig. 11. Axial Velocity Probability
Density Functions with Seeding -
2.0 cm from the Fuel Injector for
Iso-Octane, S - 0.3.

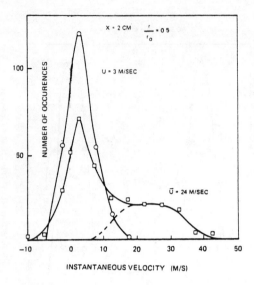

Fig. 12. Axial Velocity Probability Density Functions with and without Seeding - 2.0 cm from the Fuel Injector Iso-Octane, s = 0.3

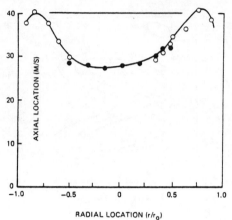

Fig. 14. Mean Axial Velocity Profile 20.0 cm from the Fuel Injector - Solid Symbols Denote Seeding

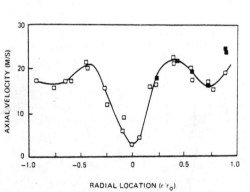

Fig. 13. Mean Axial Velocity Profile 5.0 cm from the Fuel Injector - Solid Symbols Denote Seeding

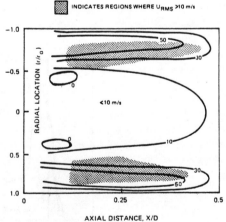

Fig. 15. Mean Asial Gas Phase Velocity Contours - Iso-Octane, S = 0.3

Mean axial velocity profiles obtained with and without seeding are presented in Figs. 13 and 14. At x=5cm there is agreement between measurements obtained with and without seeding which could lead to the mistaken conclusion, that, except in the wall region, the fuel droplets are following the local gas flow. However, away from the wall, examinations of velocity data acquisition rates showed no significant changes when seed particles were introduced; thus, it was apparent that fuel droplets were dominating the velocity probability density distributions at this axial location. Indeed except in the region close to the wall where a sufficiently high seed particle/fuel droplet number density ratio could be achieved, there was no evidence of bimodal distributions. Further downstream (x=20cm), droplet concentration was reduced sufficiently by evaporation and combustion so that the addition of seed material once again influenced the data acquisition rate. Thus the agreement between the seeded and unseeded velocity measurements at this location shows that fuel droplet and particulate combustion products are convected with the local gas velocity.

From axial velocity profiles determined in this way, contours of constant gas phase and droplet velocity can be constructed. Examples of those obtained for iso-octane at 0.3 swirl are shown in Figs. 15 and 16. These data show that there are large differences between the local fuel droplet and gas velocities in the initial mixing region. Indeed, in the outer regions of the spray cone close to the injector, negative time-averaged droplet velocities occur in regions of positive local gas velocity (see Fig. 11), and rapid fuel spray radial expansion is indicated by fuel droplet number density dominance across the combustor at x=5cm. Similar overall results were found for the No. 2 distillate fuel tests.

In addition to the mean droplet and gas velocity distributions, the laser velocimeter data provide critical insight into the spray structure and dynamics in these flames. Droplet number density measurements were used to determine the mean droplet trajectories. At high threshold levels, only the velocities of large droplets are recorded and two peaks corresponding to the time averaged hollow-cone spray location are apparent. However, when the threshold level is reduced the maximum droplet densities are significantly higher and distinct double peaks appear which show that there are larger numbers of smaller droplets surrounding the main trajectory. The spray thickness and droplet number density in the hollow cone region are also increased. Measurements at a series of axial stations were used to determine the time averaged spray trajectories of Fig. 17. These data confirm droplet spray angle insensitivity to the parameters varied in these tests and provide confirmation of earlier holographic measurements [6].

In regions of the fuel spray where data acquisition rates were high (up to 80,000/sec), "real time" velocity information was obtained. Figure 18 shows three autocorrelations of the fluctuating velocities, one within the time averaged fuel spray (r/r_0 = 0.46) and two at locations in the outer and inner fuel droplet entrainment zones (r/r_0 = 0.73 and 0.1 respectively). It can be seen that there are marked variations of energy distribution with frequency across the fuel spray and that the inner and outer fuel droplet entrainment regions contain proportionately less energy in the high wave number region. Although interpretation of the data is complicated by variations in mean droplet velocities in the polydisperse spray, the data suggest significant variations in the scales of the velocity fluctuations across the spray.

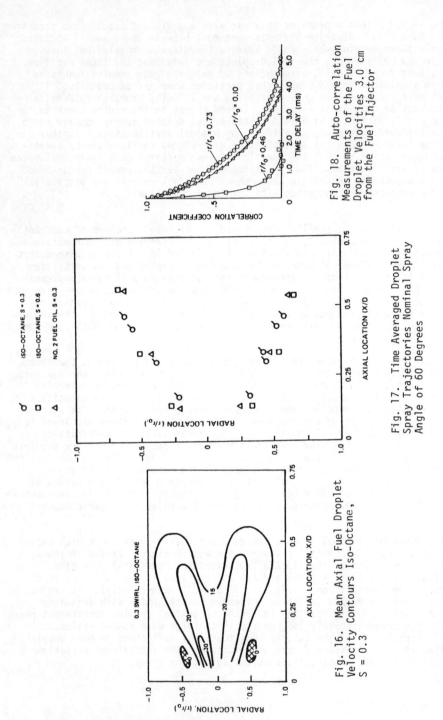

Fig. 18. Auto-correlation Measurements of the Fuel Droplet Velocities 3.0 cm from the Fuel Injector

Fig. 17. Time Averaged Droplet Spray Trajectories Nominal Spray Angle of 60 Degrees

Fig. 16. Mean Axial Fuel Droplet Velocity Contours Iso-Octane, S = 0.3

CONCLUSIONS

Laser velocimeter measurements in the initial mixing region of a confined turbulent diffusion flame burner have shown that there are large differences in the time-averaged flow fields with and without inlet air-swirl. Velocity probability density distributions and associated directional intermittency measurements indicate that there is a substantial large-scale contribution to the total rms turbulent velocity flow field. These large-scale fluctuations result in large departures from Gaussian turbulence and significant deviations from isotropy over most of the initial mixing region. Such large-scale interactions between the free shear layers and recirculating fluid have pronounced effects on the combusting flow.

Observations of the fuel droplet motions in a spray flame suggest a fuel spray picture, the principal features of which are a central core principally comprised of the larger fuel droplets whose velocities are governed by the fuel nozzle characteristics, their greater inertia enabling them to penetrate the air stream environment with little interaction. Surrounding this penetrating droplet core are clouds of smaller fuel droplets which respond relatively quickly to the ambient velocity field. This interaction leads to large scale turbulent motions, increased rms fluctuation levels, and central and toroidal droplet recirculation zones. In the outer jet regions, gas flow recirculation and swirl draw the smaller droplets well away from the mean spray trajectory.

Although care is required, particularly with respect to flow seeding, these results clearly indicate that the laser velocimeter is a practical tool for flow measurements in extremely hostile combustion environments.

REFERENCES

1.   Owen, F. K., L. J. Spadaccini, and C. T. Bowman : Sixteenth Symposium (International) on Combustion, p. 105, The Combustion Institute, 1977.

2.   Owen, F. K., L. J. Spadaccini, J. B. Kennedy and C. T. Bowman, "Effects of Inlet Air Swirl and Fuel Volatility on the Structure of Confined Spray Flames." To be presented at the 17th Symposium (International) on Combustion, Leeds, Aug. 21-25, 1978.

3.   Owen, F. K. "Measurements and Observations on the Structure of Turbulent Recirculating Flows." AIAA Journal Vol. 14, No. 11, 1976, pp 1556-1562.

4.   Owen, F. K., "Laser Velocimeter Measurements of a Confined Turbulent Diffusion Flame Burner," AIAA Paper 76-33, Jan. 1976; also AIAA Progress in Astronautics and Aeronautics: Experimental Diagnostics in Gas Phase Combustion Systems, Vol. 53, edited by B. T. Zinn, C. T. Bowman, D. L. Hartley, E. W. Price, J. F. Skifstad, 1977, pp. 373-394.

5.   Owen, F. K. "Laser Velocimeter Measurements of Turbulent Spray Flames," AIAA Progress in Astronautics and Aeronautics: Turbulent Combustion, Vol. 58 edited by L. A. Kennedy, 1978, pp. 229-245.

6.   McVey, J. B., J. B. Kennedy, F. K. Owen, and C. T. Bowman, "Diagnostic Techniques for Measurements in Burning Sprays," Paper 76-28, Western States Section/Combustion Institute Meeting, La Jolla, CA, Oct. 1976.

Laser Velocimetry in a High Velocity Combustion Flow

PIERRE MOREAU and JEAN LABBE
Office National d'Etudes et de Recherches Aerospatiales (ONERA)
92320 Chatillon, France

SUMMARY

Laser velocimetry is applied to a combustion test set-up where a premixed methane-air flow is ignited by a parallel flow of hot gases. Some new features of such a confined flame have been pointed out by the study of the velocity probability density functions, whose interpretation may depend on seeding conditions.

1. INTRODUCTION

Measurements by means of solid probes in high temperature flows, and particularly in flows with combustion, are very difficult because the probes must be strongly cooled. The presence of the probe may modify the aerodynamics of the flow and the cooling may change the reaction rate.

It is the reason why optical measurements are of peculiar interest in combustion studies.

Laser anemometry measurements have already been performed at ONERA [1, 2] and have given interesting results on the development of the flame in a high velocity premixed flow.

The aim of the new study was to complete the previous results especially about the influence of different parameters on combustion. This paper presents examples of prominent results.

2. TEST FACILITY (fig. 1)

With a view to study the development of a turbulent flame, a test facility has been designed in which the combustion in a premixed air-methane flow is initiated and stabilized by a parallel flow of hot gases produced by an auxiliary burner.

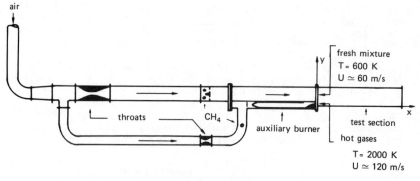

Fig. 1 — The test facility.

The air flow, preheated up to 600 K, is divided into two parts : the main flow and the pilot jet ; the mass flow are measured by means of a sonic throat. The mass flow of the pilot jet is about 15 per cent that of the total mass flow.

In the main stream (cross section 80 x 100 mm), methane is injected far upstream of the combustion chamber in order to provide a good mixing. At the inlet of the combustion chamber, the mean velocity is in the range 55-60 ms^{-1} and the equivalence ratio about 0.8, but some comparative tests were performed without methane.

The pilot gases are produced by an auxiliary methane-air combustion. Their temperature is about 2000 K and their mean velocity twice that of the main flow. The outlet cross section is 19 x 100 mm and the thickness of the separating plate reduced to 1 mm.

The combustion chamber is two-dimensional, 1300 mm long, and its square cross section is constant (100 x 100 mm). Its first part is equipped with transparent silica windows allowing measurements over the whole height of the section between the abscissae 40 and 440 mm. By changing the location of the part equipped with windows, optical measurements are allowable up to 650 mm.

By mixing with the hot pilot gases, the combustion is initiated in the premixed flow and develops downstream in the chamber. The maximum duration of each test is limited to about 30 seconds in order to avoid overheating of the set up, only the part equipped windows being cooled.

All the measurements are performed in the median plane of the combustion chamber.

3. INSTRUMENTATION

3,1. Optical system (fig. 2)

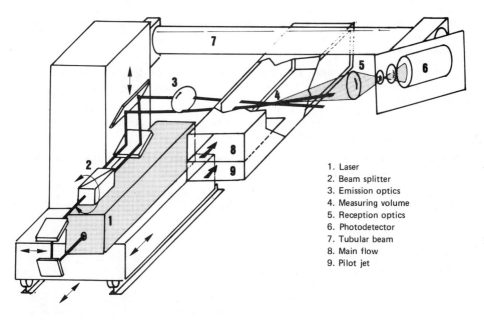

1. Laser
2. Beam splitter
3. Emission optics
4. Measuring volume
5. Reception optics
6. Photodetector
7. Tubular beam
8. Main flow
9. Pilot jet

Fig. 2 — Schematic LDV (Laser Doppler Velocimeter) set-up.

The velocimeter used for these experiments is a laser doppler unit equipped with a continuous argon ion laser whose power is about 1.4 watts on the green line (wavelength λ = 514,5 nm).

The laser output beam is split into two parallel beams 10 mm apart. They are made to cross and to focus to a common point with a lens of 500 mm focal length. The fringe spacing located in the measuring volume is 28.65 μm. The fringe patterns can been oriented by rotating the beam splitter.

For each point, the instantaneous velocity components are measured successively along 3 axes at 45° from one another. The measuring volume can been moved along three perpendicular directions by means of electric motors of 200 mm translation capability. The position is located by a digital recorder of accuracy 10 μm.

In order to explore larger domains, the whole mounting frame can be moved on two rails.

The velocimeter operates in forward scattered light. For this purpose, the collecting optics (300 mm focal length) are hung at the end of a tubular metal beam, 1.5 m length, overlapping the channel. This set up was placed successively, on the right and on the left of the incident laser beam, in order to avoid shading of the collected light and to permit measurements very close to the walls surrounding the glass windows.

3,2. Seeding

In view of the high temperature (2000 K), zirconium dioxide submicron particles are injected far upstream of the burner, at the junction between the main and the auxiliary combustion chambers, so that fresh and hot gases are uniformly seeded.

3,3. Signal processing

The digital counting device used for these experiments was described previously [3, 4] ; it was improved by connecting the signal output to a 5100 HP recorder.

Due to the short duration of each experiment (it cannot exceed 30 s) a particular experimental process must be used. After each run a one minute breaktime is necessary to let the metal cool down. As usual, the signals emitted by particles of a size outside a given range are eliminated.

The signal burst, emitted by validated particles crossing the fringe pattern, is converted into a pulse train and the frequency response signal is accumulated in the memory block of a computer. At the same time, a sample of the signal is observed on the scope of a spectrum analyser.

When a sufficient number of particles (typically 1000 particles) is reached, the turbulence and the instantaneous velocity are roughly computed. The average values are visualized on a digital voltmeter.

Before the run, the frequency filters are adjusted in the assumed range. The burner is fired and, after a few seconds necessary for the stabilisation of the combustion process, the recording is started.

If the position of the filters is correct, the data are downloaded from the computer memory toward a magnetic tape recorder, and the computer is ready for the next measurement.

This way, it is possible to work out a detailed analysis in deferred time.

4. EXPERIMENTAL RESULTS

Measurements were performed in seven sections of the burner. Because of the possibility for the combustion to be influenced by the mixing of the two flows there were 15 measurement locations in the two first sections and only 9 in the other sections.

4,1. Numerical results

Figure 3 gives an example of the velocity profiles for the seven sections analysed. We can notice an acceleration of the main flow in the upper part of the chamber and an even greater acceleration in the reaction zone. The important velocity gradient in the first section (x = 42 mm) is due to the difference of velocities of the two flows.

Figure 4 shows the proviles of the velocity vector angle. This velocity vector is always very small, angles in the range −2, +3°, so the velocity is almost parallel to the chamber axis. It is difficult to notice an evolution between the curves obtained in different sections of the combustor.

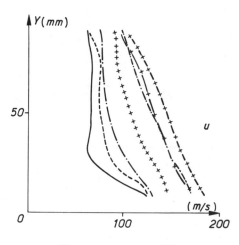

Fig. 3 — Mean velocity profiles.   Fig. 4 — Profiles of the velocity vector angle.

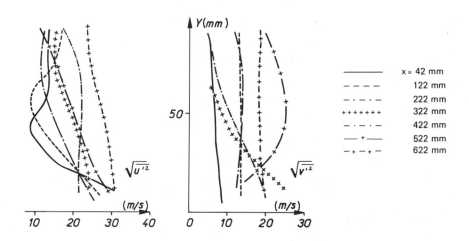

Fig. 5 — Axial velocity fluctuation profiles.   Fig. 6 — Transverse velocity fluctuation profiles.

Figures 5 to 7 represent respectively the transverse profiles of axial velocity fluctuation, transverse velocity fluctuation and Reynolds stress. As in figure 4, it is difficult to find a law governing the evolution of these profiles along the abscissae. We can however notice the important value of the velocity fluctuations.

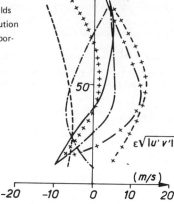

Fig. 7 — Reynolds stress profiles.

$\varepsilon = \pm\ 1$

4,2. Velocity histograms

From the recordings of instantaneous values of the velocity, it is possible to obtain the probability density function of the velocity at different locations in the combustion chamber. These locations, on figures 8 to 11, are given by the following table :

location	a	b	c	d	e	f	g	h
y (mm)	92,5	90	75	60	45	30	27.5	25
location	i	j	k	l	m	n	o	
y (mm)	22.5	20	17.5	15	12.5	10	7.5	

On figure 8 are represented the histograms obtained in the first test section. For y = 92.5 mm (fig. 8-a), the pdf has a maximum value for u = 60 ms^{-1} and shows some wrong values obtained for u = 20 ms^{-1} and u = 35 ms^{-1}. These wrong and non eliminated values modify the mean velocity, and particularly the rms velocity fluctuation which are used to calculate the angle of the velocity vector, velocity fluctuation and Reynold stress.

From locations a to h we can notice a small acceleration of the velocity corresponding to the peak of the pdf (60 to 70 ms^{-1}). For the location i, some higher values of the velocity appear. They afterwards give another peak, which increases gradually from location i to n, while the first peak disappears. On the location l, for instance, we can notice two peaks with approximately the same height, and corresponding to 70 and 130 ms^{-1}.

The locations i to n correspond to the mixing region between the two flows (fresh mixture and hot stabilizing gases) where it is possible to find particles coming either from one flow or from the other.

This is more apparent on figure 9, where the histograms corresponding to the mixing without combustion of hot gases with the main air flow are plotted. We find the same shape of histograms at approximately the same locations.

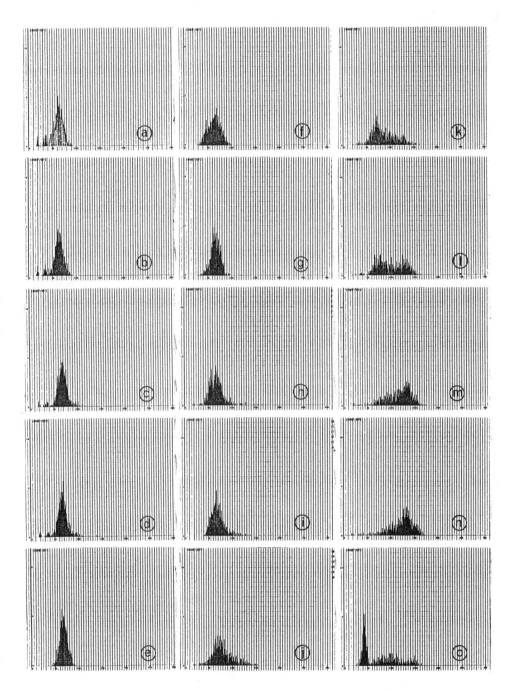

Fig. 8 — Velocity histograms with combustion in the main flow. x = 42 mm.

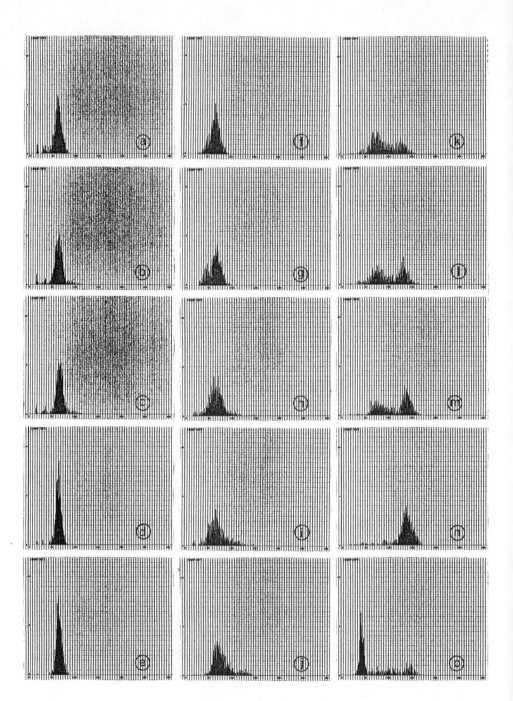

Fig. 9 — Velocity histograms without combustion in the main flow. x = 42 mm.

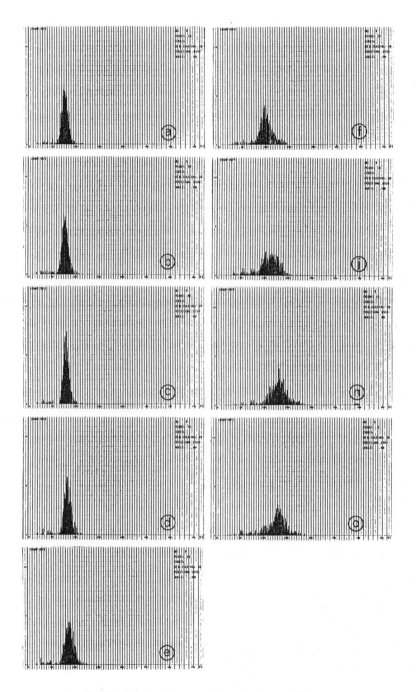

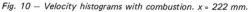

Fig. 10 — Velocity histograms with combustion. x = 222 mm.

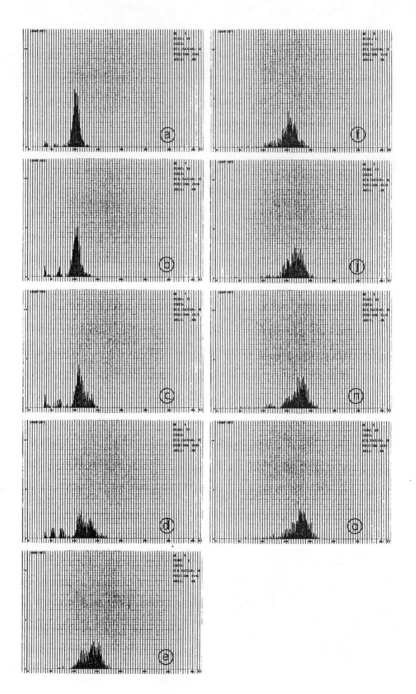

Fig. 11 — Velocity histograms with combustion. x = 422 mm.

*Fig. 12 — Shadowgraphs of the flame
(4000 frames per second).*

In both cases it seems that the pdf corresponding to location o is wrong because there is no continuous evolution from location n. The intermittent profile of the flame can explain these results. If we consider (fig. 12) four successive shadowgraphs of the flame obtained with a fast cinecamera (4000 frames per second), we can remark that a same point in the mixing zone may be successively in the fresh mixture or in the hot gases. This fact explains the velocity pdf with two peaks corresponding respectively to the fresh mixture and the stabilizing gases.

In these conditions, it is difficult to calculate a mean velocity and a rms velocity fluctuation. When the histogram presents two equal peaks, the mean calculated velocity appears in the range where the probability is almost zero and the rms fluctuation is very important.

Moreover, the height of each peak, and consequently the mean velocity and rms fluctuations, depend on the density of particles in each flow.

In our tests, the two flows are uniformly seeded at the inlet of the test facility but, in view of the combustion in the auxiliary stream, the particles density by unit volume is not the same in the two flows at the combustor inlet, and the main calculated velocity may be altered by this fact.

Figures 10 and 11 show histograms obtained further downstreams, at $x = 122$ mm and $x = 422$ mm respectively.

On figure 11-d and e we can also observe two peaks. On figure 11 a to d, the small peaks that correspond to low velocities are probably wrong.

5. CONCLUSION

Velocity measurements by means of laser anemometry were performed in a combustion chamber equipped with silica windows. In spite of the high temperature (2000 K), this method appears very efficient for flame studies. The velocity histograms reveal the mixing of the two flows with different velocities, and show that it is necessary to take care before calculating the mean velocity and rms fluctuations, as these calculated values depend on the particle densities in the two flows which, with the set-up used, are different.

ACKNOWLEDGMENTS

Authors are indebted to J.C. Guillot and A. Jerot for their assistance during the tests, and to R. Grissinger who adjusted the computer programmes to record the data and to draw the histograms.

REFERENCES

1. Boutier, A., Moreau, P. and Borghi, R. Laser anemometry in a combustion flow. ISL/AGARD workshop on Laser Anemometry. ISL Report No. 117/76 (1976), p. 313-323.

2. Moreau, P. and Boutier, A. Laser velocimeter measurements in a turbulent flame. 16th Symposium (International) on Combustion (1977), p. 1747-1756.

3.   Pfeifer, H.J. and Schäfer, H.J. A single counter technique for data processing in laser anemometry. Institut franco-allemand de Saint-Louis (ISL), Internal report (1974).

4.   Boutier, A. Data processing in laser anemometry. Conféreneence générale sur l'anémométrie laser. ISL/AGARD workshop Laser Anemometry. ISL Report No. 117/76 (1976), p. 41-53.

Two Component Laser Velocimeter Measurements in a Dump Combustor Flowfield[1]

GARY D. SMITH[2] and THOMAS V. GIEL[2]
Sverdrup/ARO, Inc., AEDC Division
Arnold Air Force Station, TN

ABSTRACT

Two component, Bragg-diffracted laser velocimeter measurements were obtained on a ducted, turbulent, hydrogen-air flow field, both with and without chemical reactions. The geometric configuration was designed to be representative of a sudden expansion or "dump" combustor. The air stream was seeded with one-micron alumina particles. Radial distributions of mean axial velocity, radial velocity, and turbulence intensity were obtained for axial stations from one-half to six duct diameters from the nozzle exit plane. A comparison of the laser velocimeter data obtained in the flow field with and without chemical reaction is presented to demonstrate the usefulness of LV measurements in determining the effect of the chemical reaction on the flow field mixing rate and the size and location of the recirculation zone within the mixing duct. Comparison of the laser velocimeter data with calculations assuming frozen or equilibrium chemistry are presented to demonstrate the self-consistency in the measured velocity, pressure and gas composition. Operational difficulties encountered in application of the laser velocimeter to a highly turbulent, chemically reacting flow field are discussed.

INTRODUCTION

Recirculating flow fields established by turbulent jet mixing of two co-axial streams in a constant-area, axisymmetric duct occur in many industrial and aerospace burner, furnace and combustor configurations. The so-called "sudden" expansion or "dump" combustors used in ramjet-rocket propulsion systems are designed on the principle of establishing and maintaining combustion in regions of recirculating flow within the combustor. Historically, combustors have been designed by "cut and "try" methods. While such methods have led to practical and thermodynamically efficient combustors, they are time-consuming and expensive. More recently, systems have been encountered in which the design criteria cannot be pragmatically achieved by purely experimental means. In order to assist in such advanced systems problems, detailed knowledge of both the fluid mechanical processes and the coupling between the turbulent flow and

---

[1]The research reported herein was performed by the Arnold Engineering Development Center, Air Force Systems Command. The sponsor of the effort was the Air Force Office of Scientific Research under the guidance of B. T. Wolfson, contract monitor. Work and analysis for this research was done by personnel of ARO, Inc., a Sverdrup Corporation Company, operating contractor of AEDC. Further reproduction is authorized to satisfy needs of the U. S. Government.

[2]Research Engineer, Technology Applications Branch, Engine Test Facility, ARO, Inc.

the chemical reactions is required. Thus radial distributions of total pressure, mean axial and radial velocity, turbulence intensity, gas composition, and wall static pressure distributions were obtained in a confined, axisymmetric, recirculating flow field typical of a dump combustor both with and without chemical reactions occurring.

SELECTION OF VELOCITY MEASUREMENT TECHNIQUE

The presence of high temperatures, chemical reactions, recirculation and high turbulence levels in the simulated dump combustor flow field precludes the accurate determination of flow characteristics by conventional methods. Conventional instrumentation systems such as hot-wire anemometers and pitot probes materially interact with the flow environment and introduce local flow disturbances which make them unsuitable for recirculation flow measurements. Such devices are subject to damage because of impact and thermal loading. The pitot probe does not adequately respond to fluid transients and can be sensitive to flow direction so that it is of little use in measurements of turbulence. Hot-wire anemometers can provide accurate turbulence information, however, they are limited to moderate turbulence levels, require an independent measurement of fluid temperature in a nonisothermal flow, and above all are fragile and incapable of survival in a hostile environment such as the simulated dump combustor flow field. These difficulties generally can be avoided by use of a laser velocimeter (LV) system which can, under controlled circumstances, obtain the required data with reasonable accuracy. If both axial and radial velocities are measured simultaneously with the LV system, it has been shown[1] that concurrent determination of shear stress and turbulence intensity is possible. Additionally, a Bragg-diffracted LV can indicate the flow directionality. Thus a Bragg-diffracted LV was selected as the velocity measuring instrument to be applied to the flow in the simulated dump combustor.

DESCRIPTION OF FLOW FIELD

The essential features of the recirculating flow field are shown in Fig. 1. For a certain range of fluid influx conditions, jet mixing of coaxial streams leads to the creation of an eddy of recirculating fluid existing on a time-averaged basis as a torodial, high vortical region with high turbulent intensities and relatively low average velocities. The pertinent features which may be used to describe the time-averaged flow field with recirculation are:

   1. The axial location of the characteristic time-averaged stagnation points on the duct wall, denoted X_{FS} and X_{RS} in Fig. 1.
   2. The time-averaged axial and radial velocity field.

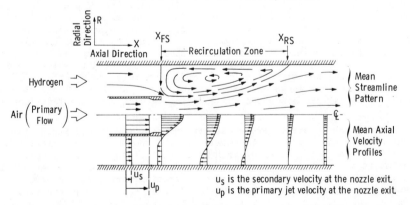

Figure 1.   Turbulent, ducted, mixing system with recirculation.

3. The time-averaged distribution of mass, energy, or temperature of the
    fluid from the primary stream which characterizes the nature of the
    mixing that has occurred between the two stations.
4. The time-averaged axial distribution of wall static pressure.

Determination of the mean velocity field is of prime importance in that the mean
velocity data can be used in conjunction with pitot pressure measurements to
provide an estimate of local fluid density and also provide a basis for determining the recirculating characteristics of the flow field.

EXPERIMENTAL APPARATUS

Test Cell

     The test cell, which is geometrically similar to the one used by Schulz[2]
and Chriss[3], is shown in Fig. 2. The test chamber consists of a 5-ft stainless
steel duct with an inside diameter of 5.24 in. A mechanically driven, axially
traversing nozzle assembly was mounted inside the duct. The nozzle assembly
was traversed in discrete steps during the testing to vary the distance between
the primary jet nozzle exit plane and the radially positionable pitot pressure
and gas sampling probe. An O-ring system provided a seal between the movable
nozzle assembly and the duct. The test chamber included a 1/4-in. wide slot
centered at the axial plane of the traversing probe tip as shown in Fig. 2. The
slot provided optical access for the laser velocimeter. A quartz window assembly was used to seal the slot.

     The primary nozzle assembly consisted of a circular pipe 2.067-in.-diam
which introduced the primary air jet at a nominal bulk flow velocity of 335
ft/sec. An annular secondary injector assembly was provided through which hydrogen entered the duct. The secondary hydrogen flow passed through two porous
plates and a screen pack shown in Fig. 2, which acted to stabilize the flow system. External water spray nozzles were positioned outside the duct to cool the
walls. A spark plug was located downstream of the traversing probe to ignite
the combustible hydrogen-air mixture.

Laser Velocimeter

     Velocity measurements were made in the duct with an off-axis, individual
realization, two-component Bragg diffracted laser velocimeter operating in the
backscatter mode as shown in Fig. 3. An argon-ion laser capable of producing an
optical power of 2 watts in the 514.5-nm wavelength line was the light source.

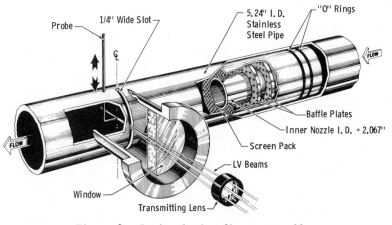

Figure 2. Recirculating flow test cell.

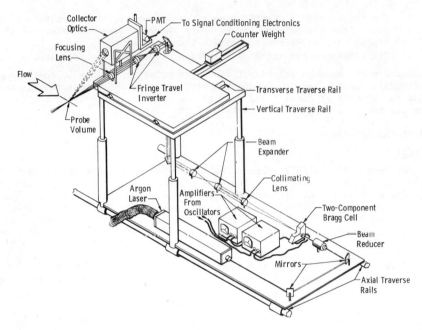

Figure 3.   Laser velocimeter and traverse system.

Light from the laser is directed by mirrors and lenses which reorient and focus
the beam at the center of a two-component Bragg cell beam splitter. The water-
filled cell has 15 and 45 MHz piezoelectric oscillators in the horizontal and
vertical branches, respectively. The two lower order beams for both the horizontal and vertical beam split were allowed to pass through a collimating lens
and a beam expander. The resulting parallel beams were then redirected by mirrors and lenses and focused together in the flow field. In the beam crossover
region, or probe volume, a set of interference fringes are formed with the dominant fringe pair aligned with the longitudinal and vertical axes of the flow
field. Since the Bragg cell imparts to each higher order beam a frequency shift
equal to the piezoelectric oscillator frequencies, the fringes are not stationary but move at a reference velocity proportional to the frequency of the ultrasonic drivers. Velocities, accordingly, are determined relative to the reference velocity so that directional ambiguity is avoided.

     Light scattered from particles passing through the probe volume is re-
ceived by a photomultiplier tube (PMT) through the collector optics. By viewing
the probe volume through an aperture at an off-axis collection angle, the effective probe volume length can be varied. The collector optics were positioned so
that operation was in the back scatter mode. The system had the capability to
measure velocities over an axial velocity range $-250 \leq U \leq 700$ ft/sec and a vertical velocity range $-250 \leq V \leq 250$ ft/sec.

     The laser velocimeter was mounted on a traverse system which allowed move-
ment of the probe volume along any of three mutually orthogonal axes. By traversing the flow along the vertical diameter of the duct, the measured vertical
fringes provide the radial velocity of the flow field. The horizontal fringes
provide the axial velocity component. By positioning the probe volume at a
point along the vertical diameter, samples of both the axial and radial velocity
components could be recorded simultaneously and averaged to provide axial and
radial velocity moments and the $\overline{uv}$ correlation at each sampled point.

In some flows natural seeding may be adequate for acquiring laser velocimeter measurements. However, at the temperatures in the chemically reacting flow field, not enough natural particles survive to obtain velocity measurements within a reasonable time period. Thus, a particle seeding device was developed with which the primary air stream was seeded. Uniform one-micron diameter particles were used in the seeder because they should accurately trace the fluid motion. It is likely however that actual particle sizes larger than one-micron were encountered because of particle agglomeration.

Two operational problems were encountered in acquiring LV data through the slot in the cell wall. Water, formed in the combustion process, deposited on the quartz window surface. It was necessary to shield the viewing port area with baffles and heat the window surface during the test to avoid condensation which could attenuate the LV signal. In addition, the test apparatus moved slightly during testing because of thermal expansion, necessitating realignment of the LV with respect to the flow field centerline at each axial position.

Precision of Measurements

The instrument uncertainty of a Bragg-diffracted laser velocimeter is dependent on the accuracy to which the fringe spacing can be determined, the stability of the crystal oscillator used to drive the Bragg cell, and the precision to which the Doppler Data Processor (DDP) can determine the signal period. The uncertainty in the LV fringe spacing, $\Delta K_u/K_u$, evaluated as described in Ref. 4 from LV calibrations, is ±0.025 for the horizontal fringe spacing K_u of 40 μm and ±0.010 for the vertical K_v of 13.4 μm. Tests of the stability of the crystal oscillators used in these experiments indicate that the uncertainty in the Bragg frequency ($\Delta f^*/f^*$) is ±6.66 x 10^{-4} and 7.69 x 10^{-4} for the horizontal and vertical measurements, respectively. A detailed study of DDP accuracy showed that the processor used to measure axial velocity had an accuracy ($\Delta \tau_i$) of ±0.005 ns, while the DDP used for vertical velocities was accurate within ±0.05 ns. For these values, Ref. 1 gives the uncertainty in the axial velocity measurements as ±4 percent of the measured value at -100 ft/sec and ±3 percent at 700 ft/sec. The uncertainty in the radial velocity measurement is ±8 percent of the measured value at 10 ft/sec and ±3.5 percent at 30 ft/sec.

In using an individual realization LV, an attempt is made to reproduce the statistical behavior of a continuously variable velocity by a finite number of measurements. The results obtained for the mean or standard deviation of the velocity are, therefore, only approximations to the true value of these parameters. The degree of approximation can be evaluated in terms of statistical confidence levels.[5] It is shown in Refs. 1 and 6 that the confidence interval for the mean velocity depends on the sample size and turbulence intensity while the confidence interval for the measured intensity depends, additionally, on the kurtosis of the velocity distribution. For a data set of 1500 samples and a 95-percent confidence level, the confidence interval for the mean velocity in the recirculating flow field is less than 5 percent, while the confidence interval for the turbulence intensity is under 6.5 percent.

In addition to estimating the statistical confidence of the individual realization LV averages, statistical bias (as predicted in Refs. 7 and 8) was also considered. Although statistical bias might be expected to be high with the highly turbulent flow being measured[9] it has been shown[10] that bias is diminished when the fluid density fluctuates as it does in this hydrogen-air flow field. Furthermore, temporal fluctuations in the particle number density, which are known to occur with fluidized bed seeders[11], can further diminish any statistical bias.[12] Therefore statistical bias was expected to be inconsequential in this study; but, to verify this expectation some data were randomly sampled as suggested in Ref. 13. Averages of all the randomly chosen data agreed with arithmetic averages of the original data indicating no significant statistical bias.

Supporting Instrumentation

A water-cooled total pressure and gas sampling probe of the type discussed in Ref. 14 was installed in the duct at a fixed axial position. The probe tip was traversed radially across the duct by a hydraulic drive mechanism. By combination of the motion of the axial traversing nozzle assembly and the radially traversing probe, axial and radial surveys were made in the recirculating flow field. The total pressure was sensed with both a variable capacitance transducer and a strain-gage transducer to obtain maximum accuracy over the wide range of pressure encountered.

The total temperatures of the gaseous supplies of air and hydrogen were measured with copper-constantan thermocouples upstream of critical-flow venturis, which were used to meter the primary and secondary mass flows.

For the data taken on the flow with chemical reactions occurring, water vapor concentration measurements were obtained using an infrared emission-absorption apparatus similar to that described in Ref. 15. A gas sample was withdrawn from the stream through the probe and passed through a platinum-wound catalytic heater to insure completion of the chemical reaction. The sample was then drawn into an evacuated chamber which had an optical viewing port on each end. The infrared emission-absorption apparatus was then used to determine the infrared transmissivity of the sampled gas, from which the water vapor mass fraction could be calculated.[15] The entire gas sampling system was steam heated to prevent condensation of the water vapor in the gas sample.

Data Reduction

To attain near maximum LV data rates, the data were recorded using a minicomputer memory as a buffer. This system could record over 2000 simultaneous fringe periods per second. The LV data were reduced using computer codes written for the IBM 370/165 digital computer. Each raw data tape was read by the digital computer and the number of axial or vertical velocity samples lying in discrete bands were computed from a relation of the form $P(U) = m/M$ where m is the number of samples found in the interval $(U \pm \Delta U/2)$ and M is the total number of samples in the data set. The velocity interval used (ΔU) was generally one-third of the standard deviation of the velocity at the point being studied. By inspecting each resulting data histogram, upper and lower limits of a digital bandpass filter were identified. These limits were purposely kept broad enough that valid samples would not be excluded.

The velocity corresponding to each valid period sample was computed from

$$u_i = K_u \left(f_u^* - \frac{1}{\tau_{ui}} \right) \qquad \text{and} \qquad v_i = -K_v \left(f_v^* - \frac{1}{\tau_{vi}} \right)$$

where u_i and v_i are the horizontal and vertical velocity components for the ith particle, τ_{ui} and τ_{vi} are the corresponding periods, f_u^* and f_v^* are reference frequencies for the horizontal and vertical fringe systems, respectively, and K_u and K_v are the horizontal and vertical fringe spacing, respectively.

Once the digital filter limits were chosen for each data point, the IBM 370/165 computer filtered the data, computed the calibration constants for the run, and obtained the desired statistical parameters of the flow by arithmetically averaging the individual measurements for each data point. It should be emphasized that in order to expedite the determination of the velocity cross correlation all data were acquired simultaneously for the two measurement directions.[3] In the simultaneous data reduction mode only those samples for which valid period measurements were obtained for both velocity components are used in the computation of averages.

RESULTS AND DISCUSSION

Experimental data were obtained in a ducted, subsonic, axisymmetric, recirculating flow field, both with and without chemical reactions. The primary
jet airflow rate was 0.580 lbm/sec and the secondary, outer stream, hydrogen
flow rate was 0.002 lbm/sec. These flow rates result in fuel/air ratio (F/A)
of 0.00345 lb H_2/lb air where the stoichiometric F/A ratio is 0.025 lb H_2/lb
air. The one dimensional velocity of the primary and secondary streams was 335
ft/sec and 3 ft/sec, respectively, and the inlet temperatures were nominally
60°F and 100°F, respectively. The fully mixed temperatures were nominally 65°F
for the nonreactive case and 1300°F for the reactive case. The static pressure
in the duct was 13.7 psia. The measurements were made at axial stations from
zero to six duct diameters from the primary nozzle exit plane. The data include radial distributions of hydrogen mass fraction, mean axial velocity, mean
radial velocity, axial turbulence intensity, radial turbulence intensity and
the velocity cross correlation term.

The radial distributions of mean axial velocity, mean radial velocity,
axial and radial turbulence intensity, and the $\overline{uv}$ cross correlation are shown in
Figs. 4 through 7 both with and without chemical reaction. The velocity data
are nondimensionalized by the one-dimensional primary jet flow velocity, u_p, and
the axial and radial distances are ratioed to the duct diameter and radius, respectively. It should be noted that only the center jet flow was seeded with
alumina particles. Since the outer jet bulk velocity was low it was felt that
by only seeding the center jet the recirculation eddy would act to distribute
the seed material throughout the flow and provide a nearly uniform particle number density. For the case of flow without chemical reaction the particle number density in the outer stream was sufficient to provide an acceptable data
rate, resulting in valid velocity measurement across the entire duct. However,
for the flow with chemical reaction, the alumina deposited on the walls so that
the recirculation eddy did not distribute sufficient seeding material across
the outer stream in the region near the primary nozzle exit plane. Therefore,
velocity measurements could not be obtained in the flow near the duct wall in a
reasonable sampling time.

The data represented in Figs. 4 through 6 show radial profiles of a property at representative axial locations both with and without chemical reactions
to demonstrate the effect of the chemical reaction on the property. It is evident from the mean axial velocity variation, Fig. 4, that chemical reaction
broadens the radial profile and reduces the axial velocity decay. The radial
velocity profiles, Fig. 5, also indicate, although less dramatically, the sustained high velocity caused by chemical reaction. The maximum value of the turbulence intensity $u^{\prime}/u_p$ and $v^{\prime}/u_p$, Fig. 6, is about the same for the two cases.
However, the location of the maximum turbulence intensity (not shown) is reached
closer to the primary nozzle exit plane in the non-reacting case because in each
case the maximum turbulence location occurs near the center of the mixing zone
which extends further downstream with chemical reaction. The turbulence intensity becomes essentially constant by $X/D_D = 4$ for the nonreactive case whereas
a significant profile still exists at $X/D_D = 6$ for the reactive case. Note by
comparing the data in Figs. 5 and 6 that the magnitude of the radial turbulence
is much greater than the mean radial velocity at nearly every point in the flow
field. The uv velocity cross correlation, $\overline{uv}/u_p$, shown in Fig. 7, also indicates a considerable effect of chemical reaction in the Reynolds stress field.
The decay of the mean axial velocity along the duct centerline is shown in Fig. 8.
As implied from the previously shown data, the measurements reveal the velocity
decays less rapidly in the reactive case than in the nonreactive case.

The velocities measured with the laser velocimeter are compared in Fig. 9
with velocities calculated from the measured gas composition and pressures assuming either equilibrium chemistry with 100 percent reaction efficiency or no

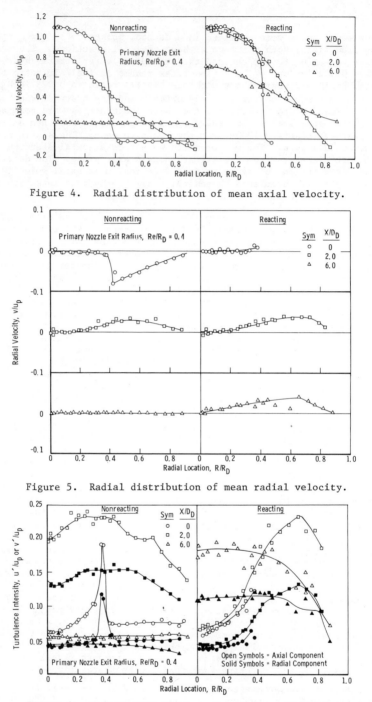

Figure 4.   Radial distribution of mean axial velocity.

Figure 5.   Radial distribution of mean radial velocity.

Figure 6.   Radial distribution of turbulence intensity.

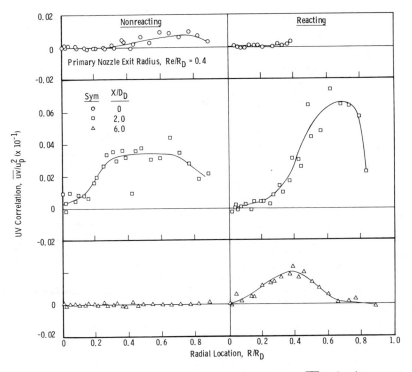

Figure 7. Radial distribution of the $\overline{uv}$ velocity cross correlation.

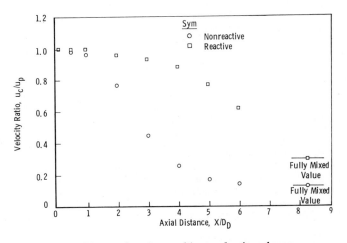

Figure 8. Centerline velocity decay.

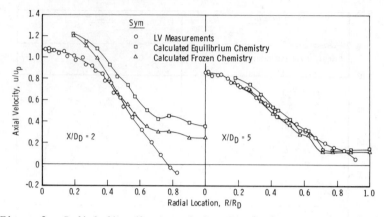

Figure 9.   Radial distribution of the UV velocity cross correlation.

reactions.   The two assumptions about the state of the chemistry allowed upper and lower limits for the velocities in the flow field to be defined.   Thus by comparing the measured velocities with the computed velocity limits, one could infer the state of the chemical reactions occurring in the gas.   The static pressure measured on the duct wall at the axial location of a velocity measurement was used in the calculation along with the enthalpies of the inlet systems. The calculation procedure is described in Ref. 16.   The LV velocity data are seen to generally agree with the calculated velocities in level and radial profile shape.   In some regions of the flow, the LV measured velocities fall outside the limits of the two calculated velocities.   This discrepancy is most probably caused by a combination of uncertainties in the measured parameters used to calculate the velocities.   The discrepancy near the duct wall results from inaccurate measurements of the small pressure differences (0.04 to 0.06 psi) and the inability of pressure measurements to indicate flow reversals. The discrepancy near the centerline results from uncertainty in the measured hydrogen mass fraction.   In general, considering all the data, the measured velocities approach the values calculated from the equilibrium assumption at locations where complete chemical reactions would be expected and with values calculated from the frozen assumption where complete reactions would not be expected.   Thus, the measured quantities appear to be self consistent.

CONCLUSIONS

Ducted, subsonic, axisymmetric, recirculating flow experiments were conducted in a simulated dump combustor with a duct diameter-to-inner nozzle diameter ratio of 2.5, both with and without chemical reactions occurring.   A two-component, individual realization, Bragg diffracted, laser velocimeter was used to obtain radial profiles of mean axial velocity, mean radial velocity, turbulence intensity, and velocity cross-correlation.   Comparison of the measured velocity with that calculated from measured pressure and composition data assuming frozen or equilibrium chemistry show that the measured velocity, pressure and gas composition data are self-consistent.   The laser velocimeter is shown to be a valuable tool to determine the effect of chemical reaction on the velocity field and hence the mixing rate and the location of the recirculation region.

REFERENCES

1.   Barnett, D. O., and Giel, T. V., Jr.   "Application of a Two-Component Bragg-Diffracted Laser Velocimeter to Turbulence Measurements in a Subsonic Jet. AEDC-TR-76-36 (ADA025355), May 1976.

2. Schulz, R. J. "An Investigation of Ducted, Two-Stream, Variable Density, Turbulent Jet Mixing with Recirculation." AEDC-TR-76-152 (ADA034537), AFOSR-TR-76-1087, January 1977.

3. Chriss, D. E. "An Experimental Investigation of Ducted, Reactive, Turbulent Jet Mixing with Recirculation." AEDC-TR-77-56, AFOSR-TR-77-0749, September 1977.

4. Barnett, D. O. and Giel, T. V., Jr. "Laser Velocimeter Measurements in Moderately Heated Jet Flows." AEDC-TR-76-156 (ADA038283), April 1977.

5. Bendat, J. S. and Piersol, A. G. Measurements and Analysis of Random Data. John Wiley and Sons, Inc., New York, 1966.

6. Cline, V. A. and Bentley, H. T., III. "Application of a Dual-Beam Laser Velocimeter to Turbulent Flow Measurements." AEDC-TR-76-56 (AD785352), September 1974.

7. Mayo, W. T., Jr. "The Development of New Digital Data Processing Techniques." USAF Contract F40600-73-C-003, Third Monthly Progress Report, Appendix; March 1973.

8. McLaughlin, D. K. and Tiederman, W. G. "Biasing Correction for Individual Realization of Laser Aneomometer Measurements in Turbulent Flows." The Physics of Fluids, Vol. 16, No. 12, December 1973.

9. Barnett, D. O. and Bentley, H. T., III. "Statistical Bias of Individual Realization Laser Velocimeters." Proceedings of the Second International Workshop on Laser Velocimetry, Engineering Experiment Station, Bulletin No. 144, Purdue University, West Lafayette, Indiana, March 1974.

10. Meadows, D. M., Whiffen, M. C., and Mayo, W. T., Jr., "Laser Velocimeter for Supersonic Jet Turbulence and Turbulence Spectra Research." AFAPL-TR-74-24, Appendix IV, pp. 163-189, (1974).

11. Glass, M. and Kennedy, I. M. "An Improved Seeding Method for High Temperature Laser Doppler Velocimetry." Combustion and Flame, Vol. 29, pp. 333-335, (1977).

12. Giel, T. V., Jr. and Barnett, D. O. "Analytical and Experimental Study of Statistical Bias in Laser Velocimetry." Proceedings of LV-III, Purdue University, July 1978.

13. Durao, D.F.G. and Whitelaw, J. H. "The Influence of Sampling Procedures on Velocity Bias in Turbulent Flows." The Accuracy of Flow Measurements by Laser Doppler Methods. Proceedings of the LDA-Symposium Copenhagen, Technical University of Denmark, August 1975.

14. Rhodes, R. P. "Probing Techniques for Use in High Temperature Reacting Flows." AEDC-TR-68-44 (AD829143), March 1968.

15. Brewer, L. E. and Limbaugh, C. C. "Infrared Band Model Technique for Combustion Diagnostics." Applied Optics, Vol. 11, May 1972, p. 1200.

16. Osgerby, I. T. and Rhodes, R. P. "An Efficient Numerical Method for the Calculation of Chemical Equilibrium in the H/C/O/N/A System." AEDC-TR-71-256 (AD741825), April 1972.

Laser Velocimetry Measurements in a Gas Turbine Research Combustor

JAMES F. DRISCOLL and DENNIS G. PELACCIO
Department of Aerospace Engineering
The University of Michigan
Ann Arbor, MI 48109

ABSTRACT

The effects of turbulence on the production of pollutant species in a gas turbine research combustor are being studied using LDV techniques. Measurements that have been made in the primary combustion zone include mean velocity, rms velocity fluctuations, velocity probability distributions, and autocorrelation functions. A unique combustor design provides relatively uniform flow conditions and independent control of drop size, equivalence ratio, inlet temperature and combustor pressure.[1]

Parameters which characterize the nature of the spray combustion (i.e., whether single droplet or group combustion occurs) have been determined from the LDV data. Turbulent diffusivity (eddy viscosity) reaches a value of 2930 cm^2/sec, corresponding to a convective integral length scale of 1.8 cm. The group combustion number, based on turbulent diffusivity, is measured to be 6.2.

INTRODUCTION

Laser velocimetry offers two exciting new capabilities useful in the study of spray combustion. First, LDV makes possible for the first time the measurement of parameters which determine the nature of spray combustion: i.e., whether single drops burn independently or group combustion occurs. A fundamental parameter that is believed to characterize the nature of spray combustion, sometimes called the "group combustion number,"[2] is proportional to the ratio of turbulent eddy dissipation time to droplet lifetime. Direct measurement of this number is possible using LDV autocorrelation and sizing techniques. In this study, LDV autocorrelation data, along with mean and rms velocity data, has been obtained for relatively uniform spray combustion conditions using LDV techniques. The turbulent eddy dissipation time, mixing lengths and eddy diffusivity in our combustor have been deduced.

A second capability of LDV is quantifying the effects of turbulence on the production of pollutant species in spray combustion. Current gas turbine research at The University of Michigan has concentrated on isolating and measuring the effects of the following parameters: (a) drop size, (b) equivalence ratio, (c) turbulence, (d) inlet temperature, and (e) pressure. The effects of independently varying drop size and equivalence ratio on the measured local NO_x, CO and UHC concentrations in our combustor have been reported elsewhere.[1]

158

To isolate and measure the effects of parameters (a) through (e) on emissions data, a unique gas turbine research combustor was developed at The University of Michigan. The following essential details have been specifically incorporated in its design.

(i)   Combustion conditions are maintained as uniformly as experimentally possible throughout the combustor can. Thirty-seven fuel needle-air nozzle pairs are arranged on an injector plate located at one end of the cylindrical can to achieve equal area distribution. The relatively uniform combustion conditions downstream can therefore be more closely modelled by the one dimensional kinetic model developed for this program. Downstream of the jet interaction region, turbulence can be characterized by two integral length scales, in the convective and transverse directions.

(ii) Independent control can be exercised over drop size, equivalence ratio, inlet temperature and combustor pressure. The Sauter mean diameter of the spray has been varied from 20 to 65 microns using concentric air blast atomization. Equivalence ratio has been varied from 0.85 to 1.5. Inlet temperature can be varied to 1100°F using an air preheater, and the combustor is designed to operate at pressure of 1-10 atmospheres. The flame is stabilized by local recirculation only, and no external flameholder is used.

(iii) The quartz combustor can allow optical access and photographic analysis, and the entire combustion region is accessible for emissions sampling and conventional probes.

EXPERIMENTAL ARRANGEMENT

The University of Michigan gas turbine research combustor is shown schematically in Fig. 1. The fuel drop generator consists of thirty-seven fuel needles of 0.01 in. diameter, each positioned concentrically with a coflow air tube. The room temperature coflow air surrounding each fuel needle is used to prevent the fuel from boiling or coking in the needle when high temperature primary air is used. Jet A aviation fuel is used. Primary air is introduced through 0.125 in. diameter air nozzles, each of which form a concentric orifice surrounding each fuel needle-coflow air tube pairs. For the present tests, the inlet temperature and combustor pressure correspond to atmospheric conditions. An air preheater enables the primary air to be heated to 1100°F and combustor pressure can be varied from 1-10 atmospheres.

By varying the primary air jet velocity from 200 to 470 ft/sec, the Sauter mean diameter of the fuel spray has been varied from 65 to 20 microns. Drop size distribution was measured and correlated over a range of Weber numbers under non-combustion conditions using direct photography. A typical drop size distribution is shown in Fig. 2.

Nominal run conditions for the LDV measurements presented are: primary zone equivalence ratio $\phi = 0.9$, primary air mass flow = 0.16 lbm/sec, droplet Sauter mean diameter = 26 microns, inlet temperature = 300°K and combustor pressure = 1 atmosphere.

A two component LDV system is used to measure velocity in streamwise and radial directions. To date, only streamwise components have been measured using the one component experimental arrangement shown in Fig. 1. The LDV is operated in the forward scatter fringe mode with fringe spacing of 15 μm. Initially, a

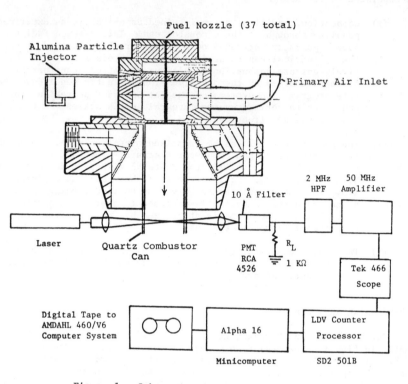

Figure 1.   Schematic of LDV System

15 mwatt Helium Neon Laser (SP124-A) was found to provide good signal to noise
ratio for gas velocities up to 500 ft/sec. A two watt Argon ion laser (SP 164-
06) is presently used for two velocity component measurements.

Alumina particles are introduced into the primary air plenum chamber using
a fluidized bed aerosol generator. Vibrations of the combustor facility are
noticeable; however, since the optical system is mounted directly to the combustor itself, no effects of vibrations on the optical alignment or on LDV
results are observed.

Signal processing is accomplished using two 50 MHz line driver amplifiers
and a digital LDV counter/processor (Spectron Development 501B). An Alpha 16
minicomputer with a 16 K byte memory is interfaced and programmed to store both
the doppler period for each validated doppler burst as well as the time interval between valid bursts. This information is passed to The University of
Michigan AMDAHL 470/V6 computer system via digital tape.

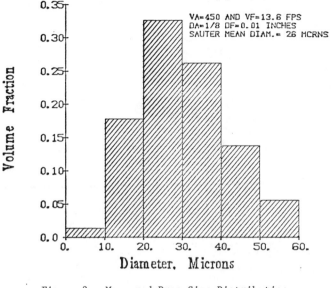

Figure 2.   Measured Drop Size Distribution
             (Ref. 1)

LDV RESULTS

    Typical LDV mean velocity profiles obtained in the combustor can at the
axial location Z = 4 in., are shown in Fig. 3. Solid symbols represent LDV data
taken in the flame for an equivalence ratio of 0.9 and Sauter mean diameter of
26 microns.

    Velocities measured under combustion conditions are approximately 2.5 times
larger than cold flow velocities measured for the same air mass flow. LDV data
obtained in the cold flow was compared with pitot probe and hot wire velocity
profiles. While the agreement in Fig. 3 between probes is reasonable, the LDV
results are believed to be the most accurate since velocity fluctuations exceeding 30% of the mean were observed. In highly turbulent flow, pitot probe readings tend to be erroneously low due to instantaneous probe misalignment with the
flow. The hot wire linearization process, which requires that fluctuations be
small, may also be a source of error.

    The turbulence levels measured with the LDV under combustion conditions were
30-35%, as shown in Fig. 4. This is only slightly higher than the corresponding
turbulence levels which were measured with no combustion using LDV and hot wire
probes. Turbulence levels in our combustor are consistent with those typical of
turbulent jets.[3] It is therefore concluded that the combustion itself has only a
secondary effect on the rms turbulence levels.

    A typical velocity probability distribution, obtained at Z = 4in. is shown
in Fig. 5. The single peaked curve is centered at the mean velocity of 160 ft/sec,
with no contribution from velocities below 60 ft/sec. This indicates that at
Z = 4 in., reverse flow is negligible and frequency shifted LDV optics are not

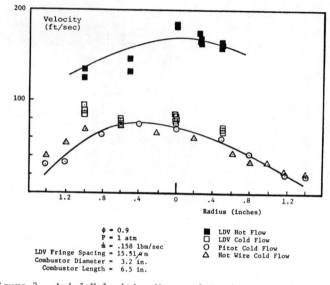

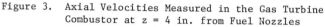

Figure 3. Axial Velocities Measured in the Gas Turbine
Combustor at z = 4 in. from Fuel Nozzles

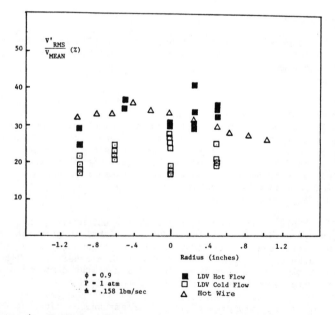

Figure 4. Turbulence Levels in the Gas Turbine Combustor
at z = 4 in.

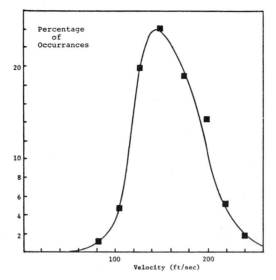

Figure 5.   Axial Velocity Probability Distribution at
            z = 4 in., r = 0.25 in.

necessary. Under certain conditions, a double peaked probability distribution
can occur, as reported by Owen[4]; such conditions allow drop velocity and gas
velocity to be measured independently.

     To determine turbulence length scales, the time of occurrence of each valid
LDV velocity sample was stored in the minicomputer, enabling the autocorrelation
function and power spectrum to be determined. Because of the random nature of
the LDV sampling process, a computer routine similar to that of Mayo[5] was developed
to compute the autocorrelation function. For every sample pair (u_i, u_j) the time
difference of k units was determined and the k th element of the autocorrelation
array R(k) was incremented by $u_i u_j$. Similarly, the k th element of the time histogram array H(k) was incremented by one. The normalized autocorrelation function
is then obtained by dividing R by H. Physically, this procedure increases the
inherent frequency response of the LDV technique. The high frequency information
contained in data sample pairs whose time difference is much less than the mean
sample interval is normally lost in the averaging process because such sample
pairs occur rarely. By weighting these sample pairs by the inverse probability
of their occurrence, high frequency information is recovered. Typical autocorrelation data deduced in this manner is shown in Fig. 6.

EFFECT OF TURBULENCE ON SPRAY COMBUSTION

     LDV techniques offer, for the first time, the capability of measuring the
fundamental parameters that characterize spray combustion. Two such parameters
are the turbulent eddy dissipation time and the droplet lifetime. The ratio of
these quantities is a measure of the ability of turbulent diffusion to provide
oxygen to drops imbedded within the fuel cloud. This process is instrumental
in determining whether single drop or group combustion occurs.

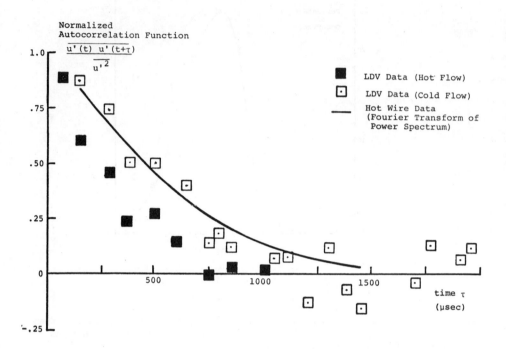

Figure 6.   Autocorrelation of Velocity Fluctuations

Chiu[2] has obtained solutions to equations governing group combustion, without specifically including a turbulence mechanism, and has defined a group combustion number G:

$$G = \frac{3(\pi R^2/D_T)(N\ r^3/\pi R^2 L)}{(r^2/\alpha\ Nu)} = 3\ Nu\ Le\ N(r/L) \tag{1}$$

where D_T is the turbulent diffusivity, r and R are drop and spray cloud radii, respectively, L is spray cloud length, α is the thermal diffusivity and N is the number of drops. The effective Nusselt and Lewis number are denoted Nu and Le. For sufficiently dense sprays characterized by large G numbers, external group combustion is predicted to occur with turbulent diffusion flames surrounding clouds of droplets. In dilute sprays corresponding to low G numbers, enough oxygen is entrained to permit single droplet combustion. LDV measurements of

the parameters in Eq. (1) are required at the present time in order to determine the critical G number that separates these two combustion regimes.

For our experimental conditions, the group combustion number G is determined as follows. Turbulent diffusivity D_T is related to the Lagrangian integral scale of turbulence Λ_L by[3]:

$$D_T = v'\Lambda_L \qquad (2)$$

Approximating Λ_L by the Eulerian integral scale and applying Taylor's Hypothesis:

$$D_T = (u'/\bar{u})\ \bar{u}^2\ T_{11} \qquad (3)$$

where T_{11} is the integral time scale obtained by integrating the normalized
autocorrelation function from t=0 to t = ∞ . The right hand side of equation
3 is determined entirely from LDV measurements, yielding: D_T = 2930 cm^2/sec
from autocorrelation measurements for which a typical data is shown in Figure
6. The value of the first factor in equation 1, $(\pi R^2/D_T)$ the time characterizing diffusion over a length scale R, is 18 msec.

The factor $(r^2/\alpha\ Nu)$ in Equation 1, which is proportional   to the droplet
lifetime $(4r^2/K\ Nu)$, has a value of .21 μsec. The diluteness factor $(Nr^3/\pi R^2 L)$
in Equation 1 is simply 3/4π times the measured ratio of fuel-air volume flow
rates; this factor is 2.4 x 10^{-5}. Spray density, based on Sauter mean diameter droplets, is 10^4/cm^3.

Therefore, the group combustion number for our spray condition at location Z = 4 inches, r = 0 inches is 6.2. Previous theoretical estimations[2]
of G cannot be accurately compared with this value unless empirical turbulent
diffusion data has been included in such theories.

ACKNOWLEDGMENTS

This research was supported under NASA Grant NSG 3148 with the NASA Lewis
Research Center, Cleveland, Ohio. Dr. Larry Cooper is the NASA technical
representative and the project director is Professor J. A. Nicholls of the
University of Michigan. Assistance in the combustor operation was provided
by Omer Kitaplioglu and Thomas Migala.

REFERENCES

1.   Anderson, R. W., Patil, P. B., Chin, J. L., Nicholls, J. A., Mirsky, W.,
     and Lyons, V., "The Effect of Drop Size on Emissions from the Primary
     Zone of a Gas Turbine Type Combustor," 16th Combustion Symposium, M.I.T.,
     1976.

2.   Chiu, H. H., and Liu, T. M., "Group Combustion of Liquid Droplets," Comb.
     Science & Tech. 17, pp. 127-142, 1977.

3.   Hinze, J. O., Turbulence, McGraw-Hill, New York, 1959.

4.   Owen, F. K., "Laser Velocimeter Measurements of the Structure of Turbu-
     lent Spray Flames," AIAA Paper 77--215, 1977.

5.   Mayo, W. T., Proc. 2nd Int'l. Wkshp. Laser Velocimetry, Purdue, 1974.

Laser Doppler Velocimetry as Applied to the Study of Flame Spread over Condensed Phase Materials

R. J. SANTORO, A. C. FERNANDEZ-PELLO,
F. L. DRYER, and I. GLASSMAN
Princeton University
Princeton, NJ 08540

ABSTRACT

In the early stages of a fire, the two mechanisms by which heat transfer occurs are conduction and convection ahead of the flame through the gas and fuel phases. Measurement of the convective transfer has been difficult in the past because velocities are characteristically low [0(10 cm/sec)] and change magnitude and direction over small distances with accompanying sharp temperature changes. With the advent of LDV techniques a non-perturbing means of determining the velocity exists. In the presentation the current work of the authors relating to LDV measurements of flame spread phenomena will be reviewed.

INTRODUCTION

The problem of fire in America is an important and serious one involving the loss of thousands of lives and millions of dollars in property. Yet our understanding of fire, its ignition, spread, growth and extinguishment, is quite limited. Fire involves the interaction of many physical and chemical phenomena and presents a complicated problem for the researcher.

As part of the research efforts aimed at reducing fire loss in America we are studying the mechanisms of flame propagation over the surface of condensed phase materials. We have been particularly interested in the measurement of convective energy transfer ahead of the flame for both induced and forced flow situations.

For a flame to propagate over a surface, sufficient heat must be transferred ahead of the flame to pyrolyze the fuel, creating a fuel/oxidizer mixture within the lean flammability limits. The mechanisms to transfer energy include radiation, solid and gas phase conduction and convection. For large fires, radiation is the dominant energy transfer mode, while for small fires radiation is not usually the most important transfer mode. Since all fires begin as small fires an understanding of the spread of small fires can have important practical ramifications.

LASER DOPPLER VELOCIMETER FACILITY

At Princeton we have centered our attention on measuring convective effects using an LDV technique. These induced convective flows are characterized by low magnitude velocities (of the order of 10 cm/sec) with changes in magnitude and direction occuring over small distances accompanied by large changes in temperature. The non-perturbing, localized measuring capabilities of the L.D.V. make it ideal for this application.

   A two component LDV utilizing a two dimensional bragg cell to achieve
frequency shifting has been constructed to allow for measurement of velocity
fields containing regions of flow reversal. Since this facility has been
described elsewhere[1] in great detail only a brief description will be presented
here. A schematic representation of the LDV facility is shown in Figure 1.
The facility employs a two watt argon ion laser tuned to the 5145A line. This
beam is projected into a two dimensional bragg cell which splits the beam into
four components while frequency shifting the output beams with respect to the
incident beam. The bragg cell utilizes 15 and 22.5 MH_z frequencies and at a
later point electronic bandpass filtering techniques are used to separate each
channel. Thus the bragg cell essentially provides the two component resolution
capability. Following the bragg cell a 6:1 beam reducer and focusing lens
provide the optical processing to form the LDV probe volume. Typically a beam
crossing angle of 5.8° for the 15 MH_z channel is employed resulting in a probe
volume whose dimensions are .16 mm x .16 mm x 3.0 mm. Following the probe
volume the beams are blocked using beam stops and the scattered light is focused onto an RCA 931A photomultiplier using a pair of 75 mm lenses whose combined focal length is 150 mm. Following the PMT the signals are input into a
signal separator (Spectron Development Lab Model DSS 1522) where the 15 and
22.5 MH_z signals are separated and heterodyned with an appropriate signal. The
signals for each channel are output to burst processors (Spectron Development
Lab Model PDAP 501A) for measurement of the doppler period. The result of
this determination is digitally output to a mini-computer along with an interval
timer reading to time mark the velocity measurement. Data is then stored on a
disc for permanent record.

EXPERIMENTS AND RESULTS

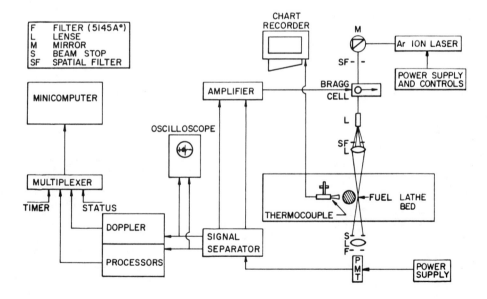

FIG. I
Schematic Diagram of the Experimental Installation

Measurements of convective effects for both liquid and solid fuels have been undertaken and the results have validated the usefulness of the LDV technique to flame spread research. This paper will treat briefly the results of one of the studies recently completed as an example of the insights that the accurate measurement of convective energy transfer lends to flame spread research.

Previous studies[2,3] of flame propagation over solid surfaces have reached differing conclusions as to the dominant mode of energy transfer. These studies have generally employed different materials, orientations or geometrics and thus a consistent interpretation has been difficult to obtain. In addition because of limitations in available techniques accurate measurements have been limited to solid and gas phase temperature. In the present study we have chosen to measure a single material (PMMA) in a single orientation (downward flame spread) for a single geometry (cylindrical rods). In these experiments two thicknesses for the material were chosen: one (5 cm in diameter) to represent the thick fuel case and a second (.16 cm in diameter) to represent the thin fuel case. Measurements were made for the gas and solid phase temperatures using thermocouple techniques while the LDV was employed to measure the induced gas phase convection. Thus the importance of the fuel thickness on the energy transfer process could be investigated.

In Figure 2 we see a plot of the measured induced gas phase velocity for both the 5.0 cm and 0.16 cm PMMA rods. In this figure the velocity has been plotted versus a non-dimensional distance from the surface while each curve corresponds to a different location ahead of the flame.

As one can see the velocity monotonically increases as one approaches the flame front. Near the flame front the flow moves slightly away from the solid surface for distances of 2 mm or more from the fuel surface. A factor of 2 to 4 is seen to be characteristic of the difference between the induced gas flow

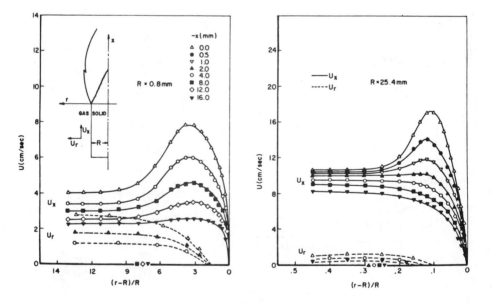

FIG. 2

Profiles of the Axial and Radial Components of the Gas Velocity Ahead of the Flame for Downward Spread Along PMMA Rods of 5 and 0.16 cm in Diameter

in the case of the thick versus thin fuel. Thus one might expect gas phase conduction to be less important in the thick fuel case than in the thin fuel.
     To better examine the relative quantitative importance of different modes
of heat transfer the gas velocity and temperature profiles were used to construct energy balances for the solid and gas phase. The energy equation in integral form was applied to the control volume shown in Figure 3. In this analysis effects due to radiation have been neglected since the flame view angle and
surface temperatures are small.
     Although the energy balance for the gas phase is not necessary to determine
the pathways of heat transfer from the flame to the unignited fuel, they can
produce useful information about the effect of flame induced air flow on streamwise gas phase heat conduction. This last aspect may prove important in understanding flame propagation in cases where convective effects play a more or less
important role.
     The solid and gas phase regions were considered and the energy balances were
calculated for each fuel rod diameter. The results of these calculations are
shown in Figure 3. The general features of the heat transfer process can be
identified from this figure. In the thick rod case nearly 90% of the total
energy is transferred via heat conduction through the solid while in the thin
case only 20% of the energy is due to solid phase conduction. Thus the initial
suspicion that the reduced convective cooling for the thin case would result in
a greater importance of heat conduction through the gas seems well grounded.
     Somewhat surprisingly the magnitudes of the convective heat transfer in
both cases is quite striking. These results clearly show that a substantial
amount of energy goes into heating the induced flow. The magnitude of the
difference between the conductive and convective terms can be interpreted to
support the occurence of a chemical ignition zone ahead of the flame which is
exothermic in nature. A full examination of the results of this work will be
given elsewhere[4].

CONCLUSIONS

     The results of the present work show for this mode of flame propagation the

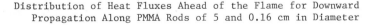

Fig. 3
Distribution of Heat Fluxes Ahead of the Flame for Downward
Propagation Along PMMA Rods of 5 and 0.16 cm in Diameter

dominant mode of heat transfer is a function of the thickness of the fuel. For thick fuels heat conduction through the solid is dominant while as the fuel approaches the thin case gas phase conduction plays a more important role. The results also show that convective heat transfer plays an important role on limiting the ability of heat to be conducted ahead from the flame through the gas phase. We believe that because the gas flow field induced by the flame plume will be different for other flame spread orientations, the dominant mode of heat transfer will also differ and thus no generalizations of the controlling mechanisms can be made.

Additionally we have shown that the LDV represents a powerful tool for the measurement of convective effects in flame propagation. Studies of the importance of force flow on flame propagation have been completed recently by Pello[5] and we are presently undertaking a study of horizontal flame propagation. We feel that the inherent experimental problems encountered in previous convective studies are circumvented by LDV. The advent of LDV as a diagnostic tool is opening up a number of combustion oriented problems for study. As the instrument base and applications grow, LDV will have a profound effect on the analysis of complex fluid-mechanics-combustion problems.

REFERENCES

1. Santoro, R.J. 1977. A Laser Doppler Velocimeter for Measurement of Flows Induced by Flames Propagating Over Condensed Fuels. Aerospace and Mechanical Sciences Report #1361, Department of Aerospace and Mechanical Sciences, Princeton University, Princeton, New Jersey.
2. Lastrina, F.A., Magee, R.S. and McAlevy, R.F.,III 1973. Flame Spread Over Fuel Beds:Solid Phase Energy Considerations. Thirteenth Symposium (International) on Combustion, 935, The Combustion Institute.
3. Fernandez-Pello, A.C. and Williams, F.A. 1975. Laminar Flame Spread Over PMMA Surfaces. Fifteenth Symposium (International) on Combustion, 217, The Combustion Institute.
4. Fernandez-Pello, A.C. and Santoro, R.J. 1978. On the Dominant Mode of Heat Transfer in Downward Flame Spread. Seventeenth Sympsoium (International) on Combustion, to be held Aug, 1978, University of Leeds, England.
5. Fernandez-Pello, A.C., Ray, S.R. and Glassman, I. 1978. Downward Flame Spread in an Opposed Forced Flow, to be published Combustion Science and Technology.

Motion of Particles in a Thermal Boundary Layer

R. W. SCHEFER, Y. AGRAWAL, R. K. CHENG,
F. ROBBEN, and L. TALBOT*
Lawrence Berkeley Laboratory
University of California
Berkeley, CA 94720

ABSTRACT

In the course of using LDV to study combustion in a thermal boundary layer, the particle count rate was found to decrease abruptly to zero inside the boundary layer. Experimental and theoretical investigation of this phenomenon was carried out. The motion of the particles may be due to the combined effects of thermophoresis and radiative heating.

INTRODUCTION

In the course of using laser Doppler velocimetry to study combustion in a thermal boundary layer, the particles seeded in the flow were found to move away from the heated surface, and as a result, the particle count rate decreased abruptly to zero inside the boundary layer, making it impossible to obtain measurements near the wall region. Such a disappearance of particles near the plate surface was not noted with the plate unheated. In order to understand this phenomenon and hopefully to find a remedy for this problem, an experimental and theoretical investigation was carried out.

EXPERIMENTAL AND LDV SYSTEM

The experimental configuration for studying a heated boundary layer with combustion consisted of a sharp leading edge quartz plate with vacuum deposited platinum heating strips. The plate was mounted vertically in an upward directed, open atmospheric pressure jet of air or premixed fuel and air. The plate was a 1.5 mm thick, 75 mm square quartz sheet with a sharp leading edge with 2° included angle. The heating strips were oriented perpendicular to the flow and were of varying width in order to improve temperature control near the leading edge. The flow originated from a 5 cm diameter nozzle mounted on a stagnation chamber 20 cm in diameter. The stagnation chamber had internal screens to suppress turbulence, and the open jet produced was of uniform velocity with low turbulence.

This work was supported by the Department of Energy, Division of Basic Energy Sciences.

*
Also Department of Mechanical Engineering, University of California, Berkeley.

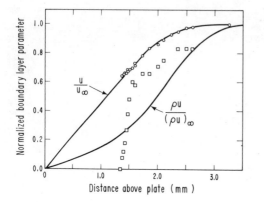

Figure 1.

Comparison of measured
velocity, particle count
rate and normalized density-
velocity product profiles
for a heated plate at 1200 K
U_∞ = 2.62 m/s.

The laser Doppler velocimeter was of intersecting dual beam type with
real fringes. A 4 watt argon ion laser operated at 4880 Å, a cube type beam
splitter and a 100 mm focal length lens were used. The beam separation is
variable, but was set at 20 mm for all the experiments. The scattering was
detected at about a 30° angle from the forward direction with a 55 mm focal
length camera lens, which imaged a pinhole on the intersection region, and an
RCA type 931 A photomultiplier. The scattering burst was both observed directly on an oscilloscope and the frequency of the burst was determined by a
Thermo Systems Model 1090 frequency tracker.

Aluminum oxide particles of nominally 2.0 μ diameter were used. The particles were suspended in water and dispersed into droplets by a collision type
atomizer. This technique was satisfactory; particle count rates varying from
300 to 1000 s^{-1} were attained.

EXPERIMENTAL RESULTS

With an unheated plate, the particle count rate in the boundary layer was
proportional to the velocity, as would be expected for a uniform particle
density in the original flow. With the plate heated to temperatures in the
range of 1000°K, visual observation of the light scattered from the laser beam
indicated a dark region near the plate surface.

The observed particle count rate and flow velocity as a function of
distance above the plate surface is shown in Figure 1. Also shown are the
normalized velocity distribution (U/U_∞) and the expected decrease in particle
count rate due to variation in velocity and density across the boundary layer,
$\rho u/\rho u_\infty$. It can be seen that the particle count becomes larger than expected
at the outer region of the boundary layer but decreases to zero abruptly at a
point approximately half way through the boundary layer, creating a well
defined particle-free region which can be called the dark boundary layer.
Such behavior is consistent with the presence of a force acting on particles
near the plate which moves the particles away from the hot surface.

A comparison of the hydrodynamic and dark boundary layer thickness is
shown in Figure 2. The dark boundary layer thickness appears to scale with
the hydrodynamic boundary layer thickness. Similar results were obtained over
a range of temperatures from 670 K to 1280 K and at velocities of 1.2 and

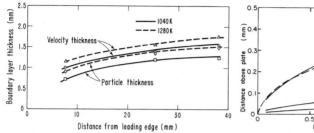

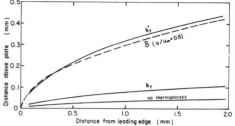

Figure 2. Comparisons of hydrodynamic boundary layer thickness ($U/U_\infty = 0.7$) and dark boundary layer thickness.

Figure 3.

Calculated dark boundary layer thickness.

2.6 m/sec. The ratio of the boundary layer thicknesses was also found to be approximately independent of the plate temperature, free stream velocity and the distance downstream from the leading edge, over the range of conditions investigated.

Additional experiments were performed to determine whether the particle disappearance was due to evaporation or burning up of the particles. The particles were passed through a CH_4/Air conical flame ($\phi = 1.1$, $U_\infty = 0.6$ m/s). It was found that the decrease in particle count rate agreed with the decrease in density across the flame front. Since the boundary layer temperatures are considerably lower than the flame temperature, the loss of particles near the plate surface is not due to their destruction by the elevated temperature.

PROPOSED EXPLANATION AND COMPUTATIONAL RESULTS

One possible explanation for the observed particle disappearance that may be consistent with the above observations is a phenomenon known as thermophoresis. It is a relatively well understood phenomenon with extensive description in the aerosol and particle research literature. Thermophoresis is the result of radiometric forces exerted by a gas on a nonuniformly heated object which it surrounds [1,2]. Under such conditions a flow is induced along the surface of the particle in the direction of increasing temperature, due to a temperature gradient along its surface. This gas flow, known as thermal creep, exerts an equal but opposite force on the surface in the direction of decreasing temperatures. For the case of a sphere in which the diameter is much greater than the mean free path, the thermophoresis force is given by [3]:

$$F_t = - \frac{6\pi M g D_p \lambda_g \sigma}{P(2\lambda_g + \lambda_p)} \frac{\partial T}{\partial x} \tag{1}$$

where D_p is the particle diameter, λ is the thermal conductivity, μ is the viscosity, P is the gas pressure, σ is the thermal slip factor and subscripts g and p denote gas and particle properties. The resulting motion of a particle in a fluid flow with a temperature gradient then represents a balance between thermophoresis forces (Eq. (1)) and Stokes drag. This is given by the following expression:

$$\frac{dv_p}{dt} = - \frac{k_T}{2\alpha} \frac{\partial T}{\partial y} + \frac{1}{2\alpha}(v_p - v_g) \tag{2}$$

where

$$k_T = \frac{2\sigma \; \lambda_g^2}{P \; \lambda_p} \tag{3}$$

$$\alpha = \frac{\rho_p \; D^2}{9\mu_g} \tag{4}$$

Equation (2) was introduced into a finite difference computer code pre-viously developed to solve the governing differential equations for gas flow over a heated flat plate [4]. Particle trajectories were calculated for several different cases. Particles were introduced at 40 different locations across the boundary layer to determine the sensitivity of particle movement to distance from plate. The particles were also introduced at two different loca-tions slightly downstream from the leading edge. Figure 3 shows the predicted dark boundary layer thickness as a function of distance along the plate. This is comparable to the approximate point at which the particle count decreased to zero in the experimental measurements. Also shown is the calculated hydro-dynamic boundary layer thickness based on $u/u_\infty = 0.5$. The results presented are for a wall temperature of 1300 K and a free stream velocity of 3 m/s. The curves labeled "no thermophoresis" and k_T represent results for the case of particle motion under the influence of Stokes drag alone and for motion with the thermophoresis force. It can be seen that the calculated particle movement is significantly less than that found experimentally even when the thermo-phoresis effect is included. Variations of the x location of particle injec-tion from x = 0.016 mm to x = 0.0016 mm had negligible effect on these results.

The coefficients k_T and α in Eq. (2) are based on properties of the bulk material, and there is some uncertainty in the values of these properties in medium sized particles. A set of calculations was made to determine the effect of variation in these constants. As can be seen in Figure 3, k_T must be increased by a factor of 25 ($k_T' = 25\ k_T$) before approximate agreement is obtained with experimental observations. On the other hand, the dark boundary layer thickness was found to be relatively insensitive to variations in α.

The particle trajectories were studied assuming that $k_T' = 25\ k_T$. The thermophoresis force results in initial acceleration of the particles away from the surface until the Stokes drag force balances the thermophoresis force. Further downstream there is a small deceleration of the transverse particle motion due to both the decreased rate of thickening of the boundary layer and the reduction in the temperature gradient. Calculations were carried out with surface temperatures from 800 K to 1600 K, and the results were in reasonable agreement with the experimental observations of increased particle density at approximately half the hydrodynamic boundary layer, independent of surface temperature and distance from the leading edge.

As was mentioned above, it was necessary to increase the thermophoresis force by a factor of approximately 25 to obtain agreement between calculated and experimental results. In the calculations it was assumed that the particle temperature gradient was equal to that existing in the gas at the same loca-tion. It is possible, however, that due to radiative heating the temperature gradient in the particle is greater than that of the gas. As a limiting case, the calculation was repeated with the additional assumption that the hot side of the particle was at the same temperature as the surface. The particle motion is significantly different than was found previously. In this case a rapid increase in dark boundary layer thickness occurs near the plate leading edge to a value somewhat greater than the hydrodynamic boundary layer thickness. Farther downstream, the rate of increase diminishes but the thickness ratio continues to increase with distance along the plate. Thus the predicted thickness ratio is significantly greater than that found experimentally.

CONCLUSIONS

The results of this investigation indicated that it may be possible to explain the particle motion normal to the streamlines in a heated boundary layer by the concept of thermophoresis. A much better understanding of the process would be obtained if the properties of the particles were known and the proper magnitude of the radiative heating were taken into consideration. Towards this aim, further experimental investigation is planned.

REFERENCES

1.   Fuchs, N. A. 1964. The Mechanics of Aerosols, p. 66, Pergamon Press, New York.

2.   Loeb, L. B. 1961. The Kinetic Theory of Gases, p. 353, Dover Publications, New York.

3.   Waldman, L. 1961. Proc. Second International Symposium on Rarefied Gas Dynamics (L. Talbot, editor).

4.   Schefer, R. and F. A. Robben, 1977. Catalyzed Combustion in a Flat Plate Boundary Layer II. Numerical Calculations. Paper presented at the Western States Section of the Combustion Institute, Stanford, October 17-18.

MEASUREMENTS IN TURBULENT FLOWS

Chairman: **S. N. B. MURTHY**
Purdue University

Measurements in Steady and Unsteady Separated Turbulent Boundary Layers

INVITED PAPER

ROGER L. SIMPSON and Y. T. CHEW
Department of Civil and Mechanical Engineering
Southern Methodist University
Dallas, TX 75275

ABSTRACT

Highly turbulent flows are those in which the velocity fluctuations are substantial as compared to the time-averaged velocity, e.g., separated flows and intense mixing regions. Prior to the development of laser anemometry, no accurate measurements in separated turbulent flows were possible. Some recent results from fundamental experiments on steady and unsteady freesteam separated flows are presented.

Details on the laser anemometer system used to acquire these data and the method of signal time-averaging are presented. The multi-velocity component LDV uses frequency shifting to relieve velocity direction ambiguity and place the Doppler signals in relatively low noise portions of the spectrum. Minimal particle seeding is used to produce good signal-to-noise ratio signals and a good data acquisition rate. Fast-sweep-rate spectrum analysis is used to process Doppler signals.

NOMENCLATURE

δ	boundary layer thickness
E	voltage output from LDV processor
f, f_B	Doppler signal frequency and Bragg cell frequency
N	number of signal bursts
$P(U)$	velocity probability distribution, equation (9)
ϕ	phase angle lead
S_u, K_u	skewness and flatness factors for U probability distribution
t, t_1	time; and period of oscillation cycle
T	total record sample time
U, V, W; U, V, W; u, v, w	streamwise, normal to wall, and transverse velocities; instantaneous, time-averaged, and fluctuation values, respectively; subscript E denotes freestream

$\overline{u^n}$, $\overline{v^n}$, $\overline{w^n}$ time-averaged fluctuation velocity moments, $n=2,3,4...$

$\hat{U}_j$, $\hat{u}_j^2$ ensemble-averaged velocities of jth bin of periodic
             oscillation cycle, defined by equations (12) and (13)

x, y, z   streamwise, normal to wall, and transverse co-ordinates

INTRODUCTION

Highly turbulent flows are those in which the velocity fluctuations
are substantial as compared to the time-averaged velocity, e.g., separated
flows and intense mixing regions. Prior to the development of laser anemometry, no accurate measurements in separated turbulent flows were possible
and substantial uncertainties were present in some mixing layer measurements
because of the instantaneous flow direction ambiguity. Simpson (1976)
has pointed out that when $\sqrt{\overline{u^2}}/U$ is locally greater than 1/3, then flow
reversal occurs part of the time. Therefore directionally ambiguous hot-
wire anemometer techniques cannot be used in this flow regime.
During this same era in which laser Doppler anemometry has been
developed, there has also been an increased understanding of such flows.
It is reasonably well accepted that an orderly sequence of fluid motions
occurs quasi-periodically. Roshko (1976) presented a review of what is
known about these seemingly coherent structures for free shear flows.
Simpson, Strickland, and Barr (1977) presented LDV measurements on a separating turbulent boundary layer. In both types of flow the most probable
frequency of passage of large-scaled coherent structures varies with U_E/δ,
where δ is a characteristic shear layer thickness. In the case of separated boundary layers, U_E/δ may be an order of magnitude smaller downstream of separation than in the attached upstream flow. When freestream
unsteadiness is present (Simpson, 1977), as in many practical cases, strong
interaction between the turbulent and non-turbulent unsteady motions can
occur. The nature of turbulent boundary layer separation is fundamentally
different from laminar boundary layer separation. It appears that because
such flows are dominated by the natural unsteadiness within the shear
layer, that the applicable governing equations are not elliptic – but
hyperbolic! Considering that the limiting factor in the performance of
many practical machines is turbulent boundary layer stall, it is easy
to see the large role that LDV must take in improving the understanding
of such flows for improved designs.
Here we will discuss the LDV system used at SMU in the last several
months to obtain data on steady and unsteady separated flows. The optical
system and particle seeding requirements for such flows are presented.
Fast-sweep-rate sampling spectrum analysis of Doppler signals is reviewed.
This type signal processing provides a velocity versus time signal, in
contrast to conventional slow-sweep-rate spectrum analysis. It is shown
that proper time-averaging of available signals can eliminate biased results
for ergodic flows. The signal processing does not produce velocity-biased
results – the averaging method does! Finally, some of our results are
discussed.

OPTICAL SYSTEM REQUIREMENTS

All of the design features required for LDV systems used in low
turbulence flows are desirable for applications to highly turbulent flows
(Durst, et al. 1976). For a fringe system of measuring a velocity component, the fringe visibility should be optimized, laser beam path-lengths
and polarization directions equalized, and the focal volume well defined
at the photodetector aperture. A sufficiently powered laser should be
used to allow an adequate signal-to-noise ratio for sufficiently small
seeding particles.

Optical frequency shifting of one of the two beams required for
each velocity component is absolutely necessary for highly turbulent flows.
This shifting moves the zero velocity frequency away from the zero frequency part of the signal spectrum. In the case of a fringe system, the fringes
will move into and out of the focal volume at a frequency equal to the
frequency shift f_B. With a sufficiently large frequency shift, the interference from the zero frequency pedestal and noise spectrum can be separated completely from the signal spectrum. The LDV is then directionally
sensitive, e.g., particles moving with equal speeds in opposite directions
produce frequency signals equally spaced above and below f_B. If the frequency shift $f_B \geq 2 f_D$, where f_D is the maximum Doppler frequency due
to the flow, then signals from all particles passing through the focal
volume can be detected (Durst, et al. 1976).

In addition, the frequency shifting device such as a Bragg cell
can serve as an efficient beamsplitter. By using different shift frequencies for measuring different velocity components, a single photomultiplier tube can detect unambiguous signals from them all simultaneously.
Of course the photomultiplier must be selected for sufficiently high
frequency response.

The SMU laser anemometer shown in Figure 1 is a typical example
system. The optical arrangement is a backscattering fringe-type arrangement for the U and V velocity components, which are in a plane perpendicular to the axis of the focusing lens. A slightly different earlier
version was reported at the Second International Workshop on Laser Velocimetry (Simpson and Barr, 1974). The major difference with the current
version is that a higher powered Spectra-Physics Model 164 Argon-Ion laser
is operated at a wavelength of 4880 Å with 1.5 watts of power. The beam
passes through a dual Bragg cell constructed after the design by Hornkohl,
et al. (1972).

The Bragg cell is contained in an aluminum housing to prevent rf
radiation leakage to the signal processing electronics. The horizontal
first-order diffracted beam is shifted 25 MHz. The Bragg cell is about
85% optically efficient. Both transducers are operated simultaneously
since two sets of signal processing electronics are available. An inexpensive Heathkit Model DX-60B CW transmitter is used for each transducer,
easily providing the about 1.5 watts of power required for a transducer.
Impedence matching circuitry was built according to accepted rf electronics
practice.

Mirrors are located to separate the Bragg cell output beams, which
are about 1.40° apart for the horizontal beams and about 0.9° for the
beams in the vertical plane. Other mirrors are used to maintain like
polarization directions and equal path lengths of the beams within 0.1
inch. The transmitting-receiving pair of lenses are mounted next to each
other with the same centerline. The 6.25 inches diameter transmitting
lens has a focal length of 37.25 inches, allowing measurements over most
of the 3 feet wind tunnel width. The second lens, 5.25 inches in diameter,

Figure 1. Laser anemometer optical system.

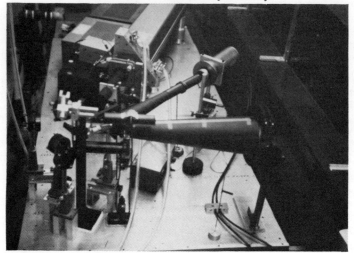

and the center portion of the transmitting lens form a receiving lens
pair which focuses the backscattered signal by way of an adjustable mirror
onto the plane of a variable opening diaphragm in front of the photomultiplier tube.

        This lens combination is mounted at an angle of 4.4° to the vertical
plane so that the centerline of the transmitted unshifted and 25 MHz
shifted beams make an angle of 0.8° with the horizontal wind tunnel wall.
Thus when a horizontal component of velocity perpendicular to the lens
center line is being measured near the wall, there is no beam interference
with the wall prior to the focal volume and the entire receiving lens
is utilized in collecting signal. The unshifted and 15 MHz shifted beams
form a fringe pattern such that the measured velocity is actually V cos
4.4° + W sin 4.4°, where V is the vertical velocity component and W is
the third component. A third signal around 10 MHz is obtained from a
fringe pattern between the 25 MHz and 15 MHz shifted beams and measures

$(U - V \cos 4.4° - W \sin 4.4°)/\sqrt{2}$.

        The probe volume size, which is directly proportional to the trans-
mitting lens focal length and inversely proportional to the laser beam
diameter was found to have little effect on the received signal quality.
A focal volume of 0.0125 inches diameter and 0.140 inches length was produced from a laser beam 1.1 mm in diameter at the $1/e^2$ locations. The
incident beams intersect at an angle of about 5.84°, producing interference
fringes 1.88×10^{-4} inches apart, λ_p. Consequently there are about 70

fringes in the focal volume at one time. Note that because these fringes
move through the vocal volume due to the Bragg frequency shift of one
beam, a given particle encounters many more fringes, e.g., at 60 fps a
particle encounters 430 fringes while at 5 fps about 5200 are crossed.

        The W component is obtained by a confocal backscattering reference
beam arrangement similar to that described by Kreid and Grams (1976).
The polarization of the 30 MHz diffracted beam from the 15 MHz Bragg cell
is rotated 90° by a half-wave plate before focusing through the center
of the receiving lens combination onto the focal volume. Thus this beam
and the unshifted, 25 MHz shifted, and 15 MHz shifted beams do not interfere in the focal volume. A portion of the 25 MHz shifted beam is used
as a reference beam after the polarization of this beam is also rotated

90° by a half-wave plate. The path lengths between this beam and the
30 MHz beam are equal but are different from the other three beams. Thus,
the laser noise associated with unequal path lengths also prevents unwanted
stray signals from occuring between the depolarized scattered light of
the three other beams and the 25 MHz reference beam. The Doppler signal
from the confocal arrangement is around 5 MHz and measures W cos 4.4° +
V sin 4.4°.

   After focusing onto the plane of the variable aperture diaphragm,
a received signal passes onto the face of a EMI 9813B photomultiplier.
The aperture alone does not define the focal volume. The signal height
discriminator in the signal processing only permits processing of the
most intense signals that occur in the center of the focal volume. With
stationary smoke in the focal volume, a signal-to-noise ratio of about
35dB is obtained.

   In all versions the optics were mounted on a single mobile cart
which allows movement along the wind tunnel test section. To locate the
position of the focal volume with respect to the horizontal test wall,
the entire optical table is adjusted vertically until a symmetrical X
is formed on the wall. The most intense portions of the beams are then
focused on the wall, so a strong 25 MHz signal is produced.

PARTICLE SEEDING FOR HIGHLY TURBULENT FLOWS

   Particle seeding for highly turbulent flows does not appear to have
any more severe limitations or requirements than other flows. Durst,
et al. (1976) outline the most important considerations. It should be
noted that it is impossible to seed a highly turbulent flow in any prescribed manner. Highly turbulent flows are characterized by intense mixing within the flow. In the case of boundary layer and free shear flows,
there is also significant entrainment of freestream fluid into the turbulent motions. This would progressively dilute the particle concentration
if only the shear flow has been seeded. Instead of needless worry over
prescribed particle concentration, we have been concerned with proper
averaging of available signals as described in section 5 below, with enough
particles to provide a high data rate, and with sufficiently small particles to accurately follow the flow. In fact, without any seeding we
were able to obtain signals from ambient dust. However, we used minimal
seeding to produce a signal data rate of about 400 per second.

   The boundary layer was seeded in the following manner. The test
wall boundary layer in the SMU wind tunnel is tripped by the blunt leading
edge of the plywood floor, the height of the step from the wind tunnel
contraction up to the test wall being 1/4 inches. 1-1/4 inches upstream
of the blunt leading edge, 33 smoke ports, 1/8 inches in diameter, are
located spanwise on 1-1/16 inches centers in the wind tunnel contraction.
A baffle plate deflects the smoke in the free-stream direction and tends
to produce a uniform spanwise distribution of smoke. More details on
this wind tunnel are given by Simpson, et al. (1977) and Simpson (1977).

   The smoke generator is essentially the same design presented by
Echols and Young (1963), with the numerical values of particle size,
flowrates, and pressures being taken from their work. The smoke is produced by six adjustable nozzles each of which blows high speed air through
4 orifices 0.04 inches in diameter into the liquid smoke material, which
in this case is dioctal phthalate or "DOP". The DOP is atomized by the
shearing action of the compressed air jets. In the data presented here
only four nozzles with a pressure drop of 10 psi were required to produce
the best LDV signals possible. The resulting mixture of air and DOP particles is blown perpendicular toward the bottom of a 5 gallon impacter

can, removing any large particles which may have been entrained in the
mixture. The mixture is then blown out of the top of the impacter can
into a manifold which distributes the smoke uniformly to the smoke ports
in the wind tunnel contraction.

In the experiments reported here, 2 cubic feet per minute of smoke
at a concentration of about 0.3×10^{-3} lbs. of smoke particles per cubic
foot of blown air was used. The density of the undiluted smoke was only
0.4% greater than that of air alone. In the test boundary layer near
the separation region, the density of the diluted smoke was only about
0.0004% greater than that of air alone, making smoke-induced density effects
negligible. Mean particle size of this stable room temperature (77°F)
smoke is approximately 1 micron. If a particle this size is accelerated
from near rest to the free-stream velocity during time δ/U_E, as it may
do when contained in some coherent structure, it still will follow the
flow velocity within 0.1% or about 0.05 fps using the analysis of Brodkey,
et al. (1969). Mazumder, et al. (1974) report the frequency response
of such a particle to be down 10% at 10 kHz when subjected to sinusoidal
oscillations. The spectral inertial subrange was below 10 kHz in the
high velocity outer part of the boundary layer for these flows, so the
particles were following the lower frequency range highly turbulent
oscillations found nearer the wall.

SIGNAL PROCESSING BY FAST SAMPLING SPECTRUM ANALYSIS

Durst, Melling, and Whitelaw (1976) reviewed the use of frequency
trackers and counters for highly turbulent flows. The current authors
will not repeat their recommendations but instead describe a third signal
processing technique that has been used at SMU for over four years.

All measurements obtained at SMU have used fast-sweep-rate sampling
spectrum analysis, as described by Simpson and Barr (1974, 1975). A
rapidly-swept-filter spectrum analyzer and some peak detection and sampling
electronics are employed in this method. The spectrum analyzer filter
of bandwidth B is swept through a given range of frequencies D up to 8000
times per second, i.e., a sweep frequency f_s of 8KHz. If a signal is
detected on each sweep, then up to 8000 velocity signal samples per second
can be obtained. Thus a velocity versus time output signal can be produced
that contains information about turbulence frequencies of up to 4KHz.
The advantages, simplicity, and accuracy of this signal processing method
have not been widely discussed. Some LDV users appear to confuse fast-
sweep-rate spectrum analysis with conventional slow-sweep-rate spectrum
analysis that requires considerable corrections to the results (Durst,
et al. 1976). Thus, a brief review of this method is given here.

The signal processing logic is as follows. As shown in Figure 2,
the signal from the photomultiplier tube or detector is put into a swept
filter spectrum analyzer. For each sweep of the filter when a signal
is detected, a vertical voltage distribution proportional to the filter
output is displayed as shown in Figure 3(a). The simultaneous horizontal
sweeping voltage, which is proportional to the signal frequency, is shown
in Figure 3(b). The peak of the vertical voltage distribution marks
the frequency of the passing particle signal and can be used as a gating
signal to allow the instantaneous value of the horizontal sweep voltage
E to be sampled. This instantaneous voltage value is related to the instantaneous velocity U of the particle through

Figure 2. Block diagram for fast-
sweep-rate spectrum analysis: A -
photomultiplier tube; B - spectrum
analyzer; C - peak detection circuit;
D - sample-and-hold circuit.

Figure 3. Signal shapes
during a single sweep of
the swept filter.

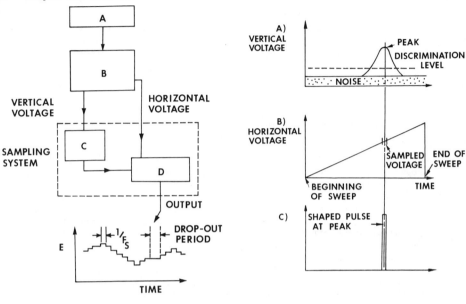

$$ U = \lambda_f \left(f_o - f_B + \left(\frac{df}{dE}\right)E \right) = \lambda_f (f - f_B) \tag{1} $$

where λ_f is the fringe spacing, fo is the analyzed frequency at the beginning of a sweep, and df/dE is a calibration constant relating the
analyzed frequency to the horizontal sweep voltage.

Prior to gating the horizontal sweep voltage, the vertical voltage
distribution is fed into a peak detector circuit which produces a pulse
simultaneously with the occurrence of the first peak value (Figure 3(c)).
In the circuit used here a 1 μsec wide pulse is produced. This output
pulse is used to trigger a sample-and-hold circuit, into which the horizontal sweep voltage has been fed. The sampled sweep voltage E is held
by the sample-and-hold circuit until a new signal from another particle
is detected. The lower part of Figure 2 shows this output voltage E versus
time, which can either be fed to a probability analyzer or long time true
integrating dc and rms voltmeters to determine the mean and rms velocities
as described in section 5 below.

It is clear from Figure 3 that the vertical voltage signal peak
must be distinguishable above the wideband noise level for detection.
The discrimination level must be above the highest noise level present
in the range of Doppler frequencies for the turbulence present. Since
the signal is processed in the frequency domain, the signal-to-wideband-
noise ratio need not be as good as signals processed in the time domain
with counters. Note that the shape of the vertical voltage signal peak
and variations in signal height above the adjustable discrimination level
do not influence the signal processing. This means that skewing and attenuation of the signal peak associated with high sweep frequencies f_s are

of no importance to the laser anemometer user as long as sufficient vertical
voltage signal remains. The E versus f relation is still linear, i.e.,
the relative location of the peak produced by a given input frequency
is the same with respect to the peak produced by any other input frequency.
Thus for a given sweep frequency f_s and bandwidth B, the spectrum analyzer
is easily calibrated with an accurate signal generator.

Several fundamental characteristics of Gaussian filter spectrum
analyzers, such as the HP 8558B used in the SMU work, are worth noting
to further explain this method. The vertical voltage output (Figure 3a)
from a rapidly-swept-filter spectrum analyzer is proportional to the energy
of the signal which passed through the filter during the time

$$\tau = \frac{BT_s}{D} \qquad (2)$$

where B is the 3dB resolution bandwidth of the filter, the dispersion
D is the range of frequencies over which the filter is swept, and T_s is
the sweep time which is slightly less than $1/f_s$. Thus this voltage is
proportional to the amplitude squared and the duration of the Doppler
signal burst detected.

It is important to note that the frequency f of the detected signal
peak is independent of the number of Doppler signal cycles in a signal
burst and the duration of a signal burst. This can be rigorously proved
by Fourier transform theory (Engelson, 1971). Durão, et al. (1975) observed
this for signal durations between 2 to 20 µs and for numbers of cycles
between 1 and 150! Thus even if the bandpass filter during its sweep
catches only a leading or trailing few cycles of a signal, or encounters
a signal burst with amplitude modulation, the correct frequency will be
measured if the vertical voltage output is above the discrimination level.
The longer the Doppler signal is detected, the less attenuation of the
vertical voltage output. The loss factor of the vertical voltage output
α_p is expressed (Engelson, 1971) approximately as

$$\alpha_p = \frac{3}{2} t_o B \qquad (3)$$

for $\alpha_p < 1$, where t_o is the time that the signal is detected. For longer
detection times, $\alpha_p = 1$. Durão, et al. (1975) also obtained data on their
Gaussian filter spectrum analyzer that agree with this equation.

To insure a sufficiently large vertical voltage output above the
discrimination level, one can choose a sufficiently large bandwidth.
The amplitude loss factor for Gaussian filters is given by (Engelson, 1971)

$$\alpha_d = \left[1 + 0.195 \left(\frac{D}{T_s B^2} \right)^2 \right]^{-\frac{1}{4}} \qquad (4)$$

and confirmed experimentally by Simpson and Barr (1974). In practice
B was less than 0.03D, with D being adjusted between 500 KHz to 5 MHz
for varying turbulence levels. For example with B = 30 KHz, D = 1 MHz,
and $T_s = 1 \times 10^{-3}$ sec, then $\alpha_d = 0.95$. There is only a 5% attenuation
of the vertical voltage peak in this case.

As long as 1/B is less than the time between successive Doppler
signal bursts, then the detected signal from each individual Doppler burst

produces an output which is independent of other Doppler bursts (Engelson, 1971). This condition is easy to achieve since $B \geq 10$ KHz and $f_s \leq 8$ KHz.

The peak detection circuit used here locates the first vertical peak above the discriminator level and does not locate additional peaks until the vertical signal has dropped below the discriminator level. This prevents sampling of signals from more than one burst at a time. After falling below the discriminator level, the peak detection circuit is automatically reset and accepts the next peak as a valid signal. Thus it is possible to obtain several valid velocity peaks on the same sweep of the filters through the frequency range.

The advantages, disadvantages, and the optimum use of fast sweep spectrum analysis are worth summarizing. The method is relatively inexpensive, uses a commercially available spectrum analyzer that can process the highest frequency signals available from a photodetector, and processes poor signal-to-noise ratio (SNR) signals. The scope display of the spectrum analyzer allows the user to see the signal and noise and to set the discrimination level just above the noise. The only disadvantage experienced by the current users is that many available bursts of Doppler signal are lost since all signals outside the narrow bandwidth B at a given time of a sweep are disregarded.

The optimum use of this method requires only a few considerations. First, a sufficiently large vertical voltage signal peak must be obtained, so α_p and α_d should not be much below unity unless large SNR signals are available. The sweep frequency f_s should be as large as possible with a given SNR in order to produce the largest output velocity data rate. The time that a Doppler signal passes through the filter τ should be less than the duration of a Doppler signal in the focal volume in order to minimize "finite transit time" signal broadening. Finite transit time signal broadening occurs when light is scattered to the photodetector from several particles simultaneously (Durst, et al. 1976). If only light particle seeding is used, then this consideration can be ignored.

TIME-AVERAGING OF SIGNALS

As pointed out above the particle number density in a highly turbulent flow cannot be made uniform. Thus the time between the passage of successive signal generating particles will be unequal. This effect alone presents no particular signal processing problem if the time intervals between successive signal bursts are small compared to $1/f_{max}$, the time period of the highest flow oscillation frequency f_{max} to be detected, i.e., if the signal is almost continuous. One can simply treat the signal as a continuous hot-wire anemometer signal to obtain the averages

$$U = \frac{1}{T} \int_o^T U(t)dt \qquad (5)$$

$$\overline{u^n} = \frac{1}{T} \int_o^T (U(t) - U)^n \, dt \qquad (6)$$

where $n = 2, 3, 4 \ldots$. When the time intervals between successive signal bursts are long compared to $1/f_{max}$ (high signal dropout rate) and are unequal, these equations should also be used in the fashion explained below.

First, let us look at the commonly used method of particle averaging for individual particle velocity measurements. The averages are made over the number of signal bursts N obtained during the time period T:

$$U_N = \frac{\sum\limits_{i=1}^{N} U_i}{N} \tag{7}$$

$$\overline{U_N^{\,n}} = \frac{\sum\limits_{i=1}^{N} (U_i - U_N)^n}{N} \tag{8}$$

where n = 2, 3, 4.... These averages are not made with respect to time and are biased unless the time intervals between signal bursts are equal. McLaughlin and Tiederman (1973) proposed a biasing correction that is based upon the idea that higher velocity flow carries more particles through the focal volume per unit time. Thus, more high velocity signal bursts will be obtained and U_N will be too high. However, high velocity particles

spend less time in the focal volume so that in the case of sampling spectrum analysis signal processing, the chance of detecting a given signal burst varies as $(U^2 + V^2 + W^2)^{-\frac{1}{2}}$. Thus, this effect tends to cancel the above mentioned bias for particle averaging. Durão and Whitelaw (1975) showed that if the Doppler bursts are randomly sampled before particle averaging, the bias effects are reduced significantly. Even so, particle averaging is not fundamentally a time average.

Consider now time-averaging of signals according to equations (5) and (6), even though the signal dropout rate may be large. Only ergodic flows, whose averaged quantities in equations (5) and (6) become independent of time for large T, are considered. This restriction is also required for particle averaging. Figure 4 shows a typical output voltage signal from an LDV signal processor which is proportional to $U(t)$. Note that the last sampled signal must be held by a sample-and-hold circuit until a new signal is detected for time-averaging. With exception of the instant at which a new signal is detected, the sampled-and-held voltage does not correspond to the actual instantaneous velocity. However, the voltage value at each instant corresponds to the instantaneous velocity at some instant during a record time T for an ergodic flow. Since any averaging process removes time domain dependency, it does not matter when during the time period T that it is averaged. It is unlikely, even for a long time T, that many particles will have the same velocity. Thus it is highly unlikely that a given signal voltage will be averaged too long.

This method of averaging eliminates the need for the high velocity flow bias correction. If the signal rate is lower from low velocity fluid than higher velocity fluid, then the number of changes in the output voltage is lower but does not influence the time integrals. For example, in Figure 4 signals A-E occur in high velocity flow while signal F occurs in low velocity flow a short time later. Signals A-E each contribute to the integrals less of the time than does signal F, but the flow passing through the focal volume just after signal F must also have been at a low velocity until the occurance of signal G because of the lower signal rate. Again the signals G-I each contribute less to the integral than does J. Note that signals F and J are generally not equal.

The mechanics of evaluating equations (5) and (6) can be carried out several ways. One way is to determine U from a true integrating volt-

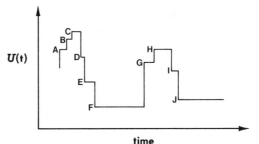

$U(t)$

Figure 4. Typical fast-sweep-rate sampling spectrum analysis output.

time

meter as done with hot-wire anemometer signals. The quantities $\overline{u^n}$ can also be obtained in this manner as long as information is not lost by high-pass filtering. Otherwise digital evaluation of these averages is preferable. The averaging time T should be selected to make the results time independent. Bates and Hughes (1977), for example, suggest that over 8000 data points are needed for convergence of $\overline{u^3}$ and $\overline{u^4}$ values. Clearly though, considering the possible particle density variation in highly turbulent flow, one should experimentally determine the minimum T required for ergodicity.

Another way of evaluating averages makes use of a velocity probability histogram, such as shown in Figure 5.

$$1 = \int_{-\infty}^{+\infty} P(U)\,dU \qquad (9)$$

$$U = \int_{-\infty}^{+\infty} P(U)\,dU \qquad (10)$$

$$\overline{u^n} = \int_{-\infty}^{+\infty} (U - U)^n P(U)\,dU \qquad (11)$$

where n = 2, 3, 4.... The histogram $P(U)$ is constructed by sampling the $U(t)$ signal such as shown in Figure 4 at equal intervals in time Δt for the period T. Thus the histogram reflects a true time integral and the results from equations (9-11) will be equivalent to those from equations (5-6). The time interval Δt between digital samples should be no larger than the shortest time between signal bursts, otherwise some data will be lost. For example, in the work at SMU, $\Delta t = 10^{-4}$ sec for about 400 new signals per second. The averaging time T was at least a half minute,

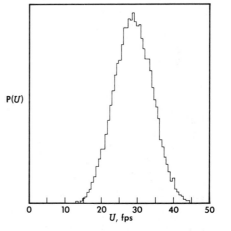

$P(U)$

U, fps

Figure 5. Typical velocity probability histogram: U = 29.1 fps. $\overline{u^2}$ = 27.0(fps)^2, S_u = 0.047, K_u = 2.66. Discrete velocity bins due to probability analyzer.

so at least 12,000 new
data signals and 3×10^5
equal time interval sam-
ples were involved for
one histogram. An added
benefit of the histogram
approach is that noise
can be detected while $P(U)$
is being constructed.
If one has an oscilloscope
display, the noise will
cause the base level of
$P(U)$ to grow. Thus, the
resulting $P(U)$ can be
corrected for noise or the
discriminator level in the
signal processor can be
adjusted on-line and a new
$P(U)$ constructed.

SOME EXAMPLES OF MEASUREMENT RESULTS

The highly turbulent
flows that are currently being examined at SMU with LDV
are separating turbulent
boundary layers with both
steady and unsteady free-
stream velocities. In both
cases the separation is induced by adverse pressure
gradients, in contrast to
backward facing steps. Earlier measurements on a different
separating flow are given by
Simpson, et al. (1977). Fast-
sweep sampling spectrum analysis was used for signal processing in each case. Histo-

Figure 6. Mean velocity profiles
at several streamwise locations for
steady freestream separating flow.
Solid lines for visual aid only.

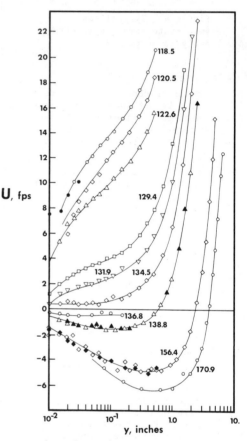

grams such as shown in Figure 5 were used to determine mean and fluctuation
averages from equations (9-11) for the steady freestream velocity case.
Equal time interval averaging for the unsteady freestream case was used
as described below.

Figure 6 shows the reduced results for the steamwise velocity U
at locations along the test wall in the vicinity of separation. In this
semi-logarithmic plot one can see that the scatter in U values is of the
order 0.20 fps, which is the estimated uncertainty due to all sources.

Figure 7 shows U, V, $\sqrt{\overline{u^2}}$, and $\sqrt{\overline{v^2}}$ results at one steamwise location.
The uncertainty in V is about the same as for U while $\sqrt{\overline{u^2}}$ and $\sqrt{\overline{v^2}}$ are
about $\pm$ 2% uncertain. This latter uncertainty level is primarily due
to the finite width of each of the 100 voltage bins into which the sampler
output voltage E is divided by the digital probability analyzer. The
results obtained on different days at the same location indicate a high
level of data repeatability.

These results are not believed to suffer strongly from bias errors. First, α_p = 1, so there is no bias in the duration of a detected signal due to the flow velocity. In other words, the time that the highest velocity particle spends in the focal volume is always large enough to produce a sufficiently large vertical voltage output from the spectrum analyzer. Minimal particle seeding was used for the best SNR and data sample rate, so significant finite transit time broadening is unlikely.

Velocity gradient broadening is not significant for any data presented here (Simpson, 1976). The focal volume diameter and length are small compared to the boundary layer thickness. In addition, signals from the center of the focal volume are the most likely since the scattered signals are the most intense. Large-scaled motions, which scale on the boundary layer thickness, appear to dominate the structure of highly turbulent flows, so strong instantaneous spatial velocity variations are unlikely. In any event, these results compare favorably with hot-wire anemometer data obtained in regions that do not contain significant time variation of the flow direction.

Figure 7 also shows $-\overline{uv}$ results. The $(U - V \cos4.4^\circ - W \sin4.4^\circ)/\sqrt{2}$ signal, as discussed in section 2 above, was used to produce

$$\overline{u^2} - 2\,\overline{uv}\,\cos4.4^\circ - 2\,\overline{uw}\,\sin4.4^\circ + \overline{(v\,\cos4.4^\circ + w\,\sin4.4^\circ)^2}$$

Since $\overline{u^2}$ and $\overline{(v\,\cos4.4^\circ + w\,\sin4.4^\circ)^2}$ were measured independently and $\overline{uw}$ was presumed very small, $-\overline{uv}$ resulted. The uncertainty in $-\overline{uv}$ is about +6%, although there is amazingly little scatter in the results in this case.

The skewness factors $S_u = (\overline{u^3})/(\overline{u^2})^{3/2}$ and $S_v = (\overline{v^3})/(\overline{v^2})^{3/2}$ are shown in Figure 8 while Figure 9 shows the flatness factors $K_u = (\overline{u^4})/(\overline{u^2})^2$ and $K_v = (\overline{v^4})/(\overline{v^2})^2$ for the same streamwise location as the data shown in Figure 7. S and K are about ±0.1 and ±0.2 uncertain. Data obtained on different days are in close agreement, with the scatter being within these uncertainty levels.

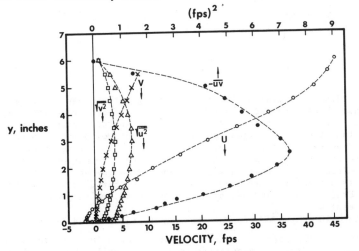

Figure 7. Mean and fluctuation velocity profiles for the 138.8 inches location. Dashed lines for visual aid only.

Figure 8.   Skewness factors at   Figure 9.   Flatness factors at
138.8 inches.   138.8 inches.

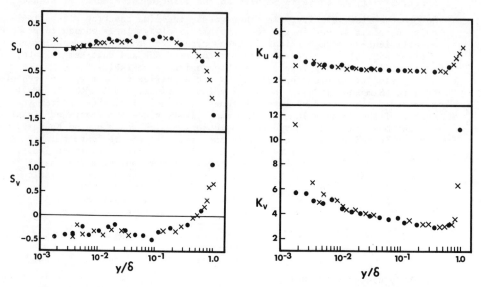

Some results at the same streamwise location for the unsteady freestream case are shown in Figures 10, 11, 12, and 13.   In this case, the
period of oscillation t_1 , 1.677 seconds, is subdivided into 96 individual
bins, each differing by 3.75° in phase angle.   At the time of the same
phase of each of 200 oscillation cycles, the LDV signal is sampled by
an A/D converter, i.e., the LDV sampler output voltage E is sampled.
The ensemble-averaged "mean" velocity for a given jth bin is

$$\hat{U}_j = \frac{1}{200} \sum_{i=1}^{200} U_j \qquad (12)$$

and

$$\hat{u}_j^2 = \frac{1}{200} \sum_{i=1}^{200} (U_j - U_j)^2 \qquad (13)$$

Note that this is equivalent to time-averaging rather than particle averaging since the A/D converter samples are made at equal time increments.
This requires that the LDV signal processing data rate be sufficiently large that the latest signal processing output voltage E is obtained
since the last A/D converter sample was taken.   This insures that there
is no more than one bin uncertainty in the phase information.   In our
case the minimum LDV signal data rate required is 58/sec, but since these
new signals are not equally spaced in time, a higher data rate is necessary.
Here we obtained about 400 new signals per second, which produced satisfactory results.
Figure 10 shows $\hat{U}/U_E$ versus phase angle for various distances away
from the wall.   The data are relatively smooth even for the freestream,
which is shown at the 8 inches location from the wall.   Figure 11 is
a semi-logarithmic plot of these data versus y for six different phase

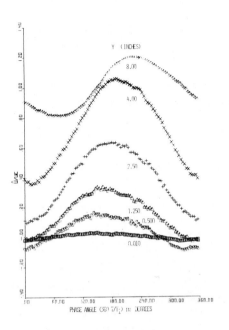

Figure 10. Ensemble-averaged
velocity $\hat{U}_j/U_E$ vs. phase angle
for several y locations at 138.8
inches streamwise location.

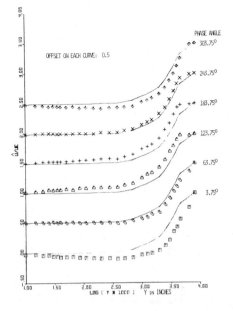

Figure 11. Ensemble-averaged
velocity $\hat{U}_j/U_E$ vs. y for several
phase angles at 138.8 inches streamwise location. Note displaced
ordinate values.

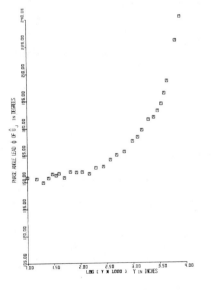

Figure 12. Phase angle lead ϕ of
$\hat{U}_j$ vs. y ($\hat{U}_j \sim \cos(2\pi t/t_1 - \phi)$)
at 138.8 inches streamwise location.

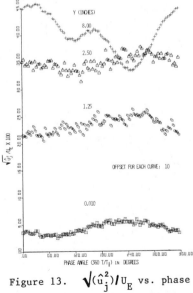

Figure 13. $\sqrt{(\hat{u}_j^2)}/U_E$ vs. phase
angle for several y locations at
138.8 inches streamwise location.

angles. The solid line is the long time average of $\hat{U}_j/U_E$ over all 96

bins for each y location. These figures and Figure 12 show that the flow oscillation near the wall leads the freestream oscillation by about 90° No high mean backflow velocities are observed, just as in the steady freestream case. The ensemble-averaged $\hat{U}_j$ velocity dips slightly below zero

for each y location near the wall and rises to a progressively higher absolute value 180° away for increasing y.

Figure 13 shows results for $\sqrt{\hat{u}_j^2}/U_E$. While there is some scatter

in these results, it is clear that the bin-averaged turbulence fluctuation $\sqrt{\hat{u}_j^2}$ oscillation lags the ensemble-averaged velocity $\hat{U}_j$ shown in Figure 10 by about 90° near the wall. Near the freestream, the boundary layer thickness is oscillating, so $\sqrt{\hat{u}_j^2}/U_E$ oscillates almost 180° out of phase with the freestream velocity.

A serious question is whether 200 samples are sufficient to closely define $\hat{U}_j$ and $\hat{u}_j^2$ for each bin. Clearly the averages of these quantities

over all 96 bins (19,200 samples) are well defined. Figures 10 and 13 demonstrate that the scatter of $\hat{U}_j$ and $\sqrt{\hat{u}_j^2}$ values in adjacent bins reflect the effects of an inadequate number of signal samples and possibly a too low signal data rate. Thus $\hat{U}_j$ and $\sqrt{\hat{u}_j^2}$ for an individual bin may be uncertain 3% of the mean freestream velocity. However, a Fourier decomposition of $\hat{U}_j$ and $\hat{u}_J^2$ for the first harmonic (0.6Hz) has the effect of smoothing through all 19,200 samples and should be considerably less uncertain. Comparison of LDV and hot-wire anemometer data at spatial locations of common validity indicate agreement well within 1% for the mean values and amplitude and phase of the first two harmonics of the oscillatory part.

While further analysis of all these data is currently underway, it is clear that new information on the structure of steady and unsteady freestream separating turbulent boundary layers is being made possible only by the use of LDV.

ACKNOWLEDGMENTS

The authors would like to thank P. W. Barr for trouble-shooting the LDV system and J. J. Sallas and Dr. R. E. Nasburg for keeping the data acquisition computer running. The steady freestream measurements were supported by Project SQUID, an Office of Naval Research Program, while the unsteady freestream measurements were supported by the U. S. Army Research Office.

REFERENCES

Bates, C. J., and Hughes, T. D. 1977. The Effect of Both Sample
    Size and Sampling Rate on the Statistical Fluid Flow Para-

meters in a High Reynolds Number, Low Turbulence Intensity
Flow. Proc. Fifth Symposium on Turbulence in Liquids.
Univ. of Missouri-Rolla.

Brodkey, R. S., Hershey, H. C., and Corino, E. R. 1969. Turbulent Measurements
in Liquids. Symp. Turbulence in Liquids., Univ. Missouri-Rolla (ed.
J. L. Zakin and G. Patterson), p. 129.

Durão, D. F. G., and Whitelaw, J. H. 1975. The Influence of Sampling Procedures
on Velocity Bias in Turbulent Flows. Proceedings of LDA Symposium
Copenhagen, pp. 138-149, P. O. Box 70, DK-2740 Skovlunde, Denmark.

Durst, F., Melling, A., and Whitelaw, J. H. 1976. Principles and Practice of
Laser-Doppler Anemometry. Academic Press. N.Y.

Echols, W. H., and Young, J. A. 1963. Studies of Portable Air-Operated Aerosol
Generators. NRL Report 5929.

Engelson, M. 1971. Spectrum Analyzer Measurements, Theory and Practice.
Tektronix, Inc.

Hornkohl, J. O., Crossway, F. L., and Lennert, A. E. 1972. Signal Characteristics
and Signal Conditioning Electronics for a Vector Velocity Laser Velocimeter. pp. 396-444. The Use of the Laser Doppler Velocimeter for Flow
Measurements, Stevenson, W. H. and Thompson, H. D., ed. Proc. Workshop
on LDV, Purdue Univ.

Kreid, D. K., and Grams, G. W. 1976. Confocal LDV Utilizing a Decoupling Beam-
splitter Combiner. Applied Optics, 15, pp. 14-16.

McLaughlin, D. K., and Tiederman, W. G. 1973. Biasing Correction for Individual
Realization of Laser Anemometer Measurements in Turbulent Flow. Physics
of Fluids, 16, pp. 2082-2088.

Mazumder, M. K., Hoyle, B. D., and Kirsch, K. J. 1974. Generation and Fluid
Dynamics of Scattering Aerosols in Laser Doppler Velocimetry. Proc.
Second International Workshop on Laser Velocimetry, II, pp. 234-269,
Stevenson, W. H., and Thompson, H. D., ed.

Roshko, A. 1976. Structure of Turbulent Shear Flows: A New Look. AIAA Journal,
14, pp. 1349-1357.

Simpson, R. L. 1976. Interpreting Laser and Hot-film Anemometer Signals in
a Separating Boundary Layer. AIAA Journal, 14, pp. 124-126.

Simpson, R. L. 1977. Features of Unsteady Turbulent Boundary Layers as Revealed
from Experiments. AGARD-CP-227; paper 19.

Simpson, R. L., and Barr, P. W. 1974. Velocity Measurements in a Separating
Turbulent Boundary Layer Using Sampling Spectrum Analysis. Proc. Second
International Workshop on Laser Velocimetry, II, pp. 15-43, Stevenson,
W. H., and Thompson, H. D., ed.

Simpson, R. L., and Barr, P. W. 1975. Laser Doppler Velocimeter Signal Processing
    Using Sampling Spectrum Analysis. Rev. Sci. Inst., 46, pp. 835-837.

Simpson, R. L., Strickland, J. H., and Barr, P. W. 1977. Features of a Separat-
    ing Turbulent Boundary Layer in the Vicinity of Separation. J. Fluid
    Mech., 79, pp. 553-594.

Design of LV Experiments for Turbulence Measurements

M. C. WHIFFEN, J. C. LAU, and D. M. SMITH
Lockheed-Georgia Company
Marietta, GA

1 INTRODUCTION

The state-of-the-art in Laser Velocimetry has developed to the point where mean velocity measurements are more or less routine. However, the measurement of turbulence parameters, which are of primary importance to many experiments, is subject to bias errors, misinterpretation and other assorted problems. Resulting from a number of years of theoretical and practical investigation into the causes of these problems, this paper will present some empirical techniques that have been used to design reliable turbulence measuring experiments in both model jets and wind tunnels.

2 EVIDENCE FROM MEASUREMENTS IN A JET

In a previous attempt to validate the LV as an instrument for turbulence research, corresponding measurements were conducted in a round jet at a low subsonic Mach number (0.28) using an LV and a hot-wire anemometer[1]. Although the two sets of results showed comparable trends and the mean velocity distributions were identical, it was found that magnitudes of the turbulence intensities measured by the LV were systematically higher than those of the hot-wire, even after corrections were made to the hot-wire results. This trend was also noted in another independent study of a subsonic jet[2]. It would appear, therefore, that this is an inherent feature of the LV measurements.

Figure 1 shows the autocorrelation curve for axial velocity fluctuations obtained with the LV[3] in the potential core of the jet where very low turbulence is expected. Because of the quasisinusoidal nature of the velocity signals in this region, the correlograms, obtained with hot-wires, have always had the appearance of weakly-damped cosine curves. On this basis, it would seem that the spike at $\tau = 0$ does not reflect the true value of the autocorrelation. This could be explained if erroneous velocity samples were being recorded whose velocity deviations from the mean were not part of the turbulence being measured. A more appropriate value of the correlation coefficient at $\tau = 0$ would be indicated by the extrapolation of the sinusoidal curve back to $\tau = 0$ shown by the dashed line. The adjacent sample at $\tau = 10\mu s$ is also in error. This is caused by a coincidence time of $10\mu s$ allowed in the 4 channel processor for this series of experiments. It is, therefore, appropriate to consider extrapolation of the curve for τ less than $20\mu s$.

Figure 2 shows the autocorrelation diagram for velocity signals in the mixing region where higher turbulence levels exist. A spike is also visible at $\tau = 0$, but the magnitude of the spike is significantly smaller than that in the

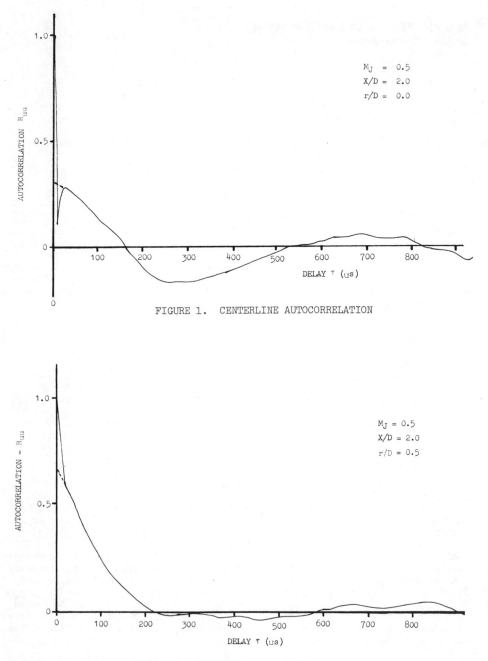

FIGURE 1.   CENTERLINE AUTOCORRELATION

FIGURE 2.   MIXING REGION AUTOCORRELATION

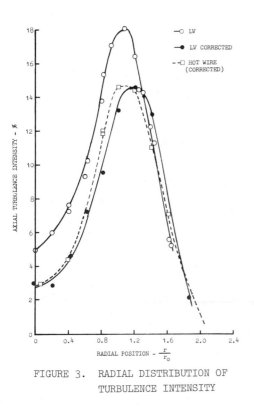

FIGURE 3.   RADIAL DISTRIBUTION OF
TURBULENCE INTENSITY

potential core. The curve at τ less
than 20μs has also been extrapolated
as shown by the dashed line.

Since the ordinate intercept of
the autocorrelation curve represents
the mean square value of the velocity
fluctuations, $\overline{u'^2}$, it would appear
reasonable to "correct" turbulence
intensity measurements by multiplying
the measured intensity ($\sqrt{u'^2}/U_J$) by
the corresponding $\sqrt{R_{uu_0}}$ shown in
Figures 1 and 2, R_{uu_0} being the
extrapolated value of correlation
coefficient (R_{uu}) at $\tau = 0$ in the
correlogram. This method of correction was applied to the radial distribution of axial turbulence intensity
obtained in a Mach 0.28 jet ($x/D = 2.0$)
as shown in Figure 3. Corresponding
hot-wire results (which have been
corrected for the tangential insensitivity of the hot-wire and the recti-
fication of the hot-wire signals[4])
are also shown. The fact that the
corrected LV results follow the
corrected hot-wire results reasonably
well indicates that the LV is indeed
measuring the flow phenomena accurately except for the additional
uncorrelated noise data mentioned
above. It would seem that this provides a viable approach to the
correction of turbulence data obtained
with LV's. The disadvantage, however,
is that a somewhat lengthy process of correlation is required to correct the data.

3   ERROR ANALYSIS

The first step to investigating the source of these erroneous turbulence
intensity measurements was a detailed examination of the LV processor. The
system currently used at Lockheed-Georgia is the 5th generation of a processor
first designed several years ago[5]. It is arranged in a four channel configuration which allows the accumulation of data from a pair of independent two-
vector optics systems from which spatial correlation information may be derived.
It is a burst counter system with threshold level triggering and a zero crossing
detector. Two sources of error have been identified as those due to signal-to-
noise ratio and digital resolution. The measured turbulence intensity is,
therefore, the sum of the actual turbulence and the contributions of these two
errors.

3.1   Signal-to-Noise Ratio   (SNR)

Mayo has derived[6] an estimate of the apparent RMS fluctuation in a zero
crossing detector as a function of SNR: $E_n = 1/2\pi c\sqrt{SNR}$ where c is number of
zero crossings. He assumed band limited noise, stationary processes, a reasonable SNR and sufficient signal validation to eliminate signals with dropped cycles.

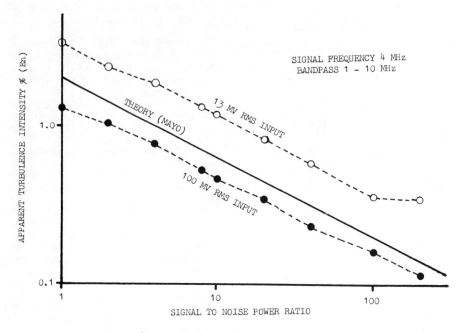

FIGURE 4.   APPARENT TURBULENCE DUE TO PMT NOISE

This estimate is plotted in Figure 4 along with measurements taken from the processor. The measurements were obtained by feeding a known signal to a Light Emitting Diode (LED) and allowing it to excite a photomultiplier tube (PMT). The LV processor which was used to measure and record the fluctuation error produced more error at maximum sensitivity (13 MV RMS) than at lower sensitivity (>100 MV). It is seen that the actual performance is in general agreement with Mayo's theoretical curve even at low SNR. The apparent improvement of the measured error (>100 MV RMS) over the theoretical is attributed to the additional signal validation afforded by the processor.

Figure 4 represents the contribution of PMT noise to the measurement executed by the LV. It in effect dictates the minimum turbulence which the LV can measure. There is a temptation to remove this contribution by assuming an RMS addition of this apparent turbulence to the actual flow turbulence. However, such a technique would require the assumption that there is one representative SNR for all particles passing through the measurement volume regardless of their size, angle of approach and point of entry into the volume.

Clearly, since the signal level varies from particle to particle the SNR will also vary and a reliable quantitative determination of SNR cannot be made. However, the average minimum turbulence intensity measurable in a particular system/experimental configuration can be estimated.

3.2 Digital Resolution

Another contribution to the measured turbulence is the limited digital resolution of the measurement. An estimate of this error can be based on a period count accuracy of ±1 count. If equal probabilities are assumed for the three counts, i.e., N-1, N & N+1, where N is the actual count, the equivalent standard deviation of the error, E_d can be derived:

$$E_d = \sqrt{.6667}/N$$

In our case N is directly related to the reciprocal of the processor input frequency f_{in}:

$$N = \frac{1}{K\,f_{in}}$$

Since f_{in} is proportional to the sum of the effective fringe velocity, V_f, (induced by the use of a Bragg Cell) and the particle velocity, V_p, the 8 fringe period count is given by

$$N = \frac{1}{K'(V_f + V_p)}$$

and

$$E_d = K' \sqrt{.667} \, (V_f + V_p)$$

the equivalent turbulence (in FPS), σ_d, is:

$$\sigma_d = E_d \, (V_f + V_p) = K' \sqrt{.667} \, (V_f + V_p)^2 \tag{1}$$

where K' is the constant of proportionality.

3.3 Estimation of Error Contributions

By proper design of the experiment it is possible to minimize the effect of biases leaving these three contributions to the measured turbulence: actual flow turbulence, σ_A, signal broadening due to PMT noise, σ_n, and digital resolution, σ_d. Assuming each process is stationary, independently random, and Gaussian (all of which may be argued) we can conclude that our measured turbulence σ_m will be the RMS sum of the three contributions:

$$\sigma_m = \sqrt{\sigma_A^2 + \sigma_n^2 + \sigma_d^2}$$

where

$$\sigma_n = E_n \, (V_f + V_p) \tag{2}$$

These effects may be seen graphically in Figure 5. The measured turbulence from a constant velocity, low turbulence flow is plotted as a function of the total input velocity, $V_f + V_p$. The flow velocity V_p was constant at about 51.6 fps and the hot wire measured turbulence was about 1.35%. A PMT noise level was chosen to allow the RMS sum to fit the measured data. The resulting PMT noise level is about .35% which, from Figure 4, represents a SNR of 20 or 30. Again there is a temptation to use this approach to correct for the effects of PMT noise and digital resolution and thereby obtain a more accurate measurement. The problem is, however, that the PMT noise does not appear to be constant for the duration necessary to accomplish an experiment. In addition to the particle size distribution discussed above, laser stability and PMT stability also affect SNR and tend to change as a function of temperature, humidity and time.

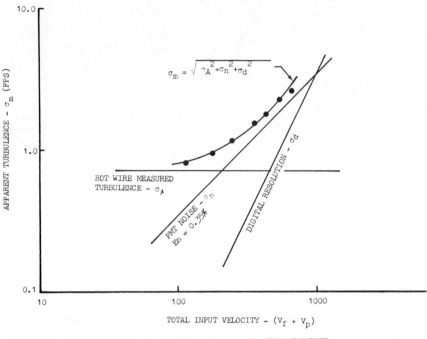

FIGURE 5.   APPARENT TURBULENCE MEASUREMENT

4   DESIGN OF EXPERIMENTS

From Figure 5 it appears that since the apparent turbulence converges asymptotically to the actual level at low input velocities, the optimum operating region for the measurement of turbulence is with minimum fringe velocity. However, the actual operating conditions are also affected by the ability of the measurement volume to detect all incident particles and the frequency limitations of the electronics. The following paragraphs present a summary of data designed to enable an experiment to be planned with predictable characteristics of angular particle capture and error contributions.

4.1 Polar Response of the Measurement Volume

It is well known that an LV system without a velocity offset is not capable of resolving the direction of flow since the particle generates the same frequency burst regardless of the direction it passes through the measurement volume. In addition, it is also obvious that a particle whose path is nearly parallel to a set of fringes would not cross the eight fringes necessary for a velocity measurement. In general, the probability of detection of a particle passing through a fringe pattern varies as some function not only of the system parameters, but also of its angle of approach to the measurement volume[7]. In an effort to describe this probability of particle detection, PPD, the polar response of a measurement volume has been derived in Appendix A as:

$$PPD(\alpha) = \sqrt{1-\left(\frac{N/N_T}{\cos\alpha + V_f/V_P}\right)^2} \qquad (3)$$

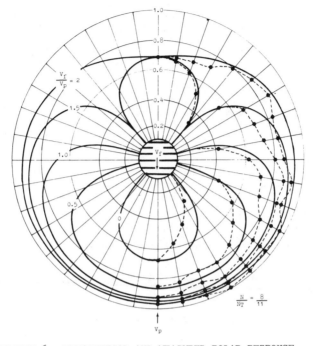

FIGURE 6. THEORETICAL AND MEASURED POLAR RESPONSE

Figure 6 is a plot of the PPD as calculated from Equation (3). On the right side of the figure is a plot of the corresponding data measured from the LV system by rotating the fringe pattern and changing the effective fringe velocity. The data were taken in a relatively low turbulence flow at about 60 fps. They represent the ratio of accepted particles to incident particles as measured at the processor. Using this ratio the value of N/N_T was calculated and the measured curves were then normalized at $\alpha = 0°$ and $V_f/V_p = 0$. Although small variations in the measured response due to anomolies in the fringe patterns can be seen, the overall agreement is good for practical applications. Note also that the increasing axial length of the lobes as V_f/V_p increases reflects an increasing probability of particle capture which represents an increase in the effective measurement volume cross-section. This means higher data rates as well as the isotropic response.

4.2 Velocity Range Chart

Figure 7 shows a chart summarizing operating regimes of the Lockheed LV system which has been successfully used to plan experiments. It summarizes the effects of processor frequency limitations, and measurement volume polar response upon the calculation capabilities of the instrument. The horizontal axis represents the velocity of the measured particles and the vertical axis the effective fringe velocity introduced by Bragg cell frequency shifting and local oscillator mixing. It is well known that a combination of these two can produce any effective fringe velocity. The chart provides a technique for selecting a fringe velocity needed to properly measure particles in a given velocity range. For example, data from a 800 fps flow with 15% turbulence intensity would have particles in the range 400 to 1160 fps (3σ bounds). The chart indicates that with a fringe velocity of about 300 fps, these could be measured with a capture angle varying from 100° at the high velocities to 360° at low. The erroneous turbulence due to SNR would range from under 0.75% at the high end to 1.5% at low velocities and that due to digital resolution would always be less than 0.5%. In order to function at higher capture angles, increased fringe velocity would be required at the cost of greater error due to SNR and digital resolution.

The curves plotted are an accumulation of ordinary processor characteristics, the polar response chart of Section 4.1 and the expressions for σ_n and σ_d of Sections 3.1 and 3.2. The diagonal lines of constant $(V_f + V_p)$ indicate processor frequency characteristics. Limitations on this are "pedestal removal" filters at the low end and the effective functioning of validation circuits at the high end. The radial dashed lines showing V_f/V_p reflect two different data biases. Firstly, the diagram indicates the polar response characteristic curve from Figure 6, and the estimated "capture angle" to be expected. Secondly, they provide an estimate of the error due to SNR found by dividing Equation (2) by V_p:

$$\frac{\sigma_n}{V_p} = E_n \left(1 + V_f/V_p\right)$$

where E_n, the effective SNR of the system, was chosen to be 10 for the purpose of illustration.

The curved lines show the expected error due to digital resolution found by solving Equation (1) for σ_d/V_p:

$$\frac{\sigma_d}{V_p} = \frac{K'}{V_p} \sqrt{.6667} \, (V_f + V_p)^2$$

4.3 Experimental Evidence

Experimental confirmation of a set of data may be obtained by plotting the mean particle velocity measured against fringe velocity on Figure 7 as the fringe velocity is varied. This has been done for two sets of data taken in the turbulent mixing layer of Mach 0.5 and 0.9 jets at locations where the local turbulence intensity is the same for both cases.

If all the measurements were unbiased, the locus of measurements would be a vertical line through the true particle velocity. However, if one or more of the bias phenomena are affecting the measurement, the locus will tend to be altered in the direction of the bias. As the fringe velocity was decreased, one may observe that little variation in mean velocity is seen until a point near a V_f/V_p ratio of 1 is reached. Thereafter, the slope becomes negative and successively higher mean velocities are recorded. Since the inflection points occur at the same value of V_f/V_p, this shows that the phenomena are related to the V_f/V_p ratio. The bias indicated by this inflection could be due either to polar response or to PMT noise. Since the bias increases at lower fringe velocities and yields the correct particle velocity at higher fringe velocities, it is concluded that the effect is due to polar response. (The bias error would occur at higher fringe velocities if it were due to PMT noise.) Consideration of the jet flow mechanics at the measurement point support this conjecture since a significant contribution of the flow is at high angles to the measurement vector. Turbulence intensity measurements also show the same inflection tending to lower intensity at lower fringe velocities suggesting that the lower vector velocity particles are not being detected when the fringe velocity is reduced.

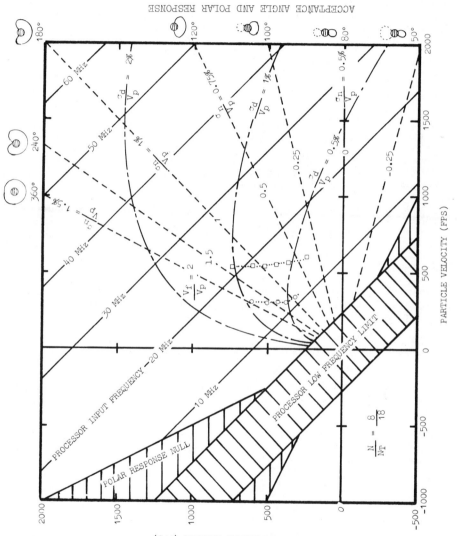

FIGURE 7. OPERATING RANGE OF LV

5   CONCLUSIONS

In planning LV experiments for turbulence measurements, the major contributors to turbulence bias errors are processor frequency limitations, measurement volume polar response, PMT signal-to-noise ratio and digital measurement resolution. In the Lockheed LV System the major controlling influence available to the experimenter is the apparent fringe velocity. By identifying the sources of these bias errors, this paper has presented a practical technique for selecting fringe velocities which minimize the effects of these biases and to some extent quantifying the remaining errors. It also provides guidelines for technical development of LV systems which will be able to make accurate measurements in very low-turbulence flows.

6   REFERENCES

1. Lau, J. C., Morris, P. J., and Fisher, M. J. 1976. Turbulence Measurement in Subsonic and Supersonic Jet Using a Laser Velocimeter. AIAA Paper 76-348.

2. Barnett, D. O. and Giel, T. V. Application of a Two-Component Bragg-Diffracted Laser Velocimeter to Turbulence Measurements in a Subsonic Jet. Arnold Engineering Development Center, AEDC TR-76-36.

3. Smith, D. M. and Meadows, D. M. 1974. Power Spectra from Random-Time Samples for Turbulence Measurements with a Laser Velocimeter. Proceedings of the Workshop on Laser Velocimetry, Purdue Univeristy.

4. Tutu, N. K. and Chevray, R. Cross-Wire Anemometry in High Intensity Turbulence. *Journal Fluid Mechanics,* 71, 785-800.

5. Whiffen, M. C. and Meadows, D. M. 1974. Two Axis, Single Particle Laser Velocimeter System for Turbulence Spectral Analysis. Proceedings of the Workshop on Laser Velocimetry. Purdue University.

6. Mayo, W. T., Jr. Spectron Development Corp., Private Communication. To be published.. Final report Naval Underwater Systems Center. Contract N00140-77-C-6670.

7. Whiffen, M. C. 1975. Polar Response of an LV Measurement Volume. Proceedings of Minnesota Symposium on Laser Velocimetry. University of Minnesota.

APPENDIX A

POLAR RESPONSE CALCULATION

Assuming a homogeneous distribution of particles within the flow, we may express the probability of particle detection as a function of the effective cross-sectional area of the measurement volume as seen by the approaching particles. Only the two-dimensional case need be considered for the LV optics being used for this study, since the measurement volume is described by the intersection of the optical paths of the transmitting and receiving optics. This may be approximated by two cylinders intersecting at a 30° angle and is, there-fore, essentially cylindrical throughout most of its *useful* volume. We consider only the particle trajectory as projected onto the circular plane normal to the LV optical axis. Since the length-to-diameter ratio of the measurement volume is more than three, the change in detection probability due to velocity compon-ents perpendicular to this plane is small compared to those due to components in the plane. A particle's probability of detection is, therefore, assumed to be limited entirely by its velocity component in the plane normal to the optical axis.

Since the axial length of the measurement volume is considered constant, the effective cross-sectional area varies in proportion to the effective width, W, which is a function of the number of fringes required for a measurement, N, the total number of fringes in the measurement cross section, N_T, the spacing of the fringes, S, and angle of incidence of the particle. Since we are interested in the probability of particle detection, all areas are normalized by the maximum possible area, i.e., the apparent diameter of the measurement volume.

Consider a particle approaching the measurement volume from some angle, α, as in Figure A-1. As long as the particle has some velocity relative to the fringes it will scatter light bursts whose frequency is the input frequency to the processor, f_p. In the burst detector system the particle must cross at least N fringes to perform a measurement so the minimum time the particle must be in the volume, (transit time) is $t_{min} = N/f_p$

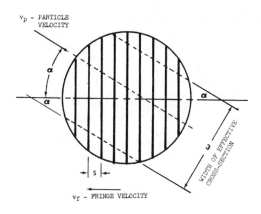

FIGURE A-1.   MEASUREMENT VOLUME

CROSS-SECTION

but $f_p = (V_p \cos\alpha + V_f)/S$

so $t_{min} = NS/(V_p \cos\alpha + V_f)$ (A-1)

Since the transit time, t, may also be expressed as a function of the volume geometry:

$$t = \sqrt{(N_T S)^2 - W^2}/V_p \qquad (A-2)$$

we derive the probability of particle detection, PPD, by solving equations (A-1) and (A-2) for W:

$$W = N_T S \sqrt{1 - \left(\frac{N/N_T}{\cos\alpha + V_f/V_p}\right)^2}$$

and by normalizing the volume diam-eter, $N_T S$, we obtain the desired expression

$$PPD(\alpha) = \sqrt{1 - \left(\frac{N/N_T}{\cos\alpha + V_f/V_p}\right)^2}$$

Laser Doppler Velocity Measurements in Subsonic, Transonic, and Supersonic Turbulent Boundary Layers

PAUL E. DIMOTAKIS
California Institute of Technology
Pasadena, CA 91125

DONALD J. COLLINS
Jet Propulsion Laboratory
Pasadena, CA 91103

DANIEL B. LANG
California Institute of Technology
Pasadena, CA 91125

INTRODUCTION

Recent measurements by Johnson and Rose (1973, 1975), Yanta and Lee (1974, 1976), and Abbiss (1976) have used laser Doppler velocimetry techniques to make direct measurements of the Reynolds stress in turbulent boundary layers in the Mach number range of 1.5 to 3.0. A serious anomaly, however, is exhibited by these measurements in that the maximum of $-u'v'$ occurs much further from the wall than is reasonable for flow at constant pressure. The anomaly has been discussed by Sandborn (1974), who supports the conjecture by some of the authors cited that density fluctuations may contribute substantially to the turbulent stresses near the wall. This conjecture is in direct opposition to the conclusion by Morkovin (1961) that effects of density fluctuations should be small compared to effects of variations in mean density for Mach numbers up to 4 or 5.

The purpose of the present experiments is to obtain redundant data over a substantial range of Mach numbers (M = 0.1 to 2.2), in an effort to resolve the anomaly in turbulent shearing stress. The flow was documented by conventional means using a Pitot tube, which was traversed through the boundary layer, to measure the mean flow. In addition, surface-friction measurements were made using both a floating-element balance and a Preston tube. The mean-flow scaling suggested by Van Driest was applied to the data, to test the adequacy of a single similarity formulation for both compressible and incompressible flow. The shearing stress was then computed from the mean flow as part of the analysis. These results were documented by Collins, Coles and Hicks (1978).

In the work presented here, the instrumentation, data acquisition and analysis for laser Doppler velocity measurements in the same flow are discussed. Measurements of u, v, u'^2, v'^2 and the Reynolds stress $u'v'$ were made over the full Mach number range.

The low speed (M $\sim$ 0.1) measurements were carried out in the GALCIT[*] Merrill wind tunnel. The high speed measurements (0.6 $\leqslant$ M $\leqslant$ 2.2) were conducted in the JPL[†] 20-inch wind tunnel.

The present paper is a preliminary report of a more extensive discussion of this work to appear by Dimotakis, Collins and Lang (1978).

HIGH SPEED LASER DOPPLER VELOCIMETRY

The first obvious problem that arises in making high speed laser

---

[*]Graduate Aeronautical Laboratories of the California Institute of Technology
[†]Jet Propulsion Laboratory

Doppler velocity measurements is that the resulting signal frequencies are very high. In principle, it would appear that this problem could be controlled by increasing the fringe spacing. In practice, however, conflicting requirements of a minimum number of Doppler cycles needed by the processing electronics, coupled with the spatial resolution considerations inescapably dictate the higher frequencies.

The restrictions on signal frequencies, above and beyond the problem of handling the high frequencies per se, arise for several reasons. The relative accuracy $\delta u/u$, with which the velocity of a single particle can be measured, is never less than the ratio of the processor clock period τ_c, to the total flight time Δt that is used for the measurement, i.e.

$$\delta u/u \gtrsim \tau_c/\Delta t . \tag{1}$$

In addition to these considerations which limit the accuracy even if the measurement environment was noise free, other factors become important at high velocities which decrease the signal-to-noise ratio of a single reading. As the flow velocities become higher, we are forced to use smaller particles to minimize problems of particle lag with rapidly decreasing scattering cross-sections and resulting signal intensities. In addition, as the velocity increases, the scattering particle spends less time in the focal volume. Even though the number of scattered photons per second remains constant (proportional to scattering cross-section), fewer total photons are collected by the detecting optics.

The extent to which the measurement of particle velocity represents the fluid velocity is, of course, a separate issue. Small particles can generally be expected to track the flow quite well if the (Lagrangian) frequency of fluctuations in the fluid velocity is less than the reciprocal of a characteristic particle response time τ_p. If the flow relative to the particle can be described by Stokes flow we have

$$\tau_p = (d_p^2/18)(\rho_p/\mu) = (d_p^2/18\nu)(\rho_p/\rho_f), \tag{2}$$

where d_p is the particle diameter, ρ_p and ρ_f are the particle and fluid densities, respectively, and μ is the (absolute) viscosity. It can be argued, and substantiated from direct spectral measurements (Klebanoff 1955, Perry and Abell 1975, 1977), that the expected fluctuation frequency in the interior of the boundary layer should be given by

$$\langle\omega\rangle \simeq \text{const}[U(y)/y], \tag{3}$$

where $U(y)$ is the local streamwise mean velocity and where the constant is of the order of unity. Consequently, the requirement for good particle tracking in a turbulent boundary layer becomes

$$\tau_p[U(y)/y] \lesssim 1 . \tag{4}$$

By way of example, using a dibutyl phthalate aerosol, a flow of $M \sim 1$ requires particles $\lesssim 1.5$ µm diameter to track the fluctuations at $y/\delta \gtrsim 0.1$. For a more detailed discussion, see Dimotakis, Collins and Lang (1978).

SCATTERING PARTICLES

The particle generator for the present experiments utilized a Laskin nozzle type construction to generate a polydisperse aerosol of dibutyl phthalate. Particles greater than 1 µm in diameter were effectively removed by an impact plate incorporated in the design. The resulting particle size distribution, measured using both a cascade impactor and a multichannel particle analyzer, is shown in figure 1.

Two different methods were used to introduce particles into the flow depending on the flow facility that was used. For the high speed measurements ($0.6 \lesssim M \lesssim 2.2$), which were conducted in the JPL 20" wind tunnel, it was necessary to seed the flow by introducing the particles into the settling chamber between the last turbulence screen and the contraction section. The particles were introduced through a tube which protruded 5 cm into the settling chamber. Holes were drilled into the tube along the stagnation line as well as the rear side at ±30° with respect to the flow vector.

For the low speed measurements in the Merrill wind tunnel ($M \simeq 0.1$), although some naturally occurring particles were present,

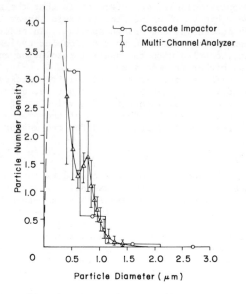

Fig. 1. Particle size distribution

the flow was seeded using the same aerosol generator in order to increase the data rate and to better control the scattering particle size distribution. For these measurements, the particles were injected through a tube spanning the test section located *downstream* of the boundary layer plate. The particles were convected around the wind tunnel circuit and over a period of a few minutes filled the whole tunnel. In practice, balancing the rate at which particles were introduced with the rate that they were lost, in order to keep the particle number density approximately constant in time, proved very difficult because the characteristic time lag was of the order of several tens of minutes. As a consequence, for the Merrill tunnel measurements the particle number density varied during the time required for the traverse through the boundary layer.

OPTICS AND MEASUREMENT GEOMETRY

The present experiments utilized the laser Doppler velocimeter in the single particle, dual (forward) scatter mode. For measurements in the JPL 20" wind tunnel, the instrument was mounted on a vibration isolated traversing mechanism on top of the wind tunnel test section, as depicted in figure 2. For measurements in the GALCIT Merrill wind tunnel, the traverse was suspended from the ceiling of the laboratory.

The light source for this instrument was a Coherent Radiation Model 52B, 4 watt Argon ion laser. The laser was operated single line at 0.5145 µm and etalon stabilized to provide a single mode beam. It was mounted on the tubular support structure for the optics as shown in figure 2.

The laser beam is directed into the transmitting optics cell where it is split into three parallel beams, of approximately equal intensity, which form a right isosceles triangle, whose base is nominally parallel to the wall. The three output beams are then focused in the center of the tunnel by a 1 m focal length lens, to a common intersection volume ∿ 0.8 mm in diameter. The optical axis intercepts the test plate from below at an angle of ∿ 1:100 in order to

permit measurement of the
boundary layer down to
the surface of the pol-
ished test plate.

The resulting inter-
section volume contains
three independent sets of
(virtual) fringe planes
correspondingly perpen-
dicular to the $\underline{u}$,
$(\underline{u} + \underline{v})/\sqrt{2}$ and $(\underline{u} - \underline{v})/\sqrt{2}$
velocity vectors (see
figure 3). By selective-
ly blocking one of the
three beams, any one of
the three velocity compo-
nents can be examined
without a change in the
focal volume geometry.

The optical axis for
the receiving optics is
aligned at an angle of $\sim$
7.5° with respect to the
test plate surface. This

Fig. 2. Measurement assembly

allows viewing the focal volume down to the test plate surface and, recalling
that the transmitting optics axis is in turn inclined with respect to the test
plate, reducing the spanwise extent of the focal volume. The collected light
cone is focused by a 1:1 imaging system into a 0.05 cm diameter pinhole which
spatially filters the collected light accepted by the photomultiplier receiving
optical assembly. The effective measurement volume diameter is defined by the
pinhole in the receiving optics and therefore is equal to $\sim$ 0.05 cm.

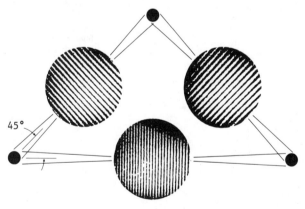

Fig. 3. Measurement fringes

SIGNAL PROCESSING AND DATA ACQUISITION

The output of the photomultiplier was amplified by a direct coupled amplifier installed at the base of the photomultiplier housing. That output was
in turn amplified by a low-noise, wide bandwidth, AC coupled preamplifier.
The resulting signal was low-pass filtered to remove the substantial shot
noise above the Doppler frequencies in each case. *Note that the pedestal, resulting from the dual scatter optics, is not removed by this processing scheme.*

The filtered signal is fed into a counter type processor (Dimotakis and Lang 1974, 1977, and Dimotakis, Collins and Lang 1978).

The processor uses a four level comparison scheme on every cycle of the Doppler burst. A TTL signal DIN is defined whose negative slope is driven by the crossing of the reference level V_o from above by the analog input (see figure 4a). DIN can be prescaled by $m = 1,2,4,8$ to yield DIN' which is used by the timing circuitry. DIN' is subsequently checked to ensure that every period is between two limits, τ_1 and $\tau_1 + \tau_2$. If τ was the period of the Doppler signal, we must have, as a consequence,

$$\tau_1 < m\tau < \tau_1 + \tau_2 . \qquad (5)$$

The time intervals τ_1 and τ_2 are front panel selectable and cover the range 60 nsec $< \tau_1,\tau_2 <$ 3 msec. The processor updates the flight time register with the time between the first negative slope of DIN' and every successive acceptable negative slope of DIN' using a 100 MHz ($\tau_c = 10$ ns) crystal. Every negative slope is also recorded in the cycle counter.

The data were recorded in two different modes. In the first mode (fixed fringe mode), the output corresponds to the (truncated) flight time for a fixed number of fringes. All the data for $M \geqslant 0.6$ (JPL 20" wind tunnel) were recorded in this mode. Eight cycles were counted for the beam pair at 0° (u_1 data) with the prescaler m set to 2. Four cycles were counted for the ±45° (u_2, u_3 data) beam pairs with the prescaler set to unity.

In the second mode (free fringe mode), the processor only requires that a fixed number of cycles be *reached or exceeded by the burst* but records the flight time and number of cycles (fringes) *of the whole* burst. Both n and (the truncated) Δt were recorded in this mode and the velocity is

$$u_i = ns/\Delta t, \qquad (6)$$

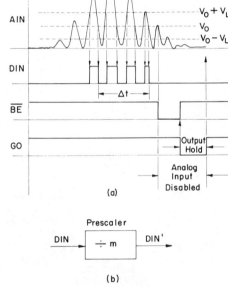

(b)

Fig. 4. Signal burst processing

where s is the fringe spacing. The data recorded at $M \sim 0.1$ (GALCIT Merrill wind tunnel) were acquired in this mode requiring at least four fringes for the ±45° data (u_2, u_3) and at least eight for the 0° data (u_1) with the prescaler m set to 2. Thus, in both modes, the same (minimum) number of fringes had to be crossed.

There are several important differences between the two modes. First, using the facility of the processor to recognize the end of the burst while retaining the previous valid measurement, the total flight time can be used, instead of an arbitrary fraction, in determining the velocity component of interest. This improves the measurement accuracy in an obvious way. Second, using the whole burst cancels the phase errors that are made in assigning equal phase to the crossings of the reference level V_o by the signal. Third, the sampling statistics are different between the two modes. In the first mode (fixed number of fringes), all we know is that the particle crossed *at least* that many. In the second mode (number of fringes unrestricted provided it exceeds a certain minimum) we know that the particle crossed *exactly* that many. That alters the sampling bias, as will be discussed later on.

The data were formatted into records of 1,024 particle measurements each and written on a Kennedy 9100 digital tape deck. Several records were recorded at each measurement location. Data rates up to 50 KHz were observed in the high speed flow.

DATA PROCESSING

In order to process the high speed data $(0.6 \leqslant M \leqslant 2.2)$, which were recorded in the fixed fringe mode, a histogram *of each record* (1,024 measurements) was formed as a function of the flight time Δt. The histograms were pruned in two passes. First, the data in any bin that contained only one count and did not have neighboring bins with more than one count were discarded. In the second pass, any data isolated from the main body of the histogram by at least one zero were also discarded. If, as a result of these two operations, more than 24 measurements out of the total of 1,024 had to be rejected, the whole record was rejected.

For the low speed data $(M \simeq 0.1)$, which were recorded in the free fringe mode and for which the actual number of flight time clock counts was large, an alternative scheme was devised. The data were sorted in velocity bins as follows. The velocity of each particle was computed as an integer percentage of the maximum velocity that the processor settings admitted, i.e.

$$I_i = 100\tau_1 n_i/\Delta t_i, \tag{7}$$

where τ_1 is the minimum Doppler period setting, n_i is the number of fringes crossed by the ith particle and Δt_i is the time of flight corresponding to the n_i fringes. The histogram was formed based on the velocity index I_i and each record was pruned in the fashion described above.

The complexity of processing single particle laser Doppler velocity data is compounded by the fact that in such measurements the fluid velocity is sampled in a biased manner. This problem, first pointed out by McLaughlin and Tiederman (1973), necessitates that the unbiased expectation value of any function of the velocity u, be computed consistently with this bias, i.e.

$$< f(\underline{u}) > = \sum_i f(\underline{u_i})B^{-1}(\underline{u_i}) \ / \sum_i B^{-1}(\underline{u_i}) \tag{8}$$

where $B(\underline{u})$ is proportional to the probability per unit time of making the particular measurement. The dependence of the sampling bias function B on the local velocity vector $\underline{u}$, as well as the shape of the measurement volume and the minimum number of fringe crossings required by the processing electronics, has been derived elsewhere (Dimotakis 1976). If the angle θ between the intersecting beams is small then, for the fixed fringe mode measurements, the expression for $B(\underline{u})$ simplifies to

$$B_{fixed}(\underline{u};\varepsilon) \simeq u_\perp (1 + u_\parallel^2/u_\perp^2)^{\frac{1}{2}}[1 - \varepsilon^2(1 + u_\parallel^2/u_\perp^2)], \tag{9}$$

where $u_\perp$ is the velocity component perpendicular to the fringe planes, $u_\parallel$ is the velocity component parallel to the fringe planes and perpendicular to the long axis of the measurement volume (beam bisector), and ϵ is the ratio n_{min}/n_T of the minimum number of fringe crossings (Doppler cycles) required by the processor to the total number (maximum) in the measurement volume. If $(u_\parallel/u_\perp)^2 \ll 1$ equation 9 reduces to the familiar form

$$B^{-1}_{fixed}(\underline{u}) \propto u_\perp^{-1} \propto (k + \tfrac{1}{2}), \tag{10}$$

where k is the (integer) output of the flight counter.

In the free fringe mode, the bias function is given by

$$B_{free}(\underline{u};n) \propto u_\perp^{-2}(u_\perp^2 + u_\parallel^2)^{3/2}(n + \tfrac{1}{2}), \tag{11}$$

or, equivalently,

$$B^{-1}_{free}(k;n) \propto (k + \tfrac{1}{2})n^2/(n + \tfrac{1}{2}). \tag{12}$$

See Dimotakis (1976) and Dimotakis, Collins and Lang (1978) for derivations and a more extensive discussion. Note that equation 12 is in disagreement with the predictions of Hoesel and Rodi (1977) who suggest that $B^{-1}_{free} \propto \Delta t \propto (k + \tfrac{1}{2})$.

RESULTS AND DISCUSSION

The mean streamwise velocity profiles are compared to the corresponding Pitot tube measurements in figures 5 and 6. The agreement can be seen to be quite good. The momentum thickness θ for the profiles was computed from the laser Doppler measurements. In figure 5 four profiles in the Mach number range $0.6 \leqslant M \leqslant 2.2$ are plotted as indicated. The agreement for the lower speed

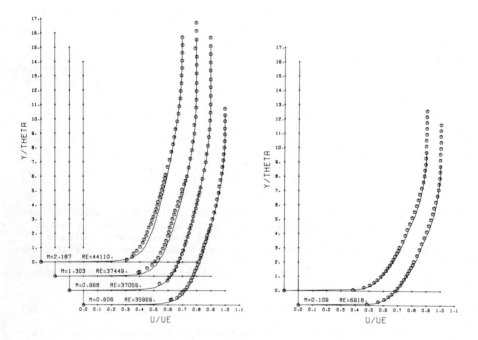

Fig. 5. $0.6 \leqslant M \leqslant 2.2$ velocity profiles   Fig. 6. $M \sim 0.1$ velocity profile

profiles is very good. We do not have an explanation for the small discrepancies that can be seen in the higher Mach number profiles. The top data points
in figure 6 correspond to a computation of the mean velocities using the harmonic mean (bias function of equation 10), the bottom points correspond to the
more exact bias compensation of equation 12. The differences are small as are
also the differences between these and mean values computed with no compensation
for the bias (not plotted). In figures 7 and 8 the root mean square fluctuations, normalized by the edge velocity, are plotted for the same flow. The
upper set of points in figure 8 are computed using equation 10 for the bias.
Again the differences between the two are small. The filled squares in figures

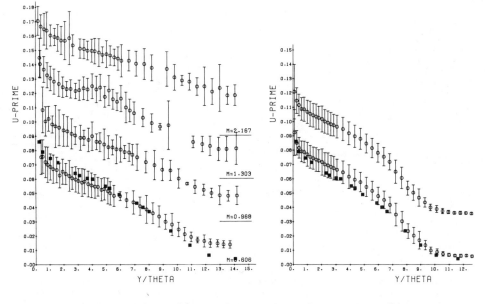

Fig. 7. $0.6 \leqslant M \leqslant 2.2$ u'/U_e profiles   Fig. 8. $M \sim 0.1$ u'/U_e profile

7 and 8 are taken from Klebanoff (1955). The apparent fluctuation level in
the free stream for the higher Mach numbers is largely the result of the
finite clock period (see equation 1), an effect that was not removed from the
data. The error bars are computed on the basis of estimates from different
records. The differences between the present data and the Klebanoff (1955)
data are within our confidence limits. The normalized Reynolds stress profile,
$-\rho_w \overline{u'v'}/\tau_w$, measurements are presented in figures 9 and 10. The solid lines
represent the Reynolds stress computed from the corresponding mean velocity
profiles by integrating the boundary layer equations in the manner described
in Collins, Coles and Hicks (1978). The filled squares are taken from
Klebanoff (1955). The upper points in figure 10 are computed on the basis of
equation 10 for the bias.
        Several questions can be answered immediately by the present data.
First, the anomalies reported in previous supersonic boundary layer measurements of the Reynolds stress are not a consequence of compressibility effects.
This can be ascertained directly from figure 9 in which no systematic variation with Mach number of the Reynolds stress measurements in the vicinity of
the wall can be seen. *Second,* the difficulty which does exist for $y/\theta \leqslant 2$,
cannot be associated with local particle lag, in the context of equation 4,

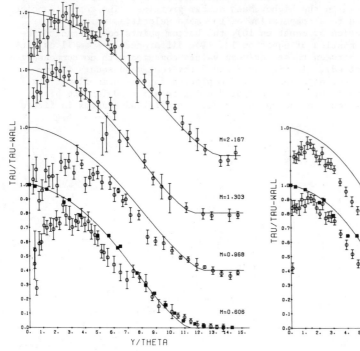

Fig. 9. $0.6 \leqslant M \leqslant 2.2$ $-\overline{u'v'}$ profiles   Fig. 10. $M \sim 0.1$ $-\overline{u'v'}$ profile

since the deviation from the expected behavior would have occurred at increasing distances from the wall as the Mach number increased (the boundary layer thickness was roughly 2.5 cm for all the measured profiles). *Third,* the previously reported anomalies cannot be attributed to the sampling bias. The differences between the correctly computed Reynolds stress data (lower data points of figure 10) and the simpler compensation by means of equation 10 are small, even though the bias function of equation 12 yields results that are closer to the expected values.

Except for the $M \sim 0.1$ stress measurements, the agreement between the computed stress profiles and the directly measured values is quite good, for $y/\theta \gtrsim 2$. The discrepancy near the wall, however, remains an open question as of this writing. The problem, we feel, is related to the behavior of the stress producing flow in the neighborhood of the wall. It is a common observation that the walls of the tunnel become coated with a thin film of the aerosol used for seeding. This indicates that the flow inside the viscous sublayer is such that a particle that enters the sublayer has a very low probability of leaving. The scaling law for particle behavior in a turbulent boundary layer, as expressed by equation 4, is not applicable in the viscous sublayer. A particle entering the viscous sublayer becomes caught in the motion of the longitudinal vortices, whose transverse extent is approximately $20 \, \nu/u_\tau$, and ends up on the wall. Consequently, a fluid element coming *from* the wall is less likely to carry particles than a fluid element moving *towards* the wall. This causes several problems that show up in different measurements.

The scattering particle number density profile for the JPL data $(0.6 \leqslant M \leqslant 2.2)$ should look approximately like a half Gaussian, resembling the concentration of a passive contaminant introduced far upstream at the wall. The recorded data rate was divided by the local streamwise velocity and plotted in figure 11. To the extent that correlations between the scattering particle number density fluctuations and the *streamwise* velocity can be neglected, the ratio is proportional to the local particle number density. The decrease as the wall is approached, extending to a large fraction of the boundary layer thickness, is quite conspicuous. The velocity component perpendicular to the wall is measured to be *negative* in the vicinity of the wall, a consequence of the fact that the positive half of the probability density of that quantity is underrepresented in the laser Doppler measurements. The underrepresentation of that motion also results in a lower measured value for the $-u'v'$ correlation since the *upswelling* carries fluid with *lower* streamwise velocity. This motion (bursting), which has been observed to be intermittent and quite violent, is held responsible for a very large fraction of the total stress near the wall (Blackwelder and Kaplan 1976).

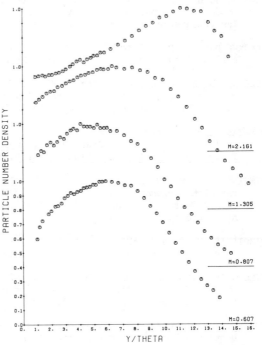

Fig. 11. Particle number density profiles

It appears that the particle transport to the wall via the viscous layer is so effective, even at the lower velocities, as to make the wall look like an infinite sink for particles. *The conclusion is then that the discrepancy in the measured Reynolds stress near the wall arises because the scattering particle number density is very strongly correlated with the instantaneous stress and not as a result of any local failure of the particles to track the flow.*

ACKNOWLEDGMENTS

We are indebted to Professor Donald Coles of the California Institute of Technology for his assistance in this project. We would also like to acknowledge the early participation of Professor Steven Barker, presently at UCLA, who contributed substantially to the final optics design.

This paper presents the results of one phase of research carried out at the Jet Propulsion Laboratory under Contract NAS 7-100 and at the Merrill Wind Tunnel of the California Institute of Technology. The work was supported by the United States Air Force, Office of Scientific Research, under contracts F44620-75-C-0007 and F44620-76-C-0046; by the Arnold Engineering Development Center, under contracts EY 7483-76-0003 and EY 7483-76-0009; and by the

California Institute of Technology, President's Fund, under Grant PF-075.

REFERENCES

ABBISS, J. B.   (1976) *Development of photon correlation anemometry for appli-*
   cation to supersonic flow.   AGARD CP-193, Applications of Non-Intrusive
   Instrumentation in Fluid Flow Research, 11.1 - 11.11.

BLACKWELDER, R. F. and KAPLAN, R. E.   (1976) *On the wall structure of the*
   turbulent boundary layer.   J. Fluid Mech. 76, 89-112.

COLLINS, D. J., COLES, D. E. and HICKS, J. W.   (1978) *Measurements in the*
   turbulent boundary layer at constant pressure in subsonic and supersonic
   flow. Part I. Mean flow.   AEDC-TR-78-21.

DIMOTAKIS, P. E.   (1976) *Single scattering particle laser-Doppler measure-*
   ments of turbulence.   AGARD CP-193, Applications of Non-Intrusive
   Instrumentation in Fluid Flow Research, 10.1 - 10.14.

DIMOTAKIS, P. E., COLLINS, D. J. and LANG, D. B.   (1978) *Measurements in the*
   turbulent boundary layer at constant pressure in subsonic and supersonic
   flow. Part II. Laser Doppler velocity measurements.   To appear as an
   AEDC report.

DIMOTAKIS, P. E. and LANG, D. B.   (1974) *Single scattering particle laser*
   Doppler velocimetry.   Bull. Am. Phys. Soc. (II) 19, 1145.

DIMOTAKIS, P. E. and LANG, D. B.   (1977) *Signal responsive burst period timer*
   and counter for laser Doppler velocimetry and the like.   U.S. Patent
   4,051,433 (September 27, 1977).

HOESEL, W. and RODI, W.   (1977) *New biasing elimination method for laser-*
   Doppler velocimeter counter processing.   Rev. Sci. Instrum. 48(7), 910-
   919.

JOHNSON, D. A. and ROSE, W. C.   (1973) *Measurement of turbulence transport*
   properties in a supersonic boundary-layer flow using laser velocimeter and
   hot-wire anemometer techniques.   AIAA Paper 73-1045.

JOHNSON, D. A. and ROSE, W. C.   (1975) *Laser velocimeter and hot wire anemom-*
   eter comparison in a supersonic boundary layer.   AIAA J. 13(4), 512-515.

KLEBANOFF, P. S.   (1955) *Characteristics of turbulence in a boundary layer*
   with zero pressure gradient.   NACA Report 1247.

MCLAUGHLIN, D. K. and TIEDERMAN, W. G.   (1973) *Biasing correction for indi-*
   vidual realization of laser anemometer measurements in turbulent flows.
   Phys. Fluids 16, 2082-2088.

MORKOVIN, M. V.   (1961) *Effects of compressibility on turbulent flows.*   Proc.
   Colloq. Mechanique de la Turbulence, CNRS (1962), 367-380.   (Proc. re-
   printed as Mechanics of Turbulence, Gordon & Breach 1964).

PERRY, A. E. and ABELL, C. J.   (1975) *Scaling laws for pipe flow turbulence.*
   J. Fluid Mech. 67, 257-271.

PERRY, A. E. and ABELL, C. J.   (1977)   *Asymptotic similarity of turbulence structures in smooth- and rough-walled pipes.*   J. Fluid Mech. 79(4), 785-799.

SANDBORN, V. A.   (1974)   *A review of turbulence measurements in compressible flow.*   NASA TM X-62, 337.

YANTA, W. J. and LEE, R. E.   (1974)   *Determination of turbulence transport properties with laser Doppler velocimeter and conventional time-averaged mean flow measurements at Mach 3.*   AIAA Paper 74-575.

YANTA, W. J. and LEE, R. E.   (1976)   *Measurements of Mach 3 turbulence transport properties on a nozzle wall.*   AIAA J. 14(6), 725-729.

Turbulent Flow in an Asymmetrically Roughened Duct of Square Cross-Section

JOSEPH A. C. HUMPHREY[*]
Princeton University
Princeton, NJ 08540

ABSTRACT

The extent to which turbulence models can be tested for accuracy and generality in their application is limited by the type and quality of experimental data presently available. This work presents some experimental measurements obtained in an asymmetrically roughened straight duct flow where cross-stream motions appear at an earlier stage than in a corresponding smooth duct. A forward scatter, frequency shifted laser-Doppler velocimeter was used to measure velocity components and Reynolds stresses. The presence of a cross-stream flow and evidence of high levels of anisotropy in the vicinity of the duct walls preclude the use of turbulence models based on isotropic turbulence viscosity considerations. Models based on the direct solution for the Reynolds stresses are, in principle, more appropriate. The present results and those in the references quoted should be valuable for turbulence model testing and development.

INTRODUCTION

The experimental investigation of turbulent flows can become exceedingly laborious and time consuming (and expensive!) so that predictions by digital computation using numerical methods based on mathematical models of turbulence are highly desireable. It is important, therefore, to have a full understanding of the limitations inherent in the various turbulence models in order that confidence may be placed in their extrapolated applications. It is obvious that the accuracy of turbulent flow predictions by a specific model depends on the adequacy of the approximations embodied in the model to overcome the problem of closure. Approximations may be required to make a calculation problem tractable but can also arise simply from the inability, at present, to cope satisfactorily with certain terms; pressure diffusion in the Reynolds stress equations is an example for closure at the stress equations level. Useful approximations depend on physical insight, theoretical reasoning and, especially, empirical fact. Thus, the advancement of turbulence modelling goes hand-in-hand with an improved understanding of turbulence and this is generally achieved through detailed experimentation. At present there exists a

---

[*]Present address: University of California, Berkeley, CA. 94720.

superabundance of turbulent flow computational results (see Reynolds (1976)) in comparison to experimental data against which the predictions can be compared. The lack of detailed and accurate experimental measurements imposes a limitation on the extent to which turbulence models can be tested. This is especially the case in three-dimensional flows, where weak velocity components or recirculation zones make it necessary to use truly non-intrusive experimental techniques which, even though optically based, can still be difficult to apply.

The experimental study of internal turbulent flows has been focussed almost exclusively on two-dimensional (2-D), or axisymmetric geometries. A brief review of some relevant works has been given by Hanjalic and Launder (1972). Practically important confined flows are not generally symmetrical and turbulence models formulated with reference to data obtained from these configurations can be expected to exhibit deficiencies in predicting more complex flows. Recently, Hanjalic and Launder (1972) and Melling and Whitelaw (1976) have investigated a 2-D asymmetrically roughened channel flow and a developing 3-D square duct flow respectively. These works establish clearly the importance of being able to predict correctly certain features of turbulent flows which arise from, for example, strong turbulence diffusion or which are due to gradients in cross-stream normal stresses (and responsible for secondary motion in the mean flow).

The present work, carried out at Imperial College (London) and reported here in part, is an example of continuing efforts to alleviate the situation described in the first paragraph. Thus, measurements obtained with a laser-Doppler velocimeter are presented and discussed for the turbulent flow in a straight square duct roughened asymmetrically by placing equally spaced rib-like elements on one wall. The geometry investigated was a hybrid of those explored earlier by Hanjalic (1970) and Melling (1975) respectively. The data obtained (available in full detail in Humphrey (1977)) should be valuable for further turbulence modelling testing and development.

EXPERIMENTAL EQUIPMENT AND MEASUREMENT PROCEDURE

The experimental equipment, flow configuration and measurement procedure used for the present study have been explained in detail by Humphrey (1977) and only a brief description is provided here. Water at room temperature was made to flow at .58 m/s (Re = 2.6×10^4) from a constant head tank, into a large plenum chamber and through a test section shown schematically in Figure 1. The test section was made of 1 cm thick plexiglass with the dimensions shown in the figure. Rib-like elements of square cross-section cut from cold-drawn brass were placed on one wall, transverse to the main flow direction, at intervals of .99 hydraulic diameters. This resulted in a pitch to height ratio of approximately 10:1. Careful alignment of the duct with the inlet contraction and tip plate in the plenum chamber ensured symmetry in the measurements to within experimental error.

Bulk average velocities (U_B) were monitored by means of an orifice flowmeter designed in accordance with BS 1042

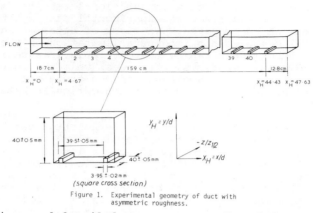

Figure 1. Experimental geometry of duct with
          asymmetric roughness.

specifications and detailed measurements of turbulent properties
at various longitudinal stations were obtained by means of a forward scatter, frequency shifted, laser-Doppler velocimeter. The
velocimeter and associated electronic components are shown graphically in Figure 2. It consisted of a 5 mW He-Ne Spectra Physics
laser, an 18000 lined bleached rotating diffraction grating with

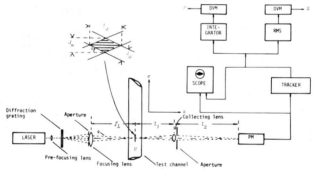

Figure 2. Laser-Doppler velocimeter and
          electronic instrumentation.

focussing lens and two additional lenses for crossing the first
order beams and focussing the measuring volume onto the pin-hole
of an EMI 9558 B photomultiplier tube respectively. The use of
rotating gratings in laser-Doppler velocimetry has been discussed
by, for example, Durst, Melling and Whitelaw (1976). In the present investigation frequency shifting was required in order to measure correctly the relatively high levels of turbulence intensity
with the tracker. The radial grating, its mounting and supporting
platform were constructed such that measurements at ± 45° to the
main flow direction could be obtained. The first order beams diffracted by the rotating grating had a half angle of 3.61° and produced a measuring volume of approximately 1.82 x 0.1 x 0.1 mm^3.

     The electronic instrumentation consisted of a DISA 55L20
frequency tracking demodulator, a Tektronix 7603 oscilloscope,
a DISA 55B30 true time integrator, and a DISA 55D35 rms meter.

In addition, two Solatron LM 1420.2 digital voltmeters were used
for displaying the integrator and rms meter results. Virtually
continuous signals of high signal to noise ratio were monitored by
the tracker at drop-out levels of less than 2%. The output from
the tracker was presented to the true time integrator and rms
meter from which mean velocities and turbulence intensities were
found. Prior to its use, the electronic instrumentation was carefully calibrated.

Two components of velocity and three Reynolds stresses were
measured in the flow of Figure 1. The data collection and reduction procedures are described by Humphrey (1977). On average 12
traverses of 25 points each were made at various longitudinal
stations corresponding to between ribs and over rib locations.
The measurements spanned the width of the duct and showed symmetrical profiles to within experimental error. Therefore, the data
over the whole cross-section was averaged with respect to the
symmetry plane. Contour plots of the measurements were obtained
by least squares polynomial regression followed by interpolation.
The quality of the polynomial fits was ascertained by reference to
the residual errors and bias due to polynomial curvature shown to
be unimportant.

The error sources and accuracy of the measurements have been
quantified by Humphrey (1977). Turbulent and non-turbulent
broadening effects and electronic noise were found negligible.
Systematic errors in the longitudinal velocity and turbulence intensity were 0%-1.5% whereas maximum random errors were approximately $\pm$ 2.5% and $\pm$ 3% respectively. These errors were due mainly
to the signal processing instrumentation and orifice meter.

RESULTS AND DISCUSSION

The results presented in this section are for the normalized
longitudinal velocity (U_x/U_B) and turbulence intensity ($\tilde{u}_x/U_B$)
only. A more complete discussion of the entire set of measurements will appear in a subsequent publication. The contours shown
in Figures 3-6 correspond to longitudinal stations of $x_H = 6.15$
and $x_H = 41.68$ hydraulic diameters which are positions halfway
between ribs 2-3 and ribs 38-39 respectively.

The mean flow entering the duct (Figure 3) shows a plug like
distribution in the core region with steep gradients towards the
walls, except for the roughened wall region where the flow moves
considerably more slowly. The peak value of U_x/U_B is slightly
displaced from the symmetry plane and is due to the early development of a normal stress driven cross stream motion of the "second"
kind. This effect is also the cause for bulging of the profiles
towards the corners similar to that observed by others in square
duct flow; see Gessner (1973), Melling and Whitelaw (1976) and
Perkins (1970). The plot of $\tilde{u}_x/U_B$ corresponding to this station
(Figure 4) shows very large values of turbulence intensity and its
spatial gradient near the rough wall ($y_H \simeq 0$) and near the side
wall ($z \simeq -D/2$). The lowest values for $\tilde{u}_x/U_B$ appear in the upper
half of the duct where, in addition, gradients of $\tilde{u}_x/U_B$ are small.
Distortion of the contours is in agreement with the circulation
sense of the transverse component of velocity (not shown here).
At this station $(u_x^2)_{rough}/(u_x^2)_{smooth} \simeq 4$ and provides an

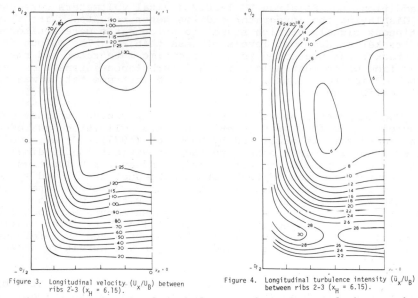

Figure 3. Longitudinal velocity. (U_x/U_B) between ribs 2-3 (x_H = 6.15).

Figure 4. Longitudinal turbulence intensity $(\tilde{u}_x/U_B)$ between ribs 2-3 (x_H = 6.15).

indication of the level of turbulence enhancement of the rough wall over the smooth at a relatively early stage in the flow.

Plots of U_x/U_B at X_H = 41.68 (Figure 5) still show bulging of the contours near the smooth wall but not towards the rough. The maximum velocity is located on the duct symmetry plane and nearer the smooth wall. At this position the flow is close to being fully developed. The longitudinal intensity contours (Figure 6) now show larger values of $\tilde{u}_x/U_B$ near the smooth wall but still smaller than at the rough wall. The distorted appearance of the

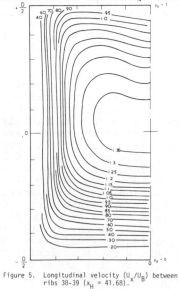

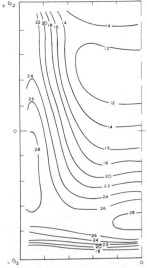

Figure 5. Longitudinal velocity (U_x/U_B) between ribs 38-39 (x_H = 41.68).

Figure 6. Longitudinal turbulence intensity $(\tilde{u}_x/U_B)$ between ribs 38-39 (x_H = 41.68).

profiles near the side wall is attributed to the effect of the transverse pressure gradient ($\partial P/\partial y$) tending to deflect fluid approaching the ribs into the bulk of the flow. The deflection appears to be stronger on the slower moving fluid near the side wall which is expected.

CONCLUSIONS

The main purpose of this experimental investigation has been to provide new and useful information for the purpose of testing and advancing turbulence models. The measurements shown here and given in detail by Humphrey (1977) indicate an early development of cross-stream flow of the second kind. This, together with high levels of anisotropy near the duct walls, precludes the use of turbulence modes based on the notion of an isotropic turbulence viscosity to predict the flow. It would appear that a model solving for the Reynolds stresses directly would be more appropriate. The implications of this suggestion will be more carefully explored in a subsequent publication.

ACKNOWLEDGEMENTS

The author is pleased to acknowledge the helpful discussions and suggestions provided by Professor J. H. Whitelaw during the course of this work. The author is indebted to Mr. R. Church for the precisely constructed test section in which the measurements were made. Financial support from the Science Research Council and Imperial Chemical Industries is also gratefully acknowledged.

REFERENCES

1. Durst, F., Melling, A. and Whitelaw, J. H. 1976. Principles and Practices of Laser Doppler Anemometry. Academic Press.

2. Gessner, F. B. 1973. The Origin of Secondary Flow in Turbulent Flow Along a Corner. J. Fluid Mech., 58, 1.

3. Hanjalic, K. 1970. Two-Dimensional Asymmetric Flow in Ducts. Ph.D. Thesis, University of London.

4. Hanjalic, K., and Launder, B. E. 1972. Fully Developed Asymmetric Flow in the Plane Channel. J. Fluid Mech., 51, 301.

5. Humphrey, J.A.C. 1977. Flow in Ducts with Curvature and Roughness. Ph.D. Thesis, University of London.

6. Melling, A. 1975. Investigation of Flow in Non-Circular Ducts and Other Configurations by Laser-Doppler Anemometry. Ph.D. Thesis, University of London.

7. Melling, A., and Whitelaw, J. H. 1976. Turbulent Flow in a Rectangular Duct. J. Fluid Mech., 78, 289.

8. Perkins, H. J. 1970. The Formation of Streamwise Vorticity in Turbulent Flow. J. Fluid Mech., 44, 721.

9. Reynolds, W. C. 1976. Computation of Turbulent Flows. Annual Review of Fluid Mechanics, 8, 183.

A New Statistical Approach to the Processing of LDV Signals of Highly Turbulent Flows

EUGEN BROCKMANN* and CHRISTO G. STOJANOFF
Technical University
Aachen, West Germany

ABSTRACT

For the case of highly turbulent flows, determination of the turbulent velocity distribution in the probe volume of a laser Doppler velocimeter requires large sampling times (due to the gaps between single particle bursts) and a huge number of data points to attain statistical confidence. In order to get rid of these disadvantages, a high scattering particle concentration which provides a stationary Gaussian random signal process can be used. A method of sampling a series of pulse pairs yields estimates of the mean frequency and the frequency spread of this process in very short times. A test of this method with computer generated LDV signals shows satisfying results.

NOMENCLATURE

N	number of pulse pairs
Q	stationary complex Gaussian process
$R(h)$	autocovariance function
$\dot{R}(h)$	first derivative of the autocovariance function
$\ddot{R}(h)$	second derivative of the autocovariance function
$W(f)$	power spectral density
b_s	true spectral width
f	frequency
h	pulse pair distance
q	random variable of Q
r_o	beam waist radius at focal point

---

* (Presently at the Applied Optics Laboratory, Purdue University, School of Mechanical Engineering, West Lafayette, Indiana, U.S.A.).

t	time
$\bar{u}$	mean velocity
w_i	weighting factors
x	normalized pulse pair distance
Δ	sum of normalized bias
Φ	normalized statistical bias
Ψ	normalized statistical bias
Ω	normalized statistical bias
δ_{fr}	distance between interference fringes
$\xi(t)$	stochastic process
μ_k	moment of k-th order
γ	normalized statistical bias
σ	standard deviation
σ_u	standard deviation of the velocity distribution
ρ	signal-to-noise ratio

Special denotations:

$\sim$	estimate
v	Hilbert transform
*	complex conjugate
$\varepsilon\{...\}$	mean of {...}
var {...}	variance of {...}
0 (...)	order of magnitude of (...)

INTRODUCTION

At the present time the application of the laser Doppler velocimeter to highly turbulent flows is partly restricted by the large amount of data which must be sampled in very short periods of time. Beyond this, it is necessary to collect data blocks over a long time scale to get statistically reliable information on the turbulent velocity spectrum. This is especially the case if LDV systems requiring a high probability of having only one particle in the probe volume at a time are used. When dealing with single particle signals, there are many pauses between signal bursts, during which valuable information is lost about the flow. The high sample rates and the large amount of data set very high requirements for the sampling and signal processing devices and make them more expensive.

From this point of view a continuous LDV signal seems to be more advantageous, provided there is a convenient method of signal sampling and processing. A continuous laser velocimeter signal can be achieved by using high scattering particle concentrations, so that all particles present in the probe volume at the same time will contribute to a continuous stochastic photodetector current. One of the aims of the present paper is to investigate the statistical properties of such a stochastic LDV signal (amplitude distribution, correlation, and power spectrum) by means of simulation. Another one is to introduce a new statistical approach for estimating the first two spectral moments of stationary complex Gaussian stochastic processes, which was originally developed for radar signals.

A study with computer generated multiparticle LDV signals has shown that this method may be applied to such signals in order to get the mean frequency (velocity) and the spectral width (velocity spread) of at least a Gaussian shaped velocity spectrum. An experimental setup is in progress to yield these first spectral moments by means of 1500 data blocks of 50 pulse pairs within 6-7 μsec.

COVARIANCE APPROACH TO SPECTRAL MOMENT ESTIMATION OF STATIONARY GAUSSIAN RANDOM PROCESSES

The photodetector current at the output of a laser velocimeter system represents a stochastic process characterized by the random phases and frequencies of the scattering particle signals and the superimposed noise along the optical and electronic channels. A common means of statistically evaluating such a stochastic process is the spectral analysis of stationary random functions [1]. As this procedure very often requires a large expense, simpler methods of estimating the spectrum often are chosen. The method presented in this chapter, developed by Miller and Rochwarger [2,3,4], estimates the first two moments of the power spectrum, rather than the spectrum itself, from knowledge of the autocovariance function or an estimate of the autocovariance function. The two moments are the mean frequency and the spectral width of the power spectrum of the signal. They correspond to the mean velocity and the velocity spread of the flow.

Let

$$Q_S = \{q^{(S)}(t) \mid -\infty < t < \infty\}$$
$$Q_N = \{q^{(N)}(t) \mid -\infty < t < \infty\}$$

(1)

be two independent stationary complex Gaussian processes with zero mean and

$$Q = Q_S + Q_N$$

(2)

Where Q may represent a measured signal, consisting of a signal process Q_S and a noise process Q_N. The random variables are $q^{(S)}(t)$ and $q^{(N)}(t)$ with

$$q(t) = q^{(S)}(t) + q^{(N)}(t)$$

(3)

The autocovariance function R(h) and the power spectral density W(f) can be defined accordingly

$$R(h) = R_S(h) + R_N(h)$$
$$W(f) = W_S(f) + W_N(f)$$

(4)

where the autocovariance and the spectral density of each single process as well as the combined process are connected through the Fourier transform, e.g.

$$R_S (h) = \int_{-\infty}^{\infty} W_S(f)e^{ifh} df$$

(5)

$$W_S (f) = \frac{1}{2\pi} \int_{-\infty}^{\infty} R_S(h)e^{-ifh} dh$$

The moment generating function

$$\mu_k (W_S) = \frac{1}{(2\pi i)^k} \frac{d^k \left(\frac{R_S(h)}{R_S(0)}\right)}{dh^k} \Bigg|_{h=0}$$

(6)

yields the first two moments of the power spectrum

$$\mu_1(W_S) = \frac{1}{2\pi i} \frac{\dot{R}_S(0)}{R_S(0)}$$

(7)

$$\mu_2(W_S) = - \frac{1}{4\pi^2} \frac{\ddot{R}_S(0)}{R_S(0)}$$

(8)

The second moment (variance) can be expressed as a standard deviation (central moment)

$$\sigma(W_S) = \sqrt{\mu_2(W_S) - \mu_1^2(W_S)}$$

(9)

Approximating the differentials in Eqs. (7) and (8) by differences for h close to zero (h≠0), the two moments of the power spectrum may be rewritten as

$$\mu_1(W_S) = \frac{1}{2\pi h} \arctan \frac{Im[R(h)-R_N(h)]}{Re[R(h)-R_N(h)]}$$

(10)

$$\sigma(W_S) = \frac{1}{\pi h \sqrt{2}} \sqrt{1 - \frac{|R(h)-R_N(h)|}{R(0)-R_N(0)}}$$

(11)

i.e. the spectral moments can be calculated if only two values R(0) and R(h) of the autocovariance function are known.

Miller and Rochwarger showed that in the case of sampling N pulse pairs from the process Q in the form

$$Q_k = \{q_k(t_1), q_k(t_2)\} \qquad \begin{array}{l} 1 \leq k \leq N \\ h = t_2 - t_1 \end{array}$$

(12)

(see Fig. 1) estimates of the autocovariance (denoted by $\sim$) can be obtained

$$\tilde{R}(0) = \frac{1}{2N} \sum_{n=1}^{N} \{|q_n(t_1)|^2 + |q_n(t_2)|^2\}$$

(13)

$$\tilde{R}(h) = \frac{1}{N} \sum_{n=1}^{N} q_n(t_1)q_n*(t_2)$$

When Eqs. (13) are introduced into Eqs. (10) and (11), estimators of the spectral moments (maximum likelihood solutions) are obtained. Beyond this the statistical behavior of the estimators of the spectral moments can be described by asymptotic formulas with respect to N where, for large N, the higher order terms may be dropped

$$\varepsilon \{\tilde{\mu}_1(W_S)\} = O(N^0) + O(N^{-1}) \ \xi + O(N^{-2}) + \ldots$$

$$\text{var}\{\tilde{\mu}_1(W_S)\} = O(N^{-1}) \ \xi + O(N^{-2}) + \ldots$$

$$\varepsilon \{\tilde{\sigma}(W_S)\} = O(N^{-1}) \ \xi + O(N^{-2}) + \ldots$$

$$\text{var}\{\tilde{\sigma}(W_S)\} = O(N^{-1}) \ \xi + O(N^{-2}) + \ldots$$

(14)

Miller and Rochwarger investigated a complex Gaussian stochastic process with a Gaussian shaped autocovariance function in the presence of Gaussian (white) noise, but indicated that a similar formulation might be derived for different types of spectra.

Stochastic Signal

a)   continuous Pulse-pairs

b)   overlapping pulse-pairs

Fig. 1.   Alternatives of Pulse Pair Spacing

The choice of a convenient pulse pair distance h should not be at random. Introducing the dimensionless statistical bias of the above defined properties

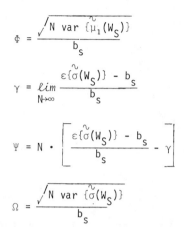

$$\Phi = \frac{\sqrt{N \ var \ \{\tilde{\mu}_1(W_S)\}}}{b_S}$$

$$\gamma = \lim_{N\to\infty} \frac{\varepsilon\{\tilde{\sigma}(W_S)\} - b_S}{b_S}$$

$$\Psi = N \cdot \left[\frac{\varepsilon\{\tilde{\sigma}(W_S)\} - b_S}{b_S} - \gamma \right]$$

$$\Omega = \frac{\sqrt{N \ var \ \{\tilde{\sigma}(W_S)\}}}{b_S}$$

(15)

best results for the spectral moments may be obtained by minimizing all of the statistical biases. One possibility, for example, is to minimize the weighted sum

$$\Delta = w_1 \ \Phi + w_2 \ \Psi + w_3 \ \Omega \tag{16}$$

For the Gaussian case, the functions Φ, Ψ, Ω and Δ are explicit functions of the signal-to-noise ratio ρ and the normalized pulse pair width

$$x = 2\pi \ b_S \ h \tag{17}$$

where b_S is the true spectral width of the signal process. Setting the weighting factor w_i equal to one, Fig. 2 shows that minima of the function Δ do exist (5). From here the procedure of estimating the spectral moments can be understood. For a known signal-to-noise ratio an optimal value of x can be found. An estimation of the true spectral width yields a good pulse pair spacing h, so that the statistical biases are minimized. Due to several estimates of b_S the results can be optimized.

APPLICATION OF THE SPECTRAL MOMENT ESTIMATION TO
LASER VELOCIMETER SIGNALS

The statistical approach, presented in the previous section, requires a stationary complex Gaussian stochastic process. In practice a measured signal usually is a real function. One of the possible methods to turn a real process $\xi(t)$ into a complex process $\overset{v}{\xi}(t)$ is the Hilbert transform

$$\overset{v}{\xi}(t) = \frac{1}{\pi} \int_{-\infty}^{\infty} \frac{\xi(\tau)}{t-\tau} \ d\tau = \xi(t) * \frac{1}{\pi t} \tag{18}$$

sometimes referred to as a quadrature filter. A stationary stochastic laser

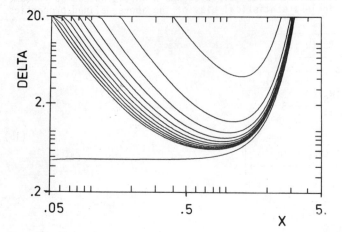

Fig. 2. Δ as a Function of $x = 2\pi b_s\, h$, Parametrized with the Signal-to-Noise
          Ratio ρ.   In Downward Reading the Curves are Drawn for ρ = -6,-3,0,3,
          6,9,12,24,1000 dB

velocimetry signal can be achieved, if always several particles are present at
one time in the probe volume.   According to the central limit theorem of probability theory, the summation of independent random processes approaches a
random process having a Gaussian distribution with a growing number of summations.
As all the particles within the probe volume are independently distributed at
random and have random phases, a Gaussian signal amplitude distribution can be
achieved in the multiparticle case.   Simulations with computer generated laser
velocimeter signals have shown that a stationary Gaussian process can be obtained
for as few as five to six particles (see Fig. 3) [5].   On the other hand a somewhat higher particle concentration would be desirable to get a better representation of the velocity spectrum of the flow.

Looking at the autocovariance function of a stationary homogeneous isotropic incompressible turbulent flow with a Gaussian shaped velocity spectrum [6]

$$\frac{R(\tau)}{R(0)} = \left(1 + \frac{\sigma_u^2\,\tau^2}{r_0^2}\right)^{-\frac{1}{2}} \exp\left\{-\frac{1}{2}\tau^2\left(\frac{\frac{\bar{u}^2}{r_0^2} + \left[\frac{2\pi}{\delta_{fr}}\sigma_u\right]^2}{1 + \frac{\sigma_u^2\,\tau^2}{r_0^2}}\right)\right\}.$$

$$\cdot \cos\left(\frac{\frac{2\pi}{\delta_{fr}}\bar{u}\,\tau}{1 + \frac{\sigma_u^2\tau^2}{r_0^2}}\right) \tag{19}$$

it appears to be a cosine function damped by a Gaussian envelope, where the
damping depends mainly on the spectral width (standard deviation of the velocity,
σ_u).   The computer simulations of laser velocimeter signals, which were performed

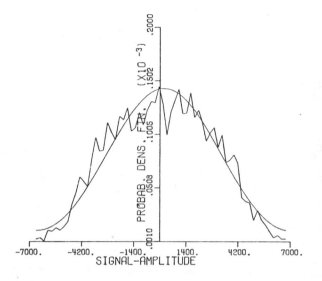

Fig. 3.   Probability Density Distribution of the LV Signal Amplitude from 4096
          Data Points (Actual and Smoothed Curves). The Signal is Simulated with
          an Average Number of 6 Particles in the Probe Volume.

to test the estimation procedure, show exactly the same statistical behavior
(Fig. 4). If the autocovariance was left in this shape, the statistical estimation approach presented in the previous section could not be applied. At this
point, however, the Hilbert transformation provides an imaginary damped sine
function corresponding to the real damped cosine function. Thus the absolute
value of the complex autocovariance function turns out to be just the Gaussian
envelope as required by the method.

     Another requirement for application of the covariance estimation is at
least a rough knowledge of the signal-to-noise ratio and the true spectral width
of the measured signal. A rough estimate of these values might be obtained with
an electronic device of low resolution before the actual measurement is done in
order to achieve an optimal pulse pair spacing. Two different types of pulse
pair sampling are shown in Fig. 1, where the distances between the pulse pairs
do not have to be equal in general.

     Tests on a digital computer have been performed including generating the
signals, the pulse pair sampling, and the spectral moment estimation. These
results were compared with those obtained by the spectral analysis. They show
quite satisfying results, even when only 100 pulse pairs are sampled out of a
data block of 4096 sample points. The relative error of the estimated spectral
moments compared to the simulation input was usually about 5-20% depending on
the first estimate of the spectral width, but could be very often improved up
to 1-5% after optimizing the pulse pair spacing by using the previous received
spectral width estimate.

     At the present time an experimental verification is in progress at the
Technical University of Aachen, in which pulse pair spacings down to 40

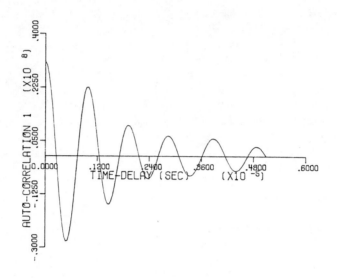

Fig. 4. Autocovariance Function of an LV Signal, Simulated with an Average
         Number of 12 Particles in the Probe Volume ($\sigma_u/\bar{u}$ = 0.1).

nanoseconds are approached. It is planned to sample about 1500 data blocks of
50 pulse pairs within 6-7 μsec by means of a diode matrix consisting of 256 by
500 elements which are illuminated by a rotating mirror.

CONCLUSION

     This investigation has shown that application of the proposed spectral
moment estimation to LDV signal analysis is possible in general. Research is
now underway to find a convenient method to optimize the pulse pair spacing in
an iterative manner. Though the analysis and simulation to date are performed
for Gaussian shaped spectra only, it seems that the estimation procedure may
also be extended to non-Gaussian spectra. Both analytical studies and computer
simulations for skewed spectra will be carried out in the near future. While
some information about the skewness of the spectrum might be gained from the
tail of the autocovariance function, in special cases an apriori knowledge of
the spectral shape could be used in a hypothesis test to improve the estimation.

REFERENCES

1. Yaglom, A. M., An Introduction to the Theory of Random Function, Dover
    (1973).

2. Miller, K. S. and Rochwarger, M. M., On Estimating Spectral Moments in
    the Presence of Colored Noise, IEEE IT-16, no. 3, pp. 303-309 (1970).

3.   Miller, K. S. and Rochwarger, M. M., Estimation of Spectral Moments of
     Time Series, Biometrica, Vol. 57, no. 3, pp. 513-517 (1970).

4.   Miller, K. S. and Rochwarger, M. M., A covariance Approach to Spectral
     Moment Estimation, IEEE IT-18, no. 5, pp. 588-596 (1972).

5.   Brockmann, E., Statistical Investigation of Laser Doppler Signals by Means
     of Numerical Simulation, Dissertation, RWTH Aachen (1977) (in German).

6.   Durrani, T. S. and Greated, C. A., Laser Systems in Flow Measurement,
     Plenum (1977).

ACKNOWLEDGMENT

     Part of this research was sponsored by a NATO Fellowship. We would like
to thank Prof. W. H. Stevenson for this support in preparing this paper.

MEASUREMENTS IN INTERNAL COMBUSTION ENGINES

Chairman: **F. K. OWENS**
Consultant

Application of Laser Velocimetry to a Motored Internal Combustion Engine

PETER O. WITZE
Sandia Laboratories
Livermore, CA

ABSTRACT

A laser velocimeter system has been developed for the measurement of mean velocities and turbulence intensities in a motor-driven non-combusting internal combustion engine. Results are presented for a high-swirl engine configuration, and a comparison is made with measurements taken with an existing hot-wire anemometer system. The purpose of the investigation was to assess the feasibility of applying laser velocimetry to engines, as well as to determine the accuracy of the hot-wire technique in a temperature-varying flow. It is shown that accurate laser velocimeter results can be obtained throughout the engine cycle, but that the correction technique developed to account for the gas-temperature dependence of a hot-wire anemometer does not give satisfactory results.

INTRODUCTION

As requirements for better fuel economy and reduced pollution emissions from automotive engines become increasingly stringent, so does the need also grow to better understand the details of engine combustion processes. Of utmost importance to improved understanding is the fluid dynamics within the engine cylinder. This is particularly true of lean-burn stratified-charge engine concepts, for which both the pre- and post-combustion fluid mechanics control the performance of the engine. In direct-injected stratified-charge engines, a liquid fuel spray is injected into a usually swirling turbulent flow such that when ignition occurs there is an optimal distribution of fuel and air within the cylinder. The goal is complete combustion of an overall lean fuel/air mixture. Parameters such as mean velocity and turbulence distributions, fuel injection timing, fuel spray pattern and direction, ignition timing and many others influence the ultimate performance of such an engine.

In dual-chamber stratified-charge engines, the combustion process is initiated in a prechamber that is connected by a passage to the main chamber. Governing the performance of this engine type is the fluid dynamics in the prechamber as ignition begins, followed by completion of combustion in the main chamber. In this configuration the relative size, shape, and fuel/air mixture of each chamber is important, but of particular interest is the size, shape, and orientation of the passage connecting the two chambers.

Even in homogeneous-charge carbureted engines, fluid dynamics is very important. Clerk[1] first recognized in 1912 that bulk gas motion and turbulence is imperative to good engine performance. If the fuel/air charge in an engine were totally quiescent at the time of ignition, the combustion process would

239

be laminar and have a fixed burn-duration. Under this latter condition an engine
that runs well at 500 rpm would probably not run at all at 2500 rpm. However,
because the bulk motion and turbulence of the fuel/air charge increase approximately linearly with engine speed,[2] the observed result is that the combustion
process also accelerates with engine speed, such that the crank-angle duration
of combustion is nearly constant.

The technology for using a hot-wire anemometer to study engine fluid dynamics has become reasonably well-developed in recent years.[2-6] However, the hot-
wire anemometer suffers from several shortcomings that greatly restrict its utility and accuracy in engine applications. Foremost is directional insensitivity
and ambiguity. In many engine configurations, such as most carbureted spark-
ignition engines, there is no true mean flow at the time of ignition. A hot-
wire anemometer can resolve no quantitative information in such a flowfield.
Another problem is the inability of a hot-wire to be used in a combusting environment. This is only partially limiting, however, since the singularly most
important fluid dynamic description of engine processes is the state of fluid
motion prior-to ignition. This condition can be well simulated by externally
motoring the engine. Of great importance to the applicability of the hot-wire
anemometer, however, is its dependence on the thermal properties of the gas
being measured. Gas temperatures in motored engines can easily exceed 500-600°C.
Because the hot wire response is extremely sensitive to gas temperature, resolving the velocity of the gas under these conditions can be both difficult and of
questionable accuracy. Since this uncertainty occurs during the moment of
greatest interest, i.e. ignition, the limitations of hot-wire anemometry are
clearly evident. Increasing awareness of these problems has provided the impetus
for the strong current interest in developing laser Doppler velocimetry for engine applications.

EXPERIMENTAL APPARATUS

The experimental measurements to be reported were taken in a single-cylinder
Wisconsin L-head engine (7.62 cm bore, 8.26 cm stroke, 1.02 cm clearance height),
modified as shown in Figure 1 to provide a highly-swirling mean velocity field
of known flow direction and low relative turbulence intensity. The oil-free engine was driven by a 5 hp variable-speed electric motor with an operating range
of 500 to 2500 rpm. The motor was able to maintain a constant average engine
speed to within 0.5 percent for typical test periods of several minutes. The
measurements to be reported were all taken at 600 rpm.

Both the engine and electric motor were rigidly mounted on a large concrete
block, isolated from the optical arrangement except for coupling through the
floor. Initially the concrete block was supported on wooden beams, but because
this permitted excessive vibration of the engine, it was necessary to change to
steel bars. Motion of the combustion chamber due to engine vibration was subsequently investigated, since this could presumably lead to a bias in mean velocity
measurements. This study was done by rigidly mounting a steady-state air jet to
the engine block in an orientation that simulated the anticipated engine air flow.
LDV measurements were then made in the core of this jet for engine-off and engine-
running conditions; no evidence of vibration-induced air motion was evident.

EXPERIMENTAL PROCEDURE

The highly transient nature of reciprocating engine fluid mechanics demands
that special procedures be developed for setting the signal processor controls
(a Macrodyne Model 2098 counter-type processor was used). Order-of-magnitude
changes in the mean velocity within a single engine stroke are not uncommon, such
that band-pass filter settings, and for some processors the velocity range, must
be tuned for a limited region of the cycle. For the present optical arrangement

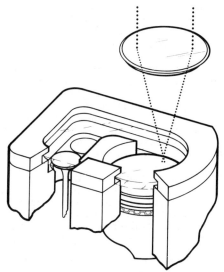

of looking down at the piston in a back-scatter mode, the signal-to-noise ratio (SNR) also varied greatly throughout a cycle, being particularly low when the piston was near top-dead-center (TDC); this condition requires varying tolerances on signal amplitude discrimination, which strongly influences the data rejection rate. Finally, because the data rate varies considerably with crank-angle position, it is often necessary to isolate the low data rate regions in order to collect enough data to satisfy the statistical requirements of sampling theory.

Signal processor setup for this experiment relied upon the use of an oscilloscope synchronized with the engine cycle. A magnetically-activated timing pulse generated at TDC by the flywheel was used to trigger the scope. Because the engine slows down slightly during compression, the pulse amplitude here is lower and thus distinguishable from the exhaust/intake pulse, which allows the scope to be triggered in-phase. The delay trigger and a faster sweep speed were then used to isolate and expand the desired region of the engine cycle.

Figure 1. High-swirl engine configuration. The tangential-velocity-component measurement location is marked by the crossing point of the laser beams.

Normal operating procedure was to display on the oscilloscope the photomultiplier output after the bandpass filters. When properly adjusted, the Doppler bursts appear as vertical lines symmetric about the ground position. The sync pulse output of the signal processor, corresponding to validation of a data point, was then added to the filtered photomultiplier signal. This combined signal gives a visual indication of the data validation efficiency, since only the Doppler bursts displayed on the scope that are immediately followed by a sync pulse have been validated. Similarly, sync pulses that do not correspond to Doppler bursts indicate erroneous validations. Using this procedure, it was possible to locate and synchronously-gate the data acquisition window so as to record valid data during one part of the engine cycle while, in effect, ignoring erroneous validations from another part. Finally, because laser induced background reflections were found to be the chief source of noise, commonly used techniques such as blocking one of the two laser beams when adjusting the signal-amplitude discrimination level was found to be unacceptable. Instead, the particle seed was turned off and the amplitude discriminator adjusted such that no sync pulses appeared within the crank-angle window being studied. The seed was then turned on, and measurement commenced. Because the SNR deteriorated with time due to seed contamination of the window and piston surfaces, at the completion of each test the seed was again turned off and a check was made for a null data validation condition.

LASER VELOCIMETRY

For this experiment the LDV optics were set up in a dual-beam backscatter mode. Approximately two watts of power from an argon-ion laser at 488 nm were typically used. During the early stages of the experiment (specifically, before engine vibration was controlled) oil-droplet particle seeding and a plastic window on the engine were tried without success. It was found that the poor optical quality of the plastic coupled with oil-wetting of the window's surface led to an unacceptable SNR when the engine vibrated. These same conditions were not run after the vibration problem was corrected, so it is not known whether the SNR would have improved enough for measurements to be possible. In any event, the plastic window was replaced with quartz, and solid particles were used as the seed.

Two different seeding techniques were used to obtain the measurements to be presented. When oil droplets were found to be unacceptable, NaCl particles of a nominal 0.6 μm diameter were tried. These were found to give good results, except during the intake stroke; here, the turbulence intensity was found to be suspiciously low, a fact that was later confirmed by hot-wire anemometer results. The problem was that the turbulence scales are very small during intake, such that smaller seed particles were required. By using nominal 0.2 μm TiO_2 particles, turbulence intensity results were achieved during intake comparable to those measured by hot-wire anemometry.

Other than during the intake stroke the dynamic behavior of the NaCl and TiO_2 particles was found to be similar. The NaCl did, however, soil the window and piston surfaces much faster, perhaps due to larger size or greater electrical charge on the particles. However, even though the TiO_2 did not visibly soil surfaces as fast, the lower signal strength from the smaller particles negated this effect, such that deterioration in SNR with test time was comparable for both seed types.

The seed particles were introduced into the engine cylinder by a steady-state air jet inserted directly into the inlet port. Thus, during those periods when the inlet valve was closed the particles continuously filled the inlet port volume, and were drawn into the cylinder when the valve was open during the intake stroke. Earlier attempts to introduce the seed into a continuously vented intake-manifold system resulted in unacceptably low data rates.

The paramount difficulty encountered in this experiment was a large laser-induced background light level, as illustrated in Figure 2a. Shown is an oscilloscope trace of the a.c. component of the photomultiplier output voltage for one engine cycle. Three zones of very high background noise level are apparent. The region in the middle of the expansion stroke is due to water-vapor condensation during the blowdown process that takes place when the exhaust valve opens prior to the piston reaching bottom-dead-center (BDC). The signal levels produced by scattering from the condensate totally swamp those from the artificially-induced seed particles; however, it was found that the Doppler signals from the water droplets are processable by the counter, although the size, and thus dynamic behavior, of the droplets is both time-varying and unknown. A second and more deleterious effect of water vapor condensation was also observed: the water droplets are believed to nucleate on the seed particles; then, during the exhaust stroke the droplets apparently grow large enough in size that they either fall out of the flow onto the piston crown, or are centrifugally driven onto the cylinder walls; electrical charge may also influence this process. In any event, when the last evidence of water droplets is gone, the LDV data rate falls to a very low level, suggesting that the water has scrubbed the seed particles from the flow. This low-seed condition persists until approximately 30° into the intake stroke, at which time a new "charge" of TiO_2 particles arrives at the measuring volume. Although this scrubbing process occurs during a part of the engine cycle that is not of great technical interest, it may be that water condensation enhances the rate at which the window and piston surfaces become soiled, such that it may be worth the effort to dry the air used in experiments of this type.

Figure 2a shows two other regions of high background noise level, both centered about TDC. These are due to scattering of the spent laser beams from the piston crown, which in the instance shown was painted flat black. It is this background light problem which makes it very difficult to make LDV measurements at the point in the engine cycle of greatest interest, i.e., the ignition and combustion phase. To overcome this problem, a first-surface glass mirror was bonded to the piston surface (a polished-aluminum mirror was found to be inadequate) and the optical plane of the incident laser beams was tilted from the

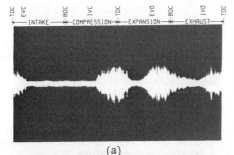

(a)

(b)

Figure 2. Background noise: a. Flat-black piston surface; b. Mirrored piston surface. (IVO = intake valve opens; EVC = exhaust valve closes.)

cylinder axis approximately 6°. The result achieved is shown in Figure 2b, for the condition of a very clean surface on the mirror. In this state the SNR was adequate; however, it took only several minutes of test time before seed-particle deposition on the mirror brought about intolerable conditions.

HOT-WIRE ANEMOMETRY

Because the response of a hot-wire anemometer is sensitive to gas temperature and pressure in addition to velocity, it is necessary to account for these effects. The increase in pressure is easily accounted-for because it only influences the density, whereas the temperature also affects the viscosity, conductivity, and temperature-differential driving force. There are three common techniques available for including the influence of varying gas temperature in hot-wire data-reduction procedures. The preferred method is to calibrate the probe over a complete range of temperatures and velocities. Since this is a very time-consuming activity, it is not practical for engine applications where probe breakage is a frequent occurrence. The technique of using a temperature-compensated anemometer system is also of questionable utility in an engine, since the frequency response generally is found to be inadequate. The remaining technique is to use an empirically-adjusted analytic model to correct indicated results for deviations from calibration conditions.

The two most often used heat transfer laws for hot-wire anemometry are those of Collis and Williams[7] and Davies and Fisher.[8] These two laws differ significantly in the reference temperatures used in their Nusselt number versus Reynolds number relationships. However, the importance of these differences may be masked by even greater variations in the dimensions of the wires used in the experiments. In order to minimize the effect of a temperature gradient along the wire due to conduction to the wire supports, Collis and Williams constructed wires of very large length-to-diameter ratio ($\ell/d > 2000$), such that conduction losses could be assumed negligible. On the other hand, Davies and Fisher were interested in studying conduction losses, and therefore used wires with an ℓ/d ratio as small as 500. The theory that they developed for partitioning the total energy loss from a wire into either convection, conduction, or radiation losses has more recently been investigated by Horvatin,[9] and it was this latter work that was used to formulate the present analysis.

The high-temperature anemometer probe used consisted of a 1.5 mm long platinum-iridium wire, 6.3 μm in diameter ($\ell/d = 238$). The probe was calibrated in ambient air for mean wire temperatures of 350 and 500°C. A heat transfer law was then developed by using this calibration data in conjunction with hot-wire anemometer measurements taken in the motored engine at both wire temperatures. These measurements were taken at the same location as the LDV data, with the wire oriented parallel to the cylinder axis such that it was responsive mainly to the tangential and radial velocity components, where the latter was assumed to be negligibly small. The air temperatures in the engine during compression were calculated from pressure transducer measurements, assuming an adiabatic process.

The temperature distribution was assumed to be uniform throughout the engine, thereby neglecting wall heat-transfer effects.

The heat-transfer law that evolved from this study was the product of the requirement that the velocity measurements made at both wire temperatures and by LDV all coincide at TDC. The Nusselt-Reynolds number relationship that resulted was $Nu = A + BRe^n$ where A, B, and n are calibration factors unique to each probe and each wire temperature. The expression found differs from the results of others, and intuition, in three ways:

1. The heat transfer coefficient in the Nusselt number was calculated by Horvatin's procedure by assuming for the boundary condition at the ends of the wire that the support needles were in temperature equilibrium with the gas. This is in direct conflict with the technique used previously by this author,[2,6] where in essence the wire supports were assumed to be at room temperature.

2. The conductivity, density, and viscosity of air appearing in the Nusselt and Reynolds numbers were evaluated at the air temperature. Collis and Williams used the mean temperature between the wire and gas as their reference temperature, whereas Davies and Fisher calculated the Nusselt number at the wire temperature, and the Reynolds number at the gas temperature. Supportive of the current result, Bradbury and Castro[10] found that the reference temperature should be weighted more heavily toward the gas temperature than the mean temperature. Dent[3] and Horvatin[9] also used the gas temperature as their reference condition.

3. The "temperature loading factor" recommended by Collis and Williams was found to be unnecessary. This result may be due more to the lack of sensitivity (or precision) of the present experiment when compared to the very controlled conditions enjoyed by Collis and Williams.

FLOW VISUALIZATION

One of the more practical limitations of LDV as a diagnostic tool is that it provides information about only one point in space for each recorded average measurement. Thus, in order to describe a flowfield completely it is necessary to perform a full three-dimensional mapping throughout space for each of three velocity components. Not only would this be a tedious task, but in most instances it would also be superfluous, since it is usually necessary to know details of only a small region of a flow. The difficulty is in knowing where to look, a priori. This is particularly true of typical internal combustion engines, for which the flow is usually complex. Flow visualization, when done in parallel with single-point measurements, can therefore be a valuable tool.

When it was found necessary to attach a mirror to the piston crown to control the background light level for LDV measurements, the opportunity to make high-speed laser shadowgraph movies presented itself. Accomplishment of this task was reasonably straightforward, the only significant difficulty being piston wobble that caused movement in the location of the image. Because of this there was a brief period of blackout during the midstroke portion of the expansion phase, but otherwise the shadowgraph movie quality was quite good. A camera speed of 5000 frames per second was used, corresponding to 0.72 degrees of crank-angle per frame.

There is a threefold difficulty with shadowgraph interpretation: (1) the observed patterns are due to density gradients, which do not directly correspond to fluid motion; (2) the observed patterns are due to the integrated effect along the line-of-sight, and thus local details are often lost or misinterpreted; (3) temperature gradients near surfaces, particularly the window and mirror, often dominate the observed patterns and yet are not representative of the bulk motion being studied. Notwithstanding these difficulties, however, shadowgraph movies can be very informative.

The shadowgraph movie taken shows the intake process to be quite asymmetric, having a helical shape as the inlet jet spirals down along the walls of the cylinder as it is drawn down by the piston. By the time BDC is reached, and then throughout most of the compression stroke, the flow has established itself as a continuous swirling structure. The axis of this spiral is not perfectly straight, such that two distinct circle patterns are observed, indicating the opposite ends of the spiral at the piston and window surfaces. At approximately 50° before TDC in the compression stroke, however, this well-structured spiral pattern appears to collapse as the flow, although still swirling, becomes noticeably more disorganized; it is not known if this is only a wall boundary layer effect. (The LDV turbulence measurements indicate no change in flow structure at the mid-plane of the clearance volume; the hot-wire results do show change although this is believed to be due to temperature, and not velocity, fluctuations.)

Also occurring during the compression stroke is an appreciable backflow from the main cylinder into the valve-area prechamber. Since most of this air is skimmed-off from the outer regions of the swirling flow, the primary measurement location indicated in the figure is not significantly influenced. During the early stages of the expansion stroke, however, the gas charge in the prechamber flows back into the main chamber, causing an increase in the mean velocity and, to a greater degree, the turbulence intensity.

Following the blackout period related to piston wobble, the late expansion phase shows swirling water droplets that appear to be thrown radially outward by centrifugal forces. Finally, the exhaust stroke is seen to be increasingly asymmetric as the axis of rotation becomes pushed to the side because of the large mass of air exiting the chamber through the off-centered exhaust port.

STATISTICS OF NON-STATIONARY TURBULENCE

Conventional time-averaging of the measured instantaneous velocity is not applicable to the transient turbulent flows encountered in internal combustion engines, where changes in the mean flow can occur on a time scale of milliseconds. The preferred procedure would be a phase average, whereby the instantaneous value at a specific crank-angle position is averaged over many cycles of the engine. Such a phase-averaged description of turbulent flow can be used to process hot-wire anemometer measurements, where the response signal is continuous; however, this procedure is not amenable to the random arrival time of LDV signals. Thus it is necessary to use a phase-time average, where the time average is over a finite crank-angle width $\Delta\theta$. If the instantaneous velocity U at engine crank angle θ is defined to consist of a mean component $\bar{U}$ and a fluctuating component u, then the phase-time averaged mean velocity can be evaluated by the expression

$$\bar{U}(\theta,\Delta\theta) = \frac{1}{N_t} \sum_{i=0}^{N_c-1} \sum_{j=1}^{N_i} U_{i,j} \ (\theta \pm \frac{\Delta\theta}{2})$$

where N_i is the number of velocity measurements recorded in the ith cycle, N_c the number of cycles, and N_t the total number of measurements. The corresponding turbulence intensity is defined as

$$u'(\theta,\Delta\theta) = \left[\frac{1}{N_t} \sum_{i=0}^{N_c-1} \sum_{j=1}^{N_i} u_{i,j}^2 \ (\theta \pm \frac{\Delta\theta}{2}) \right]^{1/2} \ , \text{ where } u_{i,j} = U_{i,j} - \bar{U}$$

The effect of approximating a phase average with a phase-time average is primarily a loss in crank-angle resolution. For most fluid dynamic motions that occur within an engine, 10° resolution has been found to be satisfactory. The primary exception to this approximation would occur in a combusting engine, where crank-angle resolution on the order of 1-2° is required.

DATA ACQUISITION AND PROCESSING

The counter-type signal processor used for this experiment was directly interfaced to a PDP-11/40 minicomputer. Most of the data recorded were taken using the computer's programmable clock and analog-to-digital converter for timing. The magnetically-activated timing pulse was sensed first, to locate TDC. Because a four-stroke engine was used, it was necessary next to sample the analog output of the pressure transducer to determine the engine phase, i.e., exhaust or compression stroke. Following this, the programmable clock was then used to locate the data-acquisition window at the desired position within the engine cycle. The window-widths used ranged from as small as 10° of crank angle to as large as 100°. During each cycle of the engine when this window was in position, the signal-processor data-inhibit control was deactivated, permitting the computer to store each validated data point within the window, as well as the clock time when the event occurred. Typically, 2000 data points were taken per test in this manner, requiring many engine cycles. These measurements were then post-processed by sorting into cells of 10° crank-angle width. The occupants of each cell were also formed into a probability distribution histogram that was displayed on a CRT to check for the presence of outlying erroneous data points.

Because the timing procedure just described relied on considerable computer software to locate the position of the window, it was necessary to calibrate the relationship between crank-angle position and the programmable-clock time as a function of engine speed. This was done by using as a reference point the instant of maximum pressure-transducer reading as the indicator of TDC of the compression stroke.

Recently, an absolute engine-crank-position indicator has been implemented, such that the programmable clock is no longer needed. A digital-output shaft encoder has been connected to the engine crankshaft; geared-down by a 2:1 ratio, the encoder gives crank position in 0.2° increments over two engine revolutions, or one cycle (i.e., 720° of absolute position). The computer software then calls upon the user to provide as input the crank-angle positions where the data-acquisition window is to open and close. When the window is open, velocity/crank-angle pairs are stored with 0.2° crank-angle resolution. This new procedure is absolute, independent of engine speed and thus requiring no calibration.

LASER VELOCIMETER RESULTS

LDV measurements were made in the direct path of the intake flow, at the intersection of the centerline of the intake channel and the cylinder radius normal to it, as indicated in Figure 1. Only the tangential velocity component was measured, for a single engine speed of 600 rpm. The mean velocity and turbulence intensity results are given in Figure 3. Each symbol shown corresponds to a unique test; tests at each crank angle were repeated a minimum of three times. The number of data points within each 10° crank-angle window for each individual test varied greatly with both crank-angle position and the particular idiosyncracies of each test setup. More will be said about data rates later, but in general each symbol plotted represents greater than 100 samples. The solid line shown in each figure indicates the average of all the measurements recorded.

The ease with which measurements were made varied greatly throughout the engine cycle, and the specific reasons for these variations tell much about the applicability of LDV to engine environments:

0-20°: No data were obtained (also for 670-720°). The cause of this condition is not well understood, but as alluded to earlier, it is speculated that water droplets that form late in the expansion stroke scrub the air in the cylinder of all but the smallest particles, such that the visibility of the remaining particles is too low to be discerned above the high background noise level caused by the close proximity of the piston surface.

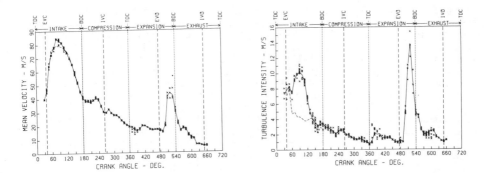

Figure 3. LDV results. Each symbol indicates a phase-time average over 10° of crank angle for greater than 100 measurements. The solid line represents the overall average for either 0.6 μm NaCl or 0.2 μm TiO$_2$ particles, except for during intake; there the solid line is just for the TiO$_2$, whereas the dashed line indicates the results using NaCl.

30-60°: Freshly-seeded air from the inlet port has reached the measuring volume. The data rate is several hundred Hz, but because the SNR is low and the time gradient in mean velocity large, the turbulence intensity results are not statistically well converged (the relative turbulence intensity in this region ranges from 15-20 percent).

70-120°: It is in this region that the problem of particle dynamics referred to earlier manifests itself. The dashed lines shown in the figures are the average of measurements taken using 0.6 μm NaCl as the seed. It is clearly evident that these particles do not follow the turbulence fluctuations when compared to the results found with 0.2 μm TiO$_2$ particles. (Hot-wire measurements reported in Reference 6 for an overhead-valve engine configuration showed intake integral time-scales to be ~ 0.1 ms, and micro time-scales ~ 0.02 ms, so perhaps this problem should have been anticipated.) The results obtained during this period were unusually sensitive to the signal level, in that the counter threshold level had to be kept close to the noise limit to prevent truncation of the high velocity fluctuations. This suggests that the visibility of the small, high-velocity-following particles was marginal. The data rate approached one kHz during this period, when the velocity was at its maximum value.

130-170°: By the end of this crank-angle region the turbulence frequency has fallen-off considerably, such that results obtained with NaCl and TiO$_2$ particles are indistinguishable. The data rate also falls as the velocity decays, to rates less than 100 Hz.

170-240°: This region is unique in that the signal was hampered by what appeared to be particle-induced background noise. Adjustment of the signal level for maximum sensitivity (just below the point where no sync pulses were observed when the seed was off) gave results that contained high frequency noise at roughly twice the Doppler frequency. This noise went away when the signal level was reduced. Because this problem first appears near BDC when the amount of seed in the cylinder is at its maximum, particularly in the volume below the measurement point, the number of particles out of the measurement volume being illuminated by the laser beams is also at a maximum. Thus the suggestion that there is particle-induced noise due to interaction with the spent laser beams.

330-390°: During this period the signal quality is very dependent on the condition of the mirrored piston surface; at TDC the data rate averaged only 40 Hz (0.1 measurements/cycle/10° window). Immediately after TDC the turbulence intensity increases dramatically, due to backflow from the valve prechamber area. The large amount of scatter in the intensity data suggests that large structures dominate the turbulence.

480-600°: At 480° water droplets begin to form due to the blowdown that
takes place when the exhaust valve opens. There is unusually wide scatter in the
turbulence measurements at this crank-angle, for unexplained reasons. The data
rate is low (~ 150 Hz) and there is a mixture of TiO_2 and H_2O particles. By 490°,
however, the water droplets totally dominate the signal, which is both frequent
(~ 1 kHz) and strong. The data rate and signal quality during the period of
rapid velocity rise are so good that real-time representation of the results would
be possible. The scatter in measurements at the top of the velocity rise is due
to the extreme time-gradients, and perhaps cycle-by-cycle variations in the flow,
but apparently not to uncertainties in the individual measurements. However, by
570° conditions become somewhat difficult again. The shadowgraph movies indicate
that the water droplets may now be very large and strongly influenced by centrifugal forces. As the number of water droplets becomes fewer and fewer due to dynamic dropout, the TiO_2 begins to become visible again.

610-660°: Only TiO_2 particles appear to be present during this region, and
signal discrimination is straightforward. However, when the background noise
level becomes appreciable after 660°, the signal is no longer discernible from
the noise. Because background light conditions here should not be appreciably
different from those encountered on either side of 360°, it must be presumed that
the water condensation process and its subsequent removal from the system detri-
mentally affected the TiO_2 particles.

LASER VELOCIMETER/HOT-WIRE ANEMOMETER COMPARISON

A comparison between hot-wire and LDV measurements is presented in Figure 4.
The hot-wire was oriented parallel to the cylinder axis. Also shown for the intake stroke is the mean velocity in the orifice as predicted by one-dimensional
continuity calculations; that is, piston speed times the piston-to-orifice area
ratio of 32.7. During intake the mean velocity hot-wire results are identical
for the two wire-operating-temperatures of 350 and 500°C, but are consistently
lower than the LDV measurements. The difference is almost 10 percent, but the
cause is not known. The hot-wire data evaluation procedure does assume an ambient temperature intake process, which is most likely not precise; however, the
error in this assumption could not be large enough to account for the discrepancy
shown.

The period of gas temperature rise due to compression lies between 240 and
480°. For the first half of this period the hot-wire and LDV mean velocity
results are in agreement because this was one of the criteria used to develop
the hot-wire temperature-correction procedure. Because the other criteria used
was that the hot-wire results at the two different wire temperatures also coin-

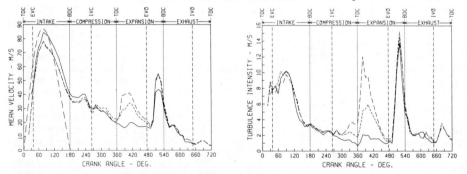

Figure 4. Comparison between LDV (solid line) and hot-wire results (long dashes,
350°C wire temperature; short dashes, 500°C). The long dashed line shown only
during intake is the mean velocity calculated from continuity for a 1.02 cm
square channel.

cide, the first criteria cannot be attributed to completely predetermining the result. Also, since the gas temperature does continually change from ambient to 230°C at TDC, it is unlikely that the agreement between the hot-wire and LDV results shown for mean velocity before TDC could be totally fortuitous. However, the large disagreement after TDC between all three measurements does indicate a very real problem. Two explanations for this difference will be discussed, both related to the observation that the responses of the two wires at different temperatures also do not agree.

The lack of agreement in turbulence results before TDC suggests that wall heat transfer effects may not be insignificant, as has been assumed. To the first order, turbulence fluctuations can be assumed to be linear perturbations from the mean (particularly before TDC, where the relative turbulence intensity measured by LDV is ~ 5 percent). If a linear behavior is the case, then the turbulence measurements for the two wire temperatures should show the same degree of coincidence as the mean velocity results. Because this is not the situation seen in Figure 4, and because the more-sensitive-to-temperature low-temperature wire shows the higher turbulence values, it appears that temperature fluctuations generated by heat transfer to the cold walls may be having an effect. Similarly, if the valve prechamber walls are cooler than the friction-heated cylinder walls, it is plausible to expect that the gas charge in the prechamber is cooler than the clearance volume air, such that the discharge from the prechamber flowing by the wire after TDC may be at a lower temperature than predicted for an adiabatic process; thus the large increase and differences in both the mean velocity and turbulence results.

Another considered explanation for these discrepancies is that thermal lag has not been accounted-for in the hot-wire temperature-correction procedure. As discussed earlier, convergence of the hot-wire and LDV results for the compression heating period prior to TDC was accomplished by assuming that the wire support needles are in temperature equilibrium with the gas. This boundary condition for the ends of the wire is used to separate conduction losses from the total heat transfer loss measured for the wire. It is not unreasonable to expect the temperature of the supports to lag the gas temperature, but if this were a significant effect the calculated conduction loss assuming no lag would be too high, leaving less heat loss due to convection, and thus giving lower than expected values for velocity. (That is, low with respect to the results for TDC assuming no lag. The absolute values of velocity would depend on knowledge of the actual lagging temperature profile.)

The agreement between hot-wire and LDV results during blowdown (500-560°) is quite remarkable, considering the highly transient nature of the process. Because the presence of water droplets in large numbers would considerably influence the response of a hot-wire anemometer, it is not evident whether quantitative agreement should even be expected in this region.

CONCLUSIONS

It has been shown that LDV measurements can be made successfully throughout the complete cycle of a motored internal combustion engine (with the apparent exception of late in the exhaust stroke and very early during intake). The same cannot be said of hot-wire measurements, since there are still unresolved discrepancies related to the wire's sensitivity to the gas temperature. However, for most engineering applications it is not necessary to include the full engine cycle, since it is usually of greater value to spatially map the turbulent flowfield during brief periods of primary interest in the cycle, such as just prior-to and including fuel injection, ignition, prechamber outflow, etc., depending on the engine type. Because of the need for spatial distribution information, optical access is probably the most restrictive limitation to the application of LDV, particularly if the engine under study closely resembles a production design.

Extension of the study described here to a combusting engine is not anticipated to produce any severe new problems, except when looking directly into combusting gases. Since this situation would occur in the vicinity of TDC, when the SNR is already very low in the motored case, combustion-related emissions can be expected to overwhelm the remaining signal. Solution of this problem would most likely require changing to a forward-scatter system, because to do so not only improves the signal strength, but perhaps even more importantly removes the background reflection problem that has been shown to be the limiting condition in a backscatter arrangement when looking directly down at the piston surface.

ACKNOWLEDGMENT

The author would like to acknowledge the assistance of T. M. Dyer for making the shadowgraph movies, and J. F. Lienhard and J. E. Fertig, who assisted as the mechanical and electrical technicians for this program. This work was funded by the Power Systems Division of the Department of Energy and the Motor Vehicle Manufacturers Association.

REFERENCES

1. Clerk, 1912. Fifth Report Gaseous Explosions Committee, British Association Report; Reported in Bone, W. A., and Townend, D. T. 1927. Flame and Combustion in Gases. Longmans, Green and Co. Ltd., London.

2. Witze, P. O. 1977. Measurements of the Spatial Distribution and Engine Speed Dependence of Turbulent Air Motion in an I.C. Engine. Paper 770220, presented at the SAE International Automotive Engineering Congress and Exposition, Detroit, Michigan, February 28-March 4.

3. Dent, J. C., and Salama, N. S. 1975. The Measurement of the Turbulence Characteristics in an Internal Combustion Engine Cylinder. Paper 750886, presented at the SAE Automobile Engineering Meeting, Detroit, Michigan, October 13-17.

4. Lancaster, D. R. 1976. Effects of Engine Variables on Turbulence in a Spark-Ignition Engine. Paper 760159, presented at the SAE Automotive Engineering Congress and Exposition, Detroit, Michigan, February 23-27.

5. Tindal, M. J., Williams, T. J., and El Khafaji, A. N. A. 1974. Gas Flow Measurements in Engine Cylinders. Paper 740719, presented at the SAE Powerplant Meeting, Milwaukee, Wisconsin, September 9-12.

6. Witze, P. O. 1977. Hot-Wire Measurements of the Turbulence Structure in a Motored Spark-Ignition Engine. Report SAND77-8233, Sandia Laboratories, Livermore, California.

7. Collis, D. C., and Williams, M. J. 1959. Two-Dimensional Convection from Heated Wires at Low Reynolds Numbers. J. Fluid Mech. 7:357-384.

8. Davies, P. O. A. L., and Fisher, M. J. 1964. Heat Transfer from Electrically Heated Cylinders. Proc. Roy. Soc. A280:486-527.

9. Horvatin, M. 1969. Some Problems Concerning the Analytical Evaluation of the Characteristics of Hot-Wires Immersed in a Fluid of Variable Pressures and Temperatures. DISA Information No. 8, Herlev, Denmark.

10. Bradbury, L. J. S., and Castro, I. P. 1972. Some Comments on Heat-Transfer Laws for Fine Wires. J. Fluid Mech. 51:487-495.

Velocity Measurements inside the Cylinder of a Motored Internal Combustion Engine

RODNEY B. RASK
Fluid Dynamics Research Department
General Motors Research Laboratories
General Motors Technical Center
Warren, MI 48090

ABSTRACT

This research is an evaluation of the capability of the LDA for making measurements inside a piston-engine cylinder. The backscatter mode of LDA operation, requiring only a single window in the cylinder, was used. Measurements of instantaneous velocity were successfully made at low speed in a motored engine. Using a single-component system incorporating a Bragg cell, non-simultaneous measurements of the two velocity components parallel to the face of the piston were made. The basic problem in making the measurements is one of minimizing extraneous light reflected back into the optics from various cylinder surfaces. Measures necessary to get a countable signal-to-noise ratio with the present commercial system are discussed.

A digital, minicomputer-based, data acquisition system recorded velocity-crankangle pairs of data. The Doppler-burst processing was done with a TSI counter and the crankangle was monitored with a digital-output optical encoder. An ensemble-average, data-processing technique has been developed to analyze the raw velocity-crankangle data and produce smooth curves of mean velocity and rms velocity fluctuation as functions of crankangle. The issue of discriminating mean velocity and turbulence in an engine cylinder is discussed briefly.

Sample results of mean velocity and rms velocity fluctuation are presented for several locations in the clearance volume. They show the basic vortical nature of the flow in this high-swirl chamber. Rms velocity fluctuation is produced during the intake process and distributes itself rather uniformly during the compression stroke.

INTRODUCTION

The fluid motions inside an engine cylinder strongly influence the performance of an engine. The instantaneous mean velocity and turbulence affect flame propagation through a cylinder. There is evidence to indicate that the speed of a flame front is dependent on absolute turbulence intensity [1]*. As engines become more sophisticated to meet economy and emissions standards, it may be necessary to improve our present understanding of how the flow field in an engine influences its operation. For instance, swirl is included in many stratified-charge engines to promote and maintain stratification. In order to properly utilize fluid motions, both mean and turbulent, one must know which motions produce what effects. Inherent in the acquisition of this knowledge is an ability to measure in-cylinder fluid velocities.

---

* Bracketed numbers designate References listed at the end of this paper.

In the past, engine-cylinder velocity measurements have been made with hot
wire anemometers. References 2-4 are some of the latest and best of these
measurements. The present work is an application of an LDA (Laser Doppler
Anemometer) system to the measurement of in-cylinder fluid velocity in a motored
engine. Researchers are just beginning to apply the LDA to engine-cylinder
velocity measurements, and references 5, 6, and 7 report their experiences.

APPARATUS

Measurements were made in an L-head, opposed-piston, 2-cylinder, Onan
engine fitted with a special cylinder head. This head contained a round window
over the whole piston, allowing an unobstructed view into the chamber. A de-
flector in the L-head region between the valves made the flow enter the cylindrical part of the combustion chamber as a nearly tangential jet. A sketch of
the cylinder-valve-deflector configuration, as it was seen by the LDA, is shown
in Figure 1.

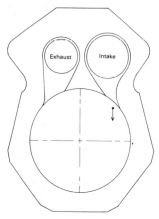

All of the measurements reported in this paper
were made in the clearance volume between the piston and the inner window surface. The measurements
were made 6.5 mm from the window surface; this is
about two-thirds of the distance from the window
to the piston (at top dead center). All measurements reported were made at approximately 270 rpm,
although some measurements were also made at
800 rpm. Measurements have not yet been made at
greater engine speeds.

A standard TSI (Thermo Systems Inc.) one-
component, backscatter system was used. The laser
was generally operated at 0.65-0.9W in the 514.5 nm
line. All measurements were made with a 250 mm
focal length focusing lens. Signal processing was
done by a TSI 1096 counter. The absolute position
encoder (Baldwin Electronics Inc. model 5V233CGXL)
provides 0.1° resolution and has BCD digital output. It was geared down 2 to 1 to distinguish
between the four strokes of the engine cycle.

FIGURE 1. CYLINDER SCHEMATIC.

When activated, the minicomputer was ready to receive data from the counter and
the crankangle encoder. The counter sent a pulse indicating that a measurement
had been validated and outputted, and the minicomputer inhibited both the
counter and the crankangle encoder. The crankangle was checked to see if it was
within the selected window (the crankangle window was put in the minicomputer
before the run). If it was outside the window, the minicomputer removed the
inhibits and another measurement was taken. If the crankangle was within the
window, the counter and crankangle encoder readings were stored in memory and
the inhibits then removed. The minicomputer stored data until its memory was
full, 7500 raw velocity-crankangle pairs. The data was sent from the minicomputer to an IBM 370/145 system over an ordinary phone line using a 1200 baud
acoustic coupler. The raw data was then transfered to an IBM 370/168 system
where the binary was converted to decimal and the appropriate conversion factors
applied to give a file of 7500 velocity-crankangle (m/s-degrees) pairs. This
ended the data acquisition phase.

OPERATION OF THE SYSTEM

The primary measurement problem was caused by too much spurious light scattered back into the optical system. The plane of the beams with respect to the
window was skewed so that the beams reflected off the window surface did not
come back into the optics. The focusing lens was masked to a rectangular

opening, 20 by 35 mm, just outside of the lens itself. It was found that optical quality quartz was the best window material; it was nearly impossible to make measurements through a plastic window. The Bragg cell-beam splitter module was separated from the backscatter module. When the beams reached the backscatter module, the various orders produced by the Bragg cell were displaced and could easily be blocked off.

In order to get good signal-to-noise ratio, great care had to be taken to ensure that everything was properly grounded. It was extremely important to have good grounding and shielding for the line connecting the power supply to the Bragg cell. Otherwise, this line broadcasted and the signal was picked up in the photomultiplier electrical wiring.

A salt-water solution was nebulized and dried to produce seed particles for the flow. The salt solution was mixed to give a 1μ particle [8] after the droplet dried. The intensity (voltage to crystal) of the nebulizer used to generate the small water droplets and the air flow rate through the particle generator were adjusted to give "good particles." To determine if the generator was working properly, the Bragg cell was removed from the system and the particles measured as they left the hose from the particle generator. The voltage and flow rate were adjusted so that nearly all of the particles had close to unity visibility when observed on the scope. The particle-sizing work of various investigators (see, for example, References 9 and 10) indicates that we should have seen close to unity visibility for a 1μ diameter particle and a 2.6μ fringe spacing. It appeared that significantly higher data rates could have been achieved if there were more particle generation capacity.

Initially there was a problem with water droplet condensation on the expansion stroke (motored engine) which prevented measurement for a short crankangle increment. When a polaroid filter was installed outside of the focusing lens, this data gap was nearly eliminated. A slight gap can be seen near 430° in the data of Figure 2. The existence of a gap also depends on where in the cylinder the measurement is made and, presumably, on the humidity of the intake air.

Figure 2 shows a sample data set after conversion of velocity to m/s and crankangle to decimal numbers. The data was taken over the full 720° of crankshaft rotation (no crankangle window). It was measured in the clearance volume at the location shown in Figure 1; the short arrow indicates the positive direction of velocity. The data of Figure 2 show modest scatter in most regions. Certainly there is no evidence that a significant number of bad data points (measured noise) were produced. A striking feature is the gaps in the data. Near TDC the piston came close to the control volume and scattered too much light back into the receiving optics. This precluded velocity measurement, leaving the two gaps near TDC.

Data rates ranged from 200-1500 measurements/second (no crankangle window) depending on how well the particle generator was working and the cleanness of the window. A typical rate was 800 measurements/second. The window was usually cleaned after 2-4 runs depending on how fast the window got dirty and where the measurement was being made (the window did not get dirty uniformly).

DATA ANALYSIS

Analyzing data from engine measurements is not as straightforward as one might first expect. The basic problem stems from the time-dependent nature of the fluid motions in a cylinder. The analysis of hot wire measurements in a motored-engine cylinder has been discussed by Lancaster [2] and Witze [3,4], who differ on the "best" data analysis procedure.

In a typical steady flow situation it is conventional to decompose the instantaneous fluid velocity into a mean velocity and a turbulent fluctuation.

$$U(t) \equiv \overline{U} + u(t) \tag{1}$$

Here U(t) is the instantaneous velocity, $\overline{U}$ the mean velocity and u(t) the turbulent or fluctuating portion of the velocity. For the steady flow case the
mean is a simple time average which is unambiguous and physically meaningful;
no real problems of concept or definition exist.

Within a piston-engine cylinder the fluid motions are time dependent
because of the inherent cyclic nature of the device. Mean velocity is now a
function of time rather than a constant. Suppose we switch from time to crank-
angle for the engine case; then,

$$U(\theta) \equiv \overline{U}(\theta) + u(\theta) \tag{2}$$

is analogous to the steady case (equation 1) except that $\overline{U}$ is no longer a constant. Remembering that $U(\theta)$ is the instantaneous velocity which actually
exists in a flow, it is clear that the problem is a matter of assigning some of
this velocity to $\overline{U}$ and the rest to u. As it is virtually impossible to make a
meaningful split by considering only an instant in time, we must average in some
manner.

Here θ is the angle through which the crankshaft has turned since the start
of the velocity measurement; it increases monotonically from zero to its value
at the end of the measurement, which may be many cycles later. We will distinguish between θ and ϕ, where ϕ is the crankangle within any cycle relative to
some chosen reference point in that cycle (in this paper, ϕ is measured from TDC
on the intake stroke). If we let the subscript i designate the engine cycle
(720° = 4π radians per cycle), counted from the time measurement begins, we can
write

$$\theta = \phi + 4\pi(i-1) \tag{3}$$

where i goes from 1 to N, and N is the number of cycles that are measured.

It is common to use ensemble averaging to define the instantaneous mean
velocity as a function of crankangle [2,3,4]. To ensemble average we write

$$\overline{U}_E(\phi) \equiv \frac{1}{N} \sum_{i=1}^{N} U(\phi,i) \tag{4}$$

where $U(\phi,i) \equiv U[\phi+4\pi(i-1)]$ is the instantaneous velocity at crankangle ϕ during
the ith cycle of the measurement; each cycle will be called one velocity record.
The subscript E indicates that $\overline{U}$ is an ensemble average.

The direct definition of absolute turbulence intensity then follows as

$$u'_E(\phi) = \sqrt{\frac{\sum_{i=1}^{N}[U(\phi,i) - \overline{U}_E(\phi)]^2}{N}} = \sqrt{\frac{\sum_{i=1}^{N}[u_E(\phi,i)]^2}{N}} \tag{5}$$

where $u_E(\phi,i) \equiv U(\phi,i) - \overline{U}_E(\phi)$.

By subtracting the ensemble average from each velocity record in this way,
all velocity differences or fluctuations from the ensemble average are included
in turbulence.

It is important to examine what effect a particular averaging scheme has
on splitting the instantaneous velocity into its mean and turbulent components,
and whether this split is the best choice for engine measurements. As seen
above, the ensemble average never really exists in one cycle, it only exists in
the average. It is important to realize that what happens in each combustion
cycle is determined only by the conditions which exist in that particular cycle.
The fluid motions in an engine cylinder are essentially restarted with each

cycle. Thus, looking back at equation (2), we realize that at a given ϕ, $\overline{U}(\theta) \equiv \overline{U}[\phi + 4\pi(i-1)]$ may not be the same for each cycle. Consequently, the use of an ensemble-averaged $\overline{U}(\phi)$ to separate $u(\theta)$ out of $U(\theta)$ introduces a certain degree of approximation into the analysis of turbulent fluctuations in an engine.

The problem, as outlined, is that it is difficult to get a perfect averaging scheme that will tell us what $\overline{U}(\theta)$ is for every single cycle. Ideally, the scheme should also allow $\overline{U}(\phi,i)$ to vary from cycle-to-cycle. One can accommodate the latter by defining a cycle-to-cycle variation in velocity, $\hat{U}(\phi,i)$, where

$$\hat{U}(\phi,i) \equiv \overline{U}(\theta) - \overline{U}_E(\phi) \equiv \overline{U}(\phi,i) - \overline{U}_E(\phi) \tag{6}$$

Here the cycle-to-cycle term at a given ϕ is the difference between the instantaneous average velocity in a particular cycle and the average velocity over many cycles obtained by ensemble averaging.

If consideration is taken of cycle-to-cycle variation, the instantaneous velocity in an engine cylinder can be expressed as

$$U(\phi,i) \equiv \overline{U}_E(\phi) + \hat{U}(\phi,i) + u(\phi,i) \tag{7}$$

Lancaster [2] used this equation along with the assumption that for each cycle $\hat{U}(\phi,i)$ was a constant over the 45° incremental crankangle range used in his data reduction, i.e. $\hat{U}(\phi,i) = \hat{U}(i)$.

While it is most desirable to use equation (7), there is the problem of defining the cycle-to-cycle term or, alternatively, of defining $\overline{U}(\phi,i)$. Probably the best method is to use a smooth continuous curve through the data of any cycle to get $\overline{U}(\phi,i)$. The smooth continuous curve preferred by the present author is either a finite Fourier series or a cubic spline. In order to curve fit properly one should have a continuous or nearly continuous velocity record for each cycle. This is not the case in the present study where the LDA data rate is rather low. In order to account for cycle-to-cycle variations one must be able to process every single cycle by itself.

At the present time, the magnitude and importance of cycle-to-cycle contamination of turbulence are unclear. In any event, the current LDA data are not dense enough to allow data processing for each cycle. Consequently, an ensemble-averaging scheme has been used for the present data analysis, and the term "rms velocity fluctuation" is used to indicate the calculated result. The scheme is outlined below.

The first step was to split the data into 1° windows. A velocity was then calculated for each window by simply arithmetically averaging the data points which fell in that window. Next, an average-velocity curve was fit to the average window values. The cubic-spline smoothing routine ICSSCU [11] from the IMSL (International Mathematical and Statistical Libraries, Inc.) subroutine library was used to fit the data. It is specially tailored for data that is known to be significantly random in nature, and it gives a continuous curve whose smoothness is controlled by input parameters. One of the required inputs to ICSSCU is a weighting function; it should be the standard deviation of the window values. This standard deviation was estimated by considering both the scatter in the data and the number of points which went into making a window average.

This mean-velocity curve was fit by iteration; each time through, the estimate of standard deviation was improved. After the mean-velocity fit was accomplished, the rms velocity fluctuation in each window was determined from the difference between the measured velocities and the fitted mean-velocity curve. The rms velocity fluctuation for all the windows was then fit with the cubic spline routine. At this point the data were screened to throw out points beyond 5 standard deviations from the fitted mean-velocity curve.

Using the reduced data set, the mean velocity and rms velocity fluctuation were again fit with the spline curves. This time, all points lying beyond 4 standard deviations were thrown out. Finally, the mean velocity and rms velocity fluctuation were again fit to the remaining data by an iterative scheme in which the standard deviations used as weighting functions were taken from the curve-fit values of rms velocity fluctuation. The final results are smoothed curves of both mean velocity and rms velocity fluctuation as functions of crankangle.

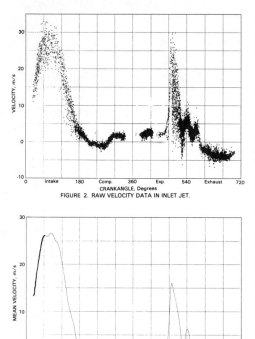

FIGURE 2. RAW VELOCITY DATA IN INLET JET.

FIGURE 3. MEAN VELOCITY IN INLET JET.

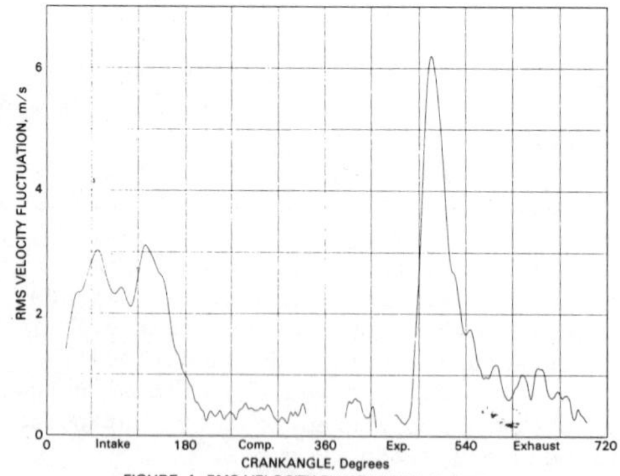

FIGURE 4. RMS VELOCITY FLUCTUATION IN INLET JET.

A nominal amount of smoothing was prescribed for the fitted curves, with statistically "expected values" used for the smoothing parameters. One could use more or less smoothing than this, as desired. This nominal smoothing and 1° window size give frequency resolution which should be more than adequate for mean-velocity fluctuations (fluctuations with a 10° period are resolved with minimal attenuation). The amount of smoothing used is important because it affects what gets prescribed as mean velocity and what gets prescribed as rms velocity fluctuation.

Figure 3 presents the mean velocity determined from the data pairs shown in Figure 2. Note the large peak in the velocity curve during the intake stroke. There is a slight region of negative flow just as the intake valve is closing. The velocity during compression and during expansion is rather small. When the exhaust valve opens there is a strong reverse flow past it (motored engine) into the cylinder, causing the velocity peak near 490°. The velocity then drops smoothly through the rest of the cycle. There is significant negative flow late in the exhaust stroke. The rms velocity fluctuation is shown in Figure 4. There is high intensity during the time of inlet flow and the time of reverse flow past the exhaust valve. This engine has considerable reverse flow past both the inlet and exhaust valves. During compression the rms velocity fluctuation is rather low and constant in magnitude.

Figures 2-4 illustrate the operation of the data analysis scheme. As indicated in Figure 1, the measuring point was in the inlet jet so that, at least during the intake stroke, one knows the nature of the mean flow. This set of data was chosen to be presented for just this reason.

VELOCITY INTERPRETATION

There exists a general problem of interpreting velocity measurements which merits discussion at this time. The basic hope is to reconstruct a characterization of the velocity field in an engine cylinder from measurements at a limited number of points. If the flow field is simple, a modest number of points may provide an understanding. However, if the flow field in a cylinder is complex, measurements at a large number of points may be necessary. Present results indicate that the flow field in a motoring engine can be very complex during certain parts of the cycle when the valves are open.

The present data are non-simultaneous measurements of two orthogonal components of velocity at the given locations. When one is looking at averaged quantities it is not a significant disadvantage that the measurements are not simultaneous. However, to recognize certain coherent structures it may be necessary to look at simultaneous measurements of two components.

It is not clear which parameters are most important in influencing engine operation. The velocity field characterization might include mean velocity, turbulence intensity, length scales, time scales, frequency distributions, etc. Perhaps one might be interested in different parameters for controlling different functions. For example, mean-velocity may be of most interest in a Diesel or stratified charge engine because of its effect on fuel/air mixing, whereas turbulence intensity may be of more interest in a homogeneous-charge engine because of its influence on flame speed.

MEASUREMENT UNCERTAINTY

If it is assumed that the system is measuring on a real Doppler burst, the error in the counter measurement of time is 0.25% at 20 MHz [12]. The digital data system uses 10 binary bits, for approximately 0.25% accuracy. The beam angle was measured by TSI and specified to 0.1%. The Bragg-cell frequency shift introduces some error due to the frequency uncertainty of the shift and the downmix circuits; this error is approximately 0.1%. The total error of a Doppler burst measurement is therefore approximately 0.7%, an extremely small value. The absolute crankangle is accurate to approximately 2° with 0.2° resolution. The dynamometer controlled engine speed to 5%; any speed variation shows up as velocity fluctuation.

There is also some influence of velocity bias on the measurements. Velocity bias occurs because more high-velocity particles go through the measurement volume than low-velocity particles (see references 13 and 14). The present data have not been corrected for bias.

The source of error with the greatest potential for causing large uncertainty is the measurement of noise instead of Doppler bursts. Operation of the system and counter setup were checked by confirming that no-particle-seeding corresponded to zero data rate. It was also found useful to block one beam and make sure that the data rate went to zero. As has been described, the data were screened at 5 and 4 standard deviations, and only about 25 points were usually rejected out of 7500. Generally, measurements made on noise fall considerably away from Doppler burst measurements. Non-screened data, such as that shown in Figure 2, confirm that there is little problem with bad, noise-based measurements in the present data.

The ensemble averaged mean velocity and rms velocity fluctuation were calculated from a limited number of velocity measurements and are, hence, subject to statistical sampling uncertainty. Generally, the error in the mean is proportional to the ratio of the relative rms fluctuation to the square root of the number of samples. Similarly, the error in the rms fluctuation is inversely proportional to the square root of the number of samples. It is not clear exactly how the smoothing routine affects the statistical uncertainty.

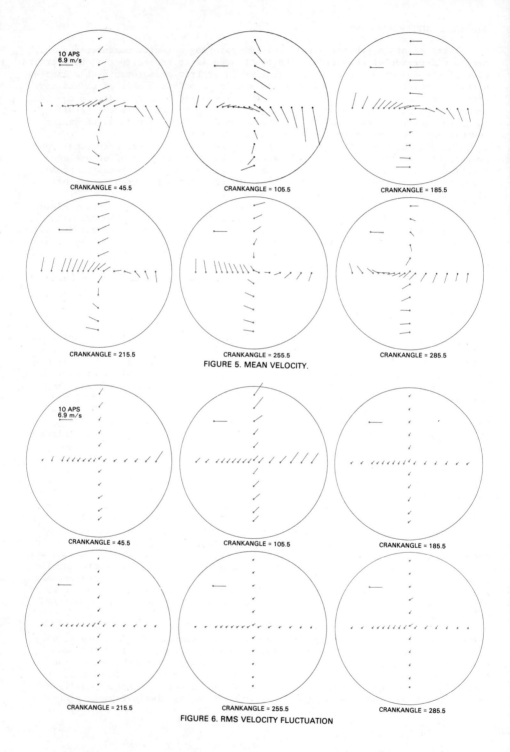

FIGURE 5. MEAN VELOCITY.

FIGURE 6. RMS VELOCITY FLUCTUATION

It appears that a relevant parameter should include samples/crankangle and maximum frequency of fluctuation in the velocity. As can be deduced from Figure 2 the sample rate varies with crankangle. At the present time a detailed analysis of the statistical uncertainty has not been done. Based on experience with artificial data it would appear that the statistical uncertainty is a few percent.

MEASURED VELOCITIES

Velocities were measured along vertical and horizontal cylinder diameters as seen in Figure 1. Both components of velocity were measured (non-simultaneously) at 26 measurement stations. The measurements were made in the clearance volume 6.5 mm from the window. The engine was operated at 270 rpm for all 52 data runs. Mean velocity and rms velocity fluctuation are presented in Figures 5 and 6.

The figures are intended to show what kind of measurements can be made in an engine cylinder, and the type of data processing/handling which is necessary to attempt a meaningful characterization of the flow. The curves presented are for sample/representative crankangles; similar curves have been produced at 1° crankangle increments over the intake and compression strokes.

Figure 5 shows mean velocity at the 26 locations in the cylinder for 6 crankangles during the intake and compression strokes. The cylinder is oriented as in Figure 1 with the inlet valve in the upper right. The circles indicate locations where velocities were measured, and the lines indicate the magnitude and direction of the local velocity. The short line in the upper left quadrant gives a reference velocity scale of 10 times the average piston speed.

The figures reveal a type of vortical flow in the cylinder. The apparent center of rotation moves clockwise during the intake stroke and counterclockwise during the compression stroke. The velocity distribution is neither solid body nor free vortex rotation, but is something more complicated. The apparent center of rotation can move around the cylinder rather rapidly as illustrated by its movement from the 4th quadrant to the second quadrant in the 215.5° to 285.5° sequence.

Figure 6 presents the rms velocity fluctuation (for two components) for the same locations and crankangles as the mean-velocity results shown in Figure 5. The plotted lines are the vector sum of u' and v' where U and V designate the horizontal and vertical components of velocity as referenced to Figure 1. u' and v' are the rms values of the turbulent part of the velocity.

After the intake process is over, the fluctuations throughout the cylinder do not vary greatly from one location to the next or from one crankangle to the next. In a low squish engine, the common assumption is that turbulence is generated primarily by the intake flow. The turbulence is then convected around the cylinder by the mean motion. Certainly the results of Figure 6 are in general agreement with this idea.

All of the measurements shown in Figures 5 and 6 are for a single plane in the cylinder. In order to better characterize the flow one should have measurements at several planes. Also, there is a third component of velocity parallel to the cylinder axis which has not been measured. Certainly the existence of any complex, three-dimensional flow pattern can confound the interpretation of two-dimensional results such as those in Figures 5 and 6. The flow in an engine cylinder can be very complex, and an extensive set of measurements is necessary to characterize even the mean velocity field.

REFERENCES

1. D. R. Lancaster, R. B. Krieger, S. C. Sorenson, and W. L. Hull, "Effects of Turbulence on Spark-Ignition Engine Combustion," SAE paper 760160, Detroit, Michigan, February 23-27, 1976.

2. D. R. Lancaster, "Effects of Engine Variables on Turbulence in a Spark-Ignition Engine," SAE paper 760159, Detroit, Michigan, February 23-27, 1976.

3. P. O. Witze, "Measurements of the Spatial Distribution and Engine Speed Dependence of Turbulent Air Motion In an IC Engine," SAE paper 770220, Detroit, Michigan, February 28-March 4, 1977.

4. P. O. Witze, "Hot-Wire Measurements of the Turbulence Structure in a Motored Spark-Ignition Engine," Sandia Laboratories Energy Report SAND77-8233, May 1977.

5. P. Hutchinson, A. Morse, and J. H. Whitelaw, "Velocity Measurements in Motored Engines: Experience and Prognosis," SAE paper 780061, Detroit, Michigan, February 27-March 3, 1978.

6. G. Wigley and M. G. Hawkins, "Three Dimensional Velocity Measurements by Laser Anemometry in a Diesel Engine Cylinder Under Steady State Inlet Flow Conditions," SAE paper 780060, Detroit, Michigan, February 27-March 3, 1978.

7. P.O. Witze, reported in "Application of Advanced Laser Diagnostics to Combustion Studies in Automotive Engines," Progress Report 5, Covering the Period December 1, 1977 through January 1, 1978, Sandia Laboratories Case Number 0760.020, January 20, 1978.

8. "Operating and Service Manual, Model 3060 Particle Generator," Thermo-Systems Inc., St. Paul, Minnesota.

9. W. M. Farmer, "Measurement of Particle Size, Number Density, and Velocity Using a Laser Interferometer," Applied Optics 11, 2603-2612 (1972).

10. R. J. Adrian and K. L. Orloff, "Laser Anemometer Signals: Visibility Characteristics and Application to Particle Sizing," Applied Optics 16, 677-684 (1977).

11. Subroutine ICSSCU, International Mathematical and Statistical Libraries, Inc. (IMSL) Subroutine Package. May 1976 Description. See Also C. H. Reinsch, "Smoothing by Spline Functions," Numerische Mathematik, 10, 177-183 (1967).

12. "Laser Anemometer Systems," Thermo-Systems Inc., St. Paul, Minnesota (1976).

13. D. K. McLaughlin and W. G. Tiederman, "Biasing Correction for Individual Realization of Laser Anemometer Measurements in Turbulent Flows," The Physics of Fluids, 16, 2082-2088 (1973).

14. W. Hoesel and W. Rodi, "New Biasing Elimination Method for Laser-Doppler Velocimeter Counter Processing," Review of Scientific Instruments 48, 910-919 (1977).

Velocity Measurements in the Manifold
of an Internal Combustion Engine

M. L. YEOMAN and A. TAYLOR
UKAEA, Materials Physics Division
AERE Harwell
Oxfordshire, England

ABSTRACT

Time averaged mean and root mean square fluctuations of velocity have been sampled for selected time intervals at predetermined points in the pulsed flow through manifolds attached to motored and firing petrol engines. Timing and data storage systems were developed for synchronizing signals produced by the LDA technique to the engine cycle. Eight histograms of velocity were accessed for one engine rotation. Silicon droplets were used as scattering centers in the presence of substantial wall films. Averaged mean velocities show strong engine cycle dependence but little spatial structure. Velocity fluctuations were unrelated to the mean velocity.

INTRODUCTION

The composition and behaviour pattern of air-fuel mixtures prior to combustion are of primary importance to the efficient and sanitary operation of a petrol engine. As the first phase of a study of pre-ignition fuel characteristics laser doppler anemometry techniques have been applied to pulsed air flows in a motored engine manifold. The objectives were to obtain accurate data sets of $\bar{V}$ and $\bar{V}^2$, identify areas of rapid flow and acceleration, compare the relative merits of several ways of operating laser anemometry techniques applied to this situation and identify possible problem areas introduced by wet-wall conditions.

TRIUMPH MANIFOLD ASSEMBLY

A 2 litre Triumph Dolomite engine was fitted with a manifold consisting of a metal frame with perspex top and bottom plates. Silicon oil droplets were used for seeding purposes and introduced through a carburettor tailored to reproduce the internal geometry of an operating carburettor. The engine was motored with a 40 horse power motor at a constant revolution rate of 1460 rpm. Coupling the motor to the engine through the gearbox resulted in speeds of from 1460 rpm in top gear to 4000 rpm in bottom gear. A Ricardo-Cousins baffled orifice plate with thermocouples measured the flow rate and temperature of air entering the carburettor. The manifold vacuum was measured by a mercury manometer via a pressure tapping on the manifold frame. An encorder was connected to the crank shaft by a pulley with a 2:1 reduction in revolution rate. The single pulse from the encorder was matched to the TDC point of the engine cycle with a known valve opening sequence.

    Detailed commissioning of the manifold assembly and an integrated optical
unit indicated that best quality signals were obtained in the reflected forward
scatter mode. For scanning purposes the two parallel beams produced by the
optical unit were deflected by a high quality mirror into a plane parallel to
the plane of the manifold, Figure 1. After passing through a lens and
deflection by a second mirror the converging beams entered the manifold. Forward scattered light was reflected by a mirror secured on the bottom surface of
the manifold into the scanning system. After the imaging lens the scattered
light was almost parallel and was finally collected by a prism arrangement in
the optical unit. The collection system of the optical unit focused the image
of the probe volume in the manifold air stream onto a fast photodetector.

    A frequency shift of either sign was produced by a Kerr cell shifter placed
immediately after the Argon laser. The 514nm line of the laser was shifted by
± 5, ± 8MHz and focused to a probe volume measuring approximately 0.1mm x 0.1mm x
1.0mm containing 40 fringe pairs. The collection prism assembly in the optical
unit was scanned in a plane perpendicular to the plane of the laser beams to
improve the spatial resolution of the system, and discriminate against flare
from oil films and manifold surfaces.

OPTICAL SYSTEM

Figure 1. Optical Scanning System

TIMING AND SIGNAL PROCESSING

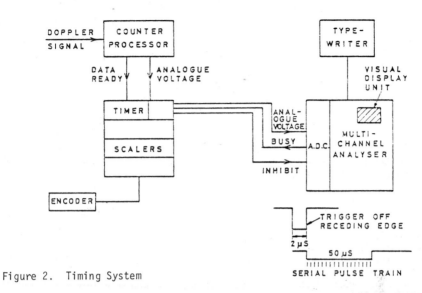

Figure 2.   Timing System

Velocity information was sampled for a selected time interval at a pre-determined point in the window cycle with the aid of an encorder attached directly to the drive shaft of the stepping motor. The encorder outputs a single pulse and two 360 pulse trains per revolution. The single pulse, called the TDC pulse, was used to trigger a delay unit and a set of scalers arranged so that the delay time could overlap the next TDC position. The two 360 pulse trains were used for timing purposes and to drive a revolution rate counter.

Figure 2 illustrates the system employed to gate the signal processing electronics. When a Doppler burst has been analyzed and a voltage is about to appear as an analogue voltage proportional to the Doppler frequency the Counter Processor outputs a 'data ready' signal which primes the Timer to receive the analogue signal from the Counter. The Timer selects a 2μs sample of the signal and feeds the sample to the A.D.C. of a Laben multichannel Analyzer. The Analyzer is inhabited by the Timer until the required point in the cycle relative to the T.D.C. pulse is reached when the Analyzer is allowed to collect data for a predetermined window period. As the 2μs pulse begins to turn over a 'busy signal' is produced by the Laben which is fed to the Timer inhibiting the system from accepting data from the Counter. When, after about 50μs, the 'busy signal' disappears the Timer waits for the next 'data ready' signal from the Counter and the process is repeated.

DATA COLLECTION

Single subgroup operation of the Laben multichannel analyzer accesses a total of 512 channels. The data is displayed as a velocity histogram for a predetermined channel width. Each point on the histogram corresponds to an air flow velocity since the Counter Processor outputs a voltage proportional to the instantaneous frequency which is proportional to the velocity of the scattering particle. The peak of the histogram for a symmetric distribution gives the mean velocity and the width of the curve is related to the variance or RMS value of

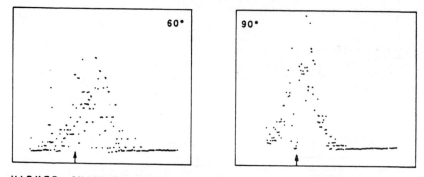

MARKED CHANNEL CORRESPONDS TO A FREQUENCY SHIFT OF
5MHz (10.4ms⁻¹) AND REPRESENTS AN AIRFLOW OF ZERO
VELOCITY

Figure 3.   Velocity Histograms of Pulsed Flow in a Manifold Assembly

the velocity fluctuations. An electronic typewriter was employed to print out
a hard copy of the number of counts in each channel at the end of a run. For
testing purposes a video screen provided a visual display facility. Figure 3
illustrates tracings of the visual display for a window interval of 10° in arm
4 of the manifold. The arrow at the base of each histogram represents a frequency shift of 5MHz and is essentially the point of zero velocity.

VELOCITY AND TURBULENCE MEASUREMENTS

The timing system was employed to cover a complete engine rotation with
the 360 pulse train from the shaft encoder. Information during the following
rotation was suppressed and the system accessed during the next repeat cycle of
the first engine rotation. A delay unit enabled the timing sequence to be started
at any point in the engine cycle in 5° steps. Information was collected during
time intervals of from 5 to 45 degrees of engine rotation and the 512 channels
of the Laben Analyzer were split into 8 subgroups of 64 channels. During a
complete piston cycle 8 histograms were recorded with a predetermined window
duration for each spatial location.

Data was recorded in arms 3 and 4 and in the top section of the manifold
with conditions corresponding to a firing engine speed of 50 mph, i.e. a manifold vacuum of 15 inches of mercury and a throughput of 34 cubic feet/min. To
obtain these conditions a speed of 2600 rpm was required. Quoted variance
values have been corrected for instrumental broadening and the contribution due
to engine vibrations.

Figure 4 illustrates the mean velocity as a function of engine cycle opposite the outer weir in the top section of the manifold. The velocity shows a
strong time dependence with a large variance.

The spatial dependence of mean velocity, Figure 5, shows relatively little
structure with a high variance, Figure 6, apparently unrelated to mean velocity.

The results quoted were obtained with a 10° window and the effect of larger
and smaller sampling intervals is shown in Figure 7. This is consistent with
previous findings on the effect of crank angle broadening on LDA measurements
in reciprocating engines.

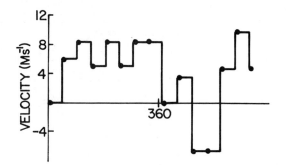

Figure 4.   Engine Cycle in Manifold Top Section

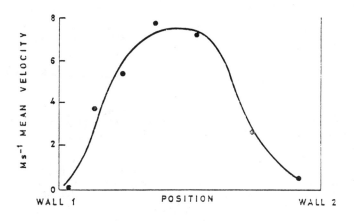

Figure 5.   Arm to Cylinder 4 Horizontal Scan

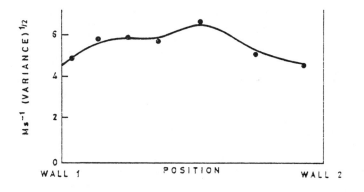

Figure 6.   Arm to Cylinder 4 Horizontal Scan

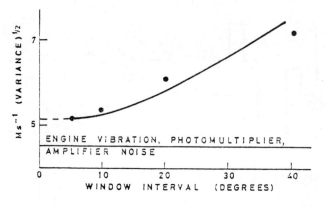

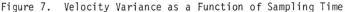

Figure 7.   Velocity Variance as a Function of Sampling Time

Laser doppler signals were obtained in backscatter with the engine in the firing mode at 2000 rpm scattering directly from petrol droplets.

SUMMARY

A timing and data storage system has been developed for synchronizing velocity information produced by the LDA technique to the cycle of a reciprocating engine. Eight histograms of velocity were accessed for one engine rotation.

Averaged air-flow velocities in the inlet manifold of a motored Triumph Dolomite engine indicate a strong dependence on position in the engine cycle but show little spatial structure. The velocity fluctuations are unrelated to the mean velocity which suggests that the variance does not originate in high shear regions.

Seeding with silicone oil resulted in adequate signals with the engine in the motored mode. In the firing mode the carburettor provided petrol droplets as seeding particles.

Laser Velocimeter Measurements in a Constant Volume Internal Combustion Engine Simulator

T. MICHAL DYER and PETER O. WITZE
Sandia Laboratories
Livermore, CA

ABSTRACT

The transient gas motion in a constant volume combustor has been studied with a dual-beam backscatter laser anemometer system, using a counter-type signal processor. This combustion 'bomb' closely simulates the combustion process near top-dead-center in a conventional automotive spark-ignition engine. The thin, cylindrical combustion chamber has pyrex windows on each end. A highly-swirling methane-air mixture was introduced into an initially evacuated chamber through a tangentially-directed valve in the side of the cylinder. Time-resolved tangential component mean velocity measurements have been made in both the pre- and post-combustion gases. High-speed shadowgraph movies have also been made, to illustrate both the fluid motion and the combustion process.

INTRODUCTION

Computer modeling of the combustion process inside an operating internal combustion engine is envisioned to be a valuable tool in the automotive design process. Detailed understanding of phenomena occurring in the combustion chamber will allow advanced engine concepts to be explored in search of higher efficiency, lower emissions, and greater fuel flexibility. Such models are currently being developed; however, to be viable tools they must be validated with experimental data taken from a variety of well-characterized, realistic experiments. One such experiment is a single-shot, constant volume combustor, designed to closely simulate the combustion processes occurring near top-dead-center in a conventional automobile engine. Because of its simplicity of concept and operation it is readily amenable to implementation of optical diagnostic methods, although the single-event aspect of the device does make the statistical measurement of turbulence time consuming. The velocity measurements reported here in the combustion bomb form an integral part of any complete data set aimed at computer modeling validation. Such measurements help to relate combustion bomb tests to the actual engine combustion process that is being simulated, and, when coupled with pressure and temperature distributions, provide the necessary initial conditions for time-dependent calculations.

EXPERIMENTAL APPARATUS AND PROCEDURES

The combustion bomb apparatus was designed and fabricated by Volkswagen Research in Germany and has been lent to Sandia as part of an international

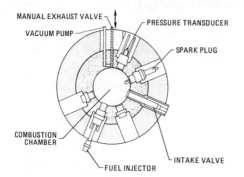

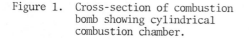

Figure 1.   Cross-section of combustion
            bomb showing cylindrical
            combustion chamber.

cooperative program. A schematic
cross-section of the device is shown
in Figure 1. The cylindrical combustion chamber is 80 mm in diameter;
two pyrex windows separated by 29 mm
form the ends of the chamber, providing complete optical access.

There are several features incorporated into the design of this
particular bomb which allow a close
simulation of actual engine combustion processes. The premixed fuel/
air charge is prepared in a well-
stirred, high pressure reservoir. A
fast response, shrouded poppet valve
is used for induction of the pressurized charge into the initially-
evacuated preheated vessel, producing
a highly swirling turbulent flow
within the combustion chamber. The
walls of the bomb are temperature controlled to simulate the operating wall
temperature in normal engines, taken in this instance to be 400 K. Timing for
the various events such as intake valve operation, ignition, and diagnostic
timing is preprogrammed on digital counters. Thus, once the experiment is
initiated, all events are automatically sequenced at their preset times.

A dual-beam backscatter laser velocimeter system was used together with
a counter-type signal processor (Macrodyne Model No. 2098). One watt of argon-
ion laser power was typically used. Because of the presence of combustion-
generated blackbody radiation and chemiluminescence, a narrow bandpass color
filter of 10 Å half-width was placed before the photomultiplier. The entire
optical system was mounted on a milling machine table, allowing three-dimensional
positioning of the measurement volume. The flow was seeded with nominally 0.2
μm titanium dioxide particles which were deposited onto the walls of the chamber
near the intake valve prior to each test. The inflowing gas jet then dispersed
the seed particles throughout the volume. Deposition of seed on the glass viewing windows made it necessary to clean them after each test. Because the seeding procedure was both non-quantitative and locally applied, the data rate was
found to vary significantly between tests and with position in the chamber, but
was typically 10^3-10^4/second.*

The signal processor was interfaced to an on-line PDP 11/34 minicomputer.
The computer was used to record velocity/time pairs, statistically process the
data, and display velocity histograms for each data set. This latter capability
allows a quick check on the consistency of the probability density distribution,
and is particularly useful in detecting the presence of statistically outlying
data points which can erroneously affect the turbulence measurements. The
velocity measurements were statistically processed using a time-ensemble averaging procedure. Specifically, the velocity data recorded for each individual
test was time-averaged in 10 ms increments (2 ms during combustion), and then
the test was repeated until the time-ensemble average for each window reached
a statistically meaningful value (typically between 100 and 1000 individual
realizations were used).

---

* Although it was realized that seeding of the flow upstream of the valve
  (perhaps within the fuel/air mixer) would have been preferable, this was
  not done for a specific reason. Molecular Rayleigh scattering is used
  as a standard diagnostic technique in this device to measure flame thicknesses, and because its application demands a very particle-free system,
  contamination of the apparatus was kept to a minimum.

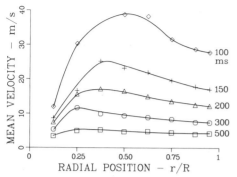

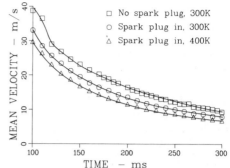

Figure 2.   Radial distribution of tan-
            gential velocity component.
            Curves show the decay of
            velocity with time. Radial
            position is normalized by
            bomb radius, R = 40 mm.

Figure 3.   Decay of tangential veloc-
            ity at the half-radius
            location. Curves show the
            effect of installing the
            spark plug and heating the
            walls.

EXPERIMENTAL RESULTS

    PRECOMBUSTION FLUID MOTION--Only one timing sequence was used for these
tests: the events were initiated at time zero, the intake valve opened at 16
ms and closed at 84 ms, and ignition occurred at 200 ms. Since the presence of
small percentages of fuel gas does not significantly affect the mixture gas

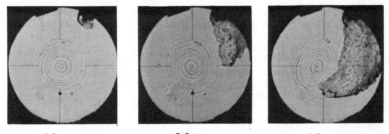

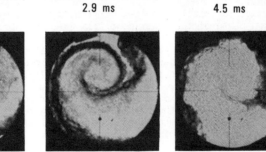

Figure 4.   Premixed methane combustion test sequence as visualized with a laser
            shadowgraph technique. The time relative to ignition is indicated
            on each high speed movie frame. The small dot at the half radius
            location indicates the measurement point.

properties, much of the precombustion
testing was done with air instead of
with combustible mixtures. Figure 2
shows radial profiles of mean tangential velocity with time as a parameter.
For these air-only tests, the bomb walls
were held at 300 K, the gas pressure was
8.5 atmospheres, and no spark plug was
installed, i.e., the walls were nearly
smooth. The results show that there is
a relatively small central-core flow
characterized by solid body rotation
(tangential velocity proportional to
radius); however, the bulk of the fluid
motion is more nearly characterized as
constant tangential velocity. An axial
survey for the same test conditions was
also taken at the half-radius position.
This survey traversed 24 mm between the
two glass windows, coming to within 2.5
mm of each window. The resulting velocity profile was found to be very uniform, indicating that the wall boundary
layer is thin and has no influence on
the bulk motion in the chamber.

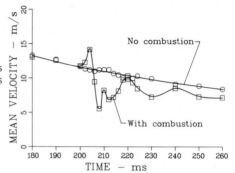

Figure 5.   Velocity history at the
half-radius point. The
curves compare the velocities with and without combustion. Ignition occurs
at 200 ms and the flame
arrives at the measurement
location at about 205 ms.

The influence of the spark plug and heating the bomb walls is graphically
demonstrated in Figure 3, which shows mean velocity versus time at the half-
radius location. Intrusion of the spark plug into the flow (approximately 8
mm) causes a distinct reduction in velocity. The increase in fluid viscosity
resulting from increasing the wall temperatures from 300 to 400 K further diminishes the gas motion.

COMBUSTION CHARACTERIZATION--The last curve in the figure just discussed
provides the baseline for the combustion testing performed. The mixture was
premixed lean methane/air with an equivalence ratio of 0.75. The flame propagation behavior is illustrated in Figure 4, which shows a sequence of frames
from a high speed laser shadowgraph visualization movie. At 4.5 ms after ignition the flame front reaches the measurement location indicated by the small
dot. Peak combustion pressure of 48 atmospheres occurs approximately 12 ms
after ignition.

An important but easily resolved difficulty was encountered during the
post-combustion phase of the experiment. It was found to be necessary to keep
the walls of the bomb hot (> 375 K) to prevent condensation of the water vapor
combustion products onto the viewing windows, an effect which seriously degrades
the quality of both the incident laser beams and the Mie scattered light.

Figure 5 shows the velocity history at the half-radius location for ignition at 200 ms. The data have been averaged in 2 ms increments for the period
from 200 to 224 ms. As one would expect, the unburned gas velocity is increased
ahead of the flame, whereas the burned gases show a reduced velocity. For comparison the curve without combustion is shown, indicating that the post-combustion gases recover to nearly the same bulk velocities they would have had if
combustion had not occurred.

ACKNOWLEDGMENT

The authors would like to acknowledge the support of Volkswagen Research,
especially Drs. W. Brandstetter and W. Lee, for loan of the apparatus and their
technical support for this program. This work was funded by the Power Systems
Division of the Department of Energy and the Motor Vehicle Manufacturers Association.

SESSION VI
DATA ANALYSIS II

Chairman: **W. G. TIEDERMAN**
Oklahoma State University

Data Retrieval in Laser Anemometry by Digital Correlation

A. E. SMART
Spectron Development Laboratories, Inc.
3303 Harbor Blvd., Suite G-3
Costa Mesa, CA 92626

ABSTRACT

In hostile environments the price of good signal quality is often very high in terms of engineering or complexity. It is therefore more cost effective to use a subtle approach to interpret a small signal than to guarantee a larger signal. Acceptance of this view gives other advantages such as suitability of smaller particles, tolerance of smaller windows and easier use of backscatter. The satisfactory retrieval of data from signals which are reduced from the classical plus noise to shot noise dominated or even photon resolved may be achieved by using correlation. The high-speed electronics now available make this technique suitable for real fringe or transit anemometry. Digital correlation is an optimal method for acquiring measurements from a transit system whose ability to operate successfully close to walls is now widely accepted. Data are presented to show the advantages of this retrieval method in some difficult measurement cases, such as rotating machinery and aero-engine exhausts.

NOTATION

c	Hyperbolic attenuation with z
d	Spot separation
ℓ_o	$1/e^2$ intensity radius for fringe system
r	Fringe contrast
s	Sampling space
x,y,z	Cartesian coordinates
θ	Half angle of beam intersection for fringes
W_o	$1/e^2$ intensity radius of spot a $z = 0$
λ	Wavelength of illumination
$I_{f(s)}$	Intensity in fringe system
$I_{t(s)}$	Intensity in two-spot system

INTRODUCTION

Laser anemometry is very broadly the measurement of velocity by detecting and decoding the light scattered from small particles. In the early days of this technology, heavy stress was placed on reference beam and then fringe optical design: as this has become more completely understood, emphasis has shifted to signal retrieval. Many signal processing schemes have been used and it is only recently that these fragmented approaches have been seen to emphasize different aspects of a more optimum detection approach [1,7].

For many good reasons, the engineering difficulty or inconvenience of obtaining a classically large signal is often great and a more efficient way of transmitting and receiving scattered light and electronically detecting it has potential advantages. These may be taken as easier backscatter, greater obliquity leading to closer approach to surfaces or smaller particles. This last has further implications for the fluid dynamicist as more accurate following of the flow and more frequent estimates of velocity yield the potential for time history, as well as statistical averages in turbulence studies. This paper addresses the considerable signal sensitivity advantages of using photon correlation techniques and the further increases which are possible by using transit anemometer optics instead of fringe optics.

THE SIGNAL

The smooth classical view of an electromagnetic wave can never be realized below infinite energy. All finite powers may be regarded in theory as a train of non-overlapping quanta (photons) and the mean rate and intervals carry all the information which may ever be extracted. A consequence of reducing signal power is the qualitative change in the nature of the pulse modulation, and this may not be reversed by later amplification back to the former power. In any real experiment, there is a limit to the smallest interval between two photons which may be resolved, and this is equivalent to low pass filtering. Reference 1 puts mathematical and quantitative rigor into these ideas and here we will review some of the new ideas which may be explored with such a generalized approach. In Figure 1, we show a computer simulation of the appearance of the signal as the energy contained within it is reduced. It shows behavior which is similar to that observed in a practical case previously published as Figure 3 of Reference 2. The signal looks dramatically different, even for the same top cut frequency of the filtering device as signal power is reduced.

Noteworthy is that the two extremes, often quoted as though they were experimental cases (Figure 1a, d) are never achieved. The 'classical' signal is never noise-free, however high the power, and the 'photon resolved' is never completely resolved because there is a certain probability of photons infinitely close together. Real signals always lie in the 'gray area' of inhomogeneous Poisson processes [1], and the only completely satisfactory approach to signal retrieval must consider how closely its assumptions approach this.

Most experiments are assumed to yield signal lying in a well defined narrow band of the above regime, but the truth is that for any real experiment, especially with polydisperse particles, the input signal may cover a very wide dynamic range in Figure 1. Many early references discussed the signal on a classical plus noise basis and more recent work has considered a quantum limited signal [3,4,5]. Both these extreme areas are amenable to good theoretical analysis and hardware conforms well to this, tracking and counting in the former case and photon correlation for very low light levels. Extensive work has been done to condition real signals into one regime or the other with questionable attention to the consequences of the conditioning.

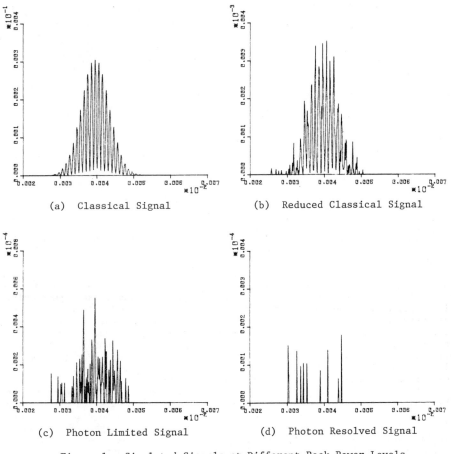

(a)   Classical Signal   (b)   Reduced Classical Signal

(c)   Photon Limited Signal   (d)   Photon Resolved Signal

Figure 1.   Simulated Signals at Different Peak Power Levels.

FILTERING

        Filtering is the removal of information. It may take the form of a con-
volution in the signal domain or a multiplication in frequency space. Its
efficacy for laser anemometry relies on the separability of required information
from useless 'noise' effects in the relevant space. There is always a top cut
imposed by electronic component limitations and any attempt to view the signal
on an oscilloscope must be after this effect. For classical interpretation, a
band pass filter is applied to the signal and conventionally this must be sufficiently greater than the signal width to allow for unbiased velocity fluctuation. The alternative of a moveable filter, matched to the signal, can only be
realized if a continuous estimate of center frequency is available. In the case
of photon correlation, no filter is used on the signal, but correlation separates 'signal' and 'noise' effects in the correlation domain as a complex non-
linear transform of the 'low frequency' signal and a smooth flat background from
the Poisson noise. Again the assumption of separability is applied, this time
in the correlation space and the 'noise' on the transformed data is now both
photon number fluctuations and stochastic failure of the stationarity condition;
i.e., if the sample of data is too short, the correlation is still not easy to
extract from the variability error.

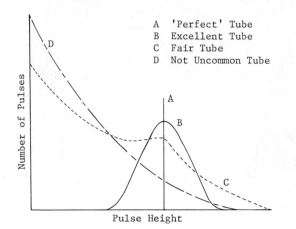

Figure 2.   Photomultiplier Output Behavior from Single Photon Input.

REAL DEVICES

The foregoing discussion has assumed a photon detector which is imperfect only in its finite time of response, an equivalent dead time. Real photomultipliers fail on a number of other counts. The production of a discriminable pulse from every photon is not guaranteed because of limited quantum efficiency, and noise may give pulses which are discriminated as photon events. Figure 2 shows schematic performance of phototubes. For semi-classical assumptions, the tube performance is often considered not to give serious problems. For quantum resolved signals, it matters very much. Actually, it matters everywhere, but sometimes one can neglect the effects without serious consequences.

A second source of spurious events in experiments is light scattered from other than the measured particles. Although the photon statistics of this light are different, there is a limit to the way this may be used to eliminate flare effects. It is best to reject spurious light by optical design. Away from walls, this improves spatial resolution, and for a given acceptable amount of flare, enables one to work as close as possible to walls. It is not uncommon for those investigators who have only used burst counting techniques to be unaware of the severe detrimental effects of flare light. Such flare is common inside poorly designed or improperly used coaxial backscatter optical systems. This flare limits burst counting systems less than photon counting systems only due to the higher required signal levels of the classical technique.

FRINGE AND TRANSIT SYSTEMS

There is a transform relationship between these two configurations as shown in Figure 3. The discussion now reduces to the relative merits of A and A* as the better encoding operator for the scattered light. It might seem at first sight that for a fixed size of scattering center an analog of the Fellgett advantage might accrue from the two-spot encoding, but this is complicated by other geometrical features and constraints on the velocity direction sensed.

From the diagram, Figure 4, we define the sense and orientation of the fringes and spots, and proceed to describe the intensity which would be detected at any point in the sampling space. We will borrow (with minor changes) from the work of Abbiss in Reference 5 to illustrate the intensity in the fringe

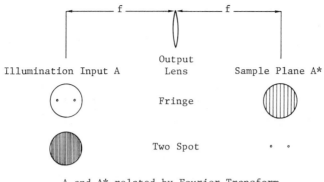

A and A* related by Fourier Transform.
Diagrams not to scale.

Figure 3.   Relationship of Fringe and Two Spot Illumination.

volume and use a change of notation to make the fringe and transit descriptions comparable.

In the case of a fringe system,

$$I_f(s) = k_f \exp\left[\frac{-2}{\ell_o^2}\left(x^2 \cos^2\theta + y^2 + z^2 \sin^2\theta\right)\right]$$

$$x \left[\exp\left(\frac{4}{\ell_o^2} x z \sin\theta \cos\theta\right) + r^2 \exp\left(\frac{-4}{\ell_o^2} x z \sin\theta \cos\theta\right)\right.$$

$$\left. + 2r \cos\left(\frac{4\pi}{\lambda} x \sin\theta\right)\right] \tag{1}$$

where k_f is a proportionality constant.

For equal intensity beams, a case which is usable for real fringe anemometry this reduces to:

$$I_f(s) = k_f \exp\left[\frac{-2}{\ell_o^2}\left(x^2 \cos^2\theta + y^2 + z^2 \sin^2\theta\right)\right]$$

$$x \left[\cosh\left(\frac{2}{\ell_o^2} x z \sin 2\theta\right) + \cos\frac{4\pi}{\lambda} x \sin\frac{\theta}{2}\right] \tag{2}$$

Because this is a complicated expression, it is quite general to restrict this volume further by the receiving aperture. For small θ, leading to a highly elongated volume, the idealized truncation in the z direction is considered to reduce the expression to:

$$I_f(s) = k_f \exp\left[\frac{-2}{\ell_o^2}\left(x^2 + y^2\right)\right] \cdot \cos\left(\frac{2\pi x \theta}{\lambda}\right) \tag{3}$$

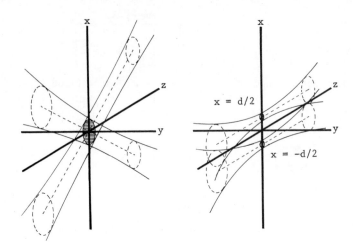

Figure 4.   Schematic of Fringe and Transit Illuminated Volumes.

We show in Table 1 some typical comparative specifications for fringe and two-spot systems. Figure 5 is the simulation of the classical fringe velocimeter signal from a 0.3 μm diameter particle with the amplitude given in photoelectrons/sec. Figure 6 is the output of the photomultiplier tube. The peak mean photoelectron rate is approximately 5×10^7, so this signal would be totally swamped by large background light. Figure 7 illustrates the classically expected pulse from one of the two spots. Figure 8 shows the PMT output. Figures 9 and 10 depict the same signal in the presence of 10^{10} photoelectrons/sec background light.

It is much more likely that one will obtain a high background from flare when using the fringe system (with a 260 micrometer diameter pinhole in the receiver) than when using a two-spot system (10.6 micrometer diameter pinhole). Thus, the two-spot system produces large signal-to-noise ratio advantages for single signals in three ways: the filter for the two-spot can be proportionately much narrower than the fringe system filter if a tracking filter is not used; the two-spot has higher intensity signals by a factor equal to the square of the ratio of the beam radii at the probe volume; the background light which will get through the pinhole spatial filter is much less for the two-spot system due to the smaller apertures. On the negative side, all optical aberrations, including those introduced by windows and turbulent index of refraction effects, will enlarge the achievable spot size and reduce the advantage of a transit velocimeter system.

For the two-spot system, the equivalent illumination intensity function may be written for Gaussian optics [6] as:

$$I_t(s) = \frac{P}{\pi W^2(z)} \left[\exp\left(\frac{-2[(x-d/2)^2 + y^2]}{W^2(z)}\right) + \exp\left(\frac{-2[(x+d/2)^2 + y^2]}{W^2(z)}\right) \right] \quad (4)$$

where P is the total power included in both beams, and W(z) is the $1/e^2$ intensity radius given by

$$W(z) = W_o \sqrt{1 + \left(\frac{\lambda z}{\pi W_o^2}\right)^2} \quad (5)$$

Table 1.   System Parameters.

Parameter	Two-Spot	Fringe Velocimeter
System		
Wavelength	514.5 nm	514.5 nm
Laser Power	0.5 watt/spot	1 watt
Optical Efficiency	0.3	0.3
Detector Quantum Efficiency	0.2	0.2
Transmitted Beam Diameter	25 mm	1 mm
Range	400 mm	400 mm
Beam Separation at Transmitter	0.5 mm $\approx$ 0	20 mm
No. of $1/e^2$ Fringes	0	25
Probe Volume Spot Diameter	10.6×10^{-6} m	N/A
Signal Frequency	N/A	38.9 MHz
Probe Volume Width	N/A	262×10^{-6} m
Single Burst Duration	26.6×10^{-9} s	0.655×10^{-6} s
Particle		
Velocity	400 m/s	400 m/s
Index of Refraction	1.5+j0	1.5+j0
Diameter	0.3×10^{-6}	0.3×10^{-6}

Background Light

Two Levels: zero and 10^{10} photoelectrons/s
              (equivalent to 20 nw optical power collected
              by phototube)

This is not simplified by assuming a truncation function in the z direction, but neither does it change its functional form. Hence, any calculations do not depend on z selectivity. This has a shade of irony in that the conventional dimensions of use for this system do mean that it is easier to restrict sensitivity away from z = 0. Again, because of Gaussian beams, the z roll-off is of quasi-Gaussian form and enormously simplifies inferences in this system.

On the basis of these descriptions of the illuminated region, we now have some computer simulations of the signal appearance for each of the two systems. As might be expected, the discriminability of the signal above the noise is dramatically better for the two-spot system. The price of improvement may be reduced data rate, depending on the size distribution of scattering particles and/or it may be significantly reduced for more than a few percent turbulence because not so many particles may cross both spots as would have crossed an equivalent fringe system. Of course, this may not be a serious problem if it is a good estimate of flow direction which is required.

To summarize, the distinction between the two systems is not very tidy, but in situations where scattered light is a problem, the transit system is superior to the real fringe system. The theoretical improvement factor is up to 10^5, but 10^2 to 10^3 is easily realized by competent optical design.

As shown by the simulations, we may use classical signal descriptions for the transit systems for much smaller scattering centers due to the concentration of light. In cases where the background flare light is kept to a reasonable value, such as 10^6 or less, much smaller signals which are photon resolved may be detected by photon correlation techniques.

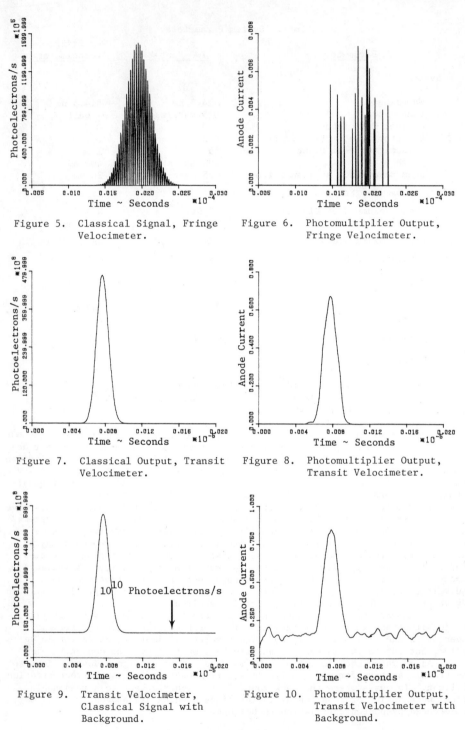

Figure 5.   Classical Signal, Fringe   Figure 6.   Photomultiplier Output,
            Velocimeter.   Fringe Velocimeter.

Figure 7.   Classical Output, Transit   Figure 8.   Photomultiplier Output,
            Velocimeter.   Transit Velocimeter.

10^{10} Photoelectrons/s

Figure 9.   Transit Velocimeter,   Figure 10. Photomultiplier Output,
            Classical Signal with   Transit Velocimeter with
            Background.   Background.

TRANSIT ANEMOMETER MEASUREMENTS

        Several pieces of excellent work [8,9] have been reported using the advan-
tages of the transit anemometer system. We enclose some simple examples here of
the type of performance observed. Figure 11 shows a boundary layer traverse at
an oblique angle onto the surface of a turbine cascade blade. The measured
points were taken in random order and no difficulty was experienced with flare
even as close to the surface as 100 μm. Figure 12 shows the angular capability
taken in a time-gated system looking between the blades in a 60" diameter fan.

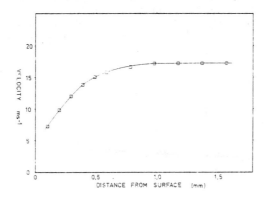

Figure 11.   Boundary Layer Traverse on Pressure Surface
             in a Turbine Cascade.

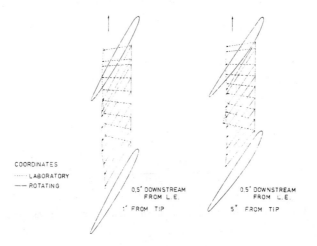

Figure 12.   Inter-blade Traverses at Design Speed.

The speed magnitude and direction were very easily measured in a few seconds. In this system, the angle is measured to high accuracy by a method discussed in Reference 2, the figure from which is reproduced here (Figure 13). Quite obviously, it is not necessary to take so many angular points -- three is usually quite sufficient for a parabolic fit on direction. The data in this figure was from a small jet.

There are cases where this transit system may be used easily where other systems may fail. It proved easy to obtain measurement in an unseeded afterburning exhaust from an aeroengine whose conditions were in excess of 800 ms^{-1} and 2000°K with steep refractive index gradients. It is possible to minimize the effect of turbulent refractive index fluctuations in this system because the projection and collection beams are co-axial and subjected to fairly similar effects. Also, the light which forms each spot travels over substantially the same path. This transit system was designed to make measurements in high-speed axial compressors, but has proved versatile and convenient for other applications. Typical accuracy is 1/2 percent on speed and 1/2 degree on angle, but this may be improved with care if the higher accuracy is necessary.

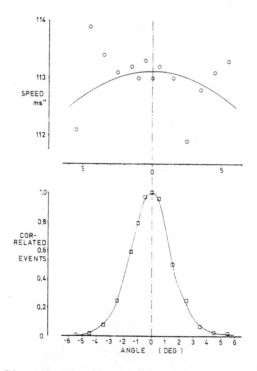

Figure 13.   Direction Identification Using a Transit Anemometer.

CONCLUSIONS

1. The application and success of photon correlation methods has increased
    the dynamic range of laser anemometry signals from which good measurement
    may be obtained.

2. Explanation of these effects shows that in conditions of flare, the change
    to a two-spot, or transit, system is beneficial, and that the same corre-
    lation machine can be used to retrieve data in a rather different way,
    but with even higher rejection of noise effects.

3. While much is yet to be said and explained with the transit system, it
    seems to offer great promise in what have been either very difficult or
    impossible situations.

ACKNOWLEDGEMENTS

        The comparative computer simulations were conducted by W. T. Mayo, Jr.,
Ms. Carolyn Greenwood and the author under U.S.A.F. Contract F040600-78-C-0002.
This portion of the research was sponsored by the Arnold Engineering Development
Center, Arnold Air Force Station, Tennessee, under the direction of Mr. Marshall
Kingery.
        The simulation programs have been developed under various research con-
tracts with NASA Langley Research Center, the Naval Underwater Systems Center,
the Advanced Research Projects Agency, and private funding by Spectron Development
Laboratories, Inc.

REFERENCES

1. Mayo, W. T. Jr., "Modeling Laser Velocimeter Signals as a Triply Stochastic
    Poisson Process", Proceedings of Minnesota Symposium on Laser Anemometry,
    October 1975.

2. Smart, A. E., "Special Problems of Laser Anemometry in Difficult Applica-
    tions", Lecture 6 in AGARD LS 90, August 1977.

3. Pike, E. R. and Jakeman, E., (ed. Goodwin, D. W.), Advances in Quantum
    Electronics, Academic Press, 1974.

4. Cummins, H. Z. and Pike, E. R. (eds.), Photon Correlation and Light
    Beating Spectroscopy, NATO Advanced Study Institute, B3, Plenum, 1976.

5. Cummins, H. Z. and Pike, E. R. (eds.), Photon Correlation Spectroscopy
    and Velocimetry, NATO Advanced Study Institute, B23, Plenum, 1976.

6. Siegman, A. E., An Introduction to Lasers and Masers, McGraw-Hill,
    New York, 1971.

7. Smart, A. E. and Mayo, W. T. Jr., "Applications of Laser Anemometry to High
    Reynolds Number Flows", presented at the Conference on Photon Correlation
    Techniques in Fluid Mechanics, Stockholm, Sweden, 14-16 June 1978.

8.   Schodl, R., "A Laser Dual Beam Method for Flow Measurements in Turbo-
        machines", ASME Paper 74-GT-157, 1975.

9.   Eckardt, D., "Detailed Flow Investigations Within a High-Speed Centrifugal
        Compressor Impeller", ASME 76-FE-13, 1976.

How Many Signal Photons Determine a Velocity?

E. R. PIKE
Royal Signals and Radar Establishment
Malvern, Worcestershire, England

ABSTRACT

The accuracy of velocity measurements in LDV depends on the number of signal photons received. Some results will be described which show that 100 photons detected is a large signal. A generalised Shannon theory will explain why this is so and a new data analysis method based on information theory will be described.

INTRODUCTION

In a previous paper at this meeting we discussed the LDV signal in terms of bursts of photons scattered into the detector by single particles crossing the probe volume. The particle velocity is assumed not to change significantly during the beam transit and thus the signal photons are bunched at the Doppler or Doppler-difference frequency for the finite duration of the transit time. The pulse rate will be further weighted by the Gaussian intensity profile of the laser beams. A method was explained for extracting the frequency from such a pulse train by autocorrelation and FCT, which was equivalent to forming the digital Fourier transform or power spectrum of the train, followed by interpolation to remove the effect of the weighting factors. This is illustrated in Fig 1 of our previous paper.

Given such a procedure to produce a velocity value we must ask what unavoidable variation to expect due to the fundamental statistical nature of photon detection. That is to say, given the same particle, following the same trajectory, with the same laser power and all other conditions remaining constant, what will be the variation in the calculated velocity due to the different realisations of the detected photon signal?

STATISTICAL ACCURACY

Calculations of this type have been carried out analytically for the case of scattering by molecular Brownian motion where Laplace, rather than Fourier, inversion is required and it is well known how many photons are required to determine a molecular radius or a translational diffusion coefficient [1]. Charts and tables exist enabling the experimenter to choose the correct laser power, for example, to give him a desired accuracy in a given measurement time. Similar results, unfortunately, do not exist for the LDV problem. The Brownian motion case took a great deal of work and is a somewhat simpler calculation. Until such time as the investment is made to push through similar "signal-to-noise" calculations for LDV we can only fall back on computor simulation. The

285

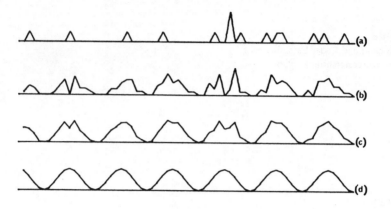

Fig 1   Typical detected signals at different mean photon
         rates.   (a) 2.5 photons/cycle   (b) 25 photon/cycle
         (c) 250 photons/cycle   (d) Infinite photon rate.

photon statistics are then generated by taking samples from Poisson distributions
with mean values proportional to the instantaneous intensity falling on the
detector.   Many realisations of a given signal are made, the velocity (frequency)
calculated from each and a mean square deviation of velocity determined.   This
variation, as a percentage error, can then be found as a function of the mean
photon flux per Doppler cycle for example.   In Fig 1 we show such simulated
signals for an ideal Doppler difference anemometer for various mean photon
rates.

   Simulation work of this kind is very tedious since there are many variables
which can be changed independently over wide ranges.   Taking signals, however,
with means of the order of 1 to 10 photons per Doppler cycle, showing some ten
cycles of oscillation in the transit time and processed with a 64 channel
correlator, such simulations as we have performed [2] showed clearly that a
total of 100 photons provides a strong signal.   This is in the sense that the
velocity estimates were within the accuracy of the interpolation procedure used.
Signals even at levels down to 3 photons/cycle gave a respectable 1% accuracy in
velocity.

   These rather unexpected but welcome results are very different from those
mentioned above for the measurement of molecular diffusion where some 10^5 photons
are required to achieve 1% accuracies.

INFORMATION CONTENT

   One is led to ask why this enormous difference in information content exists
between the two cases and, rather more subtly, how the number of fringes in the
beam width affects the information content of the signal.   Pursuit of this
question has led us into the study of the extraction of information by inversion
from the Fourier and Laplace transforms and a wide range of similar "Mellin
kernel" problems including the "single-beam" anemometer and the LDV with finite
fringe number, of which it is a limiting case.   A new analysis method has been
developed for inversion which makes use of explicit eigenvalues of such kernels
which we have found analytically for the first time.[3] With the aid of these
functions we may now not only successfully invert, in so far as is possible, the
Laplace and other difficult ill-conditioned transforms, including the analysis

of turbulence from photon correlation functions, but we may see clearly the source of the problems of such inversions and construct a generalized Shannon information measure for each one.

The general case may be appreciated from the context of LDV by considering the common problem of extracting parameters of the velocity probability distribution $p(v)$ from a correlation function $g(\tau)$ which has been allowed to accumulate over all representative velocity values. The optical geometry is assumed known so that the laminar-flow correlation function $K(v\tau)$ is determined. The problem in one-dimensional form requires inversion of the first-order Fredholm equation

$$g(\tau) = \int_0^\infty K(v\,\tau)\ p(v)dv \tag{1}$$

The Fourier sine and cosine transforms, the Laplace transform and the Gaussian single-beam anemometer equations are all cases of this equation with special values of $K(v\tau)$. For the Doppler-difference LDV [2]

$$K(v\tau) = e^{-v^2\tau^2/r^2}\ (1 + \frac{m^2}{2}\cos\frac{2\pi v\tau}{s}) \tag{2}$$

where r is the beam radius, m is a fringe visibility factor and s is the fringe spacing.

A familiar problem of this kind illustrating the difficulties is that of diffraction-limited ($\frac{\pi}{\Omega}$) imaging with a lens of finite aperture. The image is given by

$$I(x') = \frac{1}{2\pi}\int_\Omega^\Omega e^{-i\omega x'}\int_{-X/2}^{X/2} e^{i\omega x}\ O(x)\ dx$$

$$= \int_{-X/2}^{X/2} \frac{\sin\Omega(x - x')}{\pi(x - x')}\ O(x)\ dx \tag{3}$$

It is well-known that although, mathematically, given $I(x)$ the object $O(x)$ can be found exactly by inversion, physically, the lens has a finite resolution and only $S = X\,\Omega/\pi$ resolution elements can be determined along the length X of the object. S is known as the Shannon number. The difference between the mathematical and physical problems lies in the presence of noise. Even computor round-off noise will, in fact, spoil the theoretically perfect mathematical inversion. Unfortunately noise does not just make the result noisy but in general will cause the calculation to "blow-up" which is why such problems are said to be ill-conditioned.

The explanation in mathematical terms requires the introduction of the set of eigenfunctions $\phi_n(x)$ which are images of themselves

$$\int_{-X/2}^{X/2} \frac{\sin\Omega(x - x')}{\pi(x - x')}\ \phi_n(x')\ dx' = \lambda_n\ \phi_n(x) \tag{4}$$

The λ_n are a set of real eigenvalues.

Then we may form the expansion

$$I(x) = \sum_0^\infty a_n\ \phi_n(x) \tag{5}$$

where

$$a_n = \int_{-X/2}^{X/2} I(x') \, \phi_n(x') \, dx' \; . \qquad (6)$$

Substituting into the integral equation

$$O(x) = \sum_{o}^{\infty} \frac{a_n}{\lambda_n} \, \phi_n(x) \qquad (7)$$

The $\phi_n(x)$ are prolate spheroidal functions [4] and the λ_n have the property of decreasing suddenly almost to zero at the value of n equal to the Shannon number. Thus even small amounts of noise are "amplified" by the small denominator and this makes it impossible in practise to evaluate the coefficients of components with higher eigenvalue numbers than this number.

The above considerations apply to the more general Fredholm equations discussed above where, although the eigenvalue spectrum does not behave so dramatically as with Shannon's kernel, nevertheless, the eigenvalues can be arranged in decreasing order and a generalised Shannon number, depending now on the actual levels of noise present, can be defined, beyond which further information is irretrievable.

EIGENVALUE SPECTRA

We shall not repeat here the analytic derivation of the eigenfunctions and eigenvalues of the general problem but refer to a forthcoming publication [3] containing details. For the present purposes the interesting question is to compare the way the eigenvalues of different kernels fall off and hence to assess the potential information transfer capabilities of different experiments.

For the Laplace transform where

$$K(v\tau) = e^{-\alpha v\tau}$$

we find

$$|\lambda_\omega|^2 = \frac{\pi}{\alpha \, \cosh \, (\pi\omega)} \qquad (8)$$

which behaves asymptotically as ω becomes large as

$$|\lambda_\omega|^2 \sim \frac{\pi}{\alpha} \, e^{-\pi\omega} \qquad (9)$$

The exponential fall off of eigenvalues is the cause of the difficulties in inversion. Any ad-hoc approach not using these new eigenfunctions is liable to mix-in proportions of the higher "forbidden functions" which will cause unknown errors or divergence. Smoothing constraints have been used but may suffer the opposite failing of not picking up available information in the region of inter-mediate eigenvalue numbers.

For the LDV kernel written above the eigenvalue spectrum is more compli-cated but in the limit of high ω

$$|\lambda_\omega|^2 \sim \frac{\Gamma}{\omega^{\frac{1}{2}}} \, e^{\pi^2 n^2 (1 + \frac{2}{\pi})} \, e^{-\frac{\pi}{2} (\omega^{\frac{1}{2}} - 2n)^2} \qquad (10)$$

where n = r/s, the number of fringes in the beam radius. In the limit of zero fringes (single-beam case)

$$|\lambda_\omega|^2 ~\sim~ \frac{r}{\omega^{\frac{1}{2}}}~~e^{-\pi\omega/2} \tag{11}$$

and in the limit of infinite fringe number we have the Fourier transform where

$$|\lambda_\omega|^2 ~=~ \text{constant for all } \omega, \tag{12}$$

and no limits are posed on inversion.

These results demonstrate clearly the value of keeping as high a fringe number as possible and show why it is that the Brownian motion and LDV experiments require such different numbers of photons. The statistical fluctuations on the correlation function restrict inversion much more quickly with increasing eigenvalue number in the former case.

Practical inversion techniques have been evolved, using the above analysis, which extract the maximum information from the photon correlation function in light scattering experiments of this kind.[5]

REFERENCES

1. Jakeman, E. 1973. "Photon Correlation" in Photon Correlation and Light Beating Spectroscopy. Eds H. Z. Cummins and E. R. Pike. Plenum Press pp 75-150.

2. Pike, E. R. 1976. "Photon Correlation Velocimetry" in Photon Correlation Spectroscopy and Velocimetry. Eds H. Z. Cummins and E. R. Pike. Plenum Press pp 246-343.

3. McWhirter, J. G. and Pike, E. R. 1978. On the Numerical Inversion of the Laplace Transform and Similar Fredholm Integral Equations of the First Kind. J Phys A, to be published.

4. Slepian, D., and Pollack, H. O. 1961. Prolate Spheroidal Wave functions, Fourier Analysis and Uncertainty - I. Bell System Technical Journal 40, 43-63.

5. McWhirter J. G. and Pike E. R. 1978. The Extraction of Information from Laser Anemometry Data. Physica Scripta, to be published.

Comparison of Trackers and Counters

ROBERT V. EDWARDS
Chemical Engineering Department
Case Western Reserve University
Cleveland, OH 44106

ABSTRACT

Most discussions of relative merits of various detection strategies for laser anemometer have suffered from an inadequate specification of the detector properties. Further, rarely are theoretical comparisons made of the applicability of several types of detectors to the same flow. For instance, counters are usually described as useful for low scattered density flows and trackers are described as useful for high scattered density flows.

Questions such as the suitability of trackers for measurements in sparsely seeded, high turbulence intensity flows will also be discussed. Comparison will be made of trackers and counters under these conditions. Results of computer simulations as well as experimental data of various detector strategies will be presented. A critical parameter for selection of a detector is shown to be the signal to noise ratio.

INTRODUCTION

Frequency trackers are most often used for highly seeded liquid flows and counters are most often used in lightly seeded gas flows. The demands on the signal processor under these conditions are usually viewed as quite different. However, it is interesting to examine the behavior of these two types of detectors under similar conditions.   Given the paucity of experimental data, the comparisons will be mostly theoretical. Several aspects of detector performance will be analyzed:

1) Accuracy
2) Ability to discriminate signal from noise
3) Unbiased accuracy
4) Range

The tracker will be assumed to be a second order phaselock loop with:

1) A quadrature detector verification circuit
2) An analog output corresponding to the voltage controlled oscillator's drive voltage
3) A digital counter to directly measure the frequency of the voltage controlled oscillator (VCO)
4) An input bandpass filter that limits the range of the tracker

5) An adjustable tracker bandwidth (ω_N)

The counter will be assumed to:

1) Be a zero crossing detector
2) Have a verification circuit that rejects all signals less than
    a preset amplitude A_s, and/or all signals where the period
    attempts to change more than t_p during N_p cycles
3) Have an input filter that limits the range of the counter

ACCURACY

In a highly seeded flow, a noise-free laser anemometer signal can be
modeled by band pass filtered white noise [1]. Let the frequency response of
the filter be $S(\omega)$ and let its Fourier transform be $R(\tau)$.

Under these circumstances, Rice [2] showed that the mean of the output, $\bar{\omega}$,
of an ideal:

1) frequency detector should be given by

$$\bar{\omega}_T = \left(\frac{dR}{d\tau}\bigg|_0\right)/iR(o) \tag{1}$$

2) zero crossing detector should be given by

$$\bar{\omega}_z = \left(\left(-\frac{d^2R}{d\tau^2}\bigg|_0\right)/R(o)\right)^{1/2} \tag{2}$$

Lading and I [3] have shown that a properly adjusted second order
phaselock loop is an excellent approximation to an ideal frequency detector.
Therefore the output of a tracker under these conditions should be $\bar{\omega}_T$.

A counter (with the verification circuit turned off) is a zero crossing
detector and thus its mean output should be $\bar{\omega}_z$.

As an example, let

$$S(\omega) = \exp\left[-(\omega - \omega_o)^2/2\sigma_{\omega_o}^2\right] \tag{3}$$

where ω_o is the center frequency of the signal and σ_{ω_o} is the bandwidth. Then,

$$R(\tau) = \sqrt{2\pi}\,\sigma_\omega \exp\left[-\frac{\tau^2\sigma_\omega^2}{2}\right] \exp\left[-i\,\omega_o\tau\right] \tag{4}$$

$$\bar{\omega}_T = \omega_o \tag{5}$$

$$\bar{\omega}_z = (\omega_o^2 + \sigma_\omega^2)^{1/2} \tag{6}$$

$$\bar{\omega}_z = \omega_o \left(1 + \frac{\sigma_\omega^2}{\omega_o^2}\right)^{1/2} \tag{7}$$

The output of a zero crossing detector should be higher than that of a tracker. However, this will not usually be a problem. A typical laser anemometer is less than 5% wide, so let $\frac{\sigma_\omega}{\omega_0} = .05$. Then

$$\bar{\omega}_z = (1.0013) \ \bar{\omega}_T.$$

This is less than a 0.2% error, less than the stability and/or measurement error of most laser anemometer systems.

1) Original Signal
2) Our Tracker
3) Disa Tracker
4) Zero Crossing Detector

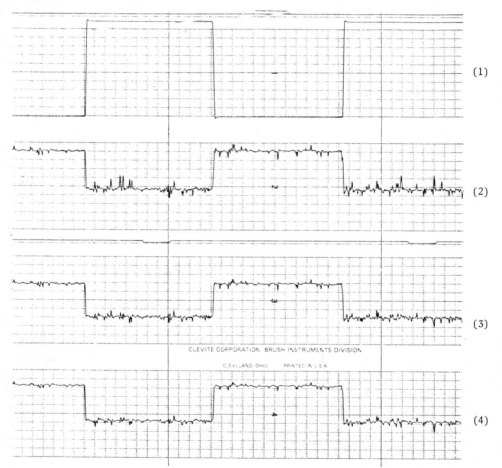

Figure 1

Lading and I tried to verify the relation between $\bar{\omega}_z$ and $\bar{\omega}_T$ for laser anemometer signals using our simulator [3]. Our system was not stable enough to quantitatively prove the relation, but we did show that the discrepancy could not be appreciably larger than that shown here.

The verification circuit in a counter is designed to eliminate parts of the signal where the number of zero crossings per second change appreciably. In a highly seeded flow, fluctuations in the number of zero crossings are an effect of "ambiguity noise" [1]. The spectrum of these fluctuations is a broadband one of width σ_ω and of power density 0.368 σ_ω at zero Hz. The effect of the verification circuit would be to reject the signal when the fluctuations exceed a frequency   determined by the tolerance of the circuit. The counter would thus be effectively off, even when a signal was present. Figure 1 shows a plot of the outputs of two different trackers and a zero crossing detector and the modulating signal obtained from a analog laser anemometer signal simulator. The mean frequency was modulated by a square wave. This is a noise free laser anemometer signal in that no noise was fed into the system. The ambiguity noise is still present, however, since it is an integral part of the laser anemometer signal. Note the "hash" on the outputs of all the detectors. This is the ambiguity noise filtered by the recorder's amplifiers. With the verification circuit on, the counter would be off during these excursions from the mean. I know of no data in the literature that were taken to confirm the accuracy of counters with two stage verification circuits in highly seeded flows.

With the information known to date, it seems that in most cases, a tracker and counter are equally accurate on noise free signals from highly seeded flows.

A sparsely seeded flow will be assumed to be one where the particle density is such that at least half the time, no seed particle can be seen in the measurement volume. Initially, the signal will be considered to be noise-free. Under these conditions, it is not reasonable to model the signal as bandpassed white noise.

A second order phaselock has the ability to lock itself to a signal anywhere in the range determined by its input filter. If a new signal appears within ω_N of the old frequency of the loop, it will lock in less than one cycle. If the new frequency is further away than ω_N , the loop will drift toward the new frequency until it is within ω_N , and then lock.

If the loop was locked to a burst, and a new burst arrives whose frequency is within ω_N of the old frequency, the loop will lock to the new burst almost instantaneously [4]. A reasonable value for ω_N is 3% of the highest anemometer frequency expected [5]. With this setting, if the flow had a turbulence intensity of less than 3%, the tracker could follow it readily. If the turbulence intensity is higher than 3%, but the probability of at least two particles passing through the sample volume in a time less than the correlation time of the flow is high, the tracker will still readily follow the flow.

When the nature of the occurence of a measurement is random such as here, it is necessary to have some method of verifying that an event in the detector is a valid measurement. Most commercial trackers have a so-called "drop-out" detector. These are usually devices for detecting the instantaneous amplitude of the envelope of the signal. If it exceeds a prescribed height, the measurement is considered valid. Obviously this

device can be fooled by a burst that the tracker never successfully locked
to. A better verification procedure is to use a quadrature circuit. When a
phase lock loop is locked, the VCO is $\pi/2$ radians out of phase with the input
signal. If an oscillator is slaved to the VCO, but is $\pi/2$ radians out of
phase with it, it will be in phase with the signal. If this slaved
(quadrature) oscillator is multiplied times the signal and low pass filtered,
the result is a signal with a positive DC component when the loop is locked.
When the loop is out of lock, the DC component disappears. A validation circuit
for low seeding flows can be keyed by the output of the quadrature detector.

A counter is designed for pulsatile signals of the kind encountered in low
seed density flows. If the signal is noise free and the signals from two or
more particles never overlap, the counter can be very accurate (better than
0.2%) [6,7]. This is ignoring particle bias errors. These are treated
extensively in the literature [8,9,10].

Ambiguity noise is a result of phase changes induced in the signal by the
presence of two or more particles being simultaneously in the sample volume
[3]. Most validation circuit reject bursts where there are measurable phase
changes during the burst. This will result in the rejection of most of the
signals where two or more particles are present at once. This can result in
the situation of the measurement rate actually decreasing as the seeding
density increases.

Let n be the mean number of particles in the sample volume. Since the
particles are usually randomly distributed in the flow system (A Poisson
distribution), the probability of exactly one particle being in the volume is
ne^{-n} . Below n = 1, the probability of a measurement increases as n
increases. Above that value of n, the probability of a measurement decreases
rapidly.

I know of no control experiments for accuracy of counters with two stage
verification circuits in the literature. There is no reason, at this point,
to believe that they will be any less accurate than the measurements reported
with the one stage verification circuit.

Unlike a second-order phaselock, a counter has no memory so that it can
measure with equal ease two successive bursts that have an arbitrary
difference in frequency. As a practical matter, all counters require an
input bandpass filter. The counter can detect bursts anywhere within these
limits. The reasons for the input filters will be discussed in the section
on noise.

NOISE

The previous section here has been predicated on noise free signals. Noise
in a laser anemometer has three primary sources:

1) The low frequency "pedestal"
2) Shot noise present in the signal from the sample volume
3) Shot noise due to stray light, not from the sample volume, getting
    into the detector

The spectrum of the "pedestal" is a low pass type spectrum with a
bandwidth on the order of the width of the desired laser anemometer signal
[11]. In order to eliminate this term, high pass filters with a cutoff
frequency about 10% of the highest frequency expected, are used in front of
trackers and counters. If frequency shifting techniques are used, the

pedestal spectrum can be moved much further from the frequencies of interest.

The definition of the signal to noise ratio is always a touchy question. We are interested, here, in the ability of the detector to discriminate between signal and noise. In principal our detectors can respond to any noise perceived by them. Therefore the signal to noise ratio will be defined as the ratio of the total signal power to total noise power during an attempted measurement.

If $\bar{n}$ is the expected number/second of photons seen by the detector, there will be a fluctuation in the numbers of photons actually received of RMS value $\sqrt{\bar{n}}$. These fluctuations are called shot noise. The spectrum of the shot noise is flat with a power density proportional to $\bar{n}$, the mean number of photons received per burst. It is easy to show that n is proportional to $W/\sigma v$ where W is the laser power, σ is a characteristic length of the sample volume, and v is the velocity of the particle at the volume. If the signal is filtered by a bandpass filter of width $\Delta \omega$, the total noise/power burst will be proportional to $\Delta \omega P/\sigma v$.

The power in the signal's spectrum is proportional to $\dfrac{\gamma^2 P^2}{v\sigma^3}$ where γ is a measure of the modulation depth of the signal compared to the pedestal. The signal to noise ratio is thus $\gamma^2 P/\sigma^2 \Delta \omega$.

This definition is equivalent to that used by Adrian et al. [6].

Background light entering the detector will cause shot noise but will add nothing to the signal. The noise power now has the form $(\dfrac{AP}{\sigma v} + \dfrac{AP}{v}) \Delta \omega$, where $\dfrac{AP}{v}$ is the background light intensity. Depending on the source of the background, A may depend on the size of the sample volume.

The same signal to noise definition will be used for highly seeded flows.

The problem that can be caused by noise will be illustrated by a crude calculation. Assume that the spectrum of a burst is a Gaussian superimposed on a bandpassed noise spectrum.

Signal $= B \ e^{-(\omega - \omega_0)^2/2\sigma_\omega^2}$, where σ_ω is the signal bandwidth. The Gaussian will be assumed to have negligible power at the edges of the filter.

Noise $= A$, $\omega_1 \le \omega \le \omega_2$, 0 otherwise

Assume that the detector's output is the mean of the spectrum seen by it.

$$\bar{\omega} = \frac{A \displaystyle\int_{\omega_1}^{\omega_2} \omega \ d\omega + B \int \omega \exp [-(\omega - \omega_0)^2/2\sigma_\omega^2]d\omega}{A (\omega_2 - \omega_1) + B \int \exp [-(\omega - \omega_0)^2/2\sigma_\omega^2]d\omega} \qquad (8)$$

$$\bar{\omega} = \frac{A [\dfrac{\omega_2^2 - \omega_1^2}{2}] + \sqrt{2\pi} \ B \sigma_\omega \omega_0}{A (\omega_2 - \omega_1) + \sqrt{2\pi} \ B \sigma_\omega} \qquad (8a)$$

The power in the signal, S, is $\sqrt{2\pi}\, B\sigma_\omega$ and the power in the noise, N, is $A(\omega_2-\omega_1)$. Thus

$$\bar{\omega} = \omega_o \frac{\left(\dfrac{\omega_2+\omega_1}{2} + \dfrac{S}{N}\right)}{\left(1 + \dfrac{S}{N}\right)} \tag{9}$$

$$\frac{\bar{\omega} - \omega_o}{\omega_o} = \frac{\left(\dfrac{\omega_2+\omega_1}{2} - \omega_o\right)}{\omega_o \left(1 + \dfrac{S}{N}\right)} \tag{10}$$

The fractional error in the measurement is a function of the fractional deviation of the actual signal frequency from the mean frequency and the signal to noise ratio. The measurements are biased toward mean frequency of the filter. Figure 2 shows a plot of $\bar{\omega}$ vs ω_o for various S/N ratios. The data was taken by Adrian et al. [6] for a low seed density flow using a

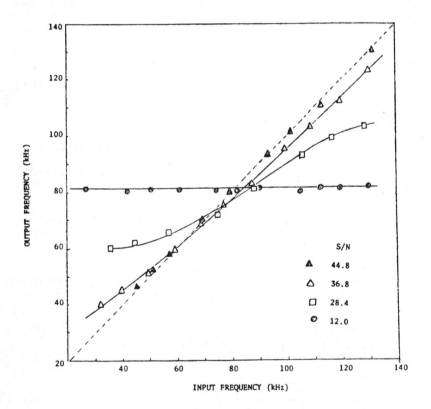

Figure 2

counter with an amplitude triggered validation circuit. Their results
qualitatively follow the form of the derivation presented here, but is
quantitatively much worse than computed. The measurements were made on a
real flow, but the signal to noise ratio was varied by adding continuous
noise to the signal. Undoubtedly the counter was occasionally being
triggered by the noise, which biased the results further toward the filter's
mean frequency. The theory presented here is in good agreement to a second
experiment performed by them, wherein they used an artificially generated
signal and triggered the detector from the actual noise-free burst. This
situation is exactly that assumed in the model. Hoesel and Rodi [7]
performed a similar set of measurements. Again the results are qualitatively
similar, but quantitatively worse than computed here. For instance, the
fractional error (for small S/N) depended on the $(\frac{S}{N})^{-11}$ rather than on $(\frac{S}{N})^{-1}$
as predicted here.

    Undoubtedly, the use of a two stage validation circuit will improve the
situation, but the results should remain approximately the same, there will be
a bias in the measurement toward the mean frequency of the bandpass filter.

    Adrian et al. [6] performed measurements on the same system using a
tracker. From their results, I am unable to perceive any measurable error
in the tracker measurements! Recall this was for a lightly seeded flow
with added noise. Their explanation for the marked difference in behavior is
one with which I basically concur. The tracker is effectively a self tuning
filter with an effective input filter of width $2\omega_N$, where ω_N is the speed
of the loop. As was mentioned before, ω_N will be on the order of .03 times the
maximum frequency expected. At any one time, the tracker cannot even see
the whole input filter.

    If the detection system is adjusted so that a 10 to 1 variation in
velocity (frequency) is expected, the bandwidth of the input filter is 0.9 x
ω_{MAX}. This bandwidth determines the signal to noise ratio for the counter.
Under these same circumstances, the trackers noise determining bandwidth is
0.06 ω_{MAX}, a factor 15 improvement in effective signal to noise. Further,
the mean frequency of the filter seen by the tracker is just the tracker's
frequency. Except at the edges of the input filter, there should be no bias
in the tracker measurements due to shot noise.

    We are in the process of making measurements of the absolute accuracy of
trackers in highly seeded flows using a properly adjusted second-order
phaselock loop for a tracker. The measurements are performed by comparing the
tracker output to the result of a spectral measurement made simultaneously.
The spectral measurements are known to be quite accurate [11] and the results
to date indicate an upper limit of 0.5% difference between the two devices.
Further, no biased errors have been noted. The signal to noise ratio is
about 10:1 in the measurements made so far.

    Again, there is a lack of data in the literature for counter measurements
of highly seeded flows.

RANGE

    The actual range of velocities detectable by a given system is determined
primarily by the setting of the input filters. The tracker and counter are
identical in that regard. The practical width of the input filter for a counter
is determined by the signal to noise ratio. When noise is a problem, (S/N < 37
according to Adrian), the input filters must be narrowed for the counter.
Since the effective noise bandwidth of the tracker is on the order of at

least 10 better than that of the counter, the tracker's effective range can
be a factor of 10 better than the counter.

Another consideration is slewing rate. How fast can the detector change
the detected frequency?

A counter can, in principle detect two adjacent bursts, one each at the
limits of the input filter range. If σ_ω is the bandwidth of the signal at
the center of the input filter, the maximum slewing rate of the counter will
be $\Delta\omega\sigma_\omega$.

A tracker can change its frequency ω_N in $1/\omega_N$ seconds. The slewing rate
is thus approximately ω_N^2 with small error. Recall that $\omega_N \doteq \sigma_\omega$ and
$\sigma_\omega \doteq .03\ \omega_o$, so the tracker slews about a factor of 30 slower than a counter.

In a low seed density flow, the counter can respond to a large change in
velocity between bursts. In any real flow, there are strong bursts which are
detected and validated and many weak bursts which are detectable but not
validated. A tracker will respond to the weak bursts by drifting toward
their frequency even if it doesn't lock to them. When a strong burst occurs,
the tracker will probably be close enough to it to lock. There seems no
obvious theoretical reason why a tracker cannot detect and validate at least
as many events as a counter. Given its superior signal to noise performance,
it may even do better than a counter.

CONCLUSION

There seems to be little question that the behavior of a tracker as a
function of signal to noise ratio is far superior to that of a counter. This
is quite important since the most convenient, indeed sometimes the only
parameters for controlling the signal to noise ratio are the laser power and
the filter bandwidth.

For steady flows, the tracker has been shown to be at least as accurate as
the counter for high signal to noise ratios and superior to it for low signal
to noise ratios. This result is true independent of the seeding density.

For fluctuating flows, the tracker should be able to perform at least as
well as a counter. It can have a higher tracking range than the counter,
when noise is a problem.

There is a need for carefully performed experiments using:

1) Trackers in fluctuating low seed density flows
2) Modern two stage validation circuit counters in high seed density
    flows

ACKNOWLEDGEMENTS

I wish to thank L. Lading for many inspirational conversations and teach-
ings. The author also acknowledges the financial support of NSF (ENG75-19185
and ENG76-22965A01).

REFERENCES

1. W.K. George and J.L. Lumley, "The Laser Doppler Velocimeter and its
    Application to the Measurement of Turbulence", J. Fluid Mech., 60,
    p. 321, (1973).

2. S.O. Rice, "Statistical Properties of a Sine Wave Plus Random Noise",
    Bell-Syst. Tech. Journal, 27, p. 109, (1948).

3. L. Lading and R.V. Edwards, "The Effect of Measurement Volume on
    Laser Doppler Anemometer Measurements as Measured on Simulated Signals",
    Proceeding LDA-Symposium, Copenhagen (1975), p. 64.

4. F.M. Gardner, Phaselock Techniques, John Wiley & Sons, Inc., (1966).

5. R.V. Edwards, L. Lading and F. Coffield, "Design of a Frequency Tracker
    for Laser Anemometer Measurements", Fifth Biennial Symposium on
    Turbulence, Rolla, Missouri (1977).

6. J. Adrian, J.A.C. Humphrey, J.H. Whitelaw, "Frequency Measurement Errors
    Due to Noise", Proceedings LDA-Symposium 75 (Copenhagen), p. 287 (1975).

7. W. Hoesel and W. Rodi, "Errors Occurring in LDA-Measurements",
    Proceedings LDA-Symposium 75 (Copenhagen), p. 251, (1975).

8. D.K. McLaughlin and W.G. Tiederman, "Biasing Corrections for Glow",
    Phys. of Fluid, 16, p. 2082, (1973).

9. W. Hoesel and W. Rodi, "New Biasing Elimination Method for Laser-
    Doppler Velocimeter Counter Processing", Rev. Sci. Inst. 48, 7,
    p. 910, (July, 1977).

10. R.V. Edwards, "How Real are Particle Bias Errors?", 3rd Internaional
    Workshop on Laser Velocimetry, Purdue University (1978).

11. R.V.Edwards, J.C. Angus, M.J. French, and J.W. Dunning, Jr., "Spectral
    Analysis of the Signal from the Laser Doppler Flowmeter: Time
    Independent Systems", J. Appl. Phys., 42, (2), p. 837 (1971).

Processing Laser Velocimeter High-Speed Burst Counter Data

JAMES F. MEYERS and JAMES I. CLEMMONS, JR.
Instrument Research Division
NASA-Langley Research Center
Hampton, VA 23665

ABSTRACT

Individual random measurements from a laser velocimeter high-speed burst counter are typically gathered and tabulated with a pulse height analyzer or a minicomputer and displayed in histogram form. This technique yields sample statistics of the measurements and can in principle yield the probability density characteristics of the velocity flow field. Additional information can be extracted by imposing coincident requirements on the individual velocity component measurements and by determining the elapse time between the random measurements.

The present paper discusses the use of a high-speed buffer interface used to gather the data from one, two, or three LV counters simultaneously. The interface will gather data at a rate up to one million measurements per second for each channel whereas a minicomputer must gather data serially via DMA and is even then limited to two components. The interface uses a triple level clock circuit to measure the interarrival times for power spectra measurements with a maximum resolution of 100 nanoseconds. The final and perhaps the most unique capability of the buffer interface is its ability, upon computer command, to establish the requirement of coincidence of measurements between any of the two input channels or all three input channels. This will allow the investigation of velocity flow field characteristics due to magnitude changes and due to flow angle changes, a unique measurement capability of the laser velocimeter.

The last section of the paper presents several examples of the data processing techniques utilizing the capabilities of the buffer interface. Also presented are several novel forms of data presentation which yield an overall view of the flow field under study. These include arrow plots of velocity, flow streamline plots, contour maps of mean velocity, standard deviation, and flow angle, and composite histograms between two components.

INTRODUCTION

Although the optical and signal processing techniques used in laser velocimeter systems have advanced greatly since the advent of laser velocimetry as a flow diagnostic tool, the processing and display of the measured data have been limited to sample statistical calculations and simple plotting techniques. These techniques are adequate to view simple trends in the mean velocity and turbulent intensity, however they are very difficult to use to obtain an understanding of the total flow field. The present paper presents another look at the statistical calculations of the mean and potential bias errors and their applicability to the randomly sampled velocity data obtained from a laser velocimeter. Also techniques for displaying the resulting data in the form of arrow plots, flow streamline plots, and contour maps are discussed.

Recently data processing techniques for obtaining turbulent power spectra from the laser velocimeter data have been developed. In order to use these techniques, a new hardware interface between the signal processing electronics and the computer has been constructed. The operation of this interface is described along with the description of additional capabilities which were built into the interface that allow the measurement of the velocity vector for each particle passing through the LV sample volume. By processing this data with extended versions of the turbulent power spectra techniques, cross spectra, interactions between velocity magnitude fluctuations and flow angle fluctuations may be investigated.

STATISTICAL QUANTITIES

An unbiased estimate may be made of an ensemble mean if the sampling process is statistically independent of the data. This is always true when the data is uniformally sampled. However, in random samples, it must be deter- mined if the sampling process is independent of the data. It has been suggested in Reference 1 that this is not the case for a laser velocimeter, and that the interdependence causes a bias error in the ensemble mean.

In order to generalize a test situation, three assumptions will be made: (1) The velocity measurements are Poisson distributed in time, (2) the particles embedded in the flow are not only randomly dispersed in space, but are also randomly dispersed in the velocity field and (3) the measurement sample taken over a finite period of time is a good representation of the stationary condition at the measurement location. Also as an aid in viewing the higher moment statistics, graphical representations of the velocity probability density functions for each measurement ensemble will be made using the histogram format.

Theory

The sample mean will be studied using three different methods: (1) simple arithmetic mean, (2) arithmetic mean with corrections for velocity bias and Bragg cell bias, and (3) time averaging. The arithmetic mean is based on the assumption that all velocities present in the flow at the measurement point have an equal probability of being measured. The computations of ensemble mean velocity, V_e, use the classical non-weighted equation:

$$V_e = \frac{\Sigma V_i}{N} \qquad\qquad (1)$$

where V_i is the ith velocity measurement and N is the number of measurements. However, from Reference 1, it was found that if the seeding particles are uni- formally distributed in the flow, the number of measurements will be weighted toward the higher velocities since more gas (and thus more particles) passes through the sample volume per unit time than at low velocities. Conversely, from Reference 2, it has been found that when a Bragg cell is used for detecting directionality in the velocity measurements, a bias occurs toward the lower velocities. Since the Bragg cell causes the fringe pattern to move, it is possible to obtain multimeasurements from the same seed particle as it passes through the sample volume. The extreme of this effect is found when a particle is stationary in the sample volume allowing an infinite number of measurements to be made.

Since the particles pass through the sample volume at an arrival rate proportional to the velocity of the flow, the bias toward high velocities can be corrected by including a multiplication factor in the standard arithmetic mean calculation yielding a new estimate of ensemble mean, V_v:

$$V_v = \frac{\sum_i a_i V_i}{\sum_i a_i} \qquad (2)$$

where $a_i = 1/V_i$

On the other hand, the measurement rate of an LV system, which is equipped with a Bragg cell for detecting directionality in the velocity measurements, is inversely proportional to the velocity of the flow. The number of multi-measurements made of a single particle as it traverses the sample volume is found by dividing the transit time of the particle passing through the sample volume by the time required for a single velocity measurement. The particle transit time, t_t, is defined as the time required for the particle to pass through the sample volume as defined by the $1/e^2$ laser power boundaries:

$$t_t = \frac{D_{sv}}{V_i} \qquad (3)$$

where D_{sv} is the sample volume diameter. The time for a single measurement, t_m, is found by:

$$t_m = \frac{10 \, L_{fr}}{V_f + V_i} + T_r \qquad (4)$$

where L_{fr} is the fringe spacing, V_f is the velocity of the moving fringes due to the Bragg effect, and T_r is the reset time of the high speed burst counter. The factor of ten is included since ten fringe crossings are required to make a measurement by the high-speed burst counter under study. Thus the number of measurements, n_i, is found by:

$$n_i = \frac{t_t}{t_m} = \frac{D_{sv}/V_i}{\dfrac{10 \, L_{fr}}{V_f + V_i} + T_r} = \frac{D_{sv}(V_f + V_i)}{V_i[10 \, L_{fr} + T_r(V_f + V_i)]} \qquad (5)$$

The correction factor then becomes:

$$b_i = 1/n_i \qquad (6)$$

and the corrected arithmetic mean for Bragg bias, V_B, is,

$$V_B = \frac{\Sigma\ b_i\ V_i}{\Sigma\ b_i} \tag{7}$$

The total correction of the arithmetic mean, $V_{V,B}$ is then

$$V_{V,B} = \frac{\Sigma\ a_i\ b_i\ V_i}{\Sigma\ a_i\ b_i} = \frac{\Sigma\ A_i\ V_i}{\Sigma\ A_i}$$

where

$$A_i = \left(\frac{1}{V_i}\right)\ \frac{V_i\left[10\ L_{fr} + T_r\ (V_f + V_i)\right]}{D_{sv}(V_f + V_i)} \tag{8}$$

$$= \frac{10\ L_{fr} + T_r\ (V_f + V_i)}{D_{sv}\ (V_f + V_i)}$$

As an example of the effect, the value of A_i is shown in Figure 1 as a function of velocity for the laser velocimeter used in Reference 2 which had a fringe spacing of 2.65×10^{-5} m, a sample volume diameter of 3.13×10^{-4} m, a Bragg frequency of 5.0 MHz, and a counter reset time of 0.4×10^{-6} s.

From Reference 3, it has been suggested that all biases are removed if the data are weighted by the amount of time elapsed between each particle arrival. The restriction of this method is that the mean particle arrival rate must be equal to or greater than the highest flow turbulence frequency that contributes to the energy contained in the overall turbulent power spectra. The method for determining the time weighted average, V_t, is:

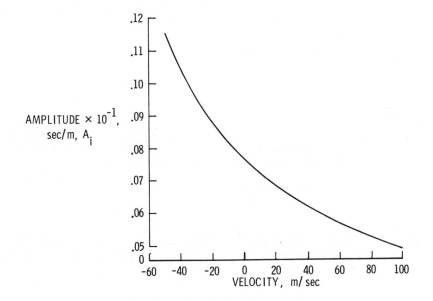

Figure 1.   Combined Velocity Bias and Bragg Bias Correction Factor

$$V_t = \frac{\Sigma \frac{\left(V_i + V_{i+1}\right)}{2} \Delta t_i}{\Sigma \Delta t_i} \tag{9}$$

where Δt_i is the time between the ith and the i+1th measurement.

In a wind tunnel test case, Reference 2, where the ensemble means were calculated using all three methods, all yielded similar results for data rates above about ten particle arrivals per second. Also, in all cases, the arithmetic mean and the corrected mean for velocity bias and Bragg cell bias yielded similar results, which infers that when a Bragg cell is used, whose frequency is large compared to the signal frequency, the bias errors are negligible. If the bias errors are found negligible for a given laser velocimeter, the simple arithmetic mean equations of arithmetic mean (V_e), standard deviation (σ), skew (S_R), and excess (E) are:

$$V_e = \frac{\Sigma V_i}{N} \tag{10}$$

$$\sigma = \sqrt{\frac{\Sigma(V_i - V_e)^2}{N - 1}} \tag{11}$$

$$S_R = \frac{\Sigma(V_i - V_e)^3}{N\sigma^3} \tag{12}$$

$$E = \frac{\frac{\Sigma(V_i - V_e)^4}{4}}{\left(\frac{\Sigma(V_i - V_e)^2}{2}\right)^2} - 3.0 \tag{13}$$

(13)

In order to determine the statistical accuracy of the mean and standard deviation obtained from each measurement ensemble, the measurement uncertainties may be calculated using the equations found in Reference 4. For a 95-percent confidence limit, the statistical uncertainty in the mean is:

$$\Delta V_e = \frac{2\sigma}{\sqrt{N}} \tag{14}$$

That is, the true stationary mean velocity of the flow lies within a radius of uncertainty about the measured mean calculated from the ensemble. Similarly, the measurement uncertainty in the standard deviation for a 95-percent confidence limit is:

$$\Delta\sigma = \sigma \sqrt{2/N} \left(1 + \frac{E}{2}\right)^{\frac{1}{2}} \tag{15}$$

It should be noted that the inclusion of excess in Equation 15 allows the measurement uncertainty to be calculated for ensembles with probability densities other than Gaussian.

A visual sense of these statistical quantities may be obtained by placing the ensemble data in histogram form. Since the high-speed burst counters output data in the form of number of reference clock pulses, the histogram interval width is established. Thus, the histograms are formed by determining the number of occurrences of each counter output pattern present in each measurement ensemble. These patterns are converted to velocity and the histograms plotted. A few example histograms and their ensemble statistics are presented in Figure 2. These ensembles were chosen to illustrate several types of probability density functions found in the tests in Reference 2.

V_e	44.5	37.4
ΔV_e	0.0	0.0
$V_{V,B}$	44.6	37.4
V_t	44.8	37.6
σ	1.3	1.4
$\Delta\sigma$	0.0	0.0
S_R	-0.37	-0.04
E	0.5	0.2
N	2136	1535
DATA RATE	36 / sec	26 / sec

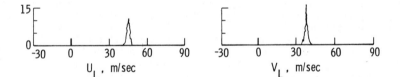

Figure 2a. Example Two Component Histograms (U_L is +45° with respect to Tunnel Centerline, and V_L is -45° with respect to tunnel centerline)

V_e	52.0	27.0
ΔV_e	1.1	0.4
$V_{V,B}$	51.7	27.0
V_t	52.5	25.7
σ	8.4	4.7
$\Delta\sigma$	0.6	0.2
S_R	0.26	0.16
E	-1.0	-0.9
N	58	159
DATA RATE	1 / sec	3 / sec

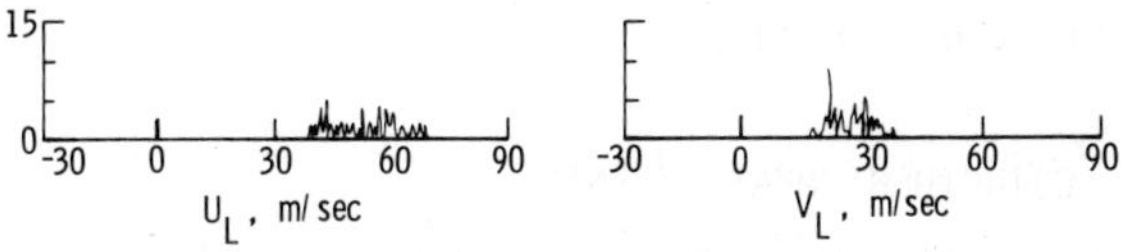

Figure 2b. Example Two Component Histograms (U_L is + 45° with respect to Tunnel Centerline, and V_L is -45° with Respect to Tunnel Centerline)

Plotting Techniques for Flow Field Display

Although the information needed for flow field diagnostics is contained in tables of the computer statistical quantities, overall understanding of the flow field is enhanced by plotting the results in the form of arrow plots, flow streamline plots, and contour maps.

The arrow plots are the simplest of the plotting techniques and basically consist of converting the mean velocities obtained from two measurement components at a point in the flow to mean velocity magnitude and mean flow angle. These two quantities are then represented by an arrow whose origin is located at the measurement location on the plot, and whose length is proportional to the mean velocity magnitude and its direction determined by the mean flow angle. An example is shown in Figure 3 for a NACA 0012 airfoil at 19.4 degrees angle of attack at a nominal Mach number of 0.148.

For the remaining two techniques, flow streamline plots and contour maps, a velocity grid must be computed. This is accomplished by using spline techniques to develope a smooth velocity profile along a vertical measurement scan, e.g., a set of measurements in the vertical direction illustrated by a set of arrows along a vertical path on the arrow plots. The spline technique is used again to develop velocity profiles in the horizontal direction, joining the vertical splines forming the grid. As an example based on the arrow plot, Figure 3, the velocity was computed every 1.6 mm along the horizontal spline fit and a horizontal spline fit was computed every 1.6 mm in the vertical direction which yield a velocity grid of 1.6 mm between grid points over the entire flow field. This technique may be used to develop grids of velocity magnitude, flow angle, two-component velocity standard deviation, etc.

The flow streamline plots are formed based on a grid of mean flow angle. A particle is located (theoretically in the computer) at an upstream location on the grid and allowed to move in the direction indicated at the grid point. When the particle intersects a vertical or horizontal line between two adjacent grid points, the flow angle for that location is determined by linear interpolation between the two grid points and the particle is sent to the next location along the computed trajectory. This process is continued until the particle leaves the grid structure. An example of this technique is shown in Figure 4 for the flow field illustrated in the arrow plot.

The contour map for mean velocity magnitude is obtained by a search through the grid map for adjacent grid points in which the velocity magnitude stored in one point is above the desired value and the value stored in the other point is below. The spatial location of the desired value is then obtained by linear interpolation between the two grid points and the computed location is stored in a two-dimensional array. When the entire grid has been searched, the program scans the resulting array to find adjacent spatial locations and places them in order based on closest spatial distance. The restriction on being considered as adjacent spatial locations is that any two spatial locations must be within two grid distances, in this example 3.2 mm. The first location point is also checked so that circular contours may be closed. When a contour is completed and plotted, the array is scanned again looking for a second contour. This process continues until all locations in the array have been plotted. A contour map for the example flow field is illustrated in Figure 5.

TIME DEPENDENT AND COINCIDENT MEASUREMENTS

In order to obtain time dependent measurements, such as time average velocity, described earlier, and turbulent power spectra measurements, additional

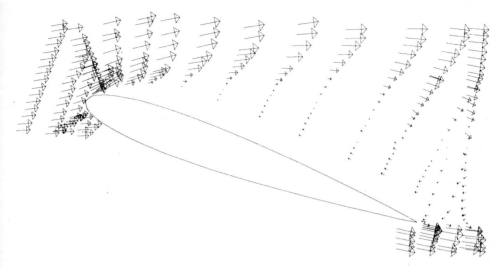

Figure 3. Velocity Vectors Computed from Measurements
Over the Wing at α = 19.4°

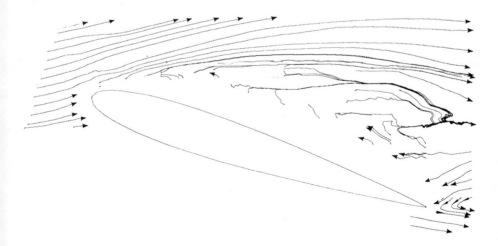

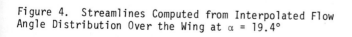

Figure 4. Streamlines Computed from Interpolated Flow
Angle Distribution Over the Wing at α = 19.4°

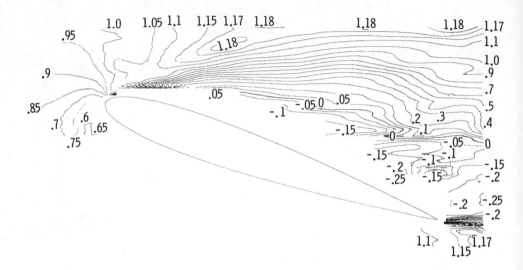

Figure 5.   Contours of Constant Local Velocity Magnitude/Free
Stream Velocity Measured Over the Wing at   α = 19.4°

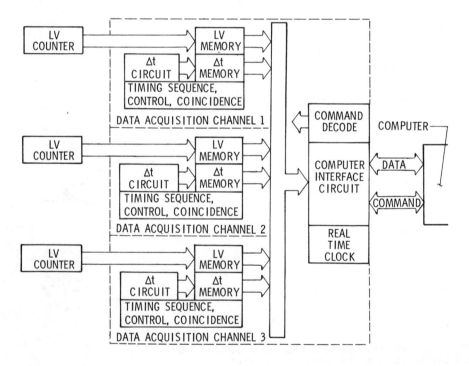

Figure 6.   Block Diagram of the Laser Velocimeter Autocorrelation
Buffer Interface

hardware must be included between the high-speed burst counter and the computer system to measure the time between successive velocity measurements (interarrival times). The hardware system developed at Langley measures the interarrival times and provides an intermediate storage of 4096 velocity - interarrival time pairs for up to three LV components and has capabilities of forcing a requirement of measurement coincidence between the components. This hardware called the laser velocimeter autocorrelation buffer interface or LVABI, will not only provide measurements of velocity time history, but allow the extraction of velocity magnitude fluctuations, and flow angle fluctuations since the velocity vector for each particle is obtained when the restriction of coincidence is applied.

The Laser Velocimeter Autocorrelation Buffer Interface

The laser velocimeter autocorrelation buffer interface (LVABI) is designed to receive data from multicomponent laser velocimeters, determine the inter-arrival times for all data acquired (or determine the interarrival times of only coincident data), and provide storage of velocity and interarrival time values at rates of up to one million data sets (LV-Δt) per second for each LV component. Circuits are provided for both rapid data transfer between the LVABI and the computer, and for computer control of the LVABI.

The LVABI has two major divisions: the data acquisitions channel(s) and the computer interface circuits. There may be one to three data acquisition channels, each having its own data capture circuit, control and timing sequence generator, interarrival time measurement circuit, coincidence circuit, and memory. The computer interface circuit serves all data acquisition channels and includes: input/output circuits for connecting the LVABI to a computer, computer command decode circuits, and a real time clock for controlling the data acquisition time. A block diagram of the LVABI is shown in Figure 6, and the specifications are given in Table 1. Two key functions, the interarrival timer and the coincidence circuit will be discussed here since each is involved in the time dependent measurements.

Table 1. LVABI Specifications

LV data acquisition rate:	1,000,000 data words/second/LV component (noncoincidence mode)
LV data word:	14-bits (10-bit mantissa, 4-bit exponent) 2-bits status (not stored), (TTL levels)
LV data valid signal:	<600 nanoseconds (TTL levels)
Interarrival time:	1.0×10^{-6} seconds to 0.65535 seconds
Interarrival time resolution:	1.0×10^{-6} s to 6.55535×10^{-3} s, 100 nanoseconds 6.5535×10^{-3} s to $65,535 \times 10^{-3}$ s, 1 microsecond 65.535×10^{-3} s to 655.35×10^{-3} s, 10 microseconds
Interarrival time data word:	18-bits time data, 2-bit status)
First interarrival time:	From initialization to the first LV data valid
Data storage:	4096 32-bit data words per channel

The interarrival timer is a three level clock whose base frequency is 10 MHz. When the system is started, or at the time of a velocity measurement, a 16-bit counter begins counting pulses from the 10 MHz clock. If a velocity measurement is not made before the counter overflows, the input clock is changed to 1 MHz and the number 6553 is loaded into the counter. The effect is that the interarrival time measurement is now as if the clock were 1 MHz from the start. If the counter overflows again, the process is repeated with a 100 kHz clock. If the counter overflows a third time, the counter is zeroed indicating that the interarrival time exceeded the maximum measureable time. This process allows an interarrival time to be measured from one microsecond to 0.655 seconds. A 2-bit status word is used to indicate the clock frequency used in the inter-arrival time measurement.

The coincidence circuit allows velocity measurements to be made of only particles which yield valid velocity measurements in both components, (or all three components in a three component laser velocimeter). Since the intensity of the scattered light from a particle will vary depending upon the character-istics of each LV component and the trajectory of the particle, an acceptance window in time for coincidence must be included. Because the transit time for a particle passing through the sample volume is approximately a microsecond for most LV configurations, the coincidence aperture window was chosen to be 1 microsecond. Once a velocity measurement is made in any component, the 1 micro-second timer is started. If the remaining components do not obtain a velocity measurement during that time, the acquired data is erased. If the remaining components do obtain measurements, the measurement values are stored in the LVABI memory along with the interarrival time between the measurement and the previous coincident measurement. In order to keep interarrival time continuous, allowing for noncoincident data, two reference clock counters are used in an alternating manner so that time is not lost if a data point is rejected due to noncoincidence.

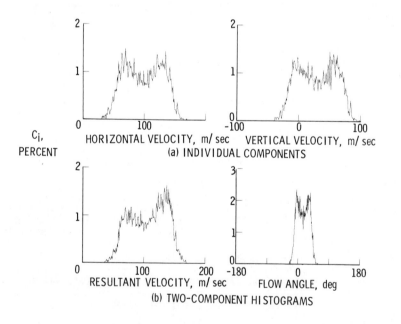

c_i, PERCENT

(a) INDIVIDUAL COMPONENTS

(b) TWO-COMPONENT HISTOGRAMS

Figure 7.   Example Two-Component Histograms

Velocity Vector Measurement

By using the LVABI in the coincident mode, the velocity vector for each particle may be measured. This type of measurement opens up a new source of information, for instance, statistical analysis may be performed on velocity magnitudes only, flow angle only, or both and the relation between velocity magnitude and velocity flow angle may be investigated. To aid in this study, composite histograms may be generated, one of magnitude and one of flow angle. An example of these histograms, along with the basic velocity component histograms, is shown in Figure 7. This technique may also open new insights in conditional sampling in high shear or mixing regions in flows, because once the data is stored in the computer, a condition may be established based on flow angle.

Turbulent Power Spectra Measurement

Techniques for measuring turbulent power spectra from laser velocimeter measurements have been developed in References 5, 6 and 7. Although the LVABI does not affect these techniques directly, it does, however, remove the problems of data packing which are a possibility in the first-in-first-out memories used in these references, and it also does not require the computer's attention during data gathering. Another problem with the present methods of data handling is the limited range of the interarrival time measurements which has been expanded with the LVABI.

Using the basic approach given in Reference 5, turbulent power spectra measurements were conducted in a wind tunnel test about a stalled airfoil in Langley's V/STOL tunnel, Reference 2. An example of the results is shown in Figure 8. Although nothing new was developed in Reference 2 with regard to turbulent power spectra measurements, the question of spectra measurement accuracy is of concern. At present, the only study of measurement accuracy was conducted in Reference 5. Although the results looked good for describing the variability error in the stop band, the variability error in the pass band is still uncertain.

A computer simulation study is now beginning at Langley which will attempt to determine the variability error and its dependence on measurement data rate, number of measurements, spectral bandwidth, and center frequency. The simulation begins with a uniformally sampled signal in time whose spectra is white. This sampled signal may now be filtered on the computer with a digital filter so as to yield the desired frequency content. The power spectra of this signal is now computed using the autocovariance approach. The resulting spectra is used as the standard in comparisons with the LV simulations. The uniform sample is further sampled based on a time generated by a random number generator whose statistics are Poisson. If the random time occurs between the times for two adjacent uniform samples, as is generally the case, the velocity value is obtained via linear interpolation between the two adjacent uniform samples. Once the simulated LV measurements are made, the power spectra is computed based on the approach in Reference 5. Figure 9 illustrates the spectra obtained from the uniform sample along with an overlay of the spectra derived from the LV simulation. This technique allows one to vary the mean Poisson data rate, the number of measurements, the frequency bandwidth of the input signal and the characteristics of the input spectra. From the results one can compare the spectra with the standard spectra to determine the extent of the variability error in the pass band, the stop band, or the composite. From this work it is hoped that an estimate of accuracy may be made on spectra obtained during wind tunnel testing.

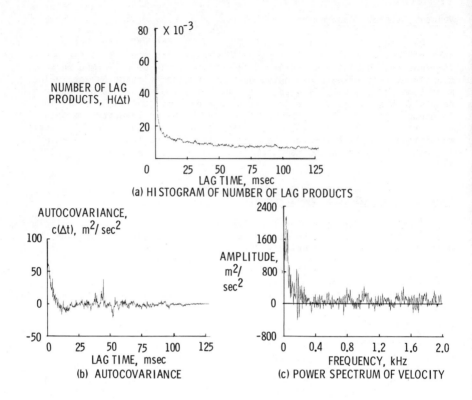

(a) HISTOGRAM OF NUMBER OF LAG PRODUCTS

(b) AUTOCOVARIANCE

(c) POWER SPECTRUM OF VELOCITY

Figure 8.   Calculation of Power Spectra for the V_L Component at the
Trailing Edge of the Wing at α = 19.4°

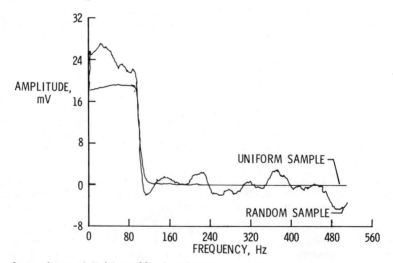

Figure 9.   Comparison of Uniformally Sampled Derived Power Spectra
with a Poisson Sampled Derived Power Spectra

Component Spectra

Although the above discussion was primarily concerned with a single component autocorrelation and spectra, and its associated uncertainty, the LVABI will allow further types of spectral measurements to be made. By applying the coincidence restriction on the data, cross spectra may be made of any two LV components. Even though cross spectra measurements of LV data are not a new idea, the technique opens other possibilities. By determining the velocity vector of each particle measured, it is possible to determine the turbulent power spectra of the flow due to velocity magnitude changes, spectra of flow angle changes and cross spectra between velocity magnitude and flow angle changes. Further, it is possible to determine the cross spectra between any two arbitrary directions.

CONCLUDING REMARKS

The laser velocimeter has been used for many years as a flow diagnostic tool for measuring one or two component mean velocities and turbulent intensities. Only within the last couple of years have turbulent power spectra been derived from the laser velocimeter measurements. However, these measurements may be only the tip of the iceburg as to the capabilities of the laser velocimeter. It has been shown that it is possible to distinguish between flow velocity magnitude and flow angle when the velocity vector is measured with a multi-component laser velocimeter. The ability to investigate the interaction of flow angle with velocity magnitude opens a new realm of information not available with classical wind tunnel instrumentation which may yield new insights into the mechanism of fluctuating flow fields.

REFERENCES

1. Tiederman, W. G., D. K. McLaughlin, and N. M. Reischmann: Individualized Realization Laser Doppler Technique Applied to Turbulent Channel Flow. Third Biennial Sym. on Turbulence in Liquids, Un. of Missouri, 1973.

2. Hoad, Danny R., James F. Meyers, Warren H. Young, Jr., and Timothy E. Hepner: Laser Velocimeter Survey about a NACA 0012 Wing at Low Angles of Attack. NASA TM74040, 1978.

3. Dimotakis, P. P.: Single Scattering Particle Laser Doppler Measurements of Turbulence. AGARD CP193, May 1976, pp. 101 to 1014.

4. Yule, G. V. and M. G. Kendall: An Introduction to the Theory of Statistics. 12th ed., Charles Griffen Co. (London), 1940.

5. Mayo, W. T., Jr., M. T. Shay, and S. Riter: Development of New Digital Data Processing Techniques for Turbulence Measurements with a Laser Velocimeter. AEDC TR7453 (Aug. 1974).

6. Smith, D. M. and D. M. Meadows: Power Spectra from Random Time Samples for Turbulence Measurements with a Laser Velocimeter, 2nd International Workshop on Laser Velocimetry, Purdue University, March 1974.

7. Wang, James C. F.: Laser Velocimeter Turbulence Spectra Measurements. Minnesota Symposium on Laser Anemometry, University of Minnesota, October 1975.

Fourier Transformation and Spectrum Analysis of Sparsely Sampled Signals

KEITH H. NORSWORTHY
Boeing
P.O. Box 3199
Seattle, WA 98124

ABSTRACT

Alternative methods for deriving Fourier transforms and power spectral estimates from sparse signal samples are evaluated by analysis and computer simulations. The results confirm that randomization of the sample interval can effectively circumvent the well known Nyquist sampling theorem and permit spectrum analysis to frequencies well above the average sample rate.

The paper compares spectral analysis by direct Fourier transformation of the time domain signal samples with spectral analysis via the Dirac comb correlation function method recently proposed by M. Gaster and J. Roberts (References 1 and 2), and introduces new random sample processing methods that are based on simultaneous equation regression analysis.

Careful consideration is given to frequency resolution (spectral window) issues and the confidence factors associated with the measured frequency spectra.

INTRODUCTION

The problem addressed in this paper is that of estimating the spectral components of a signal that is observed sparsely and at random times, see Figure 1.

The well-known Nyquist theorem teaches that periodic sampling leads to spectral folding about a frequency equal to one half the sample rate.

Recent publications by a number of authors, including M. Gaster and J. Roberts [1,2], have shown that the Nyquist theorem does not apply when a signal is observed randomly, and go on to define methods for deriving unbiased spectral estimates out to frequencies well above the average observation rate. In general, however, it is found the confidence limits of the spectral estimates derived from the random observations are substantially worse than those derived from an equal number of periodic observations.

The object of this paper is to seek random observation processing methods that will provide improved spectral estimates while retaining the desirable absence of Nyquist folding effects.

314

SPARSE SIGNAL OBSERVATIONS:

- ● PERIODIC SAMPLES:
  - ● CONVENTIONAL NYQUIST AMBIGUITY

- ● RANDOM SAMPLES:
  - ● NO NYQUIST AMBIGUITY
  - ● INCREASED SPECTRAL VARIANCE

Figure 1: RANDOMLY SAMPLED SIGNAL

PROCESSING METHODS COMPARED

Five different methods for processing random signal observations are compared, see Figure 2. These methods are briefly described as follows:

Processing Method 1 - FFT in Signal Plane

The available time observations are divided into separate time blocks of duration equal to the reciprocal of the desired frequency resolution Δf, see Figure 3.

Each time block is then further divided into equal time intervals Δt, where $\Delta t = 1/2f_{max}$, and the observations within each interval are averaged. If no observations occur in a time interval a zero is substituted for the average.

An estimate of the signal power spectrum (0 to f_{max}) is then computed by applying a conventional FFT algorithm to the averaged observations and (conjugate) squaring the spectral answers.

Processing Method 2 - Single Frequency Regression in Signal Plane

The approach here is to curve fit individual sine and cosine waves to the observations.(0_n).

1. FFT IN SIGNAL PLANE $P(f) = \overline{(FFT)(FFT^*)}$

2. SINGLE FREQUENCY REGRESSION ANALYSIS IN SIGNAL PLANE ⎫
                                                        ⎬ $P(f) = \overline{S^2(f)}$
3. MULTIPLE FREQUENCY REGRESSION ANALYSIS IN SIGNAL PLANE ⎭

4. FFT IN CORRELATION PLANE $P(f) = \overline{(FFT)}$

5. MULTIPLE FREQUENCY REGRESSION ANALYSIS IN CORRELATION $P(t) = \overline{P(f)}$
   PLANE

Figure 2: FIVE METHODS COMPARED BY COMPUTER SIMULATION

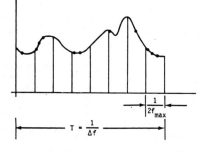

STEPS:

1. Select block size $T = \frac{1}{\Delta f}$ desired

2. Divide T into N intervals $N = \frac{2f_{max}}{\Delta f}$

3. Average observation in each interval, substituting zero if no observation

4. Perform FFT on step 3 values

Figure 3: PROCESSING METHOD 1 - FFT IN SIGNAL PLANE

The power spectrum estimate at frequency (f_k) is equated to $a_k^2 + b_k^2$, where a_k and b_k represent the least mean square error solution of the following overdetermined set of linear simultaneous equations:

$$
\left.
\begin{aligned}
0_1 &= a_k \sin 2\pi f_k\, t_1 + b_k \cos 2\pi f_k\, t_1 \\
0_2 &= a_k \sin 2\pi f_k\, t_2 + b_k \cos 2\pi f_k\, t_2 \\
&\quad\vdots \\
0_n &= a_k \sin 2\pi f_k\, t_n + b_k \cos 2\pi f_k\, t_n
\end{aligned}
\right\} \quad - - (1)
$$

Processing Method 3 - Multiple Frequency Regression in Signal Plane

This method is similar to 2, but instead of fitting one sine and cosine wave frequency at a time, the least mean square error criterion is used to fit the observations to the sum of many frequency components. The linear simultaneous equations take the following form;

$$
\left.
\begin{aligned}
0_1 &= \sum_{k=1}^{M} a_k \sin 2\pi f_k\, t_1 + \sum_{k=1}^{M} b_k \sin 2\pi f_k\, t_1 \\
0_2 &= \sum_{k=1}^{M} a_k \sin 2\pi f_k\, t_2 + \sum_{k=1}^{M} b_k \sin 2\pi f_k\, t_2 \\
&\quad\vdots \\
0_n &= \sum_{k=1}^{M} a_k \sin 2\pi f_k\, t_n + \sum_{k=1}^{M} b_k \sin 2\pi f_k\, t_n
\end{aligned}
\right\} \quad - - (2)
$$

It is found that a "stable" least mean square error solution is usually obtained if the number of simultaneous frequency components is substantially less than one half of the number of observations.

As with method 2 the power spectral estimate at frequency f_k is equated to $\left[a_k^2 + b_k^2 \right]$.

Processing Method 4 - FFT in Correlation Plane

The time observations are divided into separate time blocks, as in method 1, and autocorrelation cross products formed for each block and averaged within delay increments, $\Delta t = 1/2f_{max}$. If no cross products fall in a particular delay interval a zero is substituted for that autocorrelation coefficient. The necessity to substitute a zero occurs much less frequently than with processing method 1, because there are many more cross products per block than signal observations.

The estimate of the signal power spectrum for a given block is computed as the cosine FFT of the (averaged) autocorrelation coefficients.

Processing Method 5 - Multiple Frequency Regression in Correlation Plane

This method represents a variation to method 4. Correlation coefficients are derived, as before, by averaging observation cross products within delay increments $\Delta\tau = 1/2f_{max}$. However instead of subjecting the derived correlation coefficients to a straightforward cosine FFT, a regression analysis is used to exploit the fact that some correlation coefficients are known with higher confidence than others.

By solving the following set of linear simultaneous equations, correlation fitting errors become weighted in proportion to the half power of the number of cross products that contribute to each coefficient (C_n).

$$
\left.
\begin{aligned}
(P_1)^{\frac{1}{2}} C_0 &= (P_1)^{\frac{1}{2}} \sum_{k=1}^{M} B_k \\
(P_2)^{\frac{1}{2}} C_1 &= (P_2)^{\frac{1}{2}} \sum_{k=1}^{M} B_k \cos 2\pi f_k \; \Delta\tau \\
(P_n)^{\frac{1}{2}} C_n &= (P_n)^{\frac{1}{2}} \sum_{k=1}^{M} B_k \cos 2\pi f_k \; (n\Delta\tau)
\end{aligned}
\right\} \quad - - \quad (3)
$$

Where P_n = number of cross products averaged in C_n.
The power spectral estimates are given directly by the derived B_k values.

If the number of cross products averaged in any particular delay increment is zero, that delay value drops out of the set of equations because (P_n) is zero.

It is found that a "stable" least mean square error solution usually occurs if the number of frequency components simultaneously solved for is substantially less than one-half the number of non-zero equations.

Comparison

The Wiener-Khintchine Theorem teaches that methods 1 and 2 are precisely equivalent when the signal observations are continuous or periodic. Furthermore, simple analysis can show that methods 3, 4 and 5 are equivalent to each other, and also to methods 1 and 2, under the same assumption of continuous or periodic observations. This equivalency arises from the fact that different Fourier (frequency) components are orthogonal to each other when defined continuously or periodically.

When the observations have random timing the integral of the cross product of different Fourier components does not equate to zero so the five processing methods are not equivalent one to another.

COMPARATIVE PERFORMANCE

For purposes of comparing the performance of the different processing methods six sets of signal observations have been defined and processed through computer algorithms that implement the five different processing methods.

Timing of Observations

For convenience of comparison, the sets of observations have been derived by viewing six signal representations through a single set of observation times.

With the average (random) observation interval defined at Δt, the total observation time has been selected to equal 320 Δt, and has been divided into 16 time blocks each of duration 20 Δt.

Fourier Coefficients are sought at frequencies k/20 Δt, where $0 \leq k \leq 63$.

Table 1 shows a listing of the observation times in each of the sixteen data blocks.

Signal Waveforms Analyzed

The six time functions analyzed are listed in Table 2. The first four functions each comprise a single frequency, the fifth and sixth functions have multiple frequencies.

Example Plots

A time plot of the observations from block 1/signal 1 is shown in Figure 4 and a plot of the corresponsing autocorrelation function (averaged within each of 128 Dirac Comb bins, as described for processing method 4) is shown in Figure 5. Example spectral plots for processing methods 1 and 4 are shown in Figures 6 and 7. These spectral plots represent the average of the 16 data blocks taken from signal 1.

Two features are of primary interest; 1) from block to block how repeatable is the amplitude of the principal spectral line? and 2) how high are the spectral sidelobe responses relative to the principal spectral line?

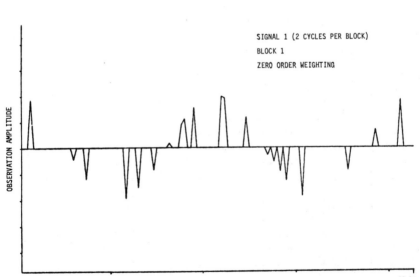

Figure 4: TIME PLOT OF OBSERVATIONS FROM BLOCK 1 OF SIGNAL 1

Table 1: LISTING OF OBSERVATION TIMES

0.5918	22.7723	40.1369	61.9640	81.4251	100.5907	120.2312	141.2445
2.8418	23.4062	40.5854	62.6040	81.8616	100.8143	120.6443	141.8697
3.5122	23.8810	40.8599	65.2013	83.0558	102.0118	121.1386	143.0139
5.4144	24.2066	43.7442	65.9504	83.4424	105.6911	121.7541	145.0297
6.1315	26.9174	43.9654	66.9335	85.4718	107.3916	122.9931	146.6089
6.8149	27.4853	44.1566	67.1182	85.8139	106.1913	123.4119	147.6083
7.6405	28.9500	45.5537	67.6504	87.0080	107.6306	124.2352	151.0504
8.2312	29.9488	46.4898	68.1075	87.2683	109.1256	124.4709	152.3436
8.4448	31.0405	46.7966	68.4575	88.4876	109.5058	125.5262	154.1335
8.9044	34.1890	47.0239	68.9429	89.0768	109.8248	125.9915	156.0431
10.2649	35.6959	48.1428	69.5921	89.6619	110.2110	126.4277	
10.4778	35.7915	48.3057	70.2876	90.6171	112.6504	128.0405	
11.5066	36.5102	48.7407	70.7096	91.2519	114.4706	128.5710	
12.7231	37.0617	49.3320	74.6230	92.3851	114.9754	128.8514	
12.9329	37.5045	50.1862	75.5072	93.0181	115.7398	129.9122	
13.2401	38.1693	51.0561	75.7481	93.7595	116.0485	130.5328	
13.5805	38.2911	53.9574	76.7546	94.1181	116.8615	131.7215	
14.3817	38.8457	56.2529	77.0072	96.5185	117.3624	131.9035	
16.7970		56.7211	77.7082	97.4486	118.4305	135.5295	
18.0629			79.2825		119.0076	136.4334	
19.3292						137.2528	
						137.5478	
						137.9528	
						139.0451	

160.3896	180.0284	201.4574	220.0827	240.2463	261.9319	280.1731	301.3934
162.3669	180.5464	202.2369	220.4314	240.5678	263.4809	280.5381	302.9868
162.6693	180.7364	203.2758	221.3815	241.8984	264.4924	280.9817	303.1214
164.9574	183.5586	205.6691	223.6946	242.4058	266.5535	284.9582	303.3518
165.7727	188.5789	207.0054	226.5111	245.0149	269.8450	286.8973	306.7908
166.8625	190.4069	210.1335	227.2122	245.5898	270.2668	287.7088	307.4319
167.4660	191.1484	211.3654	227.8392	246.8592	272.2034	289.1568	308.3959
168.4960	193.0450	212.8565	232.1080	247.2889	274.7583	289.4973	309.5537
169.3542	194.8384	216.8093	232.8338	247.6489	275.2187	291.4209	311.9656
173.1079	195.7498	217.4685	237.9107	248.2050	276.9037	291.5116	314.1286
173.6812	196.3356	218.4781	238.5456	248.9297	277.4468	296.0319	315.2402
174.1615	196.8451	219.1399	239.7091	249.3589	278.0469	296.4838	315.6484
174.3811	197.0723	219.4632		250.6518		297.2361	315.7817
175.0707	197.5915			250.9628		298.2905	316.0139
175.4106	197.9137			251.7213		298.8039	317.5793
175.7866				252.2148		299.4661	318.9539
176.0164				253.0037			319.9072
177.7497				255.1904			
179.1399				255.8654			
				257.1675			
				257.8483			
				258.7877			
				259.2412			

Table 2: SIX TIME SIGNALS ANALYZED

TEST SIGNAL	FUNCTIONAL FORM	FREQUENCY (CYCLES PER BLOCK)
1	$\cos 2\pi \frac{t}{T} \cdot 2.0$	2.0
2	$\cos 2\pi \frac{t}{T} \cdot 2.5$	2.5
3	$\cos 2\pi \frac{t}{T} \cdot 40.0$	40.0
4	$\cos 2\pi \frac{t}{T} \cdot 80.0$	80.0
5	$\sum\limits_{g=1}^{5} \cos 2\pi \frac{t}{T} \cdot g$	1, 2, 3, 4 and 5
6	$N + \cos 2\pi \frac{t}{T} \cdot 5$	Unity RMS noise plus cosine of frequency 5.0

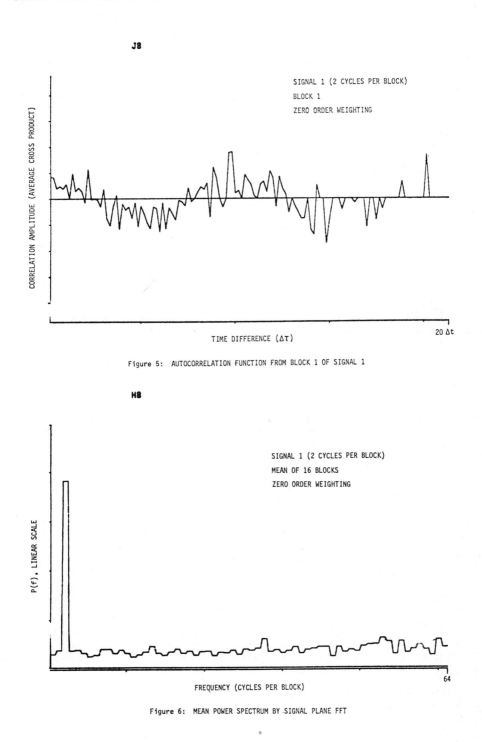

Figure 5: AUTOCORRELATION FUNCTION FROM BLOCK 1 OF SIGNAL 1

Figure 6: MEAN POWER SPECTRUM BY SIGNAL PLANE FFT

K8

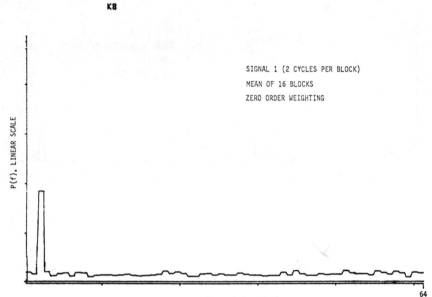

SIGNAL 1 (2 CYCLES PER BLOCK)
MEAN OF 16 BLOCKS
ZERO ORDER WEIGHTING

P(f), LINEAR SCALE

FREQUENCY (CYCLES PER BLOCK)

64

Figure 7: MEAN POWER SPECTRUM BY CORRELATION PLANE FFT

The first of these properties is handled by defining the term "spectral signal-to-noise ratio (S/N)" equal to the mean of the principal spectral line divided by its standard deviation. The second property is handled by defining the average sidelobe as a fraction of the (averaged) principal spectral line.

Comparison of Spectral Signal-to-Noise Ratios

Spectral signal-to-noise ratios derived from each of the first four processing methods are listed in Table 3 and are now further discussed.

As reported previously [1] both FFT implementations (processing methods 1 and 4) give stable spectral estimates under all conditions, with the correlation method perhaps giving slightly better results than the signal plane method at the expense of an increase in processing load. The results in Table 3 show that for these two processing methods the use of Hanning weighting is not effective in increasing the S/N values and, as expected, Nyquist folding (spectral ambiguity) occurs about a frequency equal to 1/2 Δt, or 1/2 $\Delta \tau$, see Figure 8.

Single frequency regression in the signal plane (processing method 2), when used with a zero order window, is found to be extremely effective under one special condition; namely, when the input signal has a single frequency with an integer number of cycles over the selected block length.

This condition is very restrictive and is unlikely to be frequently encountered in practice.

Table 3: COMPARISON OF SPECTRAL SIGNAL-TO-NOISE RATIOS

TEST SIGNAL (SEE TABLE 2)	MEASURED FREQUENCY (CYCLES/BLOCK)	PROCESSING METHOD (SEE TABLE 1)							
		1		4		2		3*	
		ZERO	HANNING	ZERO	HANNING	ZERO	HANNING	ZERO	HANNING
1	2	2.0	1.8	2.4	2.2	∞**	2.4	∞	∞
2	2	1.7	1.4	1.4	2.0	2.8	2.3	3.1	88.4
	3	1.6	1.4	1.4	1.6	2.7	2.2	3.5	132.2
3	40	2.0	1.5	2.6	1.9	∞	2.1	∞	∞
4	48	1.3	1.1	2.5	1.7	—	—	—	—
	80	—	—	—	—	∞	2.5	∞	∞
5	1	1.3	1.4	1.1	1.1	1.0	2.0	∞	∞
	2	1.3	1.1	1.3	1.3	0.7	1.2	∞	Zero
	3	1.3	1.1	1.1	1.2	1.1	1.1	∞	amp.
	4	1.4	1.1	1.1	1.5	1.1	1.0	∞	
	5	1.4	1.3	0.9	1.1	1.5	1.7	∞	∞
6	5	1.2	1.2	1.5	1.3	1.9	0.9	1.2	0.9

** "∞" SYMBOL USED IF S/N $> 10^6$

* PROCESSING METHOD 3 SOLVED FOR 7 SIMULTANEOUS FREQUENCIES, APPROPRIATELY CENTERED

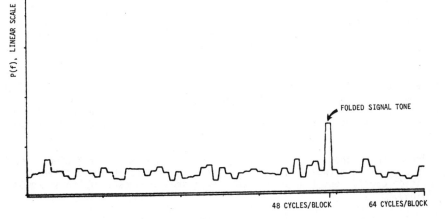

H7

INPUT SIGNAL 4 - 80 CYCLES/BLOCK

PROCESSING METHOD 1 - FFT IN SIGNAL PLANE

P(f), LINEAR SCALE

FOLDED SIGNAL TONE

48 CYCLES/BLOCK   64 CYCLES/BLOCK

Figure 8: NYQUIST FOLDING PHENOMENON

When the input signal comprises more than one Fourier component (test signals 2, 5 and 6, and test signals 3 and 4 with Hanning weighting) processing method 2 consistently gives stable spectral measurements, but the S/N values are not significantly better than those obtained with the simple FFT methods. Again, the use of Hanning weight is not effective in improving the spectral S/N values but, in this case, no Nyquist folding effects are encountered so reliable frequency analyses can be conveniently performed over a greatly extended frequency range.

Multiple frequency regression in the signal plane (processing method 3) gives truly excellent S/N results for several of the signal conditions tested. However, a "stable" least mean square error solution of equation 2 only occurs if the number of Fourier components simultaneously solved for is substantially less than one half of the number of signal observations in the selected data block. If the average random observation rate exceeds six times the maximum signal frequency ($\overline{f}_s > 6f_{max}$) it has been found that one can solve for all Fourier components simultaneously and achieve excellent accuracy on the derived frequency spectrum. If however, $\overline{f}_s < 6f_{max}$ the full frequency range, 0 to f_{max}, cannot be handled simultaneously but needs to be divided into sub-bands that are analyzed sequentially.

Test signal 4 represents an example of this condition where the full frequency range, 0 to 80 cycles per block, could not be solved for simultaneously but excellent results were obtained when solving simultaneously for only 7 Fourier components centered on the signal frequency of 80 cycles per block.

When forced to analyze sub-bands sequentially one can expect spectral S/N values that are only slightly better than those obtained with the single frequency regression method (processing method 2) because components in one sub-band show up as noise (spectral variances) when processing the other sub-bands.

When implementing multiple frequency regression method Hanning weighting of the input observations is usually found to be useful, because it serves to prevent spectral spreading by the data blocking process. With this method it is particularly important to maintain as narrow a signal spectral bandwidth as possible so that sub-band processing can be maximally effective.

As with method 2, the multiple frequency regression method does not exhibit Nyquist folding characteristics so reliable frequency measurements can be made over a very wide frequency range.

Comparison of Spectral Sidelobes from Processing Methods 4 and 5

Evaluation of the spectral S/N performance of processing method 5 (multiple frequency regression in correlation plane) has not yet been completed. However, a short study has been conducted to provide a comparison of the spectral side-lobe characteristics of the two correlation plane methods (processing methods 4 and 5).

Processing method 4 in effect weights each correlation coefficient equally, whereas processing method 5 applies greater weighting to the more reliable correlation coefficient (those derived from a greater amount of cross product averaging).

The results shown in Figure 9 show that the $(P_n)^{\frac{1}{2}}$ weighting process gives significant reduction in sidelobe amplitude providing the equation solution

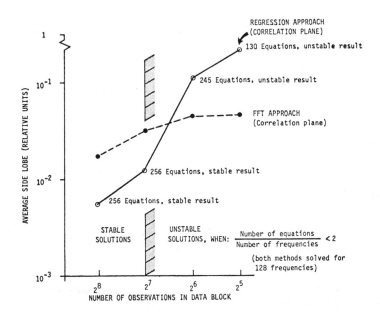

Figure 9: COMPARISON OF SPECTRAL SIDELOBE AMPLITUDES FOR PROCESSING METHODS 4 AND 5

is "stable" (number of frequency components simultaneously solved for being substantially less than one half the number of non-zero simultaneous equations). These results, Figure 9, were obtained for a single frequency cosine that had 48 cycles over a single data block observed at 32, 64, 128 or 256 random time locations in the block. In all runs, cross product averaging was applied within 256 delay increments and 128 component frequencies were analyzed.

Example spectral results for processing methods 4 and 5 are shown in Figures 10 and 11, respectively.

OVERALL CONCLUSIONS

Based on the results of the present study the following overall conclusions are formed.

1) The signal plane and correlation plane FFT methods give approximately equivalent performance, and both exhibit induced spectral folding around a frequency $1/2\Delta t$, or $1/2\Delta\tau$.

2) The single frequency and multiple frequency regression approaches both overcome the induced spectral folding phenomenon.

3) The spectral S/N performance of the <u>single</u> frequency regression method is usually no better than that of the FFT methods.

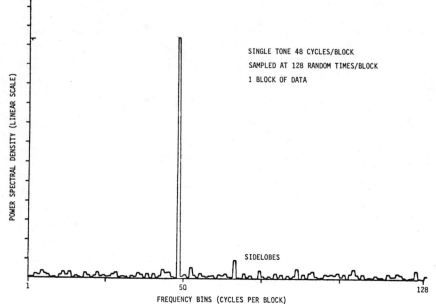

SINGLE TONE 48 CYCLES/BLOCK
SAMPLED AT 128 RANDOM TIMES/BLOCK
1 BLOCK OF DATA

SIDELOBES

POWER SPECTRAL DENSITY (LINEAR SCALE)

FREQUENCY BINS (CYCLES PER BLOCK)

Figure 10: EXAMPLE SPECTRAL SIDELOBES FOR PROCESSING METHOD 4

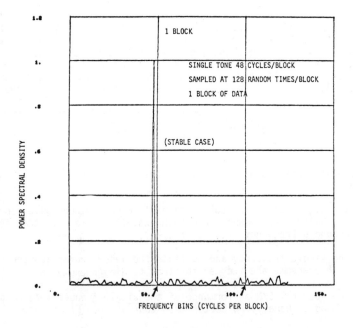

Figure 11: EXAMPLE SPECTRAL SIDELOBES FOR PROCESSING METHOD 5

4) The <u>multiple</u> frequency regression method gives excellent spectral measurements if <u>all</u> high amplitude frequency components can be solved for simultaneously, otherwise the S/N performance is only slightly better than that of the single frequency regression method.

5) Processing method 5 (multiple regression analysis applied to correlation coefficients averaged with increments $\Delta\tau$) have not been evaluated for spectral S/N performance but has been shown to have significantly lower spectral sidelobes than the correlation plane FFT method, and this can be expected to translate into improved spectral S/N performance under some circumstances.

PLANNED FURTHER WORK

It is planned that the spectral S/N performance of method 5 will be analyzed and compared with the S/N performance of other methods.

Also, in an attempt to expand the circumstances under which good spectral analysis can be obtained, processing method 3 (multiple frequency regression in signal plane) will be further investigated to determine its suitability for signals that have localized frequency bands. The approach will be to first use signal plane FFT to give an approximation of the signal spectrum, from which the frequency bands for the regression solution will be selected. It is expected that accurate spectral measurements will be obtained providing the average observation rate exceeds six times the (discontinuous) "effective bandwidth" of the signal.

ACKNOWLEDGEMENTS

The author wishes to thank Ron Murata and John Gallishaw of Boeing Aerospace Company and David Ferguson of Boeing Computer Services for their help in this paper.

REFERENCES

1. Gaster, M. and Roberts, J., "The Spectral Analysis of Randomly Sampled Records by a Direct Transform," The Proc. Royal Society London A354 (1977)

2. Gaster, M. and Roberts, J., "Spectral Analysis of Randomly Sampled Signals," J. Institute Maths. Applic. 15, 195-216 (1975).

EDITORS' NOTE

It has been pointed out by W. T. Mayo, Jr. that several papers on the subject of spectral analysis by random sampling are included in the 1974 Purdue proceedings (LV-II). See also Mayo, Shay, and Riter, AEDC-TR-74, August 1974.

Computer Simulation of Laser Velocimeter Signals

EUGEN BROCKMANN*
Technical University
Aachen, West Germany

ABSTRACT

Computer simulation of laser velocimeter signals provides a convenient means of studying various effects of the flow characteristics and the LDV system upon the signal and of testing the quality of data processing methods. The simulation programs include modeling of deterministic signal particle traces or selectable turbulent velocity distributions in the presence of several particles in the probe volume, calculation of scattering efficiencies of small and large particles, super-position with random noise, and highpass filtering of the signals. Some plotted results are demonstrated.

NOMENCLATURE

$\bar{D}$	integrated scattering parameters
$D_{k\ell mn}$	see definition following Eq. (4)
$I_{i,j}$	local intensity of beam j at particle i
$I_{sca}(\vec{r})$	scattered intensity at $\vec{r}$
M_B	number of incident beams
N_p	number of particles in the probe volume
P_i	integrated scattering parameter according to beam i
i_{PD}	photodetector current
$\vec{r}(t)$	location of a particle
t	time
$\vec{u}$	velocity of a particle
$\bar{\psi}$	integrated scattering parameter (phase)

---

* (Presently at the Applied Optics Laboratory, Purdue University, School of
   Mechanical Engineering, West Lafayette, Indiana, U.S.A.).

328

δ_i optical path length from source to particle due to beam i

η_{PD} sensitivity of the photodetector

$\xi_{i,j}$ optical path length from particle i to the photodetector due to the incident beam j

λ wavelength of laser light

COMPUTER SIMULATION

The computer simulation of signals for a dual beam type laser velocimeter is based on a modified fringe model including Mie scattering effects, where the time dependent photodetector current is generated by particles moving through the probe volume and scattering a certain amount of the incident local intensity. Assuming an optimal optical arrangement without aberrations, the probe volume is considered as an ellipsoid the boundaries of which are defined by the $1/e^2$ fraction of the central peak intensity.

At any location within this probe volume, the interference intensity can be calculated from the local intensities and phases of both beams. Following the theory given in [1] a computer code has been developed which allows the vectorial calculation of all the Mie scattering parameters of interest for any direction of propagation and polarization of the incident laser beams and their integration over any given aperture size. Thus the intensity scattered by one particle in the probe volume can be computed for any given location $\vec{r}$ as

$$I_{sca}(\vec{r}) = \bar{P}_1 I_1 + \bar{P}_2 I_2 + \bar{D} \sqrt{I_1 I_2} \cos\{\frac{2\pi}{\lambda}(\xi_2-\xi_1)-\bar{\Psi}\} \tag{1}$$

where I_1, I_2 and ξ_1, ξ_2 are the local incident intensities and optical pathlengths, respectively, and $\bar{P}_1$, $\bar{P}_2$, $\bar{D}$, $\bar{\Psi}$ are the integrated scattering parameters. For a given velocity field $\vec{u}(\vec{r},t)$ in the probe volume the position of a particle as a function of time may be expressed as

$$\vec{r}(t) = \vec{u}(\vec{r}_1 t) \cdot t \tag{2}$$

The photodetector current for a detector of sensitivity η_{PD} is given by

$$i_{PD}(t) = \eta_{PD} \cdot I_{sca}(\vec{r}(t)) \tag{3}$$

which yields the time dependent LDV signal. If more than one particle is in the probe volume, additional interference effects have to be taken into account [2,3]. Assuming for simplicity that all particles have the same size and see the same collecting aperture, and neglecting multiple scattering effects, the multiparticle interference can be calculated as

$$I_{sca}(r) = \sum_{k=1}^{N_p}\sum_{\ell=1}^{N_p}\sum_{m=1}^{M_B}\sum_{n=1}^{M_B} D_{k\ell mn}\sqrt{I_{k,m} I_{\ell,n}} \cdot$$
$$\cdot \cos\left\{\frac{2\pi}{\lambda}\left(\xi_{k,m}+\delta_k-\xi_{\ell,n}-\delta_\ell\right)-\bar{\Psi}\right\} \tag{4}$$

where the summation is carried out over all particles (N_p) and beams (M_B), δ_i is the optical path from the i-th particle to the photodetector, and $D_{k\ell mn}$ is equal to $\bar{P}_j$ for k=ℓ and m=n=j and to $\bar{D}$ otherwise.

For such a multiparticle system a simple turbulent particle flow has been simulated. With a given seeding density a certain number of particles are distributed at the time t=0 in a specially designed "random volume", which includes the probe volume. Each particle is allocated a certain velocity from a chosen distribution that carries the particle across the probe volume. A very simple movement is simulated along the one dimensional path perpendicular to the interference fringes, as only the velocity component in this direction is detected by an LDV system. Both the particle distribution and the velocity distribution are controlled by means of random number generators. In such a manner any desired velocity distribution (frequency spectrum of the LDV signal) may be obtained. Simulations have been performed which, for example, included a homogeneous density distribution and a Gaussian velocity distribution [4].

Beyond this, simulations of a single or multiparticle signal may include a highpass or a bandpass filtering process. Moreover, the computer generated signals can be added to computer generated noise of a suitable distribution and signal-to-noise ratio. When several noise sources are present in the optical and electronic channel, white noise can be used as a good approximation.

Summarizing these simulation abilities the computer model described represents a convenient method for studying and optimizing a variety of independent effects in an LDV system with regard to the optical channel as well as the signal processing. Some examples of results obtained are shown in Figs. 1-3.

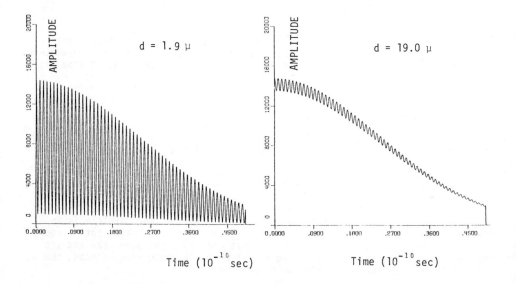

Fig. 1. Computer Simulated Single Particle Signal, Index of Refraction m - 2.7 - 0.01 i

Fig. 2. Computer Simulated Single Particle Signal, Index of Refraction m = 2.7 - 0.01 i

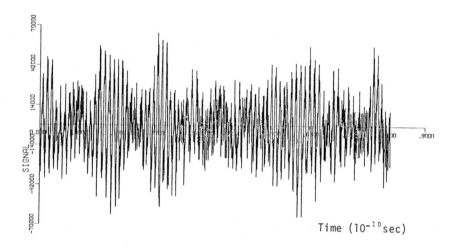

Fig. 3. Computer Simulated Multiparticle Signal, Highpass Filtered, Average Number of Particles in the Probe Volume N_P - 12 Homogeneously Distributed, Gaussian Velocity Distribution with Mean Velocity $\bar{u}$ = 2.55 m/sec and Velocity Spread σ_u = 0.255 m/sec. Superimposed White Noise, SNR δ = 12 dB.

REFERENCES

1.   Adrian, R. F. and Earley, W. L., Evaluation of LDV Performance Using Mie Scattering Theory, Proceedings of the Minnesota Symposium on Laser Anemometry, pp. 426-454 (1975).

2.   Durst, F., Scattering Phenomena and Their Application to Optical Anemometry, Journal of Applied Mathematics and Physics (ZAMP), Vol. 24, pp. 619-643 (1973).

3.   Durst, F., Melling, A. and Whitelaw, F. H., Principles and Practice of Laser-Doppler Anemometry, Academic Press (1976).

4.   Brockmann, E., Statistical Investigation of Laser Doppler Signals by Means of Numerical Simulation, Dissertation, RWTH Aachen (1977) (in German).

ACKNOWLEDGMENTS

     Part of this research was sponsored by a NATO fellowship. I wish to thank Prof. W. H. Stevenson for his support in preparing this paper.

GENERAL APPLICATIONS I

Chairman: **H. J. PFEIFER**
German-French Research Institute

Laser Velocimetry in RF Plasmas

G. GOUESBET, P. VALENTIN, and G. GREHAN
Laboratoire associe au CNRS, no. 230
Faculte des Sciences et Techniques de Rouen
BP 67-76 130, Mt. St. Aignan, France

ABSTRACT

This paper is concerned with measurements of some fluid properties in the high temperature range by means of laser velocimetry. Experiments carried out in an axisymmetric rf argon and helium plasma whose atom temperature on the axis was nearly 5000 K and electron temperature nearly 8000 K are reported.

First measurements have been made by means of a low-power laser system (5 mW). Mean velocities of scattering particles have been measured but it can be proved that the particles producing the Doppler signals were too big to follow the fluid velocity.

Further experiments have been carried out with a high-power laser system (800 mW) and mean velocities of the fluid have been successfully obtained. Although laminar, the plasma exhibited factitious fluctuations owing to a 150 Hz-modulation in the plasma generator. These fluctuations have also been measured.

Finally, by combining laser velocimetry measurements with measurements of dynamic pressures by means of a dynamic probe, the atom temperatures of the plasma were also obtained.

INTRODUCTION

Since the first results from Yeh and Cummins (1), interest in laser-Doppler velocimetry for measuring mean velocities and fluctuating velocities in fluids rapidly increased. Experimental problems became more and more complex. During recent years interest has been often focused on application of LDA to gases in the high temperature range, especially to combustion systems in order to achieve an improved understanding of processes in flames (2,3).

Plasmas are also gases in the high temperature range. For a fundamental understanding of the processes of heat and mass transfer occurring within plasmas, in particular with interest in plasma chemistry applications, it is important to get a knowledge of the local velocities in plasma flows. But there was a lack of methods for measuring local mean velocities (and fluctuating velocities) in the case of a low-velocity plasma jet. Therefore, it was decided to develop suitable LDA-systems for plasma applications.

Plasma experiments started at Rouen from 1974. A rf argon and helium plasma jet where the temperatures of heavy particles (atoms, ions) could reach 10,000 K between the turns was studied.

The plasma was firstly investigated by means of a low-power laser anemometer then by means of a high-power laser system, both using an optical interference set-up.

From the high-power laser system experiments, measurements of the mean velocities and fluctuating velocities of the plasma were obtained. The signals were processed by period-sampling. On the other hand, by coupling LDA results and dynamic pressure measurements, temperatures of heavy particles were also measured.

Other LDA studies have been made in plasmas before ours or during the same time as ours. But, in our knowledge, no plasma fluctuating velocities have been obtained in this high temperature range before the present results. Furthermore, we are not aware of other studies concerning the case of a rf plasma which is particularly important from the LDA point of view due to the presence of the radio frequency waves which constitute an important contribution to the hostile character of the environment. However, it must be pointed out that LDA-measurements were carried out by Self in the boundary layers of combustion MHD generators in a temperature range under 3000 K (4), by Barrault, Jones, Blackburn then by Todorovic, Jones and Barrault in an arc discharge (5,6) and very recently by Bayliss, Sayce, Durao and Melling in a DC transferred arc heater (7).

LDA MEASUREMENTS : MEAN VELOCITIES AND FLUCTUATING VELOCITIES

Experimental set-up with an 800 mW laser is shown on Figure 1 (see also the references 8 and 9). A very detailed description of these experiments can be found in the reference 10. Apart from the following differences the low-power experimental arrangement was basically the same as the high-power set-up (11): a low-power (5mW) He-Ne laser was used; an interference filter was used instead of a monochromator; and the signal was studied from some photographs

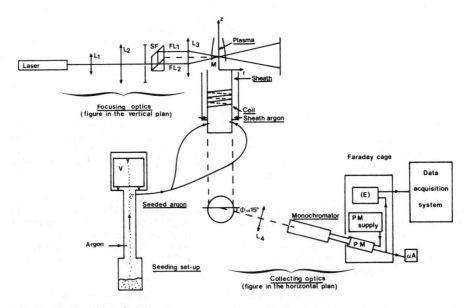

Fig. 1. Experimental Set-up

instead of by means of an automatic data acquisition and processing system.

The Plasma

The plasma was an axisymmetric laminar atmospheric induction-coupled plasma torch. The energy was supplied by an electric current of one ampere set to about MHz. The internal diameter of the silica tube was about three centimeters.

The plasma gas was an argon-helium mixture. Experiments were carried out for different combinations of the argon and helium flow rates Q_{1v} and Q_{2v} respectively, given in the table below.

Table I. Flow Conditions

Flow Rates (ℓ/min)	Q_{1v}(Ar)	Q_{2v}(He)
Low-power experiments	25.0	0
High-power experiments	21.1	0
	21.2	0.3
	21.2	0.7
	28.5	0.9
	28.6	1.3

The Optical Set-up

An interferential optics design was chosen for both low-power and high power systems. Only the high laser-power optical system is described in this paper.

The Fringe Formation

The light source was a Spectra-Physics krypton laser having luminous power equal to about 800 mW on the 647.1 nm-line.

The ($1/e^2$)-diameter of the TEM_{00} laser beam was increased from about 1.8 mm up to 5.4 mm by means of an afocal system made with the two lenses L_1 and L_2 whose focal lengths were 5 cm and 15 cm respectively. The density of electromagnetic energy was then increased in the beam cross-over region, resulting both in a better spatial resolution and an improved Signal to Noise Ratio (SNR).

The beam splitter (SF) divided the laser beam into two parallel output beams (FL1) and (FL2) with a separation of 2 cm. A convex lens L_3 (focal length 30 cm) focused all parallel input rays to a common focal point to establish a three dimensional set of planar interference fringes with a spacing i = 9.71 μm.

As a scatter center passed through the optical probe, the fringe pattern made it alternately scatter and then not scatter the light. If the velocity of the scatter in the z-direction is u_{oz} the scattered light will be modulated with a frequency f equal to u_{oz}/i. The modulation frequency, when measured, provides the velocity information.

The Collecting Optics

The focal length and the diameter of the collecting lens were 10 cm and 72 mm respectively. It collected the scattered light and imaged the fringes on to the inlet slot of the monochromator with a magnification equal to one. The width of inlet and outlet slots of the monochromator (a Jobin-Yvon H20) was 50 micrometers for both (spectral bandwidth: 0.2 nm). The monochromator removed the parasitic light due to plasma radiation and to high temperature scatter centers.

The Spatial Resolution

According to Brayton and Goethert (12) the $(1/e^2)$beam cross-over region was a nearly axisymmetric ellipsoid whose z and r dimensions were $2\Delta z$ equal to about 45 micrometers and $2\Delta r$ equal to about 1.4 mm respectively. The spatial resolution must be deduced from the intersection of the beam cross-over region and the field of view of the collecting optics. So, the length of the probe volume fell down to about $2\Delta r/4$ owing to the angle between the axis of the focusing and collecting optics equal to about 15°.

Furthermore, directions of observations must be situated in scattering lobes in order to achieve good SNR. Figures 2 show polar scattering diagrams from Grehan calculations by means of the Mie theory. A planar wave coming from the left and arriving on to a scatter center situated at 0 is considered. The polarization plane is perpendicular to the plane of the figure. The observation plane is the figure plane. Luminous intensity units are arbitrary. The position of the lens L_4 is also shown on the figures. Scatter centers are spherical alumina particles whose diameters are 1, 5 and 10 micrometers respectively. The wave length is 647.1 nm. The index of refraction is m equal to (1.65-5i) (13). Diagrams are drawn with and without taking into account the index of absorption.

From these figures, it can be seen that the choice of the angle of 15° between axes was a good balance between the need for an improved spatial resolution and the necessity to set directions of observation in scattering lobes for particles with diameter less than 10 microns.

The Seeding System

Theoretical Behaviour of Particles

The theoretical velocity behaviour of spherical alumina particles of diameter d_p is now examined. The following questions must be answered: what is the maximal diameter which can be accepted if it is wanted that the particles follow (i) the mean velocities of the fluid and (ii) fluctuating velocities of frequency f.

(i) Particles whose trajectories are parallel to the axis of the silica tube are considered. It is supposed that their velocity is zero at the time $t_0=0$. At this time it is set that the particles are just between the turns. Furthermore, it is also supposed that the plasma is instantaneously created and that it instantaneously reaches the velocity

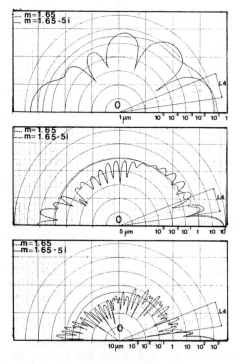

Fig. 2. Polar Scattering Diagram

u_{zf} measured at the fringe volume situated at about 7 cm from the middle of the coil.

The Table II shows for three different values of r, the velocity u_{zf} which is used in the present analysis, the atom temperature T_1 and the absolute viscosity μ. Velocities and temperatures are obtained from experimental results and the absolute viscosity is deduced from T_1 by means of the kinetic theory.

Table II.   Plasma Conditions

r(mm)	u_{zf}(m/s)	T_1(k)	μ(poise)
0	16	4500	17.0×10^{-5}
7	10	3000	10.0
12	6	2000	8.5

Then the velocity in the z-direction u_p of a particle of diameter d_p complies with the following law which is valid in the viscous regime:

$$u_p = u_{zf} (1-e^{-t/\tau})$$

where t and τ are the time and a time relaxation factor respectively.   The time relaxation factor is given by:

$$\tau = m_p/(3\mu\pi d_p) = (\rho_p d_p^2) / (18\mu)$$

where m_p and ρ_p are the mass and the density of a particle respectively.

The curves in Fig. 3 show the time t_p needed by a particle to reach the fringe system and its velocity u_p at this moment versus its diameter d_p.   The curves $t_p = f(d_p)$ show that particles whose diameters are less than 10 micrometers need an acceleration time negligible when compared with transit time. Furthermore, when they arrive at the measurement volume their velocity u_p is equal to the fluid velocity.

A more precise discussion should necessarily take into account the exact evolution of the fluid velocity along the plasma flow.   But this examination cannot be made because velocities inside the silica tube cannot be measured owing

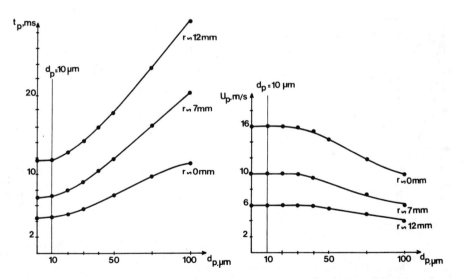

Fig. 3.   Theoretical Analysis of Particle Behaviour: Mean Velocities

to the particles deposition by thermophoresis which makes it opaque very quickly.

Nevertheless the gradients in the velocity of the fluid in the z-direction were not very important in such a way that the previous simplifying discussion remains valid. Therefore it can be concluded that particles whose diameter was smaller than 10 micrometers had the same z-velocity as the z-velocity of the fluid when they reached the fringe system.

It must be noted that these particles are big particles when compared to most of the LDA applications. This difference in required particle diameters happens because of the very viscous nature of the plasma fluid.

(ii) The following question now must be answered: Does a spherical alumina particle of diameter d_p follow a fluid fluctuation whose frequency is f (angular frequency: ω). The results of Hjelmfelt and Mockros (14) are used in this analysis.

The type III approximation of Hjelmfelt and Mockros can be used for studying the present problem. An amplitude ratio η is used. This amplitude ratio is a function of both the Stokes number and of the ratio of particles density ρ_p and of the fluid density ρ_f. So, we have:

$$N_s = (\frac{\mu}{2\pi f\, \rho_f\, d_p^2})^{1/2}$$

$$s = \rho_p\, /\, \rho_f$$

$$\eta = \frac{18\, N_s^2\, /\, s}{(1 + (18\, N_s^2\, /\, s)^2)^{1/2}} = 1\, /\, (1 + \omega^2\, \tau^2)^{1/2}$$

Figure 4 shows the amplitude ratio versus frequency for a number of particle diameters and for two different vertical trajectories.

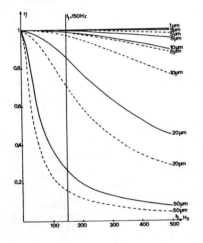

It can be deduced from these curves that particles whose diameter was smaller than 10 micrometers followed rather well "periodic fluctuations" of frequency f_t equal to 150 Hz. This frequency will play further a special role.

The Seeding Experimental Set-up

The plasma was seeded by means of dehydrated alumina particles in a counter-current system. Alumina was chosen because of its good thermal behaviour in the high temperature range according to the Engelke model giving the melting time of solid particles in a plasma jet (15).

Fig. 4. Theoretical Analysis of Particle Behaviour: Periodic Fluctuations. Full Curve, $r \approx 0$ mm, Broken Curve, $r \approx 12$ mm.

The seeding system provided a selection of various particle diameter ranges and various mass flow rates of the powder by choosing one of the three counter-current system geometries

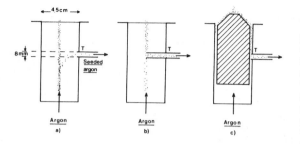

Fig. 5.   Geometries for Seeding

(Figures 5) and by varying the mass flow rates of the powder at the top of the counter-current system.

Particles were collected by means of an impaction technique at the outlet of the silica tube when the plasma was off and examined using an optical microscope. The size distributions which were obtained in this way overvalued the size distributions when the plasma was working owing to the fact that particles evaporated in the plasma jet (16) especially between the turns where the temperature was near 10,000 K on the axis.

Figure 6 shows the overvalued size distributions corresponding to the two geometries 5a) and 5b).

The geometry 5a) was used for peripheral zone measurements (r more than about 5 mm). The Figure 6a) shows that most of the particles from the overvalued size distributions had diameters less than 10 micrometers. From the Figures 3 and 4, it can be concluded that these particles followed mean velocities and 150 Hz "periodic fluctuations."

On the other hand the geometry 5b) was used for axial zone measurements. The geometry was changed during experiments by moving the tube T (Fig. 5). Using the geometry 5b), the number and the mean diameter of particles injected into the plasma were increased (see the overvalued size distributions). This process was necessary to get Doppler signals which were sufficiently numerous to obtain measurements in a reasonable time interval. In fact, using the geometry 5a) for axial zone measurements, signals were so rare that experiments practically could not be made. These increased difficulties are due to two main features:
   -the particles evaporated more quickly for axial trajectories because they traveled through hotter zones.
   -axial trajectories were not stable because of the thermophoresis phenomenon.

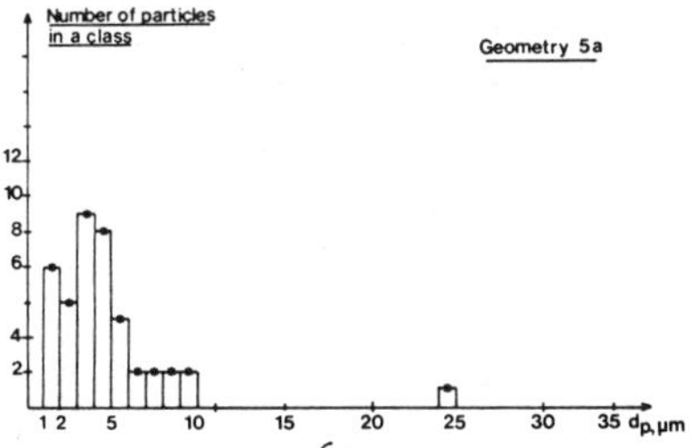

The overvalued size distribution corresponding to the geometry 5b) shows that most of the particles were

6a.

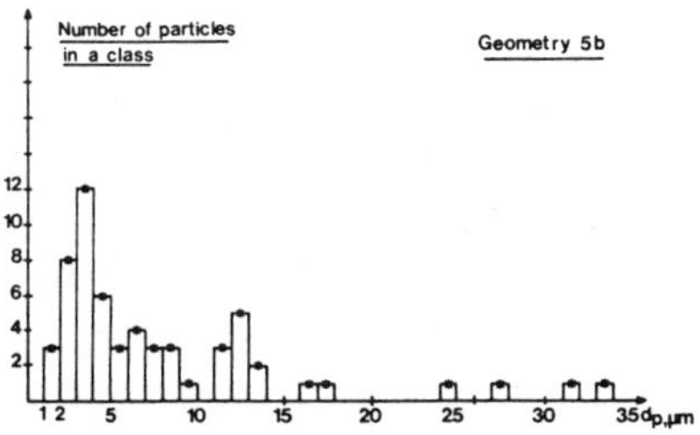

6b.

Fig. 6.   Particle Size Distributions

smaller than 10 micrometers. It must nevertheless be noted that there were two particles whose diameters are between 30 and 35 micrometers. From the Figure 3, it can be seen that such particles begin to exhibit a relative velocity which is not equal to zero, but this relative velocity is very small. Furthermore, no discontinuity effects were detected by changing the geometries. So, it can also be concluded that particles used for axial zone measurements followed mean velocities and 150 Hz "periodic fluctuations."

Finally, the geometry 5c) was used for low laser-power measurements. No Doppler signals were obtained by using geometries 5a) or 5b) in the low-power case. Measurements were made from photographs of Doppler signals. The over-valued size distribution corresponding to this geometry shows that particles as large as 100 micrometers were carried away into the plasma gas. One could then expect a relative velocity between the fluid and the particles.

It must be noted that particles of diameter 100 micrometers were larger than the fringe spacing and even larger than the z-dimension of the control volume. According to Farmer (17), a very bad visibility of the fringe modulation for such particles should be expected. However, it was observed that these particles, in fact, produced rather well modulated Doppler signals. This is in agreement with the results of Yule who experimentally showed that big particles produce visibilities and peak amplitudes increasing nearly linearly in the case (for instance) of glass spheres traversing through the center of the measurement volume (18).

The Electronic Processing System

The signal from the photomultiplier (PM), an EMI 9558B, was sent to a micro-ammeter (μA) in order to carry out mean anode current measurements or to the signal conditioner (E) where it was amplified and filtered.

The coil of the plasma generator behaved as an antenna emitting 5 MHz elec-tromagnetic radiation around it. If further refinements were not used, the emitted radiation was high enough to saturate the amplifier even when it was set on the lower band-pass. So the photomultiplier, its HV supply and signal con-ditioner (E) were put in a Faraday cage in order to solve this critical problem. Furthermore (E) contained three rejectors tuned to the fundamental frequency of generator oscillations and two harmonic frequencies in order to lower again the amplitude of the parasitic signal coming from the 5 MHz coil radiation.

The signal/noise ratios were good enough to allow processing by means of period sampling. An automatic data acquisition and processing system was used. "Instantaneous" periods measurements were carried out by means of an HP 5345 A counter with a 2 ns resolution. Data were transferred on to cassettes using a cassette recorder-reader. They were then treated by means of an HP 9810 A computer.

The Doppler Signals

In the peripheral plasma zones, signals were numerous enough to make measure-ment in a reasonable time interval. The data acquisition rate was limited by the data acquisition system. Typically between about 500 and 2000 instantaneous values accumulated during about 30 seconds were collected.

Nearer to the axis, signals became less and less numerous owing to particle evaporation and the thermophoresis phenomenon. Even using the geometry 5b), it was found that ten minutes were necessary to collect about ten values. The acquisition rate was then limited by the number of particles passing through the control volume and it is known that a bias occurs in this case. However, owing to the small fluctuations in the axial zones and taking into account the accuracy of the mean velocity measurements, that bias could be neglected.

EXPERIMENTAL RESULTS

Measurements were obtained in the free jet along the diameter of horizontal sections of the silica tube. It was necessary to work in the free jet because of the deposition of alumina particles on silica which made the tube opaque very

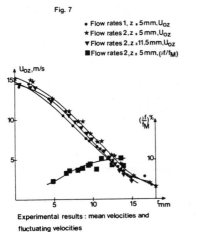

Fig. 7

• Flow rates 1, z = 5mm, U_{oz}
★ Flow rates 2, z = 5mm, U_{oz}
▼ Flow rates 2, z = 11.5mm, U_{oz}
■ Flow rates 2, z = 5mm, $(\delta f/f_M)$

Experimental results : mean velocities and fluctuating velocities

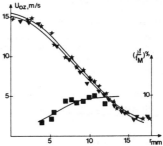

Fig. 8

★ Flow rates 3, z = 5mm, U_{oz}
▼ Flow rates 3, z = 11.5mm, U_{oz}
■ Flow rates 3, z = 5mm, $(\delta f/f_M)$

Experimental results : mean velocities and fluctuating velocities

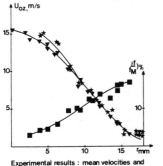

Fig. 9

★ Flow rates 4 : z = 5mm, U_{oz}
▼ Flow rates 4 : z = 8.5mm, U_{oz}
■ Flow rates 4 : z = 8.5 mm, $(\delta f/f_M)$

Experimental results : mean velocities and fluctuating velocities

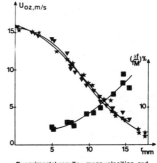

Fig. 10

★ Flow rates 5 : z = 5mm, U_{oz}
▼ Flow rates 5 : z = 8.5mm, U_{oz}
■ Flow rates 5 : z = 8.5mm, $(\delta f/f_M)$

Experimental results : mean velocities and fluctuating velocities

quickly. The r-exploration of the plasma was carried out by displacing the
plasma generator which was put on rails. The z-exploration was carried out by
displacing the LDA-system.

It must now be remembered that alumina particles which have been used followed the mean velocities of the fluid and its 150 Hz-"periodic fluctuations."
The figures from 7 to 10 show the experimental results for the various flow rates
given in the Table I and for various z-values.

The accuracy and reproduction of mean velocity measurements were satisfactory. Results are not so good if fluctuations are concerned. A first reason
is that the plasma was not so reproductible for fluctuations as for mean velocities. Besides, accuracy should be improved by means of a double-counter system.

Results from low-power measurements have also been obtained. When compared
with high-power measurements, an important discrepancy occurs (see the discussion
in "The Seeding Experimental Set-up.")

It must be recognized that the measurements of plasma velocity with a low
power laser was not successful. However, it must be pointed out that low-power
systems are sufficient to get velocities of big particles, an interesting aspect
for a lot of plasma chemistry problems, for instance fabrication of coatings
and thin-walled components.

On the other hand, velocities from high-power measurements were satisfactory. The coherence of experimental results was checked by calculating the
integrals:

$$Q = \int_0^{r_o} 2\pi r \rho_f \ u_{oz} \ dr = \int_0^{r_o} 4\pi r \ (\rho_f u_{oz}^2 \ /2) \ \frac{1}{u_{oz}} \ dr$$

where r_o is the internal radius of the silica tube. The dynamic pressures
$(\rho_f u_{oz}^2 \ /2)$ and the velocities u_{oz} were taken from experimental results. The
mass flow rates Q were then compared with mass flow rates deduced from the
Table I. Let us accept a relative uncertainty on the velocity measurements
equal to 5% and an absolute uncertainty on the dynamic pressures equal to 0.05 mm
of water. The agreement was then perfect as far as flow rates 4 and 5 were
concerned. But, concerning the other flow rates, a discrepancy is found equal
to about 12%. This discrepancy could be explained by a loss of the cylindrical
symmetry of the plasma jet when the flow rates were decreasing. This loss of
cylindrical symmetry could be visualized.

ATOM TEMPERATURE MEASUREMENTS

The dynamic pressures were measured by means of a cooled dynamic probe
linked to a micro-manometer. Coupled with mean velocities, these measurements
allowed us to deduce the density of the gas and then the atom temperatures by
using the perfect gas law.

CONCLUSION

Mean velocities and fluctuating velocities in an argon/helium rf plasma
were measured by means of a high-power LDA system. Low-power measurements were
also made but accurate mean velocities of the plasma itself were not obtained
due to velocity slip of the large particles needed to give adequate scattering.
Finally, by coupling high-power LDA measurements and dynamic pressure measurements, atom temperatures were also obtained by a new local set of methods.

ACKNOWLEDGMENTS

It is a great pleasure to thank here Mrs. J. C. Goulet and J. Mouard because of their essential contribution in the realization of the data acquisition and processing system.

REFERENCES

1. Yeh, Y. and Cummins, H.Z., Appl. Phys. Letters $\underline{4}$, 176 (1964)

2. Durst, F., Melling, A., and Whitelaw, J.H., Report ET/TN/A18, Department of Mechanical Engineering, Imperial College, London (1971)

3. Baker, R. J., Bourke, P.J., and Whitelaw, J.H., J. Inst. Fuel, p. 388, Dec. 1973

4. Self, S.A., "Laser Doppler Velocimetry in MHD Boundary Layers," 14th Symposium on Engineering Aspects of MHD, Tullahoma, Tennessee (1974)

5. Barrault, M.R., Jones, G.R., and Blackburn, T.R., J. of Phys. E: Sci. Inst., $\underline{7}$, 663 (1974)

6. Todorovic, P.S., Jones, G.R., and Barrault, M.R., J. of Phys. D: Appl. Phys., $\underline{9}$, 423 (1976)

7. Bayliss, R.K., Sayce, I.G., Durao, D.F.G., and Melling, A., "Measurements of Particle and Gas Velocities in a Transferred Arc Heater Using LDA," 3rd Synposium International de Chimie des Plasmas, Université de Limoges, France (1977)

8. Gouesbet, G. and Trinite, M., Letters in Heat and Mass Transfer, $\underline{4}$, 141 (1977)

9. Gouesbet, G. and Trinite, M., J. of Phys. E: Sci. Inst., $\underline{10}$, 1009 (1977)

10. Gouesbet, G., "Phénomènes de Diffusion et de Thermodiffusion des Espéces Neutres dans un Plasma d'Argon et d'Helium," These de Doctorat d'Etat (1977)

11. Gouesbet, G., C.R.A.S., Paris, $\underline{B280}$, 597 (1975)

12. Brayton, D.B. and Goethert, W.H., ISA Transactions $\underline{10}$, 40 (1971)

13. Handbook of Chemistry and Physics, 36th Edition, Chemical Rubber Publishing Co., Cleveland, Ohio

14. Hjelmfelt, A.T. and Mockros, L.F., Appl. Sci. Res., $\underline{16}$, 149 (1966)

15. Engelke, "Heat Transfer to Particles in the Plasma Flame," American Institute of Chemical Engineers, Los Angeles, Jan. 1962

16. Boulos, M.I., "Treatment of Fine Powders in a RF Plasma Discharge," 3rd Symposium International de Chimie des Plasmas, Universite de Limoges, France (1977)

17. Farmer, W.M., Applied Optics, $\underline{11}$, 2603 (1972)

18. Yule, A.J., Chigier, N.A., Atakan, S. and Ungut, A., J. Energy $\underline{1}$, 220 (1977)

Boundary Layer Velocity Measurements in Combustion MHD Channels

SIDNEY A. SELF
High Temperature Gasdynamics Laboratory
Mechanical Engineering Department
Stanford University
Stanford, CA 94305

ABSTRACT

Two different laser anemometer systems are described that have been developed and used to make velocity measurements in the boundary layers of combustion MHD channel flows. Such flows pose a number of severe problems for laser anemometry due to the high temperatures, high velocities, and limited optical access. The resolution of these problems and the operation of the system are detailed.

INTRODUCTION

In combustion MHD generators, the boundary layer plays a critical role in determining generator performance because the plasma electrical conductivity is much lower than in the core flow. This is due to the presence of a thermal boundary layer resulting from the fact that available materials restrict the wall temperature to values ($T_w \lesssim 2000$ K) much less than that of the core ($T \sim 3000$ K).

Sophisticated computer codes have been developed for calculating the performance of MHD generators, which are based on turbulent boundary layer models, including electromagnetic interaction effects. These models employ several empirical constants whose values are taken from experimental correlations at much lower temperatures and velocities; they also omit several potentially important factors such as 3-D effects, swirl, wall roughness, and combustor-induced fluctuations.

For these reasons it is important to make direct measurements of the profiles of the average temperature, electrical conductivity and velocity, and their fluctuations, in an operating MHD generator. This paper discusses velocity measurements by laser anemometry; other diagnostic techniques have been described elsewhere [1].

COMBUSTION MHD FACILITY

The Stanford MHD rig is shown schematically in Fig. 1. The swirl-stabilizer spray combustor burns ethanol in O_2 with N_2 diluent at a flame temperature ~ 3000 K and is seeded to $\sim 1\%$ by mass of potassium to enhance the conductivity. Typically it operates at mass flows up to 1 lb/sec at thermal

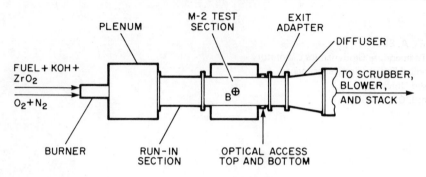

Fig. 1. MHD Flow Train.

powers up to several MW. The combustor feeds through a plenum, to promote
mixing and reduce fluctuations, and a nozzle into the MHD generator channel
which is surrounded by a large electromagnet providing a magnetic field of
~ 2.5 Tesla. The generator test section dimensions are typically 10 cm high
× 3 - 5 cm wide × 60 cm long. It has water cooled metal walls lined with MgO
firebrick, and has some 18 segmented electrode pairs in the top and bottom
walls. The electrodes are cooled to run at a surface temperature in the range
1000 - 1500 K, and each pair is connected to an external load. Under typical
conditions, P ~ 1 atm, T ~ 2800 K, the sound speed is ~ 1000 m/s, and most
experiments are run at high subsonic conditions e.g. u ~ 800 m/s, though some
experiments have been run supersonically at u up to 1600 m/s. In order to
increase the boundary layer thickness to facilitate measurements near the exit
end of the channel, a run-in section is interposed between the nozzle and the
generator test section. The test section exhausts via a diffuser, water spray
quench, scrubber and fan to an exhaust stack.

LASER ANEMOMETRY REQUIREMENTS AND PROBLEMS

It was desired to measure the profiles of average velocity and turbulence
intensity on the center lines of both the electrode (top) wall and insulating
(side) wall of the channel close to the last electrode, near the exit of the
generator test section. Of particular interest was the change in these profiles induced by the electromagnetic J × B body forces, which is predicted
to be especially significant in the case of the sidewall. Moreover, because
the boundary layer is relatively thin (~ 1 cm) it was important to achieve
high spatial resolution and make measurements as close to the wall as possible.

These requirements, combined with the hostile environment of the MHD rig
posed a number of problems, as follows.

(i) Access

Undoubtedly, the most difficult constraint is set by the access limitations imposed by the magnet, and this determined the configuration of the two
separate systems used for the electrode and sidewall boundary layers.

In the case of measurements of the electrode wall, the only possibility
for optical access is from the downstream sidewall, outside the magnet. To
measure the axial (u) velocity component, the only possibility was then to
use a single beam backscatter configuration, with the beam axis making a small

angle (~ 26°) to the flow axis. In view of the high velocities, the Doppler frequency shift is then too high (several GHz) to measure by heterodyne methods, and resort was made to measuring the wavelength shift (~ 10^{-2} Angstrom) by means of a scanning Fabry-Perot interferometer. The system devised for measuring the electrode wall boundary layer is shown in Fig. 2.

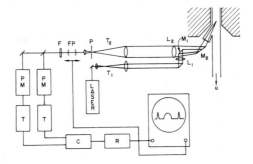

Fig. 2.   Single Beam Backscatter Anemometer.

In the case of the sidewall boundary layer, the magnet allows (with some difficulty) access vertically through the electrode walls. This permitted the use of the conventional dual beam (real fringe) configuration to measure the axial component in the sidewall boundary layer. However, to utilize such access, it was necessary to use four mirrors to route the incident and collected beams around the hardware as shown in Fig. 3. In view of the high velocities, it is necessary to use a very small angle between the incident beams to keep the Doppler frequency within limits set by the signal processing electronics. However, the use of small angles implies a long measurement volume and reduced spatial resolution as discussed below.

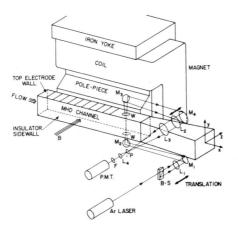

Fig. 3.   Dual Beam Forward Scatter Anemometer.

(ii) Optical Ports

The combustion flow, seeded with potassium, not only condenses water on cool windows but also rapidly condenses a white deposit of potassium carbonate on any surface less than its dew point (~ 1000°C). In the case of the electrode wall measurements, where access was obtained from the downstream sidewall, a purged and shuttered window was used on a water cooled side port. The purge gas (dry N_2) did not perturb the flow at the measurement station. For the sidewall measurements, where the access was vertically through the top and bottom electrode walls, the access holes in the channel were close to the measurement station and there was too little room to introduce purged ports and windows. For this reason access was obtained through open holes, and the pressure at the measurement station was made slightly sub-atmospheric by means of an exhaust fan, so that there was a slight inflow of ambient air through the holes.

(iii) Spatial Resolution and Scanning

In both the single beam and dual beam systems the length of the illuminated volume is much greater than its diameter. Hence, to obtain high spatial resolution, especially close to the wall it is necessary to arrange for the length of the control volume to be parallel to the wall. Then the spatial resolution in the direction normal to the wall is determined by the beam width. Of course, the length of the control volumes is also partly determined by the spatial filter in the collection optics, especially if the collection is off axis.

In the case of the single beam backscatter system, the collection was necessarily coaxial with the incident beam, and high spatial resolution was achieved by pre-expanding the incident beam to 25 mm diameter, together with the use of tight spatial filtering (50 μm pinhole) in the collection optics. By this means a spatial resolution of ~ 50 μm normal to the wall was achieved. The tight focusing of the laser beam allowed good signal/noise to be obtained in backscatter from 1 μm particles for a laser power of only 100 mW, but also necessitated heavy seeding to obtain high data rates. The optics were arranged so that the edge of the incident beam cone was at grazing incidence to the electrode wall when the focus was just at the electrode surface as shown in Fig. 4. Measurements were possible to within 300 μm of the surface, this limit

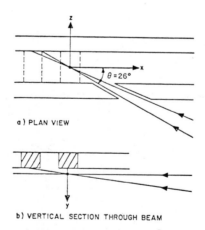

a) PLAN VIEW

b) VERTICAL SECTION THROUGH BEAM

Fig. 4.   Beam Geometry in Channel
for Single Beam Anemometer.

being set by interference produced by laser light scattered from the surface. Scanning of the measurement point was achieved by a micrometer controlled mirror, common to the coaxial incident and collection optics.

In the dual beam system the incident beams were parallel to the wall and off-axis (5°) forward scatter collection was used as shown in Fig. 5. This

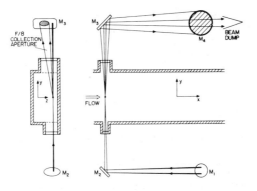

Fig. 5.   Beam Geometry in Channel for Dual Beam Anemometer.

minimized problems of interference due to laser light scattered from the wall. The beam waist was ~ 240 µm diameter, corresponding to 16 fringes with a spacing of 15 µm, and the collection optics employed a 200 µm spatial filter which imaged to ~ 300 µm diameter at the measurement point. By this means a spatial resolution of ~ 250 µm normal to the wall by ~ 4 mm parallel to the wall was achieved, and measurements were possible to within 0.5 mm of the wall. Scanning was achieved by translating the incident and collection optics, which were ganged together on micrometer positioned slides.

(iv)   Particle Seeding

Most particulate seeding materials either evaporate or burn in the combustion flow at 2800 K, and after various trails it was found that zirconia in the 1 µm size range was the only satisfactory seed under these conditions. In the case of the single beam system, where the combination of small measurement volume and scanning spectral analysis required high particle fluxes, powder was fed from a specially designed mechanical feeder in a dry N_2 flow into the plenum chamber at rates ~ 1 g/s. For the dual beam system, where the larger measurement volume combined with counter electronics permitted substantially lower particle fluxes, the powder was fed into the alcohol fuel line in the form of a paste from a syringe pump at rates ~ 1 g/min.

(v)   Other Problems

Noise and vibration levels associated with operation of the MHD system are relatively high, and problems from this source were minimized by mounting the optics on substantial support framework attached to the massive magnet iron. No significant problems were attributed to vibration or to beam bending due to refractive index inhomogeneities. This could be inferred from the steadiness of the scattered signal when the measurement point is positioned at the wall

surface, which signal was also used to determine the zero position for the
boundary layer scan.

Thermal radiation from the plasma and from the channel walls ($\sim$ 2000 K)
were adequately filtered by a 20 Angstrom wide filter in front of the detector,
and in fact the principal source of noise and interference in both systems
arises from laser light scattered incidentially from the optics, windows and
channel walls.

The considerable stray field from the magnet required that all optical
mounts and the support framework be made of non-magnetic materials. Also the
influence of the field on the laser and photomultiplier necessitated locating
them some 10 ft from the measurement point.

SINGLE BEAM BACKSCATTER SYSTEM

Details of the design and operation of this system were given in the
Second Workshop [2]. Subsequently the system was modified and used to measure
the electrode wall boundary layer profile at supersonic velocities to $\sim$ 1600 m/s.

The lens and mirror arrangement shown in Fig. 2 is used to separate the
coaxial incident and scattered beams. The Ar^+ laser beam (100 mW single mode
at 514.5 nm) is expanded to 25 mm diameter by telescope T_1, focused by lens
L_1 (f = 60 cm), deflected by the elliptic-section plane mirror M_1 and then
deflected again by the scan mirror M_2 into the port. The latter is purged
with dry N_2 to prevent water and seed condensate on the window, which is pro-
tected by a shutter when not in use. The incident beam is focused to a waist
(< 50 μm diameter) opposite the electrode, which it intercepts near grazing incidence, and the focus can be traversed through the boundary layer between
y = 0 and 30 mm by the micrometer-controlled scan mirror M_2.

Light scattered from the focus is collected and collimated by lens L_2
(f = 60 cm) and compressed to 1 mm diameter by telescope T_2. The latter includes a spatial filter P (pinhole 50 μm diameter) to reject radiation from
the plasma and walls, and particularly to reject laser light scattered from the
walls. After compression, the collected beam enters the confocal spherical
Fabry-Perot interferometer, whose mirror separation is scanned piezoelectrical-
ly, over one free spectral range, in synchronism with the oscilloscope timebase
at $\sim$ 1 Hz. Alternate sets of mirrors, with free spectral ranges Δf = 2 GHz
and 8 GHz were used for subsonic and supersonic measurement respectively. The
Fabry-Perot finesses is $\sim$ 70 which gives a velocity resolution < 1.5% of u_{max} =
$\lambda\Delta f/2 \cos \theta$, corresponding to the free spectral range.

Scattered light pulses due to the passage of individual particles through
the focus, which typically contain $\sim$ 100 photons, then pass through a spectral
filter F (width 2 nm) and are divided between two photomultiplier detectors.
This allows the use of coincidence techniques to discriminate against dark
current pulses and shot noise pulses due to laser light scattered from the walls.
The photomultipliers PM feed trigger circuits T and a coincidence circuit C.
The rate of signal pulses in coincidence is determined by the rate meter R,
and the signal rate is displayed on the storage oscilloscope.

Subject to the requirement that the particle flux is constant, this signal
rate represents the velocity probability density function P(u), from which the
average velocity and turbulence intensity may be measured. A typical display
(Fig. 6) shows P(u) and reference signals at u = 0 and u = u_{max}, which arise
from laser light, unshifted in frequency, scattered from the channel wall.

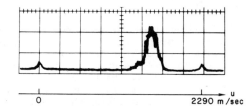

Fig. 6.   Velocity Probability Density Display
for Single Beam Anemometer.

The flow was seeded with ZrO_2 particles of ~ 1 μm diameter at a rate of
~ 1 g/sec, to yield displays of P(u) at a given y-position in a single sweep
(~ 1 sec).

Results of measurements of the boundary-layer profiles of average velocity
and turbulence intensity in a test under supersonic conditions, with free
stream velocity of 1600 m/sec, are shown in Fig. 7.

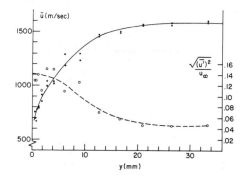

Fig. 7.   Electrode Wall Boundary Layer Profiles.

Plotted on logarithmic scales the profile u(y) approximates a straight
line corresponding to a dependence $(u/u_\infty) = (u/\delta)^{1/n}$ with n ~ 4.5.

The relatively high level of freestream turbulence is thought to consist
primarily of low frequency (< 1 kHz) fluctuations associated with the combustor.
A comparison of these measurements with calculations is made in Ref. [3], where
further velocity measurements under subsonic conditions are also detailed.

The principal limitations of this instrument are: a) Interference from
laser light scattered from the channel walls, especially for small y. This
results in unduly large reference signals which preclude measurement for
y < 0.3 mm.   b) The need to seed the flow with relatively large fluxes of
particles to obtain a satisfactory rate (~ 10^5/sec) of detected signals. This
arises from the very small scattering volume combined with the use of a single-
channel spectrum analyzer, which makes use of only ~ 2% of the available signals.
Furthermore fluctuations of particle flux due to imperfect mixing in the plenum
chamber can give rise to spurious modulation of the observed P(u).   c) The
slowness of the photographic data recording method and the need for hand reduction of data.

DUAL BEAM FORWARD SCATTER SYSTEM

This system was designed and used for measurements of the sidewall boundary layer. Access vertically through the electrode walls between the magnet
polepieces was obtained by the use of four deflecting mirrors as shown in Fig.
3, off-axis forward scatter was used, as shown in Fig. 5, to minimize problems
due to light scattered from the walls. The laser beams exit on the collection
side but miss mirror M_4 and are dumped in a beam dump constructed from black
glass Scanning is accomplished by translating the field lenses and mirrors
L_1 M_1 and L_2 M_4 together. The angle between the incident beams was ~ 2°,
yielding a fringe spacing of 15 μm and a Doppler frequency of 60 MHz for
u = 900 m/s. The collection optics aperture is F/8, and the photomultiplier
and laser (100 mW on 514.5 nm) are mounted remotely to minimize interference
due to the stray magnetic field.

Signal processing employed a TSI 1096 counter. The data acquisition
system shown in Fig. 8, and associated software was developed to process data
points (corresponding to the transit of individual particles) in batches of
10K, print out the average velocity and turbulence intensity, and display the
velocity probability density. The analog output from the signal processor is
also digitized at 1 kHz in an A/D converter and used to derive the frequency
spectrum of velocity fluctuations.

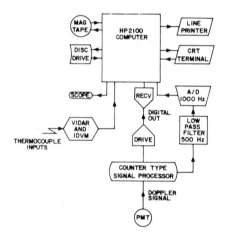

Fig. 8.   Data Acquisition System for Dual Beam Anemometer.

A typical on-line print out of the velocity probability density is shown
in Fig. 9. The central ordinate is located at the mean velocity while the

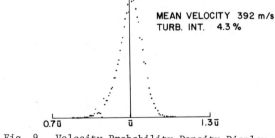

Fig. 9.   Velocity Probability Density Display.

abscissa extends from 0.7 $\bar{u}$ to 1.3 $\bar{u}$. Turbulence intensities in the range
4 - 6% were obtained depending on combustor conditions. The spectrum showed a
roughly 1/f dependence, decreasing by an order of magnitude between 5 and 500
Hz.

While the system performed satisfactorily at velocities up to 500 m/s,
the boundary layer measurements were made at rather lower velocities because
this enhances the effect of the J × B forces on the velocity profile. Figs.
10 and 11 show the measured and computed boundary layer profiles under identical conditions, except that in Fig. 10 there is no electromagnetic interaction,
while in Fig. 11, both magnetic field and current were present.

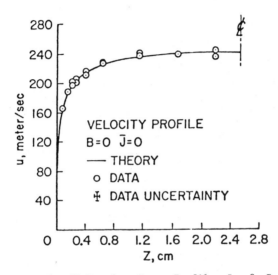

Fig. 10. Sidewall Boundary Layer Profile, J = 0, B = 0.

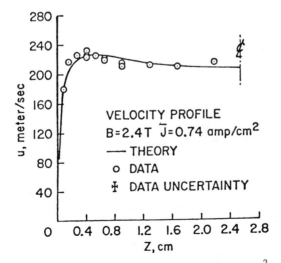

Fig. 11. Sidewall Boundary Layer Profile J = 0.74 A/cm^2, B = 2.4 T.

The marked change in profile with the presence of the $J \times B$ braking force is evidenced by the velocity maximum near the wall in Fig. 11. This is caused by the fact that because the temperature and hence electrical conductivity is lower in the boundary layer, the current is concentrated in the core and produces a greater braking force on the core. The electromagnetic interaction also produces a significant increase in the skin friction and heat transfer at the sidewall. Evidence was also found of a reduction of the turbulence intensity in the presence of the magnetic field due to eddy current damping of the fluctuations.

Further details of these measurements and their comparison with computations is available in a report [4].

CONCLUSION

The reported work has demonstrated that valuable measurements of boundary layer velocity profiles can be made in sub-scale combustion MHD generator channels despite the problems associated with high velocity, high temperature and difficulty of access. Further, more refined measurements are planned to check our ability to correctly compute the details of the turbulent channel flow including MHD interaction effects. The prospects for laser anemometer measurements in larger, pilot-scale MHD systems appear to be promising.

ACKNOWLEDGEMENTS

The contributions of Dr. R. Rankin to the development and use of the dual-beam system, and the encouragement of Professor R. H. Eustis are gratefully acknowledged. This work was supported by the U.S.A.F. under contract AFAPL F-33615-72-C-1088, by NSF under grant GK 04116, and by DOE under contract EX-76-C-01-2341.

REFERENCES

1. S. A. Self and C. H. Kruger, "Diagnostic Methods in Combustion MHD Flows," J. Energy 1, 25-43 (1977).

2. S. A. Self, Second International Workshop on Laser Velocimetry, Vol II, 44-66 (1974).

3. J. W. Daily, C. H. Kruger, S. A. Self, and R. H. Eustis, "Boundary Layer Profile Measurements in a Combustion Driven MHD Generator," AIAA Journal, Vol. 14, Aug. 1976, pp. 997-1005.

4. R. R. Rankin, "Insulating Wall Boundary Layer in a Faraday MHD Generator," Topical Report, April 1978, Stanford University.

High Speed Velocity Measurements Using Laser Doppler Anemometers

D. DOPHEIDE
Physikalisch-Technische Bundesanstalt
Abteilung 1, Braunschweig, West Germany

F. DURST
Universität Karlsruhe
Karlsruhe, West Germany

ABSTRACT

The present paper summarizes theoretical investigations carried out to optimize LDA systems for applications in high speed fluid flows. Most high speed LDA measurements have been carried out at Doppler frequencies much below the frequency capabilities of modern LDA frequency counters. This paper suggests that this might be due to the multiaxial mode output of cw lasers employed for high speed velocity measurements. This is theoretically explained in the paper. Considerations are also given to the application of photodetectors for high frequency LDA measurements. It is shown that photomultipliers permitting high anode currents are advantageous in laser Doppler anemometry, when applied to high Doppler frequencies.

INTRODUCTION

Over the last few years rapid developments in laser Doppler anemometry have resulted in commercial instruments that are nowadays widely used for fluid flow studies yielding experimental results that have important implications for new practical and theoretical developments in modern fluid mechanics. Recent experimental investigations of laminar flows by Cherdron, Durst, and Whitelaw [1] have resulted in information that permits asymmetric flows in symmetric two-dimensional sudden expansion test sections to be explained. Measurements of turbulent transport properties in free shear flows (see List and Kozovinos [2]), and studies of properties of highly turbulent recirculating flows (see Durst and Rastogi [3]), have provided new information that will influence theoretical studies in these fields. Application to fluid flows in combustion systems (see Baker, Hutchinson, Khalil, and Whitelaw [4]), and studies of aerodynamic problems (see Abbis [5]), have demonstrated the great potential of laser Doppler systems in different flow situations which have been considered inaccessible to probe-type velocity sensors.

Much of the aforementioned success has depended on the layout of LDA-systems for particular flow situations and on system optimizations for specific measurements. However, most of the optimization considerations have been carried out for low frequency LDA systems concentrating on the properties of the optical system (see Durst and Whitelaw [6] and Eliasson and Dändliker [7]), not taking into account properties of photodetectors and the bandwidth of signal processing systems. It is only recently that Durst and Heiber [8] have focused attention to the influence of photodetector properties on the signal-to-noise ratio of LDA-signals. Their study suggested that some photomultipliers might have the fast frequency response required for high LDA-signal frequencies but other detector properties might still prevent signals of good signal-to-noise ratio to be obtained and hence, permit fast LDA-measurements of high accuracy to be carried

out. Durst [9] applied this knowledge to an optical system designed for LDA-
measurements in transonic flows taking into account individual properties of important components of LDA optical systems.

The present research is based on the aforementioned investigations and is
aimed at their extension and completion. Although all considerations given in
this report are generally applicable to laser Doppler systems, the present paper
focuses on LDA-application to high speed flows. It is pointed out that all high
speed velocity measurements known to the authors have been made employing relatively small angles between the laser beam, yielding Doppler frequencies far below
the frequency capabilities of commercially available signal processors. Theoretical derivations show that this may be due to the different frequencies of the
axial modes of a laser beam. These prevent sensible laser Doppler measurements
of high LDA-frequencies in spite of the availability of suitable LDA-frequency
counters.

The paper also discusses the influence of photodetectors on laser Doppler
signals and provides information which permits the signal-to-noise ratio (SNR)
of LDA signals to be calculated.

THEORETICAL INVESTIGATIONS RELATING TO HIGH FREQUENCY LDA-MEASUREMENTS

PHENOMENOLOGICAL DESCRIPTION OF MULTIAXIAL-MODE LASER RADIATION EFFECTS IN
LASER-DOPPLER ANEMOMETRY--It is well known that most of the characteristics of
lasers that are of importance to laser-Doppler anemometry pertain to properties
of the optical resonator cavity rather than the physics of the laser action. It
is the resonator action that swamps out the random nature of the spontaneous emission from the gas molecules. The stability of the laser geometry results in the
high temporal and spatial coherence that characterizes laser radiation. However,
because of the nature of the amplifying transition and the large power requirements, most lasers have a multi-frequency output and the laser is said to operate
multimode. It has already been pointed out that the appearance of many modes is
the major cause of the "effective coherence length" reduction which requires the
use of optical arrangements with equal light path length for optimum laser-Doppler
systems (see references [10] and [11]). The behavior of Gaussian light beams
also requires path length compensation (see reference [12]).

Taking the aforementioned influences of the resonator into account, the radiation from a multiaxial mode laser light source may be expressed in the form
given by the following equation. In this equation "j" denotes the mode number
and "M" is a total number of excited axial modes. The time dependence of the
phase, $\phi(t)$, may be caused by small thermal variations in cavity length since
most available lasers are not completely stabilized. The phase variations result in a time-dependence of the magnitude of the electric field vector, $\{\varepsilon(t)\}_j$,
associated with the oscillating axial modes, j. This appears superimposed upon
magnitude fluctuations caused by mode competitions and other causes.

$$\{\varepsilon(\xi,t)\} \;=\; \sum_{j=1}^{M} \{(E_0(t))_j\} \; \exp\left[-i[K_j\xi - \omega_j t + \phi_j(t)]\right] \qquad (1)$$

In laser-Doppler anemometry the relatively small time intervals which the
scattering particles spent inside the measuring control volume justifies, especially for high speed velocity measurements, the neglect of the time dependence
of the phase, $\phi_j(t)$, and of the magnitude of the electric field vector, $\{(E_0(t))_j\}$.
This permits equation (1) to be treated assuming:

$$\{(E_0(t))_j\} = E_0 = \text{constant}, \qquad \text{and} \qquad \phi(t) = \text{constant} = \phi_j \qquad (2)$$

The actual spacing of the different axial laser modes can be computed according to the following relationship:

$$\Delta f_M = \frac{c}{2L} \tag{3}$$

If the laser is operated near threshold, the width of the effective gain curve is not known, and hence, the number of excited modes cannot be computed but has to be taken from the laser specifications. However, most available lasers are operated well above threshold in order to obtain high light power output. This causes the width of the effective gain curve to approach the line width ΔF_D of the atomic transition which participates in the laser action. Using this information, the number of excited modes can be approximately computed from the following relationship:

$$M = \frac{\Delta F_D}{\Delta f_M} = \frac{\Delta F_D \cdot (2L)}{c} \tag{4}$$

In the LDA-system, the output of a multimode gas laser is usually split by a path compensated beam splitter which produces the two light beams shown in Figure 1 that are focused into the measuring control volume by the front lens of the LDA-optic. If a particle penetrates this measuring control volume it will scatter light from each of the two incident beams and this light will interfere in space to provide a beat signal that contains the required velocity information. Taking the aforementioned radiation into account, and neglecting depolarization effects by the scattering process, permits the scattered radiation from each of the two light beams to be expressed as follows:

$$(E_s)_1 = \sum_{n=1}^{M} \frac{a_{1n}(t)}{R_p} \exp\left[-i\left[\frac{2\pi R_p}{\lambda_n} - \omega_n t - \phi_n\right]\right] \tag{5a}$$

$$(E_s)_2 = \sum_{m=1}^{M} \frac{a_{2m}(t)}{R_p} \exp\left[-i\left[\frac{2\pi R_p}{\lambda_m} - \omega_m t - \phi_m\right]\right] \tag{5b}$$

The aforementioned equations can be taken as mathematical descriptions of the scattered light fields detected at a point "Q" of the photodetector surface. The distance R_p describes the distance from the particle to the detector and $a_{1n}(t)$ is the amplitude of the scattered laser radiation produced by beam $k = 1,2$.

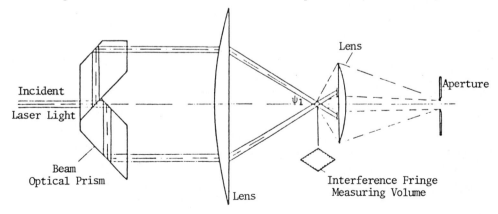

Figure 1. Example of path length compensated LDA-system.

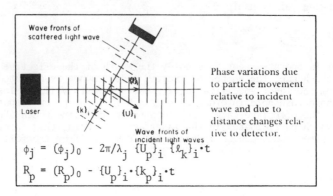

Figure 2. Time dependence of photodetector signals.

The particle motion permits the phase of the scattered light wave and the distance between particle and photodetector to be expressed. This yields the following relationships for the two scattered light waves (see Durst, Melling, and Whitelaw [13]); shown in Figure 2.

$$\phi_j = (\phi_j)_0 - \frac{2\pi}{\lambda_j} \{U_p\}_i \{\ell_k\}_i \cdot t$$

$$R_p = (R_p)_0 - \{U_p\}_i \cdot \{k_p\}_i \cdot t \tag{7}$$

Hence, the scattered field may be analytically described as follows:

$$(E_s)_1 = \sum_{n=1}^{M} \frac{a_{1n}}{R_p} \left[\exp -i \left[\frac{2\pi}{\lambda_n} (R_p)_0 - (\phi_n)_0 - \omega_n \cdot t - \frac{2\pi}{\lambda_n} \{U_p\}_i \cdot \left(\{\ell_1\}_i - \{k_p\}_i \right) \cdot t \right] \right] \tag{8a}$$

$$(E_s)_2 = \sum_{m=1}^{M} \frac{a_{2m}}{R_p} \left[\exp -i \left[\frac{2\pi}{\lambda_m} (R_p)_0 - (\phi_m)_0 - \omega_m t - \frac{2\pi}{\lambda_m} \{U_p\}_i \cdot \left(\{\ell_2\}_i - \{k_p\}_i \right) \cdot t \right] \right] \tag{8b}$$

The intensity at the surface element of the photodetector may be expressed in the following way:

$$I_Q = \frac{1}{T} \int_0^T \left[R[(E_s)_1] + R[(E_s)_2] \right]^2 dt \tag{9}$$

Taking into account the fact that the integration time of the photodetector is much larger than the inverse of the light frequency permits the following expression to be derived:

$$I_Q = \sum_{j=1}^{M} \frac{a_{1j}^2 + a_{2j}^2}{R_p^2} + \sum_{n=1}^{M} \sum_{m=1}^{M} \frac{a_{1n} a_{2m}}{R_p^2} \cos \left\{ 2\pi (R_p)_0 \left[\frac{1}{\lambda_m} - \frac{1}{\lambda_n} \right] + \left[(\phi_m)_0 - (\phi_n)_0 \right] \right. $$

$$\left. + 2\pi t \left[(\nu_m - \nu_n) - \{U_p\}_i \left[\frac{1}{\lambda_m} \left(\{k_p\}_i - \{\ell_2\}_i \right) - \frac{1}{\lambda_n} \left(\{k_p\}_i - \{\ell_1\}_i \right) \right] \right] \right] \tag{10}$$

This equation can be simplified if the frequency difference between the two beams is small such that the following relationship holds:

$$\lambda_m = \lambda_n \quad , \quad \text{i.e.,} \quad (\nu_m - \nu_n) \ll \frac{1}{2} (\nu_m + \nu_n) \tag{11}$$

The final expression therefore is:

$$I_Q = \sum_{j=1}^{M} \frac{(a_{1j}^2 + a_{2j}^2)}{R_p^2} + \sum_{n=1}^{M} \sum_{m=1}^{M} \frac{a_{1n} a_{2m}}{R_p^2} \cos \left\{ [(\phi_m)_0 - (\phi_n)_0] \right. $$

$$\left. + 2\pi t \left[(\nu_m - \nu_n) + \frac{1}{\lambda} \{U_p\}_i \left(\{\ell_2\}_i - \{\ell_1\}_i \right) \right] \right\} \tag{12}$$

Introducing the sensitivity vector:

$$\{n\}_i = \{\ell_2\}_i - \{\ell_1\}_i \tag{13}$$

permits the photomultiplier current for a single axial mode laser to be written as:

$$i_a = \text{const} \sum_{j=1}^{M} \frac{(a_{1j}^2 + a_{2j}^2)}{R_p^2} \left[1 + \frac{a_{1j} a_{2j}}{(a_{1j}^2 + a_{2j}^2)} \cos \left(2\pi t \frac{1}{\lambda} \{U\}_i \{n\}_i \right) \right] \tag{14}$$

The same expression for multiaxial mode laser radiation becomes:

$$i_a = \text{const} \left[\sum_{j=1}^{M} \frac{(a_{ij}^2 + a_{2j}^2)}{R_p^2} + \sum_{n=1}^{M} \sum_{m=1}^{M} \frac{a_{1n} a_{2m}}{R_p^2} \cos \left\{ (\Delta\phi_{m,n}) \right. \right. $$

$$\left. \left. + 2\pi t \left[(\nu_m - \nu_n) + \frac{1}{\lambda} \{U\}_i \{n\}_i \right] \right\} \right] \tag{15}$$

where $(\nu_m - \nu_n)$ may be written as:

$$(\nu_m - \nu_n) = (m-n) \frac{C}{2L} = \Delta_{m,n} \frac{C}{2L} \tag{16}$$

This yields signal frequencies as given by the following equation

$$\nu_{s,\Delta} = \left| \pm |\Delta_{m,n}| \frac{C}{2L} + \frac{1}{\lambda} \{U\}_i \{n\}_i \right| \tag{17}$$

Measurements of high frequency LDA-signals requires the employment of high frequency response photodetectors and preamplifiers, and hence, the detection electronics might not only respond to the Doppler frequency:

$$\nu_D = \nu_{s,o} = \left| \frac{1}{\lambda} \{U\}_i \{n\}_i \right| \tag{18}$$

but also to higher frequencies such as:

$$\nu_{s,1} = \left| \pm \frac{c}{2L} + \frac{1}{\lambda} \{U\}_i \{n\}_i \right| \tag{19}$$

If Doppler frequency measurements are attempted using counter techniques the frequencies given in equations (18) and (19) should be clearly separated in order to be able to apply bandpass filters prior to frequency counting. This experiment may be expressed as:

$$\nu_D \leq B \cdot \nu_{s,1} \quad , \quad \text{where } B \leq 1^* \tag{20}$$

Hence, for laminar flow the following relationship may be derived:

$$\nu_D = \frac{1}{\lambda} \{U\}_i \{n\}_i \leq \frac{Bc}{2L(1 + B)} \tag{21}$$

Assuming a turbulent flowfield that can be described by a Gaussian velocity probability density distribution, permits the following relationship to be derived:

$$\bar{\nu}_D(1 + 3T_u) \leq B \left[\frac{c}{2L} - \bar{\nu}_D(1 + 3T_u) \right] \tag{22}$$

where T_u is the local turbulence intensity defined as:

$$T_u = \frac{\sqrt{\overline{\Delta\nu_D^2}}}{\nu_D} \tag{23}$$

Hence, the highest mean frequencies that can be sensibly measured in turbulent flowfields are:

$$(\bar{\nu}_D)_{max} = \frac{B \cdot c}{2L(1 + B)(1 + 3T_u)} \tag{24}$$

Figure 3 shows plots $(\bar{\nu}_D)_{max}$ as given by equation (24) with a value for $c/2L$ = 150 MHz which approximately holds for average power argon-ion lasers. Figure 4 made use of the above general relationship to show that there is an inherent requirement to reduce the angle between the laser beams of an LDA-optic. This requirement is imposed by laser properties rather than the frequency capabilities of electronic processing systems. Figure 4 shows the following relationship derived from equation (24):

$$U_\perp \leq \frac{B \cdot c \cdot \lambda}{4L \sin \psi (1 + B)(1 + 3T_u)} \tag{25}$$

---

* The factor B will depend on the filter characteristics. B will decrease with decreasing steepness of the filter.

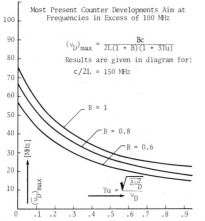

Most Present Counter Developments Aim at
Frequencies in Excess of 100 MHz

$$(\nu_D)_{max} = \frac{Bc}{2L(1 + B)(1 + 3Tu)}$$

Results are given in diagram for:

$$c/2L = 150 \text{ MHz}$$

B = 1

B = 0.8

B = 0.6

[MHz]

$(\nu_D)_{max}$

$$Tu = \frac{\sqrt{\overline{\Delta\nu_D^2}}}{\nu_D}$$

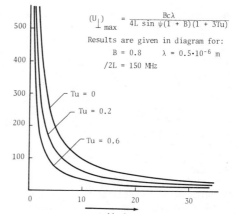

$$(U_\perp)_{max} = \frac{Bc\lambda}{4L \sin \psi(1 + B)(1 + 3Tu)}$$

Results are given in diagram for:

$$B = 0.8 \qquad \lambda = 0.5 \cdot 10^{-6} \text{ m}$$
$$/2L = 150 \text{ MHz}$$

Tu = 0

Tu = 0.2

Tu = 0.6

ψ [deg]

Figure 3. Frequency restrictions imposed on laser Doppler measurements
by multiaxial mode laser radiation.

Figure 4. Angle restrictions imposed
on laser Doppler optics by multiaxial
mode laser radiation.

It is worthwhile to mention that the application of frequency shifting devices in optical systems does not change the aforementioned relationship for
the maximum permissible Doppler frequency of LDA-signals. Hence, frequency shifting does not extend the range of application of laser Doppler anemometers if this
range is defined by the multiaxial mode radiation of a gas laser. Such extensions
will require mode selectors to be incorporated into the laser beam in order to
ensure single mode laser radiation. Such systems are used for single component
velocity measurements in high speed flows employing argon-ion lasers. They are
more difficult to incorporate into two-dimensional optical systems such as those
shown in Figure 5.

PHOTODETECTOR PROPERTIES AND THEIR INFLUENCE ON LASER DOPPLER SIGNALS--Most
laser Doppler systems employed for high speed velocity measurements incorporate
photomultipliers to detect the optical signals and to convert them into electrical

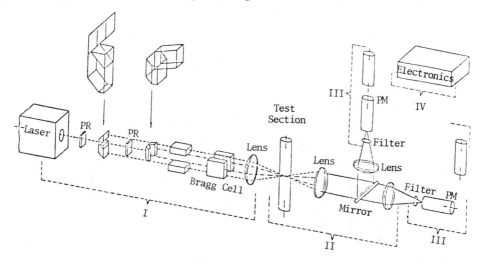

Figure 5. Dual-beam anemometer for two-component LDA measurement.

ones from which the Doppler frequency can be deduced. Up to date, the selection of photomultipliers for high frequency LDA-measurements has been entirely based on rise times of such devices but also considering those properties that have been proven to be essential for good signal-to-noise ratio in low speed flows, e.g. high quantum efficiency at the wavelength of the incident laser radiation and high internal amplification. All other photomultiplier properties have been widely considered as being insignificant for the signal-to-noise ratio of laser Doppler signals obtained in high speed flows. Durst and Heiber [8] showed in a recent study, however, that photomultipliers having the same frequency response, equal quantum efficiencies, and the same internal amplifications but different permissible maximum anode currents can yield large differences in signal-to-noise ratios of LDA-signals. Since these considerations are essential for high frequency LDA-measurements, the major points are briefly summarized in this section.

Durst and Heiber [8] studied detection systems of laser Doppler anemometers by deriving expressions for signal-to-noise ratios for different photodetectors. These are described by an equivalent circuit diagram given in Figure 6. This diagram employs current sources for the cathode current and dark current of photodetectors as well as noise contributions. The internal amplification of photodetectors like photomultipliers or avalanche photodiodes is taken into account as well as the excess noise they might introduce. Furthermore, the anode capacitance is considered and the load resistance and its noise contribution together with the noise introduced by the preamplifier. The capacitance and the load resistor define the bandwidth of the detection system and they were chosen to match the Doppler frequency to be detected.

Using the equivalent circuit diagram in Figure 6 permits the following expression to be derived for the signal-to-noise ratio at the output of the detection system:

Photodiodes:

$$SNR = i_{cath} \Bigg/ \left[2e(i_{cath} + i_D)\Delta f + 4kT\frac{1}{R_A}\left(1 + \frac{R_n}{R_A} + \frac{4}{3}\pi^2 R_n R_A \Delta f^2 C_A^2\right)\Delta f \right]^{1/2}$$

Photomultipliers:

$$SNR = i_{cath}\delta^K \Bigg/ \left[2ei_{cath}\frac{\delta^{K+1} - \delta^K}{\delta - 1}\Delta f + 2ei_D\Delta f + 4kT\frac{1}{R_A}\Delta f \right.$$

$$\left. \left(1 + \frac{R_{pa}}{R_A} + \frac{4}{3}\pi^2 R_{pa}R_A\Delta f^2 C^2\right)\Delta f \right]^{1/2}$$

Avalanche Photodiodes:

$$SNR = Mi_{cath}\Bigg/\left[2ei_{cath}M^2M^X\Delta f + 2ei_D\Delta f + 4kT\frac{1}{R_A}\right.$$

$$\left.\left(1 + \frac{R_{pa}}{R_N} + \frac{4}{3}\pi^2 R_{pa}R_A\Delta f^2 C_A^2\right)\Delta f\right]^{1/2}$$

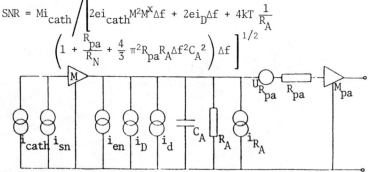

Figure 6. Equivalent circuit diagram suggested by Durst and Heiber [8] to study signal-to-noise ratios of LDA-signals.

In these equations, the expressions for the excess noise were introduced for photomultipliers and avalanche photodiodes as given in References [14] and [15] respectively.

The quantities applied in the above formulae are:

C_A = anode capacitance

e = elementary charge

Δf = bandwidth of detection system

i_{cath} = cathode current

i_D = dark current

k = Boltzmann constant

M = internal amplification

R_A = anode resistance (load resistance)

R_{pa} = equivalent noise resistance of preamplifier

SNR = signal-to-noise ratio

T = temperature

x = exponent in excess noise factor

π = 3.14159 . . .

Durst and Heiber [8] evaluated the above formulae for typical values of photo-detectors such as:

Photodiode: $C_A = 10^{-11}$ F; η_q = 70 percent

Photomultiplier: $C_A = 10^{-10}$ F; η_q = 7 percent

Avalanche-Diode: $C_A = 10^{-12}$ F; η_q = 21 percent

Source of Light: λ = 633 nm (helium-neon)

These results are shown in Figure 7 for a photomultiplier with a maximum permissible anode current of $(i_A)_{max} = 10^{-3}$ A and in Figure 8 for $i_{max} = 10^{-5}$ A.

For the photomultiplier with the higher permissible anode current, i.e. with $(i_A)_{max} = 10^{-3}$ A, the shot noise is the dominating noise contribution in the entire range yielding the given slope of the log SNR-log P_S curve of one-half. Only for the largest bandwidth Δf of the detection system, the SNR saturates. Hence, increases in scattered light power cannot be utilized in the detection system to increase the signal-to-noise ratio of LDA-signals. Figure 7 also shows that photomultipliers are advantageous if large bandwidth detection systems are employed. For small Δf-systems and high scattered light powers, photodiodes can yield better signal-to-noise ratio performances. This behavior is characteristic for the photodetectors studied in Figure 7.

Figure 8 provides the data that were computed by Durst and Heiber [8] for a photomultiplier with a smaller value for the permissible anode current, i.e. for $(i_A)_{max} = 10^{-5}$ A. Detectors of this type are employed in some commercial LDA-systems. Photodetector systems that employ such photomultipliers saturate in SNR if the bandwidth of the detection system is broad, i.e. larger than 10^7. For a detection bandwidth of $\Delta f = 10^8$ the SNR saturates at a level which is too low for most LDA-signal processing systems. Existing photodiodes, and especially avalanche photodiodes, are superior to photomultipliers if these have low values of maximum permissible anode currents.

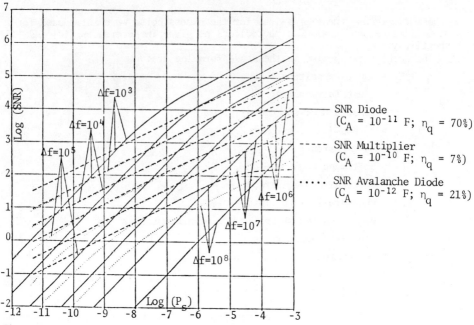

Figure 7. Signal-to-noise ratio for different photodetectors as a function of detected scattered light power for $(i_A)_{max}$ = 10^{-3} A for the photomultiplier.

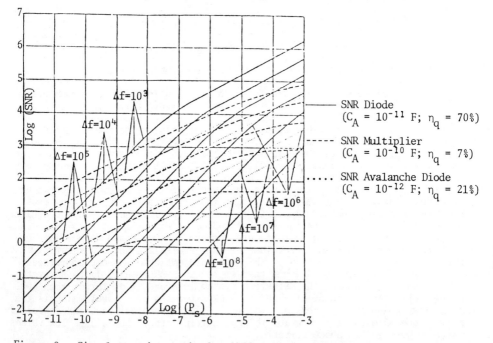

Figure 8. Signal-to-noise ratio for different photodetectors as a function of detected scattered light power for $(i_A)_{max}$ = 10^{-5} A for the photomultiplier.

CONCLUSIONS AND FINAL REMARKS

Theoretical considerations on the application of LDA-systems to high speed flows show that measurements will require limiting the Doppler frequency to values below a maximum value which is much lower than the frequencies modern counter developments aim for. The frequency limitations are due to the multi-axial mode output of high power cw lasers. High frequency measurements can only conveniently be carried out with single mode lasers.

Measurements of high frequency LDA-signals also require the consideration of the inherent limitations of the anode current of the photomultipliers. Photomultipliers with high permissible anode currents are advantageous for laser Doppler measurements.

ACKNOWLEDGMENTS

The present paper described collaborative research carried out between the Sonderforschungsbereich 80 of the University of Karlsruhe and the Physikalisch Technische Bundesanstalt in Braunschweig. The work is related to the development of a laser-Doppler system that will perform high accuracy measurements in high speed flows as a basis for accurate flow rate measurements for calibration purposes.

The authors are very grateful to Ginger Common for typing the present paper and for all her help to finish it.

The work was financially supported by the Deutsche Forschungsgemeinschaft and the Physikalisch Technische Bundesanstalt. It was completed during the time the second author held a position as Regents' Professor of the University of California at Davis.

REFERENCES

[1] Cherdron, W., Durst, F., and Whitelaw, J. H. 1978. Asymmetric Flows and Instabilities in Symmetric Ducts with Sudden Expansions. Journal of Fluid Mechanics 84:13.

[2] Kozovinos, N. E. and List, J. 1975. Turbulence Measurements in a Two-Dimensional Buoyant Jet using Laser Doppler Velocimetry. Proceedings of the LDA-Symposium, Copenhagen, Denmark.

[3] Durst, F. and Rastogi, A. K. 1978. Theoretical and Experimental Investigations of Turbulent Flows with Separation. Proceedings of the Turbulent Shear Flow Symposium. Springer-Verlag.

[4] Baker, R. J., Hutchinson, P., Khalil, E. E., and Whitelaw, J. H. 1974. Measurements of Three Velocity Components and their Correlations in a Model Furnace With and Without Combustion. Proceedings of the 15th Combustion Symposium p. 553.

[5] Abbis, J. B. 1974. Photon Correlation Measurements in Supersonic Flows. Royal Radar Establishment, Tamborough, England. Paper presented at Electron-Optic Conference.

[6] Durst, F. and Whitelaw, J. H. 1973. Light Source and Geometrical Requirements for the Optimization of Optical Anemometer Signals. Opto-Electronics 5:137.

[7] Eliasson, B. and Dändliker, R. 1974. A Theoretical Analysis of Laser-Doppler Flow Meters. Optica Acta 21:119.

[8]   Durst, F. and Heiber, K. F. 1977. Signal-Rausch-Verhältnisse von Laser-
      Doppler-Signalen. Optica Acta 24:43.

[9]   Durst, F. 1977. Möglichkeiten genauer Volumenstrommessungen mittels
      Laser-Doppler-Messverfahren. 3.PTB-Seminar, PTB-Report ME-13.

[10] Foreman, J. W., Jr. 1967. Optical Path-Length Difference Effects in Photo-
      mixing and Multimode Gas Laser Radiation. Journ. Appl. Optics 6:5:821.

[11] Rudd, M. J. 1968. Laser Doppler Velocimetry Employing the Laser as a
      Mixer Oscillator. Journ. Scientific Instr. 2:1.

[12] Durst, F. and Stevenson, W. H. 1978. The Influence of Gaussian Beam Pro-
      perties on Laser Doppler Signals. Accepted for publication in Journ.
      Appl. Optics.

[13] Durst, F., Melling, A., and Whitelaw, J. H. 1976. Principles and Practice
      of Laser-Doppler Anemometry. Academic Press, London-New York-San Francisco.

[14] RCA-Photomultiplier Manual, published by RCA as handbook.

[15] Melchior, H. 1973. Sensitive High Speed Photodetectors for the Demodulation
      of Visible and Near Infrared Light. Internal Report, Bell Laboratories,
      Murray Hill, New Jersey 07974.

Laser Velocimetry and Holographic Interferometry Measurements in Transonic Flows

W. D. BACHALO
Spectron Development Laboratories, Inc.
Costa Mesa, CA 92626

D. A. JOHNSON
NASA, Ames Research Center
Moffett Field, CA 94035

ABSTRACT

Measurements of the transonic flow about a two-dimensional airfoil have been made with holographic interferometry and laser velocimetry. Quantitative data obtained with the interferometer are compared to the laser velocimeter and surface pressure measurements to evaluate the accuracy of the technique. Good agreement in the results confirmed the two-dimensionality of the flow and the potential of the interferometer in making unsteady transonic flow measurements in the future.

INTRODUCTION

The application of optical techniques to transonic flow research has provided a greater insight into such complex flows. Because the transonic flow is perturbation sensitive, the use of nonintrusive measurement techniques is especially relevant. The development of the techniques will be very important to unsteady transonic flow research. In the past, the transonic data consisted primarily of surface pressure measurements (with a relatively long time response) and wake total and static pressure measurements. Although the pressure data are indispensable, additional information about the flow behavior is required for both the aerodynamicist and the computational fluid dynamicist. This information is necessary for understanding and predicting separated flows.

Usually the best performance of an airfoil occurs under conditions not far removed from separation. With the onset of separation, a precipitous loss of performance and attendent global effects on the flow field occurs. The occurrence of shock-induced separation in transonic flow produces a strong interaction between the boundary layer and the pressure gradient near separation. The accuracy of the numerical techniques in predicting such flows generally tends to decrease with increasing shock strength and becomes entirely unsatisfactory when the flow has separated. Poor agreement between the predictions and experiments can, in part, be attributed to inadequacies in understanding of the turbulent boundary layer separation mechanisms and in the turbulence modeling. Because the predictions depend very heavily upon empirical results, reliable turbulence and mean flow measurements remain of crucial importance.

The present investigations have two purposes: to obtain detailed measurements of the entire flow field about a two-dimensional airfoil including the inviscid, viscous, and turbulence properties, and to evaluate the applicability and accuracy of holographic interferometry in transonic flow measurements. If the accuracy of the holographic techniques could be demonstrated, a useful tool would be available to complement the laser velocimeter and pressure measurements,

369

as well as providing an excellent means of visualizing the entire flow field.
The value of the interferometer is especially important in unsteady transonic
flows which are of contemporary interest in helicopter research.

Several flow conditions were tested varying from zero degrees to 6.2° angle
of attack at which case shock-induced separation occurred. Combined measurements
allowed a complete definition of the inviscid and mean viscous flow properties
providing data on static pressures and temperatures, Mach contours, densities,
flow angles, and mean velocities. The laser velocimeter was also used to measure the Reynolds normal and shear stresses (Reference 1), but only those data
relevant to the comparison of the techniques are presented here.

EXPERIMENTAL TECHNIQUES

The experiments described in this paper were conducted in the Ames 2-foot
by 2-foot transonic wind tunnel. This facility is a closed-return, variable
density tunnel with slotted upper and lower test section walls through which suc-
tion may be applied to alleviate the wall constraints to the flow field. During
the tests, the tunnel conditions were maintained at a free stream Mach number
of 0.8 and a chord Reynolds number of 2×10^6. An instrumented NACA 64A010 airfoil with a 6-inch chord was selected for the test. The model was equipped with
surface pressure taps, embedded hot wires, and a transition strip at the 17%
chord position to ensure that the boundary layer was fully turbulent at the shock.

Laser Velocimeter

A dual-wavelength laser velocimeter system was used allowing simultaneous
measurement of the streamwise and normal velocity components. Bragg cell frequency shifting was introduced into both laser wavelengths. Frequency shifting
was essential in resolving the direction of the flow when separation occurred
and in reducing the relative frequency bandwidth resulting from the high turbulence levels. Forward scatter light collection was used with the transmitting
and receiving optics installed within the wind tunnel plenum chamber for better
light collection efficiency (Figure 1). The Argon-ion laser, color separation

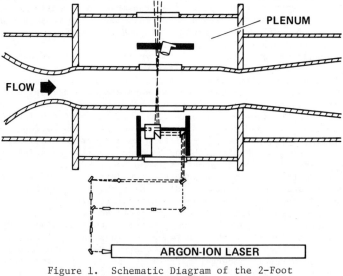

Figure 1. Schematic Diagram of the 2-Foot
by 2-Foot Laser Velocimeter

optics, and Bragg cells were located outside of the plenum. Translation of the probe volume was achieved with a system of mirrors and a computer controlled traversing system. A more detailed description of the laser velocimeter may be found in Reference 2.

In the past, research was devoted to evaluating possible flow seeding techniques to supply scattering centers for the laser velocimeter (Reference 3). In the 2-foot by 2-foot facility, naturally occurring particles were often plentiful enough for an adequate data rate (~ 1000/s). The particles are believed to consist primarily of lubricant oil that has been vaporized in the tunnel drive system and recondensed just upstream of the test section. Measurements of the particle response to the step change in velocity across a normal shock led to the conclusion that the particles were small enough (estimated at 1 μm in diameter) to have adequate response to the velocities in the transonic flow.

Holographic Interferometry

The conversion of large wind tunnel schlieren systems to holographic visualization systems was suggested by Trolinger[4]. Since that time, the technique has been developed and exploited in several aerodynamics studies[5,6]. Interferometry, the mixing of two coherent waves for the purpose of measuring distortion in one of the waves, has been widely used and is well understood. The introduction of holography as an intermediary to store the light wave information allowed a good deal of versatility in the technique and greatly extended the possible applications. In flow diagnostics, the holographic technique does not differ, in principle, from the Mach-Zehender technique. However, with holography, the interference is between two reconstructed waves that have followed the same (or similar) optical paths but are separated in time whereas in the Mach-Zehender technique, the waves follow different paths but interfere at the same time (Figure 2). This distinction is important because with the same paths the imperfections in the optical elements can be cancelled.

Although there are several ways in which holography can be used as an intermediary in interferometry[6], the dual plate technique was used in the present investigation. In this method, an exposure is made on a photographic plate, with the wind tunnel not running, and subsequent plates are made at the test conditions. After processing, the plates are positioned in the reconstruction plate

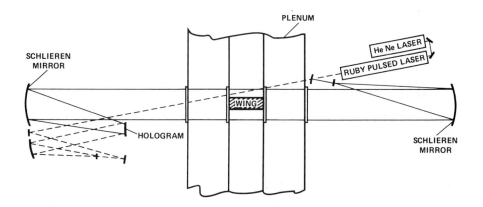

Figure 2.   Schematic Diagram of the
Holographic Interferometer

holder, illuminated with the reference beam, and aligned for the infinite fringe
interferograms. The images are then photographed and used for flow field diagnostics. The schlieren and shadowgraph techniques are also available at the time
of reconstruction. A complete description of the technique may be found in Reference 5. One reference plate can be used with several plates taken with the flow
on, saving a great deal of test time. Vibrations in the system are relatively
insignificant because the exposure duration is extremely short (~ 20 nanoseconds)
and motion between exposures is eradicated in the reconstruction process.

To reduce the interferogram data to density contours, two-dimensional flow
was assumed in the relationship

$$\rho_2 - \rho_1 = (\frac{\varepsilon}{L}) \frac{\lambda}{K_{GD}} \tag{1}$$

where ρ_1 and ρ_2 are the densities at two locations, ε is the fringe shift between
the locations, L is the geometrical width of the test section, λ is the laser
wavelength, and K_{GD} is the Gladstone-Dale constant relating the refractive index
to the specific fluid density at wavelength λ. When the interferometer is aligned
in the infinite fringe mode, each fringe corresponds to a constant density contour.
The density at a point in the flow can be determined from the wind tunnel conditions and either the measured velocity or surface pressure at that point. With
one density contour tagged, the remaining contours can be determined using Equation 1.

EXPERIMENTAL RESULTS

Representative interferograms of the flows tested are shown in Figure 3 for
0°, 3.5°, and 6.2° angle of attack taken at a freestream Mach number of 0.8 and
a Reynolds number of 2×10^6 based on the chord length. These interferograms
(when viewed in full scale) provide a graphic view of the entire flow field including the shock wave, boundary layer, wake, and even the disturbances from the
transition strip and embedded static pressure tubes. The transient flow features
are fixed by the short duration of the exposure so that the density distributions
(integrated along the beam path) in the turbulent boundary layer and wake are
visible.

The changing character of the flow with increasing angle of attack is clearly discernible on these interferograms. At 3.5° angle of attack, an increase in
the boundary layer thickness at the foot of the shock takes place as the flow is
suddenly subjected to the adverse pressure gradient posed by the shock. The upstream influence of the displacement thickness causes compression waves to form
thus producing the phenomena referred to as self-induced separation. These compression waves coalesce to a shock wave above the viscous layer. As a result of
the increasing displacement thickness, the shock is oblique, turning the inviscid
flow away from the surface. In the far field it has curved back to become a normal shock and then degenerates into weak compression waves. For the 6.2° case,
the shock is of sufficient strength to separate the boundary layer. Here, the
upstream influence has progressed well ahead of the shock location in the inviscid
region. The shock curvature is more pronounced due to the significant increase
in displacement thickness as the boundary layer suddenly leaves the surface and
takes on the character of a free-shear layer. This strong interaction between
the boundary layer and the inviscid flow has global effects on the flow that
drastically affect the lift and drag of the airfoil. No periodic oscillations
of the separated flow as seen on biconvex airfoils[7] were detected.

Although the flow visualization is valuable in detecting the significant
features of the flow and directing the laser velocimeter measurements, the interferogram can also produce quantitative information. Since the surface pressures
on the model can be measured very reliably, these data were selected for the

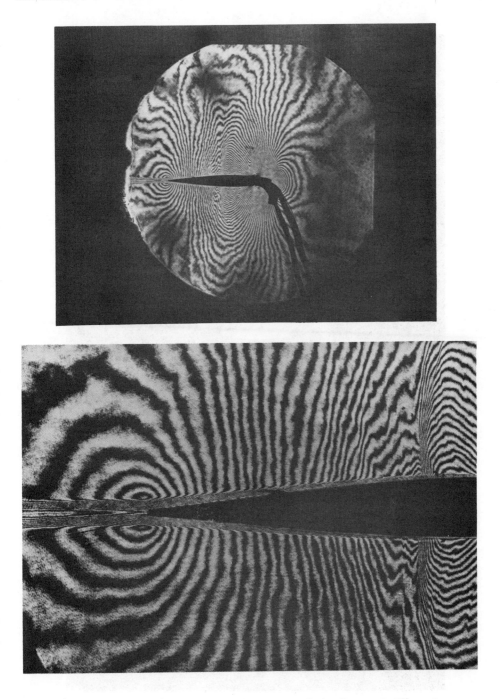

Figure 3. Holographic Interferograms of the Transonic Flow at
$\alpha=0°$; $M_\infty=.8$; $R_c=2\text{x}10^6$

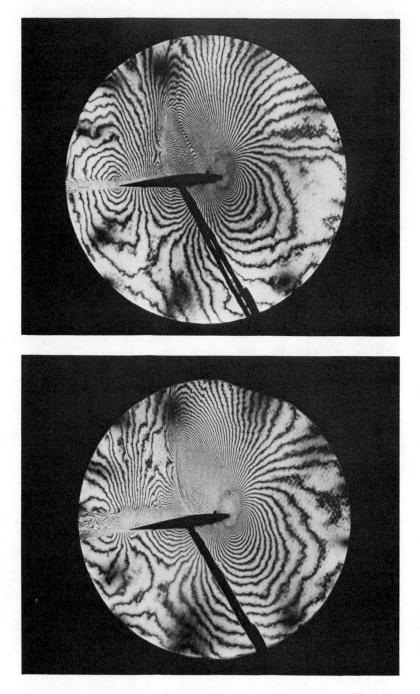

Figure 3.   Holographic Interferograms of the Transonic Flow at
            $\alpha=3.5°$ and $6.2°$; $M_{\infty}=.8$; $R_c= 2 \times 10^6$

initial comparisons. With the assumption of isentropic flow, the densities measured from the interferogram were reduced to surface pressure coefficient by the
relationship

$$C_p = \frac{2}{\gamma M_\infty^2} [(\frac{P}{P_t}) (\frac{P_t}{P_\infty}) - 1]$$

(2)

and

$$\frac{P}{P_t} = (\frac{\rho}{\rho_t})^\gamma$$

(3)

where M_∞ is the freestream Mach number, $\gamma = 1.4$, and p_t and ρ_t are the pressure
and density at the total conditions. Since the field of view does not extend
upstream to the undisturbed flow, a fringe remote from the model was identified
using the inviscid velocity measured at that point with the velocimeter. Fringes
were then simply counted from the reference to obtain the density at each point
in the flow field.

Agreement between the measured surface pressure and the interferometer data,
Figure 4, is very good confirming the relative two-dimensionality of the flow.
A discrepancy occurs in the zero degree case immediately upstream of the shock.
The possibility of shock oscillation was considered as a source for the discrepancy since the surface pressures are time averaged whereas the interferometer
measures a spatial average. However, subsequent shots showed similar results.
In the 3.5° case, the good agreement even with the increased shock strength confirmed that the loss in total pressure across the shock, and hence the change in
entropy, was negligible. The pressure drop through the compression waves just
upstream of the shock can be seen in this figure.

At an angle of attack of 6.2° wherein the flow has separated, some evidence
of a total pressure loss across the shock exists. An estimation of the shock
strength was made to obtain a total pressure loss correction. Slight improvement was realized in the agreement at the trailing edge.

Once the accuracy of the interferometer alignment to the infinite fringe
mode was verified by comparison to the pressure distributions, the Mach contours,
Figure 5, determined from

$$\frac{\rho}{\rho_t} = (1 + \frac{\gamma-1}{2} M^2)^{-\frac{1}{\gamma-1}}$$

(4)

were traced from the appropriate fringes. The Mach contours provide a quantitative mapping of the global features of the flow and are invaluable for comparisons
to numerical predictions of the inviscid flow field. For example, in Figure 5,
Mach contours determined from a solution of the Navier-Stokes equations have been
compared to the data for the zero degree case. The calculated results for this
simple case agree very well with the data.

The flow speed was calculated from the density distribution using the assumption of isentropic flow of a perfect gas and the relationship

$$\frac{V}{C_t} = [\frac{2}{\gamma-1} (1 - (\frac{\rho}{\rho_t})^{\gamma-1}]^{\frac{1}{2}}$$

(5)

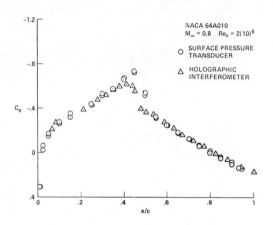

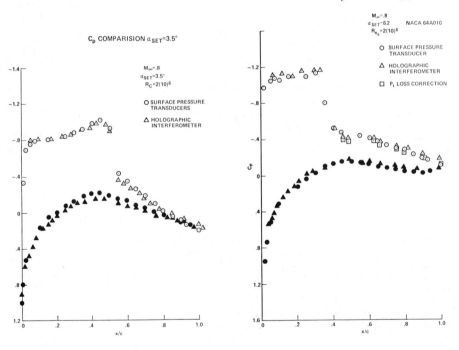

Figure 4.   Comparisons of the Measured Surface
Pressure Coefficients

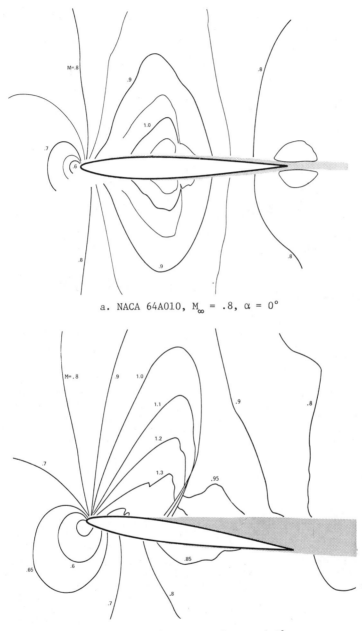

a. NACA 64A010, $M_\infty = .8$, $\alpha = 0°$

b. NACA 64A010, $M_\infty = .8$, $\alpha = 6.2°$

Figure 5. Mach Contours Measured with
the Interferometer

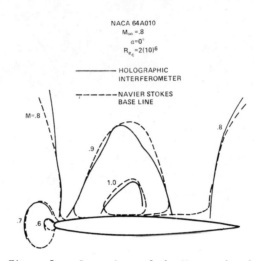

Figure 5c.   Comparison of the Measured and
Calculated Mach Contours

where C_t is the speed of sound at stagnation conditions.   Figure 6 shows comparisons to the flow speeds to those measured with the laser velocimeter.   For the zero degree case, the agreement is excellent.   However, for the 6.2° angle of attack, the laser velocimeter shows a progressive increase in the difference of the results in the regions of strong velocity gradients.   It is evident that the particles are not tracking the strong acceleration and deceleration of the flow.   The magnitudes agree well at the leading edge, peak value, and the trailing edge.

Some additional important features of the viscous flow are apparent in Figure 7.   Reversed velocities measured were relatively low indicating that the wall shear has a minimal effect on the flow in that region.   This low shear at the wall explains the free-shear layer characterization of the flow there.   It was also found that the separated region closed at the trailing edge of the model.   Extensive measurements of the turbulence intensities, Reynolds shear stress, and wake profiles have been made for this case, and these data have been reported elsewhere (Reference 1).

In Figure 8, the wake density profiles were determined from the interferograms and compared to those computed from the laser velocimeter data using the Crocco relationship given by

$$\frac{T}{T_e} = 1 + r \frac{\gamma-1}{2} M_e^2 \left[1 - \left(\frac{U}{U_e}\right)^2\right] + \frac{T_w - T_{ad}}{T_e} \left(1 - \frac{U}{U_e}\right) \qquad (6)$$

and the perfect gas law. T_e, T_{ad}, and T_w are the edge, adiabatic and wall temperatures, and r is the turbulent recovery factor equal to 0.88.   The good agreement confirms the ability to obtain accurate mean density profiles even in separated flows.   Assurance that the mean viscous characteristics of the flow can be reliably determined with the interferometer will allow the use of this technique for assessing the viscous effects in future research on two-dimensional unsteady flows.

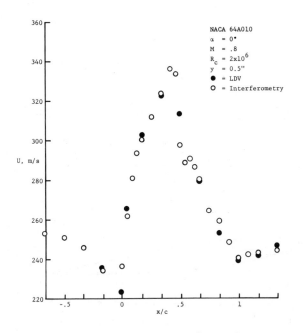

a.   Zero Degrees Angle of Attack

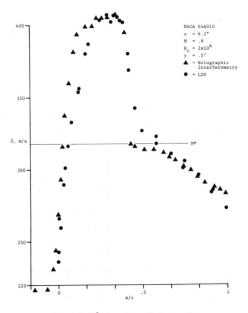

b.   6.2° Angle of Attack

Figure 6.   Flow Speed Distributions Measured with the
            Velocimeter and the Interferometer

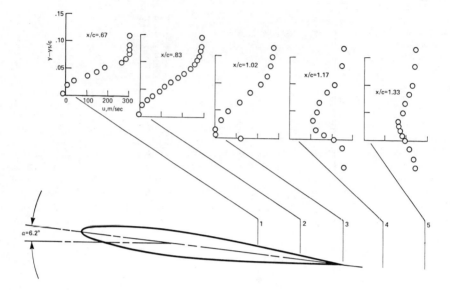

Figure 7.   Mean Velocity Profiles in the Viscous Region

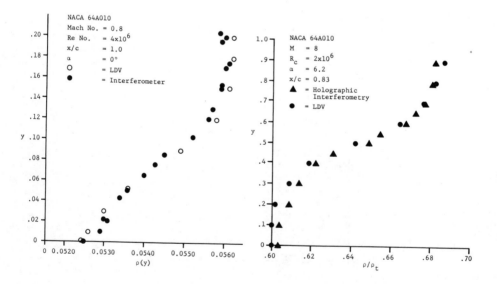

Figure 8.   Mean Density Profiles

CONCLUSIONS

It has been shown by way of comparison that holographic interferometry can provide accurate quantitative data in large scale transonic facilities. This result has several important consequences. The inviscid flow Mach contours can be established with little effort allowing the laser velocimeter to be used more effectively in making turbulence measurements. Surface pressures can be determined with very good accuracy using the interferometer so that the cost of instrumenting two-dimensional and axisymmetric models can be significantly reduced. Even more important is the ability to measure unsteady pressures in the case of flapping airfoils and other unsteady flows. The development of the holographic interferometer will make a very important contribution to the research on helicopter rotor flows.

In the future, techniques will be introduced to ensure the reliability of the holographic data. Some subjectivity is currently involved in the alignment of the holographic plates to the infinite fringe mode. With no region of undisturbed flow in the field, there is a possibility of not achieving the correct alignment. Additional information will be used at the time of the alignment to prevent this type of error.

ACKNOWLEDGEMENTS

This work was funded by the Aerodynamics Branch, NASA-Ames Research Center.

REFERENCES

1.   Johnson, D. A. and Bachalo, W. D., "Transonic Flow About a Two-Dimensional Airfoil -- Inviscid and Turbulent Flow Properties", Paper 78-117, AIAA 11th Fluid and Plasma Dynamics Conference, Seattle, Washington, July 1978.

2.   Johnson, D. A., Bachalo, W. D., and Modarress, D., "Laser Velocimetry Applied to Transonic and Supersonic Aerodynamics", AGARD Conference 193 on Applications of Nonintrusive Instrumentation in Fluid Flow Research, May 1976.

3.   Bachalo, W. D., Modarress, D., and Johnson, D. A., "Experiments on Transonic and Supersonic Turbulent Boundary Layer Separation", AIAA Paper 77-47, 1977.

4.   Trolinger, J. D., "Conversion of Large Schlieren Systems to Holographic Visualization Systems", 15th National Aerospace Instrumentation Symposium, Las Vegas, Nevada, May 1969.

5.   Havener, A. G. and Radley, R. J., Jr., "Quantitative Measurements Using Dual Hologram Interferometry", ARL 72-0085, AD 749 872, Aerospace Research Laboratories, Wright-Patterson AFB, Ohio, June 1972.

6.   Trolinger, J. D., "Laser Instrumentation for Flow Field Diagnostics", AGARDograph No. 186, March 1974.

7.   Seegmiller, H. L., Marvin, J. G., and Levy, L. L., Jr., "Steady and Unsteady Transonic Flow", Paper 78-160, AIAA 16th Aerospace Sciences Meeting, January 1978.

Definition of Dual Flow Jet Exhaust Characteristics by Laser Velocimetry

H. A. KUMASAKA
Boeing
P.O. Box 3707, M.S. 1W-03
Seattle, WA 98124

ABSTRACT

A Laser Doppler Velocimeter (LDV) system is being used to define the characteristics of model jet flows which simulate the dual flow exhausts of contemporary turbofan engines. The purpose of this study is to evaluate the noise suppression potential of various nozzle configurations which are designed to modify the velocity and turbulence profiles of jet exhausts. Mean velocity and turbulence profiles are shown for several nozzle designs and flows which have been deliberately manipulated to reduce peak velocities and turbulence while maintaining a given total flow rate.

The LDV system arrangement and support facilities for flow studies are briefly described, and results of preliminary studies to validate the use of the LDV system for flow measurements are discussed. Brief descriptions are also presented of current data acquisition and processing procedures and of proposed modifications to improve the performance and measurement capabilities of the LDV system.

BACKGROUND

The Noise Technology Laboratory (NTL) is part of the Noise Technology Staff of the Boeing Commercial Airplane Company. A major function of this laboratory is to provide facilities, instrumentation, and data acquisition and processing support to studies of aircraft noise generation and suppression (Ref. 1). LDV system development within the NTL is directed toward a capability to define the characteristics of exhaust flows for jet noise studies. This capability is required for high-temperature, high-velocity flows where conventional hot-wire/hot-film anemometry instrumentation cannot be used.

Much of the study of jet noise generation is based on model testing. Using established scaling concepts, testing allows preliminary evaluation of design concepts early in a noise suppressor development program. Installation of noise suppressors on aircraft obviously depends on other factors, primarily performance penalties due to added weight or to thrust losses. However, model testing can identify promising designs without premature commitment to the expenses of full scale hardware fabrication and testing.

MODEL JET FLOW FACILITY AND SEEDER PROVISIONS FOR LDV TESTING

Consistent with the model study approach, much of the development and application testing of the LDV system is conducted in the Model Jet Flow Facility (MJFF), Figure 1. This facility permits scaled simulation of exhaust flows of

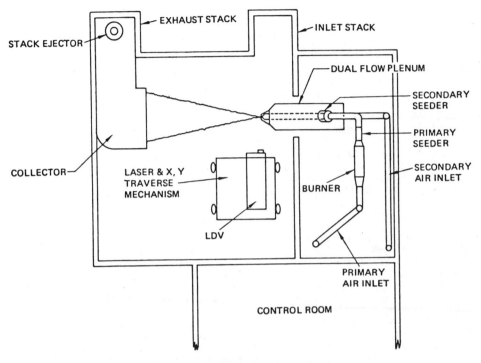

Figure 1. Model Jet Flow Facility

Figure 2. Flow Seeder Installations

turbojet and turbofan engines; separate flow ducts with propane burners are independently controlled and allow simulation of a wide range of by-pass flow conditions at different engine settings.

As shown in Figure 2, the primary and secondary flow ducts are seeded independently for LDV studies. The seeder designs are based on a fluidized-bed principle, and seed injection is by an aspirating jet. Additional air pressure inputs are used to percolate and premix the seed material.

The seeding material most commonly used is a mix of 70-80% alumina powder with 20-30% Tullanox, a hydrophobic fumed silica material. This mix has been found suitable for both cold and hot flow studies and is now used without additional drying or pulverizing. The nominal diameters quoted for these materials in supplier specifications are 0.05 micron for the alumina and 0.007 micron for the Tullanox; however, photographs of magnified seed samples disclose that the more representative seed particle is an agglomerated collection of smaller particles, characterized by a diameter in the 1-10 micron range.

LDV SYSTEM DESCRIPTION

The LDV system under development is based on the dual-beam, or Differential Doppler, arrangement. Figure 3 shows components for a two-channel, confocal backscatter arrangement enclosed in a common housing; major components of the system include a continuous wave Argon-ion laser (rated at five watts maximum output), beam projecting optics, signal receiving optics, and photomultiplier tube elements for data capture. The housing is installed on an x-y traverse table driven by remotely-controlled synchronous motors. Following initial alignment, the housing is pressurized slightly to minimize the contamination of mechanical and electro-optical components by seed material.

Basic operating principles of the LDV dual-beam mode arrangement are well-described in the literature, and a detailed description is not repeated in this discussion. Based on the simplified fringe model for this arrangement, it is simply noted that:
  (1) two equal intensity beams of laser light are introduced in parallel
       to a focusing lens, and an interference pattern of light and dark
       fringes is formed in a measurement probe volume at the intersection
       of the focused beams; and
  (2) the basic LDV system measurement of velocity is based on capture and
       processing of the modulated frequency signal information carried in
       the light signal that is scattered by a particle in flow that traverses
       the fringe pattern of known spacing.

DATA ACQUISITION

Photomultiplier outputs of the Differential Doppler frequency signal are processed through DISA Model 55L90 Data Processor modules; these are counter-type processors which employ a 5/8 fringe comparator for data validation. The modules provide panel displays of mean velocity and data acquisition rate information. These displays are used only for on-line monitoring; primary data acquisition is through the digital outputs provided by the processors. LDV data acquisition is supported by a Prime 300 minicomputer system, through a computer interface built in-house specifically for LDV data acquisition.

The digital outputs of the counter-processors are in the form of velocity-time pairs. More precisely, the outputs represent the frequency of the Differential Doppler signal from the photomultiplier and the time of capture expressed in terms of the time-since-last-sample. Both outputs are in 12-bit binary expressions - an 8-bit mantissa and a 4-bit exponent. Processor modifications have been made to yield a straight 12-bit output of time data, and the actual time output for measured data is in a 16-bit accumulated time representation from the interface.

A portion of a typical listing of data following capture, storage, and display is shown in Figure 4. Options for the resolution of time information are a 1 milli-second or 0.1 microsecond time base.

TEST PROCEDURE

Figure 5 depicts a flow survey arrangement and shows the seeded flow illuminated by the intersecting laser beams. Also shown is the corner-cube retroreflector installation commonly used to exploit the forward-scattered light signal. Since forward-scattered light is several orders of magnitude more intense than the backscattered light, the forward-scatter retroreflector is used whenever feasible to improve signal-to-noise ratios.

Following set-up of flow conditions for test, the procedure for acquiring data to define mean velocity and turbulence profiles is based on a transverse sweep in fixed position increments across the jet. At each position, a sample population of up to 4096 individual measurements is obtained and displayed as a normalized histogram in a mean $\pm$ 2 standard deviation range.

For most measurements in free jet flows, the reasonable assumption is that the sample is from a normally distributed population. Based on convention, the population mean, $\bar{V}$, defines the mean velocity (U) and the standard deviation, s, represents the normalized rms velocity ($\sqrt{u'^2}$), or turbulence, component.

A typical velocity histogram for 256 measurements at a single location is shown in Figure 6. The mean and turbulence values for each measurement are entered into a summary file whose contents are then plotted to display profiles defined by the transverse survey. Figure 7 shows the results of a flow survey for a dual flow model.

FLOW STUDY APPLICATION

A recent study to evaluate noise suppression design concepts based on the definition of mean velocity and turbulence profiles included flow surveys for the following nozzle configurations:
- (1)   Configuration A - a reference, or baseline installation combining a conical primary nozzle in a conical secondary nozzle;
- (2)   Configuration B - a multi-element primary nozzle in the conical secondary nozzle; and
- (3)   Configuration C - the multi-element primary nozzle in a multi-element secondary nozzle.

These nozzle configurations had equivalent flow areas based on a common model scale factor, and flow conditions for testing were based on matched pressure ratios and flow temperatures. The mean velocity and turbulence profiles for transverse surveys made at comparable downstream exhaust locations for each of these models are shown in Figures 8 and 9.

These measurements were obtained with a single-channel system arrangement combining a 30-inch focal length lens and a 1-inch beam spacing, using the 5145 Å wavelength of the laser source. For the 1.4-mm beam diameter (based on $1/e^2$ peak intensity limits), the nominal length of the probe volume calculated directly from equations in the literature is 0.8-inch (Ref. 2). This is obviously excessive for the resolution of sharp velocity gradients. However, the effective length of the probe volume is also governed by additional factors such as:
- (1)   an aperture plate at the photomultiplier entry which establishes an optical depth-of-field, designed to reject extraneous light signals which originate outside the measurement probe volume, and
- (2)   the data processor requirement for a minimum number of fringe crossings for data validation and acceptance, which eliminates the outer

regions of the measurement probe volume which contain fewer than the
minimum number of fringes required.

Based on direct experimental evidence, the effective length of the probe
volume for the previously described arrangement (with an aperture diameter of
0.027 in.) appears to be 0.4-inch. Although still excessive for spatial resolution of sharp velocity gradients, additional decrease of the focal length or
increase in beam spacing to reduce the length of the probe volume also results
in a change in the system calibration factor. The equivalent signal frequencies
for high velocities then approach the upper limit frequency capability of the
system, about 50 MHz. This limit is established primarily by capacitive losses
in the long signal leads from the photomultipliers to the data processors. The
system arrangement must therefore be tailored for a specific application with
regard to model size and the expected flow velocity range. The typical arrangement represents a compromise which reasonably satisfies the spatial resolution
requirements and operation within the frequency range of measurement capability.

These additional considerations notwithstanding, Figures 8 and 9 show substantial differences in the mean velocity and turbulence profiles of the three
models and disclose that the multi-element primary/secondary nozzles contribute
to progressive reduction of the peak velocities in the downstream jet exhaust.

Noise technology studies have established the strong dependence of jet
noise on flow velocity, and low frequency noise is recognized to originate in
the downstream exhaust region, characterized by large scale turbulence. Each
nozzle change which reduced peak velocities in the downstream exhaust is therefore expected to reduce low frequency noise. This is confirmed by comparisons
of relative noise power outputs for these nozzles, shown in Figure 10. The
noise power spectra in Figure 10 are based on measurements of far field noise
for the same nozzles tested at the same flow conditions in the NTL Large Test
Chamber (LTC). The LTC is a large anechoic room where free field noise testing
is performed (Ref. 3).

SYSTEM DEVELOPMENT AND PROPOSED IMPROVEMENTS

The previous example shows how even the modest first-stage capabilities
of the LDV system to define mean velocity and overall turbulence profiles
provide useful data to support a flow noise analytical study. These results
are acquired routinely using the present single-color, one-component system
arrangement supported by available software. Demonstration of this capability
and the potential for expanded LDV capabilities to provide additional insight
into flow characteristics and their relation to jet noise justifies continued
LDV development.

The LDV technique is the only available method for defining velocity and
turbulence characteristics in flows characterized by high-temperatures, high
velocities, or contaminated environments. The well-known additional advantage
of "non-interference" results from optical projection of the LDV measurement
volume, i.e., no mechanical device disturbs the flow.

Our immediate goal is to demonstrate the additional capability to define
local turbulence spectrum characteristics of jet flows. Toward this end, a
program based on the autocorrelation and transform approach has been written.
This program is being evaluated in terms of data simulations and preliminary
spectrum definition based on the current relatively small sample collections.
Major consideration requiring definition include the sample size for a valid
spectrum, the supporting software for data capture and transfer, and the upper
limit frequency of definition based on the ability of seed particles to
accurately track fluctuating flow disturbances.

Work is continuing on the development of a two-channel system based on
4880 Å and 5145 Å laser wavelengths to support two-component or two-point
measurement applications. The target capability is to achieve spatial correlation and cross-spectrum definition of the scale of flow turbulence to isolate

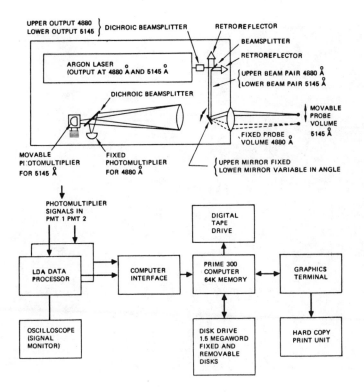

Figure 3. Laser Doppler Velocimeter System and Data Acquisition Block Diagram

```
1
HOW MANY (OCTAL) DATA POINTS? 40
LOCATION= '107761    V = 5.5454    T = '537
LOCATION= '107763    V = 5.5761    T = '1336
LOCATION= '107765    V = 5.0246    T = '2767
LOCATION= '107767    V = 6.2807    T = '13512
LOCATION= '107771    V = 5.3922    T = '17777
LOCATION= '107773    V = 5.8212    T = '27136
LOCATION= '107775    V = 4.9633    T = '73576
LOCATION= '107777    V = 4.9633    T = '77330
LOCATION= '110001    V = 6.1582    T = '106317
LOCATION= '110003    V = 5.8212    T = '115134
LOCATION= '110005    V = 6.2807    T = '117733
LOCATION= '110007    V = 5.3616    T = '137732
LOCATION= '110011    V = 6.4033    T = '142730
LOCATION= '110013    V = 4.9633    T = '37314
LOCATION= '110015    V = 5.8518    T = '40756
LOCATION= '110017    V = 6.0969    T = '47754
DO-
```

Figure 4. Partial Listing of Individual Measurements in Data Record

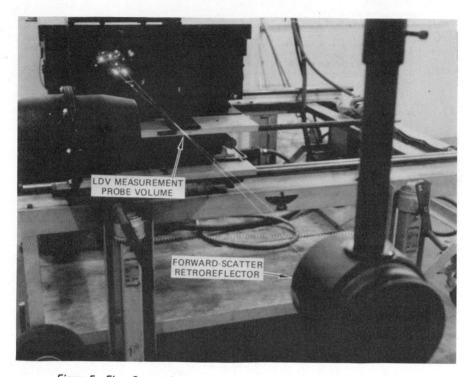

Figure 5. Flow Survey Arrangement Showing Forward Scatter Retroreflector

```
CHAN. 1:  ARITH.  <U> = 15.629 MHZ, U-RMS = 3.3483 MHZ, 100%U-RMS/<U> = 21.384
          HARM   <U> = 14.856 MHZ, U-RMS = 3.3897 MHZ, 100%U-RMS/<U> = 22.817
BIN COUNT
 1,   0   I
 2,   1   IX
 3,   3   IXXX
 4,   5   IXXXXX
 5,   4   IXXXX
 6,   4   IXXXX
 7,   8   IXXXXXXXX
 8,  11   IXXXXXXXXXXX
 9,   8   IXXXXXXXX
10,  10   IXXXXXXXXXX
11,  12   IXXXXXXXXXXXX
12,  11   IXXXXXXXXXXX
13,  20   IXXXXXXXXXXXXXXXXXXXX
14,  14   IXXXXXXXXXXXXXX
15,  12   IXXXXXXXXXXXX
16,  13   IXXXXXXXXXXXXX
17,   8   IXXXXXXXX
18,  14   IXXXXXXXXXXXXXX
19,   4   IXXXX
20,  12   IXXXXXXXXXXXX
21,  12   IXXXXXXXXXXXX
22,   3   IXXX
23,   7   IXXXXXXX
24,   8   IXXXXXXXX
25,  12   IXXXXXXXXXXXX
26,   4   IXXXX              V1 = 791.48 FEET/SEC , TURB1 = 21.384 %
27,  10   IXXXXXXXXXX
28,   7   IXXXXXXX
29,   4   IXXXX
30,   2   IXX
31,   1   IX
32,   2   IXX
TOTAL BIN POPULATION = 246, NO. PTS. OUTSIDE <U>+-2SIGMA = 9
```

Figure 6. Normalized Velocity Histogram

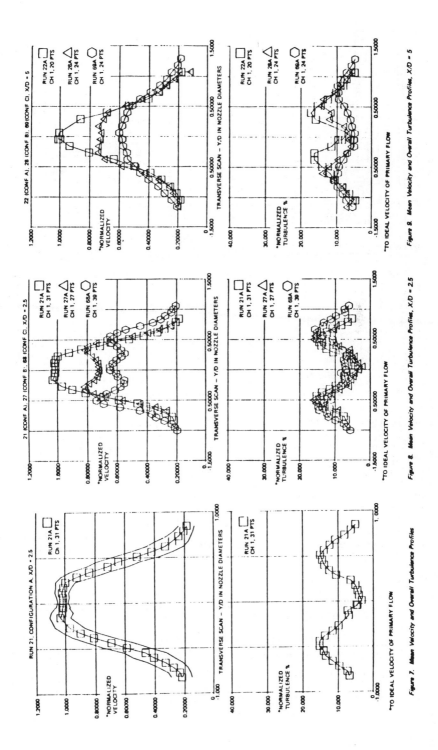

Figure 7. Mean Velocity and Overall Turbulence Profiles

Figure 8. Mean Velocity and Overall Turbulence Profiles, X/D = 2.5

Figure 9. Mean Velocity and Overall Turbulence Profiles, X/D = 5

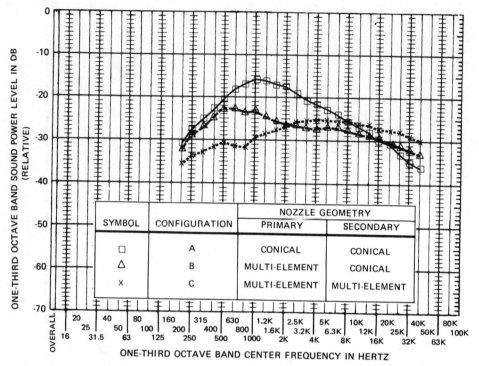

Figure 10.   Comparisons of Relative Noise Power Spectra

the source regions of noise in different frequency bands.

It should be mentioned that the NTL/LDV represents only one laser system which is being developed for a specific flow study application. Other laser systems being developed by Boeing include a flow visualization arrangement based on laser holography, and a two-color, backscatter scanning system for wind tunnel studies.

Studies are also being performed on laser signal processing based on recurrence rate correlation of event arrival times and photon correlation using a recently-acquired Malvern correlator. This latter work shows promise of ultimately providing a capability for defining turbulence spectra using single-beam, low power lasers to investigate flows at substantial distances and without any requirement for artificial flow seeding.

References

1.   Descriptive Brochure, Noise Technology Laboratory, Boeing Commercial Airplane Company, June 1977.

2.   The Accuracy of Flow Measurements by Laser Doppler Methods. Proceedings of the LDA-Symposium, Copenhagen, 1975.

3.   McGehee, B.L., "A Test Facility for Aircraft Jet Noise Reduction", presented at Annual Conference of the Institute of Environmental Sciences, Philadelphia, Penn., April 25-29, 1976.

The Use of LDV in Two-Phase Bubbly Pipe Flow

J. P. SULLIVAN and T. G. THEOFANOUS
Purdue University
West Lafayette, IN

ABSTRACT

Measurements of the liquid phase velocity and turbulent intensity in two-phase air-water pipe flow are presented.

INTRODUCTION

The bubbly flow regime of two-phase flow is characterized by a suspension of discrete bubbles in a continuous liquid. The flow of an air-water bubbly mixture in a vertical pipe was chosen to study the turbulence properties of a two-phase flow.

Most two-phase flows are highly turbulent. Turbulence directly affects heat and mass transfer rates at solid-fluid and fluid-fluid interfaces and therefore must be taken into account in physically realistic phase-interaction models. However, experimental data on the turbulent aspects of two phase flows is extremely limited. In the case of bubbly, vertical air-water flows, Sarazawa's (1) data using hot film anemometry is the most extensive. Several investigations of two-phase flows using laser Doppler velocimetry have been presented in the literature (2,3). The laser techniques have been used for measuring the velocity of the liquid phase and bubble rise velocity. The present work is aimed at obtaining the mean velocities and turbulent intensities of the liquid phase.

EXPERIMENTAL SETUP

The two phase flow system consists of a vertical glass pipe 57 mm inside diameter. Nitrogen is introduced in the plenum through ten 0.33 mm diameter holes. Tap water is used as the source, since the single pass arrangement avoids vibration and gas separation problems. The measuring station is 24 diameters downstream of the entrance to assure fully-developed flow.

The laser Doppler velocimeter (LDV) employed is a conventional dual scatter-or-fringe type of system. Rotation of a beamsplitter in the system, rotates the fringe pattern making the system sensitive to the axial component of velocity or the component ±45° to the axial. The angle between the two transmitted beams is 15.9°, which, along with beam expansion techniques, gives a probe volume size of approximately 0.05 mm in diameter and 0.25 mm long.

A TSI counter processor (MODEL 1990) is used to convert the LDV signal from the photomultiplier tube to an analog signal proportional to velocity. The analog signal is displayed on an oscilloscope and is input to a minicomputer for digital processing.

391

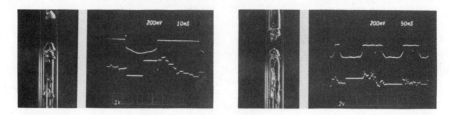

Figure 1.   Examples of LDV Data

DATA VALIDATION

Three experiments for checking system operation and data accuracy were performed. The first provided data for comparison with the classical single-phase turbulent pipe flow data of Laufer   (4). As Figures 5 and 6 demonstrate, agreement is excellent. The second and third experiments were aimed at assuring that light scattered from bubble interfaces is discriminated against and thus not interpreted as a valid liquid velocity measurement.

When a bubble passes through the laser beams, large amounts of light are scattered, reflected, and refracted. It is therefore necessary to set the discriminator level of the LDV processor correctly so that the light scattered by bubbles is not interpreted as liquid phase velocity.

To check the discrimination, single bubbles were introduced into a small tube containing a water flow. Figure 1 shows an example of velocity traces for this flow. The upper oscilloscope trace indicates the presence of bubbles when the signal is low. The lower trace is the analog output of the LDV processor. The constant level of the analog output when bubbles are in the probe volume in-dicates that light scattered from the bubble interfaces is not interpreted as

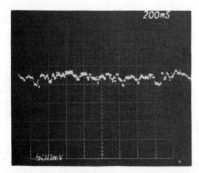

Figure 2a.
Velocity Trace Turbulent Pipe Flow
U = 46 cm/sec,   u' = .057 U
(Velocity Scale: 14 cm/sec/div)
  (Time Scale: 200 msec/div)

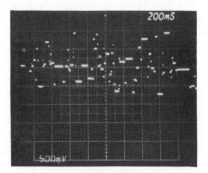

Figure 2b.
Velocity Trace at Center of
Bubbly Pipe Flow
X = 9.3 x 10^{-5}, U = 55 cm/sec,
u' = .212 U
(Velocity Scale: 14 cm/sec/div)
(Time Scale: 200 msec/div)

valid liquid velocity data. While the bubble is in the probe volume the large amount of scattered light causes an occassional data point to be accepted. However, as can be seen in Figure 1, this occurrence is insignificant compared to the data rate when the bubble is not present.

Examples of the data for bubbly flow in the pipe are shown in Figure 2. Figure 2a shows the turbulent pipe flow and Figure 2b the flow after bubbles are added.

For all cases presented the large particle discrimination level of the TSI Model 1990 processor was set to eliminate the maximum number of large signals, the comparison check was set at 1.0%, and the gain set to obtain a data rate of approximately 100 samples/sec.

The mean velocities and turbulent intensities in the axial direction and in directions ±45° to the mean were measured. Biasing corrections were not applied to this data. From the data the mean velocity and axial and tangential intensity distribution were calculated.

A strobe photograph of the bubbly flow region of interest is shown in Figure 3. Mean velocity profiles for the two-phase conditions are compared to the single phase velocity profile in Figure 4. The profiles are similar in the central portion of the flow ($\sim$ 70% radius). However, significant departures are observed near the wall. These departures may be attributed to the radial phase separation well known for low quality vertical bubbly flows. Presumably the higher void fraction observed near the wall "drives" the liquid in a chimney-like effect.

Figure 3. Strobe Photograph of
Bubbly Pipe Flow.

$X = 9.3 \times 10^{-5}$

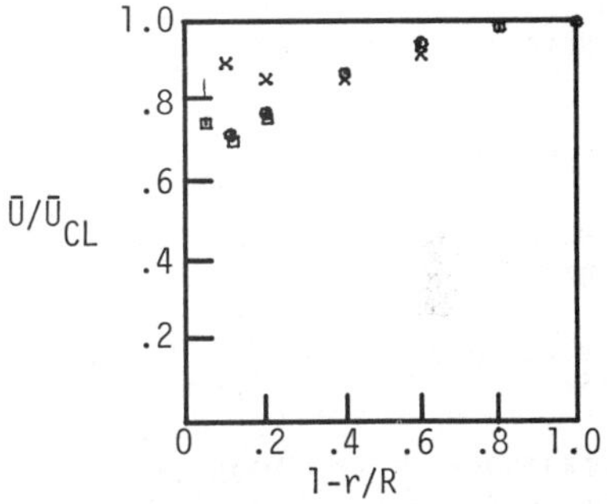

Figure 4. Mean Axial Velocity Distri-
bution.
⊘ Single Phase U = 27 cm/sec, X = 0
✕ Two Phase $\overline{U}$ = 25 cm/sec, X = 3.69×10^{-5}
⊡ Two Phase $\overline{U}$ = 56 cm/sec, X = 9.3×10^{-5}

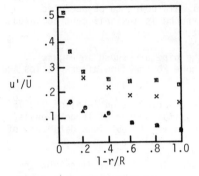

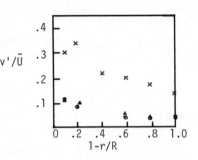

Figure 5. Axial Turbulent
Intensity Distribution

Figure 6. Tangential Turbulent
Intensity Distribution

▲ Laufer's Data
⊙ Single Phase Ū=27 cm/sec, X=0
✕ Two Phase Ū=25 cm/sec, X=3.69x10⁻⁵
⊞ Two Phase Ū=56 cm/sec, X=9.3x10⁻⁵

RESULTS

The radial variation of the axial intensity, u', is shown for two flow conditions in Figure 5. As expected, the relative intensity increases near the wall,
and the effect of the quality is seen to diminish as the wall shear controlled
turbulence region is approached. Figure 6 shows the first available data on tangential (θ-direction) intensity. These data follow the general characteristics
of the axial intensity and indicate a substantially isotropic behavior. The
axial intensities from the LDV data are approximately twice the values obtained
by Sarazawa using a hot film. The reason for this discrepancy is unknown. Additional data and discussions are contained in Reference 5.

REFERENCES

1.   Sarazawa, A. (1974), "Fluid-Dynamic Characteristics of Two-Phase Flow,"
     Kyoto Univ.

2.   Davies, W.E.R. (1973), "Velocity Measurements in Bubbly Two-Phase Flows
     Using Laser Doppler Anemometry," (Part I & II) UTIAS Tech. Note #184.

3.   Mahalingam, R., Limaye, R.S. and Brink, Jr., J.A. (1976), "Velocity Mea-
     surements in Two-Phase Bubble-Flow Regime With Laser-Doppler Anemometry,"
     AICHE Journal, Vol. 22, No. 6.

4.   Laufer, J. (1954), "The Structure of Turbulence in Fully Developed Pipe
     Flow," NACA Report 1174.

5.   Sullivan, J.P., Houze, R.N., Buenger, D.E. and Theofanous, T.G. (1978),
     "Turbulence in Two-Phase Flows," Symposium on Mechanism of Two-Phase Flow,
     Paris.

SESSION VIII
PARTICLE DIAGNOSTICS I

Chairman: **A. M. MELLOR**
Purdue University

An Optical Particle-sizing Counter for In-Situ Measurements

DON HOLVE and SIDNEY A. SELF
High Temperature Gasdynamics Laboratory
Mechanical Engineering Department
Stanford University
Stanford, CA

ABSTRACT

A particle sizing counter is described, suitable for in-situ measurements in two phase flows of laboratory scale. It employs near-forward scatter from the focus of a He-Ne laser beam, together with pulse height-analysis of the signals from individual particles. A novel and essential feature of the technique is a numerical inversion scheme to unfold the dependence of the scattered signals on particle trajectory through the measurement volume. The inversion procedure is performed by an on-line computer, and utilizes a prior calibration with monodisperse aerosols of known size. As presently configured, the instrument has a demonstrated capability of determining size distributions in the diameter range 2-25 μm, at concentrations up to ~10^5 cm^{-3}. The measured dependence of response on particle diameter agrees well with calculations from Mie scattering theory. It is anticipated that the technique can be extended to cover particle diameters up to at least 50 μm, and down to 0.5 μm and concentrations up to 10^6 cm^{-3}. It should also be adaptable to hot flows and absorbing, irregular particles.

INTRODUCTION

The measurement of particle size distributions in two phase flows is of considerable current interest, especially in connection with energy conversion devices such as liquid spray and pulverized fuel combustors, and particulate cleanup devices such as electrostatic precipitators. While the particulate characteristics, such as mean diameter, size distribution, mass loading, etc., may vary quite widely in such systems, the sizes of interest generally fall in the range 0.5-50 μm, with concentrations up to 10^6 cm^{-3} (for the smaller sizes).

The available measurement techniques [1,2] also vary widely in type and capability. Several, including microscopy, cascade impactors, Coulter Counters, mobility analyzers and commercial optical counters require a sample to be extracted from the flow. This poses problems related to obtaining a representative sample, especially for the larger sizes and in hot, high velocity flows. Also, many sampling methods are cumbersome and slow in operation. For these reasons optical techniques are of especial interest since, in principle, they are capable of making in-situ measurement with continuous and rapid readout. Moreover, by using lasers, they are adaptable to high temperature systems having high thermal radiation background.

397

Optical techniques all depend on Mie scattering [3,4], and can be broadly divided into imaging and non-imaging types. The former, including flash photography [5] and holography [6], are limited in practice to size $\gtrsim 10$ μm, and pose a difficult data reduction problem. Non-imaging methods can be subdivided into two classes: those which measure on a large number of particles simultaneously [7,8,9], and those which count and size individual particles, one at a time [10,11]. In this paper we describe the development of an in-situ, forward scatter laser particle sizing counter following the last approach. (Optical counter-sizers for gas flows using the forward scatter technique have been commercially available for some years but are limited by the need for sample collection and dilution.) The present technique has a demonstrated in-situ capability on cold, small scale, low velocity flows of sizing particles in the range 2-25 μm at concentrations up to 10^5 cm^{-3} [12]. Recent measurements on a small acetylene-air burner also demonstrate the applicability of the technique to particle-laden combustion systems. In its present form the instrument will accommodate flows up to 40 cm in width. By adjusting the instrumental parameters to suit the particle characteristics, it should be possible to extend the range downwards to ~ 0.5 μm and upwards to $\gtrsim$ 50 μm, and handle concentrations to 10^6 cm^{-3} for the smaller sizes.

MEASUREMENT VOLUME CONSIDERATIONS

The characteristics of the measurement volume are a crucial factor in the design of an in-situ optical sizing counter because the particle trajectories are not controlled as they are in a sampling-type optical counter. The effective measurement volume is determined in part by the intensity distribution in the illuminating beam, and in part by the geometry of the collection optics, including stops and apertures. For a given measurement volume there will clearly be a limit of useful operation set by the occurrence of coincidences when two particles are simultaneously present in the volume, giving rise to erroneous signals. Roughly, it can be said that for an effective measurement volume of V_m cm^3, the maximum concentration that can be measured without significant interference due to coincidence is $N_{max} \sim V_m^{-1}$ cm^{-3}. An exact treatment of coincidence effects is given in Appendix I of Ref. 12.

In the particular system investigated (Figure 1) the illumination is provided by a 1 mW He-Ne TEM$_{00}$ mode laser, which is focused by cylindrical lenses to a ribbon beam with waist widths (to $1/e^2$ intensity) of approximately 2 $w_x \sim 100$ μm x 2 $w_y \sim 300$ μm, with a waist length greater than 1 cm. The collection axis is in the x-z plane at an angle θ_c to the z-axis. The collection lens, consisting of back-to-back f/2.8, 25 cm focal length lenses, designed for infinite conjugate ratio, images the center of the beam waist to the center of a pinhole aperture of diameter 2 $w_A \sim 100$ μm diameter, with

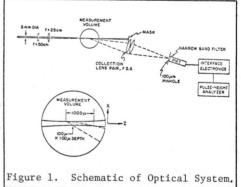

Figure 1.   Schematic of Optical System.

unity magnification. Thus the viewed volume is approximately a cylinder of diameter 100 μm intersecting the ribbon beam at an angle θ_c, while the measurement volume is given approximately by the intersection of the viewed volume and the illuminating beam. From geometrical considerations the measurement volume can be characterized roughly as a cylinder with slant ends, of diameter 2 w_A ~ 100 μm, length 2 $w_x/\sin \theta_c$ ~ 1000 μm for angles of interest, and volume V_m ~ 10^{-5} cm^{-3}. The particle flow is directed in the y-direction.

The scattered signal is trajectory dependent, and a flow of monodisperse particles yields a signal peak amplitude count distribution having a sharp cutoff at some maximum signal value, corresponding to particles traversing the center of the measurement volume, with a spread to smaller amplitudes.

The overcome this problem of nonuniformity of response within the measurement volume, a numerical inversion scheme, combined with a calibration procedure has been devised(see Inversion Technique section)to unfold the distribution of signal amplitudes and yield an indicated size distribution which eliminates the dependence on trajectory.

OPTIMAL COLLECTION GEOMETRY: SCATTERING THEORY

A necessary requirement for an effective optical sizing counter is a monotonically increasing dependence of signal amplitude on particle size, and minimum sensitivity to the complex refractive index $\tilde{n} = n_1 + in_2$ of the particle (where the imaginary part represents absorption).

Response calculations have been carried out for the collection geometry illustrated in Figure 1 using a computer code developed by Davé to generate the Mie coefficients [13].   Figure 2 shows the response characteristic of the collec-

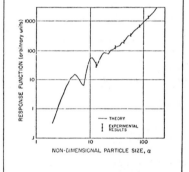

tion geometry (4°-20°) chosen for experimental study.   Resonances at $\alpha = 5$ and 11 limit the size resolution for $\alpha = 3 - 14$ but no more so than do typical cascade impactors over the entire size range.   Experimental calibrations (✦) show good agreement with the calculated results.

Calculations performed for a range of refractive indices and collection geometries show that for the larger particle sizes, small scattering angles should be used to minimize sensitivity to $\tilde{n}$,   while for the smaller sizes, larger scattering angles should be used to minimize the measurement volume and hence raise the maximum concentration that can be handled.

Figure 2.   Experimental Results for Oleic Acid Particles Compared with Theoretical Response Function F(d).

TRAJECTORY DEPENDENCE--INVERSION TECHNIQUE

In devising an in-situ particle-sizing counter, one has to face the problem that the peak value of the detected signal, which is used to size the particles, is a function not only of particle size but also of its trajectory through the measurement volume.   The principal factor controlling this trajectory dependence is the distribution of intensity $J(x,z,d)$ shown in Figure 3.   It is assumed that particles have an equal probability of passing through any element of the cross-section, and that all particles have the same mean velocity U,   irrespective of size trajectory.

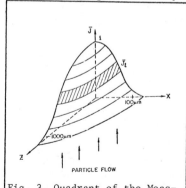

Fig. 3. Quadrant of the Measurement Vol. Function J(x,z,d)

The problem, then, is to devise a technique to unfold the measured signal amplitude count distribution in the presence of a nonuniform measurement volume function $J(x,z,d)$ to yield that which would be obtained with a uniform function, and hence to obtain the true aerosol size distribution, $N(d)$. The inversion method is described briefly in the following discussion. A complete derivation of the method is given in Ref. 12. A set of equations relating experimental pulse height analyzer data $C_i(A_i)$, measurement volume function $J(x,z,d)$, and the desired number distribution function $N(d)$ can be written as follows:

$$
\begin{vmatrix} c_1 \\ c_2 \\ c_3 \\ \vdots \\ c_i \\ \vdots \\ c_m \end{vmatrix} = U \cdot \begin{vmatrix} \Delta S_{11} & \Delta S_{12} & \Delta S_{13} & \text{---} & \Delta S_{1i} & \text{---} & \Delta S_{1m} \\ 0 & \Delta S_{22} & \Delta S_{23} & \text{---} & \Delta S_{2i} & \text{---} & \Delta S_{2m} \\ 0 & 0 & \Delta S_{33} & \text{---} & \Delta S_{3i} & \text{---} & \Delta S_{3m} \\ \vdots & \vdots & \vdots & & \vdots & & \vdots \\ 0 & 0 & 0 & \text{---} & \Delta S_{ii} & \text{---} & \Delta S_{im} \\ \vdots & \vdots & \vdots & & \vdots & & \vdots \\ 0 & 0 & 0 & \text{---} & 0 & \text{---} & \Delta S_{mm} \end{vmatrix} \begin{vmatrix} N_1 \\ N_2 \\ N_3 \\ \vdots \\ N_i \\ \vdots \\ N_m \end{vmatrix} \qquad (1)
$$

where C_i = signal count rate for normalized signal peak amplitudes in a logarithmically-scaled amplitude range, ΔA_i.

N_j = concentration of particles in the logarithmically-scaled size
parameter range, Δd_j.

ΔS_{ij} = cross-sectional area of the measurement volume (Fig. 3) normal to
the flow direction, which yields normalized signal peak amplitudes
in the range, ΔA_i, for particles in the size range, Δd_j.

The inversion of Eqn. (1) is written symbolically

$$
\underline{N} = \frac{1}{U} \underline{\underline{\Delta S}}^{-1} \cdot \underline{C} \qquad (2)
$$

Given the count rate distribution of signal amplitudes $C_i(A_i)$ in the pulse
height analyzer channels, then if the $\underline{\underline{\Delta S}}$ matrix is known, Eqn. (2) can be
solved to yield the number distribution $N_j(d_j)$. The central problem of this
approach is the determination of the cross-section matrix $\underline{\underline{\Delta S}}$.

The $\underline{\underline{\Delta S}}$ matrix is obtained by passing a monodisperse particle distribution
$N(d)$ of known diameter d_m and concentration N_m through the measurement volume at
a known mean velocity $\overline{U}$. For a specified time the signals are accumulated in
the pulse-height analyzer to yield the count rate distribution C_i. Since
$N(d_j) = 0$ for all j except $j = m$, the system of equations (1) reduces to a
set of m equations of the form

$$
C_i = U \,\Delta S_{im} N_m \, , \qquad (3)
$$

which can be solved directly for the ΔS_{im}. This yields the m elements in
the mth column of the matrix. When the amplitude response of the pulse-height
analyzer is logarithmic, it can be shown (Appendix II, Ref. 12) that the $\underline{\underline{\Delta S}}$
matrix simplifies so that all the elements on any diagonal are equal. Thus
knowledge of ΔS_{im}, $i = 1,m$ completely determines $\underline{\underline{\Delta S}}$.

Once the ΔS_{ij} matrix elements are determined from such a calibration, they
are entered in the matrix inversion algorithm of a mini-computer. Then, for an
aerosol under investigation, the count rate distribution C_i from various
channels of the pulse height analyzer are automatically entered into the
computer.

Because of the experimental uncertainty in determining $\underline{C}$ and $\underline{\underline{\Delta S}}$, the
problem $\underline{N} = (U \,\underline{\underline{\Delta S}})^{-1} \underline{C}$ cannot be accurately solved by straightforward matrix
inversion techniques. Instead, the solution $\underline{N}$ is derived by searching for a
non-negative vector $\underline{N}$ which minimizes the residual vector $R = \underline{C} - U \,\underline{\underline{\Delta S}} \,\underline{N}$. A
satisfactory solution to this problem was found through use of a non-negative
least squares (NNLS) solution procedure developed by Lawson and Hanson [14].
This procedure minimizes error propagation and also outputs values of residuals

R_j which allows one to judge the accuracy of the resulting N_j distribution. Data manipulation and display can be accomplished in less than 30 seconds on an HP 2100 mini-computer.

EXPERIMENTAL RESULTS

The primary effectiveness of an in-situ device is realized for measurements in hot flow systems. For hot systems a monochromatic illumination source coupled with spectral filtering of the scattered signal is essential to eliminate thermal background. Estimates show that background radiation is negligible for particles larger than 0.3 μm for temperatures up to 2600 K using a 10 Å narrow band filter. Measurements were performed on a small burner system. A Perkin-Elmer slot burner 4 cm long by 0.15 cm wide was mounted below the optical meas-

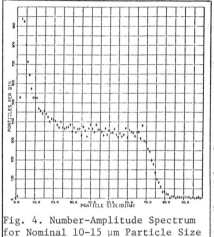

urement volume with the long dimension along the beam axis. Soda-Lime glass beads of various diameters were batch fed into a simple venturi nozzle with attached vibrator and carried by an auxilliary air stream and mixed with the methane-air flow prior to exit through the slot. Measurements were taken 2 cm above the burner slot where, under "hot" conditions, the flame temperature was measured to be 1600 K.

Comparison measurements were taken by obtaining sequential scattering spectra (e.g. Fig. 4) of the same particle size distribution in both cold and hot flows. Because of jet spread and batch feeding of the particles, the number density at the measurement volume was unknown, so that comparison data are presented in terms of particle size vs. cumulative percentage on standard log-probability plots.

Fig. 4. Number-Amplitude Spectrum for Nominal 10-15 μm Particle Size Distribution.

Figure 5 gives results for a 10-15 μm particle size distribution in both cold and hot flows. No significant difference occurs between the measurements under cold or hot conditions. An independent size distribution measurement of the particles prior to injection was obtained by use of a Coulter counter and

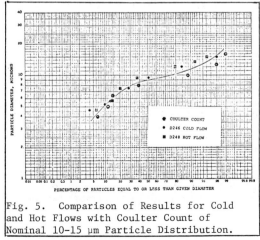

shows good agreement with our optical size measurements, giving a maximum difference of 20%. The 50% cumulative percentage (CPCT) value or d_{50} is within 10%.

A broad distribution is a more rigorous test of the inversion technique since the count rate in lower amplitude channels due to small particles at low concentrations becomes increasingly swamped by contributions from larger particles. Figure 6 shows the results for a nominal 1-30 μm size distribution which is approximately log-normal (solid line) according to the particle supplier. Agreement is good up to the d_{50} point above which optically sized results

Fig. 5. Comparison of Results for Cold and Hot Flows with Coulter Count of Nominal 10-15 μm Particle Distribution.

indicate a smaller percentage of
particles in the larger size range.
It was observed that some particles accumulated in the burner
plenum indicating that larger
particles were not all following
the flow. Thus one would expect
that the optical measurements in
the flow should qualitatively fall
below the static sample distribution results for larger particles.

CONCLUSIONS

The in-situ particle sizing
counter described herein satisfies
the principal practical requirements of such an instrument for
many important applications. The
optical access requirements are
modest and can be readily satis-

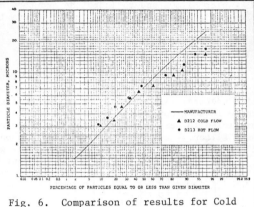

Fig. 6. Comparison of results for Cold
and Hot Flows for Nominal 1-30 μm Particle
Distribution. Solid Line in Distribution
Supplied by Particle Manufacturer.

fied for laboratory scale flow systems. Particles can be sized over at least a
10:1 diameter range with a resolution $\Delta d/d=0.15$, which is adequate for most
aerosols of interest. The instrument has been conclusively demonstrated[12]as
capable of resolving the structure of a polydispersion in the range 1.6-25 μm.
Results in both hot and cold flows are self-consistent within 10% indicating no
adverse problems with thermal background or beam bending in small scale flame
systems. The in-situ measurements compare well with independent pre-injection
sizing of the test particle size distributions.

By adjusting the instrumental geometry, it should be straightforward to
extend the size range down to 0.5 μm, although two separate light collection
geometries will be necessary to cover this entire range. Another important
feature is that the measurement volume can be made small enough to allow measurements with high particle concentrations, in the range of interest for many
applications.

The key feature of the method is the use of a computer-based inversion
scheme which allows one to unfold the effect of the random trajectory dependence of the scattered signals together with a straightforward calibration
technique which gives the necessary information on the scattering function
F(d) and the matrix elemental areas ΔS_{ij} required for the inversion.

Although the experimental work was based on the use of non-absoring
spherical droplets, the method should also be applicable to absorbing and
irregular particles. For absorbing spherical particles of known refractive
index, Mie scattering computations can be used as a guide, but if the refractive
index is unknown, or the particles are irregular, a relatively simple calibration technique [15,16] can be used to obtain the response function F(d).

ACKNOWLEDGEMENTS

Our thanks are due to Li Shing Cheng who performed the numerical calculations, and to Philip Krug for his enthusiastic work on the electronics and software. Informative discussions with Al Lieberman and Art Bates of Royco
Instruments and Takashi Nakamura of the High Temperature Gasdynamics Laboratory
are gratefully acknowledged.

REFERENCES

1. Cadle, R. D., The Measurement of Airborn Particles, Wiley & Sons, New York, 1975.

2. Allen, T., Particle Size Measurement, Halsted Press, 1975.

3. Van De Hulst, H. C., Light Scattering by Small Particles, John Wiley ¢ Sons, 1957.

4. Kerker, M., The Scattering of Light and Other Electromagnetic Radiation, Academic Press, 1969.

5. McCreath, C., Roett, M., Chigier, N., "A Technique for Measurement of Velocities and Size of Particles in Flames," J. Phys. E., 5, 601 (1972).

6. Belz, R., Dougherty, N., "In-Line Holography of Reacting Liquid Sprays," Proc. Engin. Applications of Holography, Feb. 1972, Los Angeles, Calif.

7. Hodkinson, J., "The Optical Measurement of Aerosols," Ch X of Aerosol Science, ed. C. Davies, Academic Press, 1966.

8. Holve, D., Self, S., "Optical Measurements of Mean Particle Size in the Exhaust of a Coal-Fired MHD Generator," Fall, 1976 meeting of Western States Section of the Combustion Institute, La Jolla, Calif.

9. Dobbins, R., Crocco, L., Glassman, I., "Measurement of Mean Particle Sizes of Sprays from Diffractively Scattered Light," AIAA Journal, Vol. 1, #8, Aug. 1963, pp. 1882-1186.

10. Farmer, W. M., "Measurement of Particle Size, Number Density, and Velocity Using a Laser Interferometer," Appl. Opt., 11, 2603, 1972.

11. Hirleman, E., Wittig, S. "In Situ Optical Measurement of Automobile Exhaust Gas Particulate Size Distributions: Regular Fuel and Methanol Mixtures," 16th Symposium (International) on Combustion, MIT, August 1976.

12. Holve, D., Self, S., "An Optical Particle-Sizing Counter for In-Situ Measurements," Technical Report for Project SQUID, subcontract 8960-7, December, 1977.

13. Dave, J. V., "Subroutines for Computing the Parameters of the Electromagnetic Radiation Scattered by a Sphere," IBM Palo Alto Scientific Center Report No. 320-3237, May 1968.

14. Lawson, C., Hanson, R., Solving Least Squares Problems, Prentice Hall, Inc., N.J., 1974.

15. Marple, V., Rubow, K., "Aerodynamic Particle Size Calibration of Optical Particle Counters," J. Aerosol Sci., Vol. 7, p. 425, 1976.

16. Marple, V., Rubow, K., "A Portable Optical Particle Counter System for Measuring Dust Aerosols," paper submitted to American Industrial Hygiene Association Journal, December 1976.

Simultaneous Particle Size and Velocity Measurements in Postflame Gases

SIGMAR L. K. WITTIG and KHALED SAKBANI
Institut für Thermische Strömungsmaschinen
Universität Karlsruhe (TH)
Postfach 6380
D-7500 Karlsruhe 1, West Germany

ABSTRACT

An optical arrangement is described for simultaneous particle size and velocity measurements using the Laser-Two-Focus (L-2-F) technique in combination with the Multiple-Ratio-Single-Particle Counter (MRSPC). Preliminary measurements behind sooting butane flames and comparison with conventional velocity probing and impactor-type particle counters illustrate the advantages of the technique.

INTRODUCTION

The simultaneous determination of particle size distribution, concentration and velocity is of primary importance in a wide variety of practical applications. Our own interest arises from the problems found in turbomachinery environments - specifically in gas and steam turbines. Various aspects of flow measurements in low pressure end steam turbines with relatively high Mach numbers (transonic) flow and spontaneous condensation have been discussed by Wittig et al. [1]. Also, problems associated with flow characterization in and behind internal combustion engines such as Otto engines and gas turbines have been analyzed by Hirleman and Wittig [2]. Other questions arising from new areas of application are mentioned in various papers of this workshop (see for example W.D.Bachalo [3]).

There are numerous methods for particle sizing as well as velocity measurements. Broadly, they may be devided into two categories:

1. Interferring techniques such as pitot-tube measurements and sampling by extraction

2. In-situ, non-interferring techniques generally with real time characteristics. Laser velocimeters and light scattering techniques are typical examples.

In recent years, the optical methods have been of increasing importance to flow analysis in general and specifically in turbomachinery. Stevenson [4, 5] has summarized the principles in his recent reviews of the various techniques.

As indicated earlier, our general objective is the detailed analysis based on simultaneous measurements of particle size and velocity within and behind gas turbine combustors and in low pressure end steam turbines. In this preliminary study, we therefore attempted to achieve the following goals:

i.   to develop and calibrate an appropriate system and to check on its limitations

ii.   to obtain in a first test actual measurements in and behind flames

iii. to compare the results with conventional techniques.

APPARATUS

The Multiple-Ratio-Single-Particle Counter (MRSPC) has been shown in our previous investigations to be well suited for measurements in combustion and two phase environments as described in detail by Hirleman and Wittig [2, 6]. Based on Hodkinson's suggestion [7] Gravatt [8] utilized the ratio of light intensities scattered from individual particles. As can be seen from the Mie-intensity calculations of Figure 1 and discussed in detail in references 2, 6, 7 and 8, the major advantages are found in the relatively small dependence on particle index of refractions as well as its surface and shape. Furthermore, errors introduced by the Gaussian intensity distribution of the illuminating laser beam are avoided. Ambiguities arising from highly polydisperse size distributions with a wide range of size parameter - i.e. diameter -

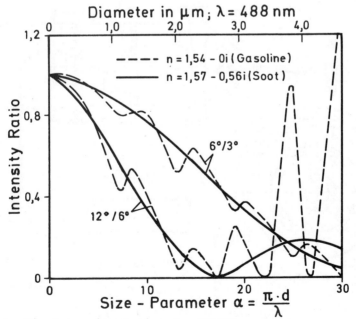

Figure 1: Mie-Intensity Ratios for MRSPC

can be eliminated by utilizing several intensity ratios as introduced by Hirleman and Wittig [6]. A schematic of the new counter
is shown in Figure 2. An in-depth analysis of various other effects and the uncertainties encountered is presented by Hirleman
and Wittig [6, 9].

The Laser-two-Focus (L-2-F) technique has been demonstrated
by Schodl [10] to be quite suitable for measurements in turbomachinery and especially in compressors. In the present study,
main flow characteristics were of primary interest. Although only a
few experimental investigations - compared for example, with
Laser-Doppler studies - have been reported using the L-2-F velocimeter, primarily because of intensity considerations it seemed
to be well suited for a combination with the MRSPC counter. The
schematic of Figure 3 and the photograph of Figure 4 illustrate
the first prototype setup with the particle counter axis offset
by 90°. On-axis design is possible as well. The spacing between
the two L-2-F beams was calculated and measured to 583 μm and the
beam diameter is 18 μm. The axial length of the probe volume
$(1/e^2 -$ intensity limit) is 4,6 mm and we were able to make measurements as close as 0,1 mm (parallel) to the wall of the flow
duct. The beam diameter of the MRSPC was 100 μm and the axial
length 8 mm.

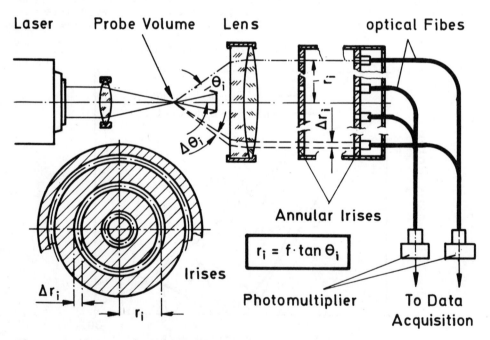

Figure 2: Schematic of MRSPC

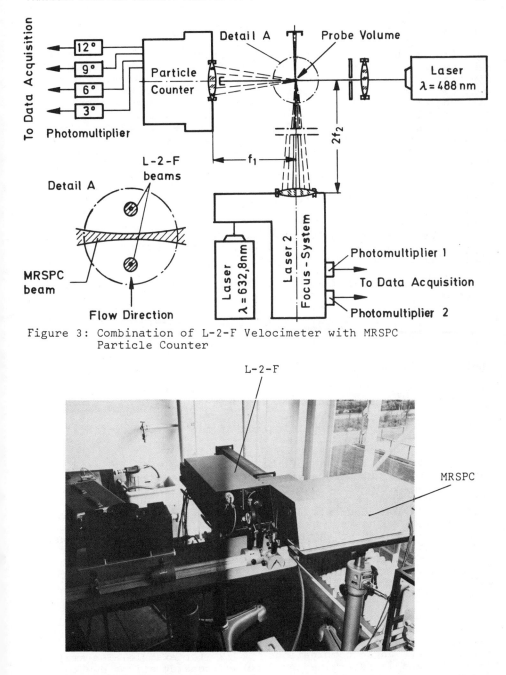

Figure 3: Combination of L-2-F Velocimeter with MRSPC
Particle Counter

Figure 4: Photograph of L-2-F Velocimeter and MRSPC Particle
Counter Combination

CALIBRATION AND EXPERIMENTAL RESULTS

For calibration and comparison purposes, cold flow measurements were made behind a burner exit of circular cross-section. Air from a central air supply system was used as carrier gas. Polystyrene spheres suspended in water were "atomized" by a Collison-type atomizer with subsequent drying of the air. Alternatively, the water droplets were used for seeding. Figure 5 demonstrates the signal quality from the L-2-F velocimeter. Mapping the burner exit flow, the profile of Figure 6 is obtained. The comparison of the optically determined velocities with small Pitot tube measurements revealed an agreement of the two techniques within 10 %. The errors can largely be attributed to the disturbance induced by the relatively large Pitot tube (1 mm in diameter). Note from Figure 6 that the burner diameter was approximately 6 mm.

Extreme care was taken in calibrating the MRSPC and in adjusting the proper signal levels i.e. system constants from the photomultipliers.

The signal ratio from the photomultipliers at two angles is given by [6]

$$\frac{S_2}{S_1} = \left[\frac{V_{D2}V_{P2}\sin\theta_2\Delta\theta_2}{V_{D1}V_{P1}\sin\theta_1\Delta\theta_1}\right] \frac{i_1(\theta_2,\alpha,n) + i_2(\theta_2,\alpha,n)}{i_1(\theta_1,\alpha,n) + i_1(\theta_2,\alpha,n)}$$

where i_1 and i_2 are the Mie-intensity functions

and V_{Di} the optical losses

with V_{Pi} amplification of the photomultipliers

After mapping the size of the probe volume with a 5 μm diameter wire from a hot wire anemometer fixed to a x-y micrometer traverse, the dynamic range was tested by spinning the wire on a rotating wheel through the sensitive volume at various speeds. As with our previous counters (see [2] and [7]), detailed calibration and the final adjustment was accomplished using Dow-Latex spheres of 1.09 μm and 0.46 μm. Adjusting the amplifications V_{Pi} to such levels that the signal ratio for a known particle size yields a value identical to the theoretical value determined from Mie theory, all other particles should match their appropriate intensity ratios. This is illustrated in Figure 7 for the intensities recorded under 12° and 6°.

The 1.09 μm spheres served for adjusting the correct levels and subsequently the 0.46 μm particles were found to match the predicted values. Similarly, intensity ratios were found for various other angle combinations - $6^o/3^o$ for example. The smallest particle size is determined by accuracy considerations due to the insensitivity of the ratio to small particle size (see Figures 1 and 7) and to the signal to noise ratio. In the present study, particle sizes smaller than 0.2 μm were not measured. For particles with an unknown index of refraction, calibration must be

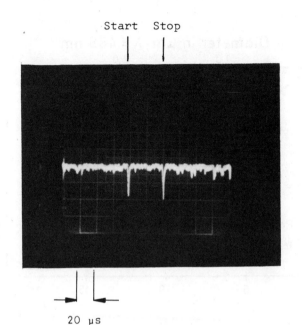

Figure 5: L-2-F Velocimeter Signals

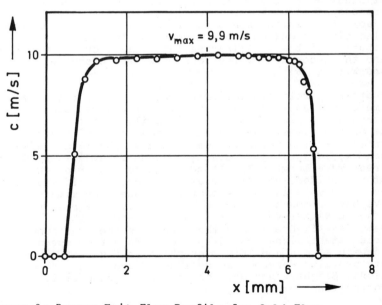

Figure 6: Burner Exit Flow Profile for Cold Flow

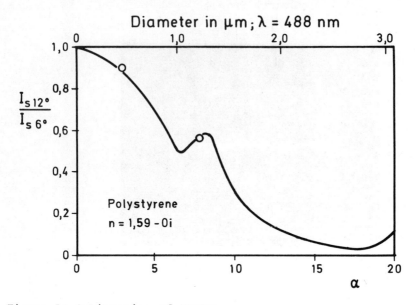

Figure 7: Calibration of MRSPC

achieved by a sequence of measurements at various diameters. In
the present study we were also to match the optical measurements
on polystyrene spheres with particle concentrations determined
utilizing isokinetic sampling and separation in an Andersen sampler. Detailed discussions on the influence of the Gaussian intensity distribution of the illuminating laser light on the particle
spectrum and the necessary corrections have been discussed by
Hirleman and Wittig [6].

    After obtaining reliable calibration data, the particle coun-
ter/velocimeter was applied to the hot gas flow of an open,
sooting butane flame. Problems arose primarily in the high temperature zone of the flame. The background radiation of the flame
strongly influenced the velocity measurements. The 15 mW He/Ne
laser chosen was too weak, a problem which can be avoided by use
of a different wave length. The 15 mW Argon-Ion laser of the particle counter provided accurate results. No optical difficulties,
however, were encountered in the postflame zone approximately
200 mm above burner mouth. Figure 8 shows an oscillogram indicating that the sensitive volume is located   between the two velocimeter beams (see Figure 3) and the signal is received from the
same particle (start signal not displayed). Main flow speeds were
recorded between 11 m/s and 16 m/s as shown in Figure 9. However,
the velocity profiles were not as well developed as under cold
gas conditions. Slightly unstable operation was characteristic
for the sooting flame.

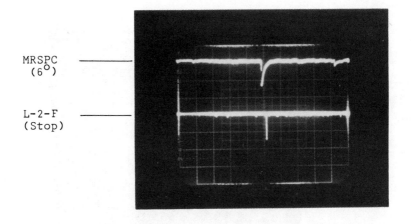

Figure 8: Simultaneous Velocity and Particle Size Determination

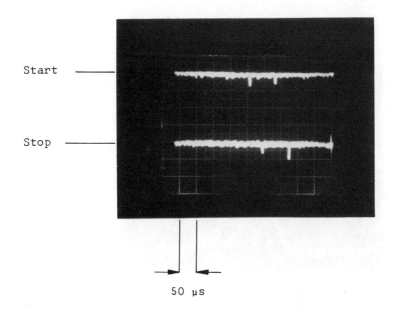

50 µs

Figure 9: Velocity in the Postflame Zone of a Butane Flame

6°
(100 mV/div)

12°
(20 mV/div)

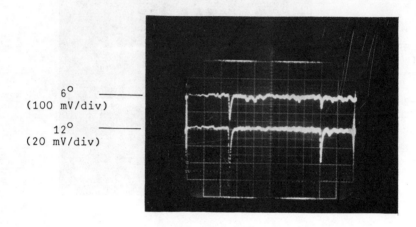

Figure 10: Particle Size in the Postflame Zone of a Butane Flame

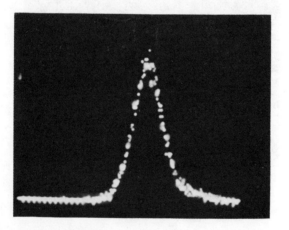

Figure 11: Typical Particle Distribution Displayed on
           Multi-Channel Analyzer

Particle - i.e. primarily soot - size determination did not pose any problems unless densities exceeded 5×10^4 particles/cm^3. This range, however, easily could be extended by reducing the size of the probe volume. Figure 10 displays the signal characteristic with Figure 11 showing the particle size distribution as recorded on a multi-channel analyzer.

The normalized size distribution with the MRSPC in-situ optical counter is shown in Figure 12 and compared with the distribution obtained by isokinetic sampling and Andersen sampler analysis. Two effects seem to be responsible for the difference in size distribution shown by the two instruments: As discussed earlier, the signal intensity from small particles frequently is below(see[6]) noise level of the MRSPC. Furthermore, the low flow speeds favour separation and fallout of particles with diameters exceeding 1 μm in the sampling system. The unsteady nature of the flow induced additional errors into the sampling procedure. It, therefore, can be expected that the actual maximum can be found between 0.7 μm and 1.1 μm. Despite the relatively broad stages of the Andersen sampler, both instruments show a rapid drop-off towards 2 μm and bimodal distributions were not observed, indicating that - as expected - the nucleation phase is closer to the high temperature zone. A detailed analysis, however, of the soot forming process was not attempted in this study.

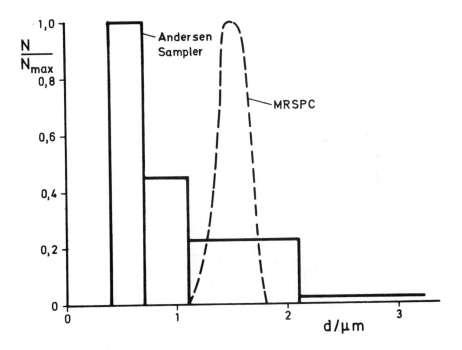

Figure 12: Comparison of Normalized Size Distribution of
          Postflame Soot Particles Determined with MRSPC
          and Andersen Sampler

CONCLUSIONS

In combining the Laser-two-Focus velocimeter with a single particle counter it was possible to achieve simultaneous in situ main flow velocity and particle size measurements in combustion environments. The technique shows major advantages in comparison with conventional probing and sampling techniques. Extreme care, however, must be take in signal processing. The ratio single particle counter may also be combined with other optical veloci- meters such as the Laser Doppler anemometer. Difficulties still exist with high particle density   flows. Attempts are presently under way to extend the range by at least an order of magnitude for application in condensing systems and specifically in steam turbine environments.

ACKNOWLEDGMENTS

Thanks are due to various students involved in the project. Messrs. F.Windschmitt, H.Schütz and F.Steinmann helped in designing the modified L-2-F velocimeter and took preliminary measurements. K.Eisele adopted the Andersen sampler and K.Schmidt assisted in redesigning the MRSPC-sampler.

REFERENCES

[1]   S.L.K.Wittig, W.H.Stevenson, D.K.Whirlow and W.A.Stewart
        "Laser Doppler Studies in the Westinghouse Wet Steam
        Cascade"
        Proceedings of the Second International Workshop on Laser
        Velocimetry, H.D.Thompson and W.H.Stevenson, Eds. Vol.II,
        p.182 (1974)

[2]   E.D.Hirleman and S.L.K.Wittig
        "In Situ Optical Measurements of Automobile Exhaust Gas
        Particulate Size Distributions: Regular Fuel and Methanol
        Mixtures"
        Sixteenth Symposium (International) on Combustion, The
        Combustion Institute, Pittsburgh, Pa., p.245 (1976)

[3]   W.D.Bachalo
        "On-Line Particle Diagnostics Systems for Application in
        Hostile Environments"
        see this workshop

[4]   W.H.Stevenson
        "Principles of Laser Velocimetry"
        Progress in Astronautics, Vol.53, Experimental Diagnostics
        in Gas Phase Combustion Systems, B.T.Zinn, Ed., p.307 (1977)

[5]   W.H.Stevenson
        "Spray and Particulate Diagnostics in Combustion Systems:
        A Review of Optical Methods" Central States Section Meeting,
        The Combustion Institute, April 3-4 (1978)

[6]   E.D.Hirleman and S.L.K.Wittig
        "Laser Application in Particulate Analysis: The Multiple
        Ratio Single Particle Counter" in Laser 77 Opto-Electronics,
        W.Waidelich, Ed., ipc-press, p.740 (1977)

[7]   J.R.Hodkinson
        "Particle Sizing by Means of the Forward Scattering Lobe"
        Applied Optics, Vol.5, No.5, p.839 (1966)

[8]   C.C.Gravatt
        "Real Time Measurements of the Size Distribution of Parti-
        culate Matter by a Light Scattering Method"
        J.Air Poll.Cntrl.Ass. 23, p.1035 (1973)

[9]   E.D.Hirleman and S.L.K.Wittig
        "Uncertainties in Particle Size Distributions Measured with
        a Ratio-type Single Particle Counter",
        Proceedings, Optical Society of America Conference on Laser
        and Electro-optical Systems, San Diego, California,
        May 1976

[10]   R.Schodl
        "A Laser Dual-Beam Method for Flow Measurements in Turbo-
        machines"
        ASME paper 74-GT-157 (1974)

Particle Sizing in Flames with Laser Velocimeters

N. A. CHIGIER, A. UNGUT, and A. J. YULE
Department of Chemical Engineering and Fuel Technology
University of Sheffield
Sheffield, England S13 JD

ABSTRACT

The measurement of particle diameters by a single particle counting LDA system is described with the objective of developing a technique for simultaneous particle velocity and size measurement in burning fuel sprays. The measured distribution of the peak values of the low pass filtered LDA signals is converted into the particle size distribution by applying equations describing the light scattered by the particles as a function of particle size, composition and position in the measurement control volume. The selection of optical arrangements to optimise instrument performance is described, and measurements are made for different arrangements. Test experiments using single particles, monosize particle streams and cold sprays indicate good agreement with results from other sizing techniques. Results are presented for a kerosene spray under burning and cold conditions. The problems and future developments of the technique are discussed.

NOMENCLATURE

A	$=$	$(b_1^2 k^2 w^2)^{-1}$
b_1, b_2, b_3	$=$	half widths of MCV based on $1/e^2$ points in x,y and z directions
d	$=$	particle diameter (see Eq. 6. etc. for subscripts).
$\bar{d}$	$=$	Rosin Rammler mean diameter
d_1	$=$	width of field determined by collection optics
E_G	$=$	gate level (light power units)
E_P	$=$	peak of mean signal (collected light power)
E_{P_o}	$=$	E_p value for particle in centre of MCV
f_1	$=$	focal length of focusing lens
I	$=$	light intensity
I_{max}	$=$	peak intensity at focus of one beam
I_{SD}, I_{ST}, I_{SR}	$=$	scattered light intensity (see Section 4 for components)
k	$=$	$2\pi/\lambda$
L	$=$	distance between collection aperture and center of MCV
m	$=$	real part of refractive index
MCV	$=$	measurement control volume

416

n'	=	number of particles with diameters between d and d + Δd
n	=	number distribution of particle diameter (see Eq. 6.)
N	=	percentage of signals greater than E_p (see Eq. 7.)
P	=	power of one beam
$S(d)$	=	area of y = 0 plane in which particle d is detected
$S_{E_p}(d)$	=	area of y = 0 plane in which particle gives signal greater than E_p
u	=	particle velocity in y direction
$\bar{u}(d)$	=	average velocity of particles, diameter d
V	=	volume % per micron
x,y,z	=	coordinate system for MCV (see Fig.4.)
α	=	off-axis angle of collection optics
β	=	Rosin-Rammler exponent
θ	=	beam intersection angle
λ, λ_f	=	light wavelength and fringe spacing
ϕ	=	collection angle
ω	=	scattering angle relative to incident beam

1 INTRODUCTION

There is a need for an instrument for making rapid 'point' measurements of the sizes and velocities of droplets in burning fuel sprays. Such a technique is valuable both as a diagnostic technique for combustors and, fundamentally, for studying the physical mechanisms of spray combustion, with the aim of improving modelling and design techniques for combustors. Primary interest is in the particle range 10μm < d < 300μm, although larger particles may occur under conditions of poor atomisation. Accurate sizing of particles smaller than this range is not of great interest, as in general, they follow the local gas flow, vaporize quickly, and contain an insignificant volume of the local unburned fuel. A technique for simultaneous measurement of particle size and velocity also allows discrimination between signals produced by droplets and signals produced by seeding particles. Thus the gas velocity field in a flow can be measured without biasing effects introduced by the droplets.[1]

Modifications of LDA techniques have been studied, with the aim of simult-aneously measuring particle sizes and these are discussed below. The particu-larly hostile environment of a burning fuel spray introduces problems of signal noise ratio and particle number density, and these restrict the types of tech-nique which can be applied. The authors here report on a signal amplitude mea-surement technique, and describe its development and application to a spray flow. Attention is focussed on the solution of problems in instrumentation and data processing and on the selection of optical geometry.

2 TECHNIQUES FOR SIMULTANEOUS PARTICLE VELOCITY AND SIZE MEASUREMENT

Double flash photography and holography are capable of simultaneously measuring particle diameters and velocities in spray flames.[2] They suffer from problems of lengthy and sometimes inaccurate data analysis, even when an automatic analysis system is used.[3] Spatially averaged particle size distri-butions can be obtained by measuring various properties of the scattered light from a single beam passing through a flow.[4] These techniques provide a size distribution for a cloud of particles, whereas in the present study measure-ment is required with good spatial resolution.

In published work relating particle sizes to LDA signals most interest[5] has focused on the relationship between particle diameter and signal visibility, which is the ratio of the ac and dc parts of the signal. It appears that the particle diameter can be related to visibility under certain conditions, e.g. the particle diameter must be smaller than the MCV fringe spacing.

The particle sizes of interest here are one order of magnitude larger than those considered in investigations[6] which have the primary objective of optimizing signal/noise ratios. Interest has been focused on particle diameters of the order of 1 μm so that particles follow the aerodynamics of the gas flow.

As particle size increases above 10 μm, computations using the Mie light-scattering theory become extremely lengthy and thus, for the present particle sizes, simplified optical analyses are required. The accuracy of a simplified theory requires to be checked by 'calibration' experiments with known particle sizes, and this forms a part of the work described here.

The visibility technique has the advantage that relative measurements are made. Particle sizes can also be derived from other relative measurements, such as the ratio of the scattered light intensities of two angles. For example Durst[7] used two photomultipliers on either side of the optical axis of a forward-scattering LDA arrangement, and Chou et al[8] used two photomultipliers to measure the ratio between the forward and back scattered light to measure particle diameters of the order 1 μm.

3   GENERAL DESCRIPTION OF 'PEAK MEAN SIGNAL' TECHNIQUE

For the problem of spray combustion, the authors have utilized a new technique in which the peak E_p of the low pass filtered, mean signal is measured for each realization.[9] The notation is shown in Fig. 1. Instruments exist which utilize the monotonic signal height/particle size relationships which can be obtained by using forward collection of scattered light and a white light source. For the present case, in which laser light is used to permit velocity measurements from the fringe modulations, the interference of the two beams produces complex particle size/signal amplitude relationships. It is thus necessary to carefully select the geometries of focusing and collecting optics to obtain an unambiguous particle size/signal amplitude relationship for the size range of interest while at the same time having sufficient visibility to measure velocity. The technique has important advantages over other possible approaches when applied to burning fuel sprays. The most important is that forward scattering optical arrangements, with small MCV dimensions, have been used successfully for velocity measurements in flames, with good signal noise ratios and moderately powered lasers (200 mW and upwards). This basic arrangement need not be changed significantly when using the signal amplitude measurement technique. Thus, a basic forward scattering LDA system, used for gas flow measurements, can be maintained in its existing arrangement.

The signal amplitude technique can be added to an LDA system with a relatively simple additional electronic module, and with the data processing carried out by the existing computer. The technique is restricted to the case $d > \lambda_f$ which is the opposite to the case for size measurement by signal visibility. Thus one can consider the signal amplitude and visibility approaches to be complementary rather than competing. For the spray situation, measurement of the largest droplets by the visibility method would require an MCV width of several mm, which is not acceptable in all but the most dilute sprays.

The development of the technique has involved tests of the light scattering analysis, the data processing routines and the electronic circuitry of the interface unit between the photomultiplier and the computer.

4   EQUATIONS FOR COLLECTED LIGHT

Figure 2 shows the arrangement in the general case where the collection system is at an angle to the LDA axis. The peak mean signal is considered

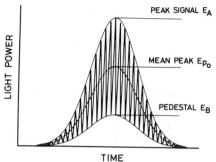

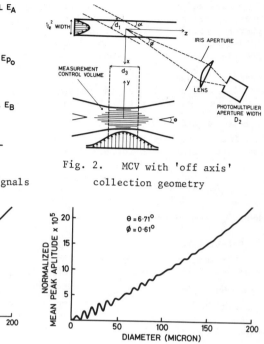

Fig. 1.   Notation for Doppler signals

Fig. 2.   MCV with 'off axis'
collection geometry

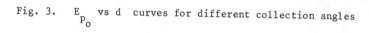

Fig. 3. E_{p_o} vs d curves for different collection angles

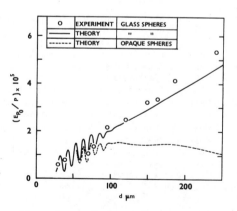

Fig. 4.   Comparison of theoretical predictions with experiment

because of the ease of calculation by using geometrical optics and also ease of measurement by using low pass filters. Calculations are made of the light power E_{p_0} collected within a collection angle ϕ as a function of particle diameter for particles in the center of the measurement volume. The mean signal is the signal averaged in time over several Doppler modulations. In the present case the particle diameter is greater than the fringe spacing and $\lambda_f \ll b_1$.

When $(\pi d/\lambda) > 1$, as for the present case, light scattering can be calculated with reasonable accuracy by using a combination of refraction, reflection, and diffraction. The scattering produced by these three components is derived for a spherical particle in the center of a Gaussian light beam:

Diffraction

For a single beam, diffraction theory gives[10] for small scattering angles ω measured relative to the beam direction:

$$I_{SD}(\omega) = \frac{4\pi^2 I_{max}}{(L\lambda k^2 \omega^2)^2} \left[\int_{\xi_o}^{\infty} e^{-A\xi^2} \xi \, J_o(\xi) \, d\xi \right]^2 \tag{1}$$

where $\xi_o = k\omega d/2$ and J_o is the zero order Bessel function.

Refraction

For small ω, internal reflection in a particle can be neglected. The scattered light intensity due to refraction is [9,10]

$$I_{ST}(\omega) = I_{max} d^2 e^{-\eta^2/b_1^2} \frac{m^4}{(m+1)^4 (m-1)^2 L^2} \tag{2}$$

where

$$\eta = \frac{dm\omega}{2\sqrt{2}(m-1)} \tag{3}$$

Reflection

The light scattered due to external reflection for small ω is [9,10]:

$$I_{SR} = e^{-d^2/2b_1^2} \left[\frac{I_{max} d^2}{16L^2} \right] \left[\frac{\omega/2 - (m^2 - 1 + \omega^2/4)^{\frac{1}{2}}}{-\omega/2 + (m^2 - 1 + \omega^2/4)^{\frac{1}{2}}} \right] \tag{4}$$

The total scattered light intensity for one beam is calculated by summing the three components including the phase effects. The total collected scattered light power E_{p_0} is calculated for the specific LDA system by summing the contributions of the two beams integrated over the collection aperture.

The range of particles which can be measured using E_{p_0} depends on the optical arrangement and particle properties.

Figure 3 shows the effect of increasing the collection aperture. It can be seen that it is necessary to maximise the value of ϕ, consistent with obtaining signals of sufficiently good visibility, in order to maximise the range of diameters with an unambiguous particle size/peak mean signal relationship. These predictions are for transparent spheres of refractive index 1.5. The analysis showed that for glass particles, and clear liquid fuels such as kerosene, with this arrangement, the collected light is composed mainly of refracted light for d>140 μm and the contribution to the measured light energy from reflected light is negligible. For opaque particles, E_{p_0} becomes insensitive to changes in d at smaller values of d than for the case of transparent particles. Thus, measurements with opaque particles are restricted to a narrow range of d for this optical arrangement. For example Fig. 4. shows the predicted peak signal/size relations for glass and opaque spheres for one optical geometry used experimentally. However, judicious choice of focusing and

collecting optics produce relations suitable for opaque particles over a useful size range. Care must be taken to verify, experimentally, that the required particles produce sufficiently visible signals to permit velocity measurements.

5   EFFECT OF PARTICLE POSITION ON SCATTERED LIGHT

5.1   Signal Peak – Particle Size Distribution, Conversion Using Cross-
      Sectional Area of MCV

      In order that the particle size distribution can be derived from a meas-
ured E_p distribution, it is necessary to know the dimensions of the region in
which measurements are made as a function of the particle size and composition.
Interest is focused here on the cross-sectional area S of the measurement
volume in the y=0 plane. This area defines the region in which particles moving
perpendicular to this plane are detected and the value of E_p is measured.
      The collection optics define a field of view in which particles are meas-
ured. For example off-axis collection (Fig. 2.) results in a slice of the
fringe-containing region in which particles are accepted for measurement. In
all cases it is required that $d_1 \gg d$. For a given diameter, S depends upon
the collection optics, the light intensity distribution in the MCV and the
gate level E_G. Generally, the larger the particle, the higher the value of E_p
for a given position in the MCV, and S increases with increasing d. Calculation
of S(d) for different collection optics is described in 5.2.
      The temporally averaged numerical particle size distribution in the spray
is

$$n(d) = \frac{(\text{No. flux of particles between d and } d + \Delta d)100}{(\text{total No. flux of all particles})\Delta d} \qquad (5)$$

with the limit taken as $\Delta d \to 0$. The computer produces a probability distribution of

$$N(E_p) = \frac{(\text{No. of signal peaks greater than } E_p)100}{(\text{total No. of signals measured})} \qquad (6)$$

The measured and required distributions are related by

$$N(E_p) = \frac{100 \int_{dE_p}^{\infty} n \ S_{E_p} \ d(d)}{\int_{dE_G}^{\infty} n \ S \ d(d)} \qquad (7)$$

where d_{E_G} is the smallest measurable diameter and d_{E_p} is the particle diameter
which produces a signal E_p at the centre of the MCV.

5.2   Measurement Cross-Sectional Area as a Function of Particle Size and
      Collection Optics

      Three cases are discussed with different collection optics. For each case
experimental results have been obtained.

(i)   On-Axis Collection Optics

      The spatial distribution of light intensity in the MCV is determined by
P, f_1 and the initial beam diameters and separation. It is useful to define
the intensity distribution in terms of a local mean, or spatially averaged,
intensity I(x,y,z), which is the average light intensity illuminating a particle with a diameter which is an order of magnitude larger than the fringe
spacing. In the region of focus:

$$I = 2I_{max} \ e^{-2\left[(x/b_1)^2 + (y/b_2)^2 + (z/b_3)^2\right]} \qquad (8)$$

Contours of equal mean intensity in the y=0 plane are ellipses described by

$$(x/b_1)^2 + (z/b_3)^2 = \ln\left[(2I_{max}/I)^{\frac{1}{2}}\right] \qquad (9)$$

This is true for a restricted length of the MCV in the z direction. When d_1 is large enough to introduce no restrictions, one can assume[9] that E_p is directly proportional to the local value of I. The area within one contour of equal intensity is

$$\text{Area} = \pi b_1 b_3 \ln(2I_{max}/I)^{\frac{1}{2}}) \;,\quad \text{so that}$$

$$S_{E_p} = \pi b_1 b_3 \ln[(E_{P_0}(d)/E_p)^{\frac{1}{2}}] \qquad (10)$$

and

$$S = \pi b_1 b_3 \ln\left[(E_{P_0}(d)/E_G)^{\frac{1}{2}}\right] \qquad (11)$$

Thus, because $E_{P_0}(d)$ is known, the relation between n and N is known from Eq. 7.

(ii) Off Axis Small Angle Forward Scatter, Collection Optics

Considering Fig. 2, one can have two cases. In the first $d_1 \ll b_1$ so that the light intensity in the slice of the MCV is a function of distance in the direction of the collection optics only and, for any particular value of z, variation of light intensity in the x direction can be neglected. This situation, with a one-dimensional Gaussian light distribution, results in the same form of Eq. 7, as for the 'large α' geometry described below. In theory the arrangement has the advantages of good signal/noise ratio and simplified data processing resulting from one-dimensional data convolution. However the arrangement was not physically possible with the optical system and spray droplet sizes used by the authors. In the second case d_1 is comparable to b_1 and the value of E_G is chosen so that the gate level, not d_1, limits the width of the MCV.

It can be deduced that equations 10 and 11 hold, but with more accuracy than for on-axis optics. This is because the Gaussian light intensity distribution assumption is reasonable for the full length of the slice of the MCV whilst, for on-axis collection, it is invalid at the extremeties of the MCV.

(iii) Off Axis, Large Angle, Collection Optics

When α is large, $\simeq 90°$, the collection P.M. aperture defines a length d_1 of the MCV in the z direction within which particles are measured. Because $b_3 \gg d_1$ the light intensity distribution within this segment of the y=0 plane is a function of x only. Thus

$$S = \sqrt{2}\, b_1 d_1 \, \{\ln(E_{P_0}/E_G)\}^{\frac{1}{2}} \qquad (12)$$

and

$$S_{E_p} = \sqrt{2}\, b_1 d_1 \, \{\ln(E_{P_0}/E_p)\}^{\frac{1}{2}} \qquad (13)$$

The experimental realization of this one-dimensional situation was achieved by maintaining the measurement optics in the forward direction, to preserve good signal/noise ratios for velocity and size measurement, and by using a separate 90° gating P.M. to select particles in the central part of the MCV.

6· EXPERIMENTS

The experimental programme consisted of three parts: (i) the use of on-axis system and various test experiments to investigate the validity of the approach and the accuracy of the theoretical analysis, (ii) the use of a one-dimensional system to study the advantages of this arrangement, (iii) the use of a near forward direction system to make measurements in burning and non-burning kerosene sprays.

An LDA signal processor has been modified so that E_p can be measured in addition to the particle velocity. An integrator amplifier has an output frequency (up to 11 MHz), which is a function of its input voltage which is derived from the low pass filtered signal. The total number of cycles from this amplifier is a measure of the integrated mean signal. Up to 17,770 signals can be measured at a maximum rate of 5000 realizations/sec. The number of cycles of a 5-MHz oscillator are counted when the mean signal is above E_G, chosen so that the smallest particles of interest are just detected when they pass through the MCV center. E_p is calculated from these two parameters. Velocity is calculated in the usual way by counting Doppler cycles. The interval between signals was measured to allow comparison of spray densities. The computer outputs $N(E_p)$; a separate programme is then used to convert this into the particle size distribution, size-velocity correlation, etc.

6.1 On Axis Collection

Glass particles of various sizes were traversed vertically by means of a rotating disk to which an optically flat microscope slide was attached. Figure 4 compares measurements of E_{p_0} versus d with theory and agreement is seen to be good. The laser light power P, in one beam, using a 5mW He-Ne laser, was measured by directing the beam into the P.M. via filters. These experiments[9] showed that the assumption of a Gaussian variation of E_p with particle position in the x and z directions was valid. Experiments were then carried out using narrow sprays of glass particles passed through the centre of the MCV. using a 200mW Argon laser. As can be seen from Fig. 5, there was reasonable agreement between the size distributions measured by LDA, and those measured by microscope analysis of slides. Thus, in general, these on-axis experiments showed the validity of the peak mean signal concept when applied to particle sizing.

6.2 90° Gating P.M. System

In order to obtain a one-dimensional dependence of signal amplitude on position, a series of experiments used a second 'gate' P.M. at 90° to the main axis. The signal processor was modified so that signals from the on-axis P.M. were accepted only when a signal was present simultaneously at the gate P.M. These experiments[11] showed that data processing was simplified and this technique is useful for certain conditions. However, for burning fuel sprays the gate P.M. signal was inevitably small and the signal/noise ratio became poor. With on-axis collection, signals were accepted from the full length of the MCV in the z direction. Although the gate P.M. reduced the effective MCV length to 250 μm, particles anywhere in the cross-beam region produced signals at the forward P.M. which could interfere with signals from the particles of interest in the centre of the MCV. This reduced the maximum local number density of particles to approx. $5 \times 10^8/m^3$. In order to measure in dense burning fuel sprays, a third arrangement was utilized.

6.3 Off Axis Light Collection, Small α

Experiments were carried out with a 'monosize' spray of kerosene and also with a kerosene spray under both burning and non-burning conditions. An argon ion laser operating at 200mW was used.

The optical arrangement, without a gate P.M., had the following parameters: $\theta=5.84°$, $\lambda_f=4.71$ μm, $\phi=0.41°$, $\alpha=6.8°$, and the magnification of the collecting optics was 0.38. The $1/e^2$ dimensions of the MCV were: $b_1=250$ μm, $b_2=250$ μm and $b_3=5$ mm. The $1/e^2$ length of the Field of View was 1.9 mm and $d_1=230$ μm.

These parameters gave good signal/noise ratio and visibility for the kerosene sprays studied. The P.M. aperture was fitted with a TFP26-9496 interference filter. The optical arrangement satisfied the conditions discussed in Section 5.2(iii) so that a two-dimensional peak signal/position relation was obtained and Eq. 7 becomes:

$$N(E_p) = 100 \int_{dE_p}^{d_{max}} n(d) \ln(E_{p_o})/E_p \, d(d) \Big/\!\!\Big/ \int_{dE_G}^{d_{max}} n(d) \ln(E_{p_o}/E_G) \, d(d) \qquad (14)$$

In order to avoid the matrix inversion problem an iterative curve fitting method was used to derive n(d), from $N(E_p)$. An equation was assumed for n(d) and $N(E_p)$ was then calculated by using Eq. 14. The degree of fit was then maximised by successive iterations. A Rosin Rammler distribution, with an upper cut-off, gave good fit to data:

$$V = \text{Vol \% per } \mu m = 100 \left(\frac{\beta}{d}\right)\left(\frac{d}{d}\right)^{\beta-1} e^{-\left(\frac{d}{d}\right)^{\beta}} \qquad (15)$$

or expressed as a number % per μm.

$$n(d) = 100 \, d^{\beta-4} e^{-\left(\frac{d}{d}\right)^{\beta}} \Big/ \int_{o}^{\infty} d^{\beta-4} e^{-\left(\frac{d}{d}\right)^{\beta}} d(d) \qquad (16)$$

This size distribution was fitted to the measurements by the iterative technique which was weighted towards the larger droplets on the high side of the n(d) peak. The distribution of the smaller droplets does not affect the volume distribution, neither can it be expected to be necessarily described well by the Rosin Rammler curve which is designed to match the volume distribution.

The average velocity of droplets in a certain size range is of interest. One can write:

$$\bar{u}(d) = \sum_{i=1}^{n'(d)} u_i(d) \Big/ n'(d) \qquad (17)$$

where $u_i(d)$ is the velocity of the ith droplet between d and $d + \triangle d$. The measurements give velocity as a function of E_p and these measurements are converted to give the required $\bar{u}(d)$. This conversion is achieved by using the integral relation between the particle diameter and the peak signal. The measurements are expressed as a cumulative oversize distribution so that,

$$\bar{u}_{cum}(E_p) = 100 \left[\begin{array}{c}\text{Sum of droplet velocities} \\ \text{for particles with peak} \\ \text{mean signals} > E_p\end{array}\right] \Big/ \left[\begin{array}{c}\text{Sum of} \\ \text{velocities of} \\ \text{all particles}\end{array}\right]$$

which is related to $\bar{u}(d)$ by:

$$\bar{u}_{cum}(E_p) = 100 \int_{dE_p}^{d_{max}} \bar{u}(d) \, n(d) \ln(E_{p_o}/E_p) \, d(d) \Big/ \int_{dE_G}^{d_{max}} \bar{u}(d) n(d) \ln(E_{p_o}/E_G) d(d) \qquad (18)$$

$\bar{u}(d)$ is also derived from this equation by an iterative curve fitting technique.

The standard deviation, $\overline{u'^2}(d)$ of the velocities of particles about $\bar{u}(d)$, as a function of size, can be calculated by the same method.

The droplet size distribution n(d) is temporally averaged and this is of practical interest in the field of combustion. The spatially averaged particle size distribution $n_{spat}(d)$ is:

$$n_{spat}(d) = \frac{100}{\Delta d} \frac{\int_{o}^{T} (\text{No./unit volume of particles between d and } (d+\Delta d)dt}{\int_{o}^{T} (\text{No/unit volume of particles of all sizes}) \, dt}$$

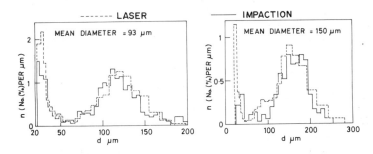

Fig. 5. LDA particle size measurements in sprays of glass particles

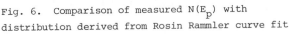

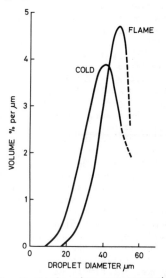

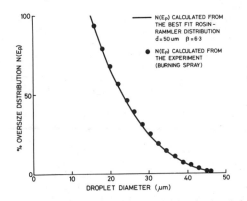

Fig. 6. Comparison of measured $N(E_p)$ with distribution derived from Rosin Rammler curve fit

Fig. 7. Particle size measurements in cold and burning kerosene spray

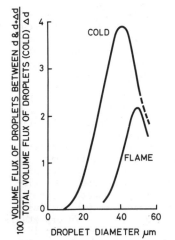

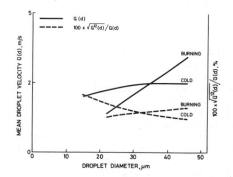

Fig. 8. Volume flowrate versus drop size in burning and cold conditions

Fig. 9. Mean and variance of particle velocity in cold and burning conditions

The relationship between $n_{spat}(d)$ and $n(d)$ is:

$$n_{spat}(d) = n(d)/\bar{u}(d) \;\; {}_{d_{E_G}}\!\!\int^{d_{max}} \; (n(d)/\bar{u}(d) \;) \; d(d) \tag{19}$$

The size distributions do not include droplets smaller than d_{E_G}. It is important to ensure that the numbers of these droplets are insignificant when data is converted to volume/size distributions.

7   APPLICATION TO FUEL SPRAYS

The optical arrangement and data processing described in 6.3 were used to investigate kerosene sprays. As a preliminary test LDA measurements were made in a near monosize spray of kerosene produced by a TS1 particle generator with numerical mean diameter 46 µm, according to a Royco particle counter. The experiments were used to check the calibration of the P.M. tube and the signal processing unit. Good agreement was found, with the LDA technique giving $\bar{d}$ = 44 µm and a narrow distribution with a Rosin Rammler exponent β = 9.

Measurements were made in a small twin fluid atomized kerosene spray, stabilized in the wake of a circular disk, under nonburning (cold) and burning conditions. Velocities and sizes of more than 1000 droplets were measured and data was analysed to calculate size and the mean and variance of droplet velocity. Figure 6 compares the measured $N(E_p)$, for one position in the spray with the distribution which gave best fit when calculated from a Rosin Rammler distribution. It can be seen that the agreement is very good. Temporal size distributions at the same position in cold and burning versions of the same spray are shown in Fig. 7. The low number of realizations measured above a diameter of 50 µm results in the iterative curve fitting method giving increasingly poor statistical representation at the upper end of the size distribution. Increasing the number of signals measured will extend the upper limit. Particles smaller than 15 µm were not measured because the E_{po}/d relationship for the optical system had a lower limit in this region for an unambiguous determination of d from E_{po}. The contribution, by volume, of these small droplets is negligible. However by modifying the collection optics, measurement of these small sizes can be achieved. As can be seen from Fig. 7, after ignition, the size distribution is shifted to the larger diameters and becomes relatively narrow. These changes are expected as they are due to the preferential evaporation of small droplets. The vaporization is clearly seen when the same results are plotted in Fig.8, in a form permitting comparison of flow rates. There was little observable difference between the temporally and spatially average volumetric size distributions for this position.

Fig. 9 shows the measured mean and variance of droplet velocity, as functions of particle size for the same position. For the cold spray at this position, the velocities of the larger droplets are slightly higher than those of the smaller droplets. This is explained by the small droplets decelerating to the gas velocity relatively more quickly because of their higher drag/inertia ratios. The variance of droplet velocity is higher for the smaller droplets and this is again expected as, the smaller the droplet, the more closely it should follow turbulence fluctuations. The velocity/diameter distribution for the burning case is interesting as it indicates that the variation of particle velocity with particle diameter is more pronounced for this case. At this stage the relative importance of the physical mechanisms involved is not known, and awaits a full mapping of the spray. The interpretation of particle size distributions in terms of combustion related mechanisms is extremely complex because of the many mechanisms which can affect the particle dynamics and vaporisation. Indeed the LDA particle sizing technique has been developed primarily to gain more information on these mechanisms. Effectively the history of droplets from the atomizer to the measurement position must be known to enable a physically meaningful comparison between the cold and burning cases.

8   CONCLUDING REMARKS

The peak mean signal technique is capable of simultaneous particle velocity and diameter measurements in both burning and non burning sprays. Providing care is taken in calibrating components of the system, and the particles are spherical with known optical properties, the size distributions measured are comparable in accuracy with those derived from image measurement techniques. The LDA method has the advantages of excellent spatial resolution and rapid data acquisition and processing. There are many possible optical geometries for the peak signal method and, although the off-axis forward scatter system functions in moderately dense burning sprays, further work may further improve the range and accuracy of the technique.

9   REFERENCES

1.   Styles, A.C. and Chigier, N.A., Combustion of Air Blast Atomized Spray Flames. 16th. Symposium (International) on Combustion, (1977).

2.   McCreath, C.G. and Chigier, N.A. Liquid-Spray Burning in the Wake of a Stabilizer Disc. 14th. Symposium (International) on Combustion, (1972).

3.   Yule, A.J., Cox, N.W. and Chigier, N.A., Measurement of Particle Sizes in Sprays by the Automated Analysis of Spark Photographs. Particle Size Analysis Conference, University of Salford (1977).

4.   Swithenbank, J., Beer, J.M., Taylor, D.S., Abbott,D. and McCreath, C.G. A Laser Diagnostic for the Measurement of Droplet and Particle Size Distribution. AIAA paper No. 76-53.(1976).

5.   Farmer, W.M., Measurement of Particle Size, Number Density, and Velocity Using Laser Interferometer. Appl. Opt., 11, 2603, (1972).

6.   Durst, F. and Eliason, B., Properties of Laser Doppler Signals and their Exploitation for Particle Size Measurements. SFB 80/TM82, University of Karlsruhe, (1976).

7.   Durst, F. and Zare, M., Laser Doppler Measurements in Two Phase Flows. SFB 80/TM/63, University of Karlsruhe, (1975).

8.   Chou, H.P., Waterston, R.M. and Pratt, N.H., A Theory of Particle Size Measurement Using Crossed Laser Beams. LDA Symposium, Technical University of Denmark, Lyngby, Denmark, (1975).

9.   Yule, A.J., Chigier, N.A. Atakan, S. and Ungut, A., Particle Size and Velocity Measurements by Laser Anemometry. Journal of Energy,1,220-228(1977).

10.   Ungut, A., Ph.D. Thesis, The University of Sheffield, May, 1978.

11.   Ungut, A., Yule, A.J., Taylor, D.S. and Chigier, N.A., Simultaneous Velocity and Particle Size Measurements in Two Phase Flows by Laser Anemometry. AIAA 16th Aerospace Sciences Meeting, Paper No.78-74,(1978).

Design and Calibration of a Laser Interferometer System for Particle Size and Velocity Measurements in Large Turbine Precombustors

W. M. FARMER and J. O. HORNKOHL[*]
Spectron Development Laboratories
3303 Harbor Blvd., G-3
Costa Mesa, CA 92626

G. J. BRAND and J. MEIER
Solar Turbines International
An International Harvester Group
2200 Pacific Hwy., P.O. Box 80966
San Diego, CA 82138

ABSTRACT

The design and calibration of an instrument to measure particle size and velocity in large turbine precombustion systems is described. This device uses backscattered light, measures two orthognal velocity components and scans the probe volume with fixed fringe period over a 1 m depth-of-field. Particle size measurement range is between 5 and 300 micrometers.

INTRODUCTION

To design a high-efficiency turbine precombustion system it is necessary to know the size of fuel oil droplets which are injected into the combustion chamber and the swirl patterns of the flows which help to disperse and mix the droplets with the injected air. Measurement of these flows with a LDV or a PSI (Particle Sizing Interferometer) system requires that the instrument design accomodate high temperature and noise levels which surround the combustion system and the bright light background associated with the combustion process (both for burning particles and for the radiant gases.) Furthermore, consideration must also be given to droplet number densities in these types of devices which can be extremely high, severely limiting the data acquisition rate. The purpose of this paper is to describe the design and calibration of a PSI system designed to function in this type of environment.

The design constraints for the instrument required that it had to operate in a backscatter mode; it had to position the geometric center of the sample space over a 1 meter scan range; two true velocity components were required and a particle size range between 5 and 290 microns had to be covered. Additionally, the optical system had to function in an environment where temperatures might exceed 150 degrees C and where noise levels might be on the order of 120 db. Particle speeds might be expected to exceed several hundred meters per second, the particles might be undergoing combustion, and number densities might exceed 10^2/cc. The electronic system was required to obtain two measurements of

---

[*]W. M. Farmer and J. O. Hornkohl are now at the University of Tennessee Space Institute, Tullahoma, TN.

particle size from orthogonal fringe sets, obtain two simultaneous orthogonal
velocity components, produce particle size and velocity histograms, compute the
mean velocity for a pre-selected velocity component, and produce a velocity
distribution histogram for a given particle size increment. The following discussion shows how those constraints were met.

OPTICAL SYSTEM DESIGN

The particle sizing interferometer optics are mounted on two shelves surrounded by an environmental housing. This housing contains a cooling water
manifold implanted in the sides, bottom, and hinged doors allowing proper temperatures to be maintained during long term tests. The housing is vibration
isolated and mounted on a support carriage which can be rolled from site to site
and adjusted in height between 1.16 m and 1.6 m. Leveling screws are provided
to stabilize the assembly and raise the carriage casters off the ground. Barry
(R) vibration isolation mounts are used between the environmental housing and
carriage to eliminate acoustic and floor vibrations.

A 4 watt Ar$^+$ laser is mounted on the utility shelf beneath the primary
optics shelf containing the transmitting and receiving optics. The utility shelf
also holds the power supplies for the laser and the 2D Bragg cell beam splitter.
This shelf is accessible at both ends through hinged doors.

The primary optical system on the top shelf is accessible through a remov-
able top panel which is covered with a noise dampening material designed to eliminate vibrations present in high acoustic noise environments. Transmission and
signal reception is achieved through a 10 cm diameter fused quartz window which
is bubble and straition free. It is AR coated on the internal surface only.
Figures 1 and 2 show the construction of the primary optical system. It is
composed of five major subsystems.

The first subsystem is the Bragg cell beam splitter. The Bragg cell contains
two ultrasonic transducers which operate in the third overtone at 15 and 22.5
MHz respectively. The transducers are mounted in a stainless steel housing which
is attached to a tilt mount for each of two rotations about axes which are orthogonal to the normals of the transducers planes. The Bragg cell housing is stainless steel in order that fluids with different acoustic properties can be used
to obtain a broad range of beam split angles for the same driving frequency.
The entire Bragg cell housing can be rotated through 90 degrees about the input
beam axis in order to establish any arbitrary orientation for the velocity
measurement coordinate system.

The second subsystem consists of lenses L2-L5. These lenses control the
value of the fringe period, the maximum sample space position for the optical
system, and the number of interference fringes in the sample space.

The scanner for the transmitting optics composes the third subsystem. The
scanner consists of a set of total internal reflection prisms which fold the
optical path parallel to a screw drive. The screw drive is such that it can
move a total distance of 50 cm. Since the beam path is folded along the screw
drive a traverse distance of 50 cm along the screw drive corresponds to a
sample space scan variation of approximately 100 cm. The position of the re-
flecting prism mounted on the screw drive is read with a potentiometer bridge to
an accuracy of $\pm$ 0.25 mm. The position of the sample space with respect to the
front face of the cabinet window has been correlated with a screw drive position
such that the readout on the scanner control panel reads directly in sample space
position.

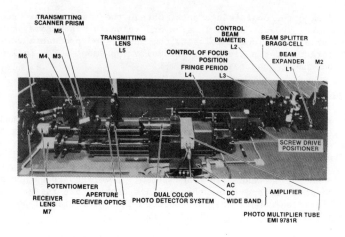

Fig. 1. Mounting shelf for transceiver system transceiver.

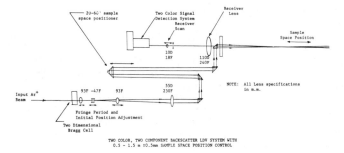

Fig. 2. Schematic of transceiver system.

The fourth subsystem consists of the ten centimeter receiver lens and the receiver scan system. The receiver scan system is designed such that a collimating lens and limiting aperture can be positioned at the image of the geometric center of the sample space in order to collimate the input to the light detection system. For typical operation the limiting aperture consists of a variable iris. However, if increased spacial resolution is required a number of fixed apertures can be mounted at the rear of the iris.

The fifth subsystem consists of the dual color light detector.[1]   The optical system has been designed such that near full power utilization of the laser can be obtained by using this system which consists of a dichroic filter color separator and two bandpass filters.

ELECTRONICS SYSTEM

The electronics system for acquiring and processing the data can be divided into seven subsystems. These subsystems consist of:

1. The transceiver scan control which provides remote asynchronous positioning control of the sample space.

2. The 100 MHz Tektronics storage oscilloscope which is used in instrument calibration and insitu analysis of signal quality.

3. The signal separator, SDL Model DSS 1522, which is used to electronically separate the velocity components and to provide a carrier frequency for identifying velocity direction.

4. The signal processors, SDL Model PMAP 550, which measure the signal time period, signal visibility, and the scatter intensity.

5. The data sorter, SDL Model DSAC 5000, which is a microprocessor system for immediate small data volume analysis of test and calibration data.

6. The Nova 2/10 minicomputer with direct memory access which is used for real time data acquisition and temporary storage of data for high data acquisition rates.

7. The Cipher Data Magnetic Tape System which is used for storage of test data transferred from the Nova computer memory and retrievable for statistical analysis on an IBM 370/158 mainline computer.

The transceiver scan control system is a controller for the asynchronous scanners in the transceiver section. Through the use of a bridge circuit, the position of the scanning prism for the transmitter in the position of the aperture and collimating lens and the receiver system can be specified with an uncertainty of $\pm$ 0.25 mm. The operation of the transceiver is such that the operator reads the position of the sample space on a digital panel meter on the front face of the controller and stops the scanner when the desired focus position is reached. He then switches over to the receiver and by use of a reference table sets the receiver to the proper receiver position reading.

Figure 3 is a block schematic of the signal processing electronics. The signal from the PMT preamp is sent first to the signal conditioner, the DSS 1522, for signal conditioning in order to obtain the signal time period. A portion of the signal is also sent directly to the visibility processor which is located in the PMAP 550. The visibility processor in this instrument is somewhat different from that described in another paper presented elsewhere in that this processor contains a high frequency rectifier which allows it to operate on Bragg cell type signals, i.e., signals of high frequency but relatively narrow band width.[2]

Note:

f_{DPX} - Doppler Frequency for V_x

f_{DPY} - Doppler Frequency for V_y

f_{DX} - Mixer Difference Frequency
for x-Component (Programmable)

f_{DY} - Mixer Difference Frequency
for y-Component (Programmable)

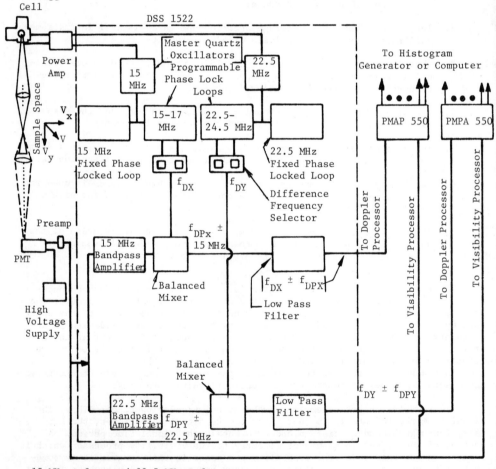

15 MHz $\pm$ f_{DPX} and 22.5 MHz $\pm$ f_{DPY}

Fig. 3. Block diagram of signal processing electronics.

The function of the DSS 1522 is to condition the signal from the PMT. First, it filters the sum and difference carrier frequencies which are present in the scattered light signal and are centered at 7.5 and 37.5 MHz. It then separates the signals corresponding to the 15 and 22.5 MHz channels. Thirdly, it introduces a switch selectable "carrier" or "differnce frequency" which can vary from 0 to 990 KHz in 10 KHz steps. The function of this carrier frequency is to optimize signal-to-noise ratio for a particular experimental condition and to introduce high accuracy into the frequency measurements. The signal conditioner also contains a low pass filter which rejects the sum and difference components introduced into the heterodyne process which uses the selected carrier frequency.

After the velocity component signals leave the signal conditioner, the signals enter the PMAP 550 signal processor where the signal time period is determined by measuring the signal for two different period averages as is a standard routine in most LDV burst counter signal processors. The number of cycles in both the long and short average is a predetermined system constant which is programmable from 0 to 99 in unit increments. These averages are used in logic tests to determine whether the signal is of acceptable quality before it is recorded by the data acquisition system. This logic is also used to control the visibility processor. Particle size measurements are accomplished by measuring the signal visibility with the PMAP 550. In addition, it measures the mean value of the scattered light signal magnitude so that the scattered intensity can also be used to estimate the particle size when a scattered intensity versus particle size calibration is available.[2-4] The signal processors have built-in time period calibrators which are front panel accessible. 12 calibration frequencies are available which range between 1.6 KHz and 8 MHz.

The data acquisition and retrieval systems consist of the microprocessor data sorter and the minicomputer and magnetic tape recording system. The microprocessor is used to acquire relatively small amounts of data and to present this data in a form which is immediately usable for on-line systems checks and instrument set-up. After the system is functioning in a data acquisition mode and data is being acquired by the minicomputer system, the microcomputer system can be used for periodic checks of data samples. The microprocessor is equipped with programs which:

1. Provide velocity histograms for either velocity channel.

2. Provide a particle size histogram for either velocity component.

3. Provide a velocity histogram for a particle size increment selected from the size histogram.

The particle size histogram is expressed in terms of percent fringe period and contains 51 size intervals ranging from less than 10% of the fringe period to greater than 100% of the fringe period. Should the operator so choose, he is at liberty to enter weighting factors for the histogram to obtain a volumetric normalized size distribution function.[3] The particle size histogram is determined from signal visibility measurements which are correlated with particle size using a library function which relates signal visibility to particle size. The microprocessor also indicates the number of events or attempted measurements by the signal processor and it records the acquisition time for each set of measurements. The primary function of the minicomputer and the magnetic tape system is to acquire data at maximum data acquisition rate. The measurements directly enter the minicomputer memory and are transferred to the magnetic tape system for processing on a large computer with much more sophisticated data manipulation and analysis programs than can practically be included in the much smaller microprocessor or minicomputer memories.

W. M. FARMER, J. O. HORNKOHL, G. J. BRAND AND J. MEIER

SYSTEM CALIBRATION

There are a number of calibrations which must be performed on an instrument of this type before particle size and velocity can be measured or estimated and it is of paramount importance that these calibrations be done accurately to minimize errors. The calibrations for this instrument can be divided into the following group:

1. Fringe period calibration.

2. Scanner calibration.

3. Sample space calibration as a function of particle size and instrument parameters.

4. Particle size calibration with respect to the assumed visibility function.

5. Data acquisition rate calibration to determine the ability of the instrument to acquire data in hot fuel sprays where particle number densities are high.

Because the fringe periods in this instrument are relatively large, significant errors can occur if only the angle between the beams is used, for example, to measure the fringe period. We have found that the error can significantly increase as the fringe period increases using this technique. Therefore, the fringe periods to be used with the instrument were determined by measuring the tangential velocity of a rotating wheel which has a known rotational frequency and an accurately measured radius. The wheel used in the calibration was designed such that its circumference was 1 meter ± 0.1 mm and the rotation rate was sensed with a magnetic tachometer. Since the wheel radius and frequency of revolution are known to high accuracy, division of the measured signal frequency into the tangential velocity of the wheel yielded a measure of the fringe period with an uncertainty of ± 0.2%.

The transceiver control readout was calibrated to read the various sample space positions as a function of prism position on the transmitter and limiting aperture position on the receiver. Because the scanner is an asynchronous device, it was necessary to calibrate the receiver position in terms of sample space position. This was done using the manufacturers specified focal length for the receiving lens in the thin lens equation to predict the location of the image of the sample space. Due to the uncertainty in this focal length, several iterative trials were necessary using different object-image distances before the correct receiver positions could be accurately predicted.   A library function using the thin lens equation was then computed for the receiver position such that an operator could select a position for the geometric center of the probe volume and then move the receiver aperture to the proper location to observe the optimum signal intensity.

To accurately measure such parameters as the particle mass concentration using this instrument, it is necessary to know the dependence of the sample space on particle size and instrument parameters. One calibration to determine the sample space length or "depth of field" used a glass bead 84 microns in diameter (attached to an AR coated optical flat by Van der Wall's attraction) placed in the geometric center of the sample space. The receiver was focused at the proper image position, the laser was placed in a light regulation mode at 100 miliwatts of light power output and the photomultiplier tube was set to a reference voltage of 600 volts. The fringe period was arbitrarily set for 98 microns on the 15 MHz velocity channel. The optical flat was mounted such that it

could be independently traversed up and down the optical axis. Acceptable signal threshold settings were chosen and the particle was scanned back and forth
along the optical axis until the signal processor ceased processing for a given
threshold setting. The signal processor logic was set so that a number of
signal cycles could be measured. A long count of 80 was chosen with a short
count of 50 and the difference frequency was chosen to be 10 kilohertz for each
velocity channel. One example of the results of this calibration showed that at
the 1 m focus position the sample space length for an 84 micron diameter glass
bead was approximatley 4.3 cm. However, this value decreased significantly
at the extremes of the scan positions. At the 0.5 m position the sample space
length was approximately .635 cm while at the 1.5 m position it was approximately
1 cm. This type of calibration provides sufficient reference for sample space
normalization should particle diameter, laser power or PMT gain change. It is
interesting to note that the geometric depth of field as computed from the size
of the limiting aperture, sample space position, and F number of the receiving
lens can be significantly greater or significantly less than those values found
for the sample space length. This calibration illustrates the necessity for
carefully specifying the scattering properties of the particles being observed
with either a PSI or an LDV system when spatial resolution is specified for a
given instrument.

It should be pointed out that for the particle used in the depth of field
calibration measurements, the signal visibility was also read using the signal
processor. This parameter remained constant throughout the depth of field scans
and for the various sample space positions. The visibilities read for the separate channels are 0.32 for the 15 MHz channel and 0.09 for the 22.5 MHz channel.
These values of the visibility give a reasonably good agreement with the size of
the bead as determined from microscope measurement. For the 15 MHz channel this
size is approximately 85 microns and for the 22.5 MHz channel the size is approximately 89 microns. It may be expected that there will be some discrepancy between the two measurements because the glass bead is mounted on an optical flat
which contributes some scatter to the light which is scattered by the particle.

A number of tests have been made to test the accuracy of the visibility described by a Bessel function. This visibility description is known to apply
for certain forward scatter measurements but there is some question concerning
its application to backscatter. In one test, glass beads with a sample size
distribution that had been determined using microscope measurements were injected
into an air flow passing through the sample space.

Figure 4 compares the size distribution results obtained with the interferometer and the microscope count when the 15 MHz channel fringe period is 98
microns. The measurements for the 22.5 MHz channel showed that nearly all the
particles were greater than 60 microns.

Additional calibration tests are being performed using a monodispersed
generator which can utilize either the fuels used in the combustion system or
water. The monodispersed generator tests are arranged in the geometry shown in
Figure 5. In this geometry, the monodispersed generator produces a stream of
droplets which are separated by approximately 4 to 5 particle diameters. Pulsed
electrostatic fields are used to charge certain droplets and then to separate
the charged droplets from the uncharged droplets. This allows the experimen-
talist to choose the spacing between the droplets such that no two droplets are
simultaneously in the sample space. In order to verify that the droplet diameter
is as expected based on calculations involving the monodispersed generator flow
rate and vibrating orifice driving frequency a stroboscopic lamp and photo-
microscope are used to simultaneously photograph the drops as they pass through
the sample space of the PSI optical system. The monodispersed generator is
suitably enclosed such that small air currents in the laboratory do not deflect

the droplets stream which pass through the sample space. This stream can then be used to scan the sample space cross section in order to determine its particle size dependence.

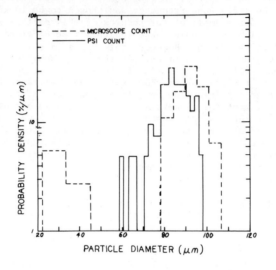

Fig. 4. Particle size histogram of glass beads obtained with interferometer system and a microscope.

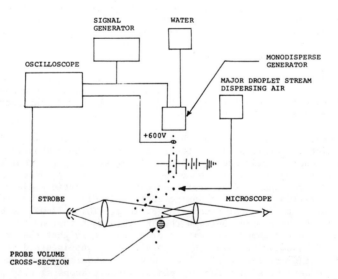

Fig. 5. Monodisperse generator test geometry.

In order to observe the qualitative and quantitative operation of the PSI system in examining flows undergoing combustion, the PSI has been used to examine a burning fuel spray centered in a 30 centimeter diameter duct. The PSI sample space was scanned across the duct and it was found that the PSI functioned better (a higher ratio of acceptable measurements to attempted measurements was found) when the spray was burning than when it was cold. It was also found that the PSI measurements were in qualitative agreement with the expectations of how the fuel spray should be distributed spacially. A reason why the PSI functions better in the hot flow than in the cold flow can be found in the fact that the hot flow immediately consumes numerous small droplets which can seriously affect the acceptable data rate in the PSI system.

SUMMARY AND CONCLUSIONS

A PSI system designed to function in a high temperature, high vibration environment in order to obtain particle size and velocity measurements in large turbine systems has been described. The use of a 2D Bragg cell allows the use of the two primary laser lines with a single photo detector system. Calibration tests performed on the instrument have shown the limitations of the instrument in terms of depth-of-field and have demonstrated its versatility in a number of measurements. The use of a small microprocessor system with a larger minicomputer system for data acquisition has been found to be an invaluable aid in instrument calibration and for on-line checks of data during test runs. Tests performed with calibrated glass beads have shown that the visibility function selected for the library is adequate for particle size measurement. The calibration tests for the PSI system suggest that the model for the estimate of the instrumental sample space is applicable. The programmable features of the signal processing electronics have been found to be a very useful aid in calibrating the system and in applying it to experiments where parameters can be reasonably well defined á priori. The calibration results suggest that depth-of-field estimates based on geometric optics used for the design of the instrument can be assumed to be reasonably accurate estimates for a specified set of operating parameters for the PSI system.

REFERENCES AND FOOTNOTES

1. W. M. Farmer, "Two-dimensional Bragg Cell LDV System Using Multiple Light Frequencies," Appl. Opt., 17, 166 (1978).

2. W. M. Farmer, J. O. Hornkohl, K. E. Harwell, and F. A. Schwartz, "Particle Sizing Interferometer Measurements In Rocket Exhausts," Presented at the Third International Workshop on Laser Velocimetry, Purdue University, July 11-13, 1978.

3. M. Kerker, The Scattering of Light and Other Electromagnetic Radiation, Academic Press, New York, 1969.

4. A. J. Yule, N. A. Chigier, S. Atakan and A. Ungut, Particle Size and Velocity Measurement by Laser Anemometry," AIAA Journal Energy, 1, 220 (1977).

Submicron Particle Size Measurements in an Acetylene/Oxygen Flame

JAMES F. DRISCOLL
Department of Aerospace Engineering
The University of Michigan
Ann Arbor, MI 48109

DAVID M. MANN
Air Force Rocket Propulsion Laboratory
Edwards, CA 93523

ABSTRACT

A new optical technique for making in situ measurements of aerosol particle size in the submicron size range has been investigated. Denoted Laser
Doppler Spectroscopy by Hinds and Reist,[1] the technique makes use of the spectral broadening of scattered laser light due to random particle motion. In
this study, LDS has been demonstrated to be suitable for application in unsteady high temperature environments of 2,200°K in flowing gases.

Particle diameters were measured in an acetylene/oxygen flame. Particle
size was determined as a function of equivalence ratio and height above a flat
flame burner through which premixed acetylene, oxygen and a nitrogen diluent
flowed. Sizes were found to vary from 40 to 250 nm as the height above the
burner varied from 0.5 to 3.0 cm. Little variation was found as a result of
equivalence ratio changes from 2.5 to 5.0. Sizes measured from electron
microscope photographs of soot particles collected from the flame show reasonable agreement with those calculated from scattering measurements. Methods
to optimize the operating parameters of the sizing technique are discussed.

INTRODUCTION

Until recently, no single technique has afforded the capability of in situ
sizing of submicron particulates of unknown composition. A new technique was
proposed in 1972 by Hinds and Reist [1] which provides this capability for
the first time. With this technique, denoted laser Doppler spectroscopy, mean
particle diameter is determined directly from a measurement of the broadening
of the optical spectrum of laser scattered light. Spectral broadening results
from the vigorous Brownian motion of the aerosol particles in the laser beam,
due to the Doppler effect, and is related to the particle diffusion coefficient
which in turn is a function of mean particle size. A measurement is made of
the homodyne power spectrum which typically extends through the 0-50 KHz range
and contains information that is discarded in conventional laser Doppler heterodyne detection. By employing both homodyne and heterodyne spectroscopy, it
is possible, in principle, to measure particle size and velocity simultaneously. In 1974, Penner, Bernard, and Jerskey [2] modified the procedure for use
in flowing gases and presented results obtained in an ethylene/oxygen flame.

438

The net displacement of a single particle undergoing Brownian motion will vary in time, resulting in an oscillation of the frequency of the Doppler shifted scattered light about ω_o. The time averaged optical spectrum will therefore be broadened by an amount depending on the average rate of particle displacement.

The halfwidth of the optical spectrum is measured using homodyne detection. For the case of many particles radiating simultaneously in the focal volume, all frequency components in the broadened optical spectrum are simultaneously incident on the photodetector. The beating of each discrete frequency interval on the optical spectrum with every other discrete frequency interval results in a spectral profile centered at zero frequency, called the homodyne spectrum, which is a Lorentzian function for a monodisperse, stationary aerosol [1]:

$$S_I(\omega) = \frac{I^2}{\pi} \frac{2K^2D}{(2K^2D)^2 + \omega^2} \tag{1}$$

with a halfwidth:

$$HW = \frac{1}{\pi} \left(\frac{4\pi}{\lambda} \sin \frac{\Theta}{2} \right)^2 \frac{kT}{3\pi\mu d} \cdot \frac{F_c}{F} \; (\ell/d) \tag{2}$$

EXPERIMENTAL CONDITIONS

Scattered radiation was measured from a 28 m watt HeNe laser beam passing through an acetylene-oxygen flame. A flat flame laboratory burner was used in this study which consisted of a core of bundled copper rods between which flowed the premixed acetylene, oxygen and a nitrogen dilutant introduced to vary flame temperature and thus flame speed. The 3.18 cm diameter burner surface was recessed and surrounded by an annular flow of nitrogen, which considerably aided flame stability.

Radiation was observed at 25° and 60° scattering angles using an RCA 4832 photomultiplier tube and a 1.6 µm interference filter. Fluctuations in the phototube current were recorded on analog tape at a speed of 60 ips using an AMPEX 1200 FM recording system. The recorded data was then analyzed using a digital spectrum analyzer (Spectral Dynamics 360-35). This unique instrument utilizes an oscilloscope screen to display the computed power spectrum of input data. The time averaged spectrum is continuously updated as recorded data is continuously digitized and Fourier transformed. Data was first high pass filtered at 1 Khz and digitized at sampling rates of 16-32 Khz. Each computed power spectrum was determined from 2.5×10^5 data points.

DISCUSSION

The results of particle size measurements, shown in Figs. 1 and 2 are encouraging because they indicate that homodyne spectroscopy can be applied in a straightforward manner to determine particle sizes in flames. Measured particle diameter is observed to increase from 40 nm to 260 nm as the height h above the flame front increases from 0.5 cm to 3.0 cm. This trend is consistent with the fact that as h increases, particle residence time increases and the probability of particle agglomoration increases.

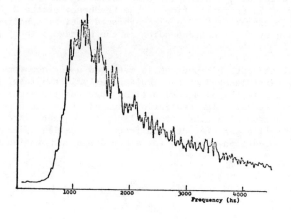

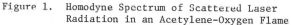

Figure 1.   Homodyne Spectrum of Scattered Laser
            Radiation in an Acetylene-Oxygen Flame

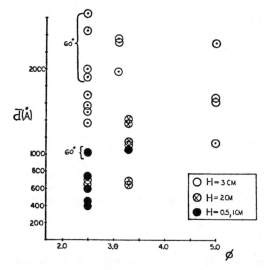

Figure 2.   Mean Particle Diameter at Various Equivalence
            Ratios in an Acetylene-Oxygen Flame

Electron microscope photographs were taken of carbon particulates sampled on a Nichrome wire at 3 cm above the burner. Particulates were observed to be in the 200 nm range, in approximate agreement with the results of Fig. 2.

The research reported in this paper was performed at the Air Force Rocket Propulsion Laboratory, Edwards, California; funds were made available by the Air Force Office of Scientific Research and administered by Auburn University.

REFERENCES

1.   Hinds, W. and Reist, P.C., "Aerosol Measurements by Laser Doppler Spec-
     troscopy," Aerosol Science, V. 3, 1972, p. 501.

2.   Penner, S.S., Bernard, J.M., and Jerskey, T., "Laser Scattering from
     Moving Polydisperse Particles in Flames, II. Preliminary Experiments,"
     Acta Astronautica, V. 3, 1976, p. 93.

GENERAL APPLICATIONS II

Chairman: **H. D. THOMPSON**
Purdue University

Laser Velocimeter Measurements in Natural Convection Flows in Triangular Enclosures

R. D. FLACK and C. L. WITT
Department of Mechanical and Aerospace Engineering
University of Virginia
Charlottesville, VA

ABSTRACT

Velocities in laminar free convection air flow fields were measured using a laser velocimeter. To test the accuracy of the measurements, velocities were first measured in the laminar boundary layer around a heated vertical flat plate. Secondly, measurements were taken in a two-dimensional triangular enclosure in which laminar natural convection prevailed. Results for the vertical flat plate measurements are compared to theoretical and previous experimental (streak photography) results; in general, the LV results compare within 2% of previous results. The triangular enclosure consisted of two isothermal side walls (one heated and one cooled), an insulated bottom, and optical glass end plates. Velocities were measured throughout the enclosure and the overall data is presented such that the flow patterns can be observed. The purpose of this paper is thus two-fold: first, and most important, to demonstrate the capability of laser velocimters in a free convection gas flow field, where they have not been successfully utilized previously; and second, to provide some quantitative data for a specific natural convection problem which has not been studied previously.

NOMENCLATURE

g = Acceleration due to gravity

Gr = Grashof number, see Equation (2)

$\hat{Gr}$ = $g\beta_m \cos\gamma (T_H - T_C) L^3 / \nu_m^2$

h = Distance from bottom of triangular enclosure

H = Height of triangular enclosure

$\overline{H}$ = h/H

L = Length of isothermal surface

T = Temperature

u = Average velocity

$\hat{u}$ = Nondimensionalized velocity, see Equation (2)

U = Nondimensionalized similarity velocity, see Equation (2)

W   = Width of triangular enclosure

x   = Dimension tangent to an isothermal surface

X   = x/L

y   = Dimension normal to an isothermal surface

Y   = $Gr^{1/4}y/L$

Z   = Length of the test section

β   Volumetric coefficient of expansion

γ   = Angle between an isothermal surface and the gravity vector

η   = $Y/(4X)^{1/4}$

θ   = Angle between laser beams

θ_1, θ_2 = Angles in triangular enclosure, See Fig. 3

ν   = Kinematic viscosity

Subscripts

H,C = Pertaining to hot and cold surfaces in triangular enclosure

m   = Pertaining to mean value of temperature

r   = Reference condition at which fluid properties are evaluated

w   = Pertaining to a wall condition

∞   = Pertaining to free stream value

INTRODUCTION

Free convection has been of interest in heat transfer applications for decades. Theoretical techniques used to solve natural convection problems yield both heat transfer rates and velocity information since the energy, conservation of mass and conservation of momentum equations are solved [1,2]. Experimentally, more data is oriented toward heat transfer rates or convection coefficients. Mach-Zehnder interferometers [3] or Wollaston prism interferometers [4] are usually used in two-dimensional and axisymmetric free convection problems. Some velocity field data is available for several specific cases. Velocity data is usually obtained from streak photography [5]. However, with all of these experimental techniques the flow must exhibit a simple geometric dependence, usually two-dimensionality, or in some cases axisymmetry. In three-dimensional free convection flow fields the above experimental techniques are not acceptable for checking three-dimensional theoretical solutions. One technique for measuring velocities in free convection gas flows which has not been exploited is a laser velocimeter (LV). For this paper, a laser velocimeter was used to measure the velocities due to free convection of air around a vertical isothermal flat plate. Data is compared to previously determined velocities. Also, the LV was used to measure the velocity field in an air filled triangular enclosure with two isothermal side walls and an insulated bottom. For this case data is presented and compared to previously presented results for inclined isothermal plates.

LASER VELOCIMETERS

To date LV's have not been used in free convective gas flows. Amenitskii
et al. [6] used a LV in free convective liquid flows (glycerin and water), in
which the impurities in the liquids were used as particles. Some doubt has
arisen, however, for LV applications in convective gas flows. Since the
driving buoyancy forces and gas velocities are relatively low, some speculation
concerning the accuracy of LV measurements in free convective flows is present,
since gravity forces may interfere with individual particles motions (particle
dynamics). The data presented in this paper indicates that LV's are capable of
accurately measuring such fluid velocities and that this subject of particle
dynamics is not a problem in free convective gas flows.

EXPERIMENTAL APPARATUS

Vertical Flat Plate

First, to demonstrate the LV's capabilities in two-dimensional free convective gas flows a vertical flat plate assembly was constructed. The plate was
152mm wide (Z) and 152mm high (L) and is constructed of aluminum. A water bath
was attached behind the assembly with an electrical heater coil and stirrer
(see Figs. 1 and 2). The plate was also instrumented with twelve imbedded copper
constantan thermocouples and the surface temperature of the plate (T_w) was uniform within 1°C.

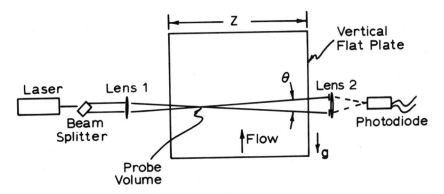

Figure 1: Laser Velocimeter with Flat Plate Assembly

The laser velocimeter which was used is a dual beam forward scatter as
shown in Fig. 1. The laser is a 0.5 watt Argon Ion laser and the 4880 Å line
was utilized. The beam crossing angle, θ, was equal to 5.14°, which resulted
in an elliptical probe volume approximately 200μm in diameter and 2900μm long.
Orientation of the receiving lens (lens 2) reduced the length of the probe
volume to approximately 800μm.   The two laser beams were in a plane parallel
to the flat plate. Thus, the vertical velocity was measured. Velocities in
any other directions could be measured by rotating the beam splitter. Smoke
(which filled the room) and entrained dust were used as the scattering particles.
Particle sizes are estimated to be approximately 1μm for which valid signals
were obtained. Smaller particles were also present as evidenced by a considerable level of noise.

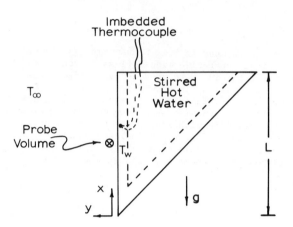

Figure 2:   Vertical Flat Plate Assembly

Scattered light was focused onto a United Dectector Model PIN 020B silicon photodiode. The output of the photodiode was amplified by a factor of 100 and each "burst" signal was displayed on a storage oscilloscope. The Doppler frequency was determined for each particle by counting ten cycles of each burst signal.

For the presented data only laminar flow was examined. Thus, very little "scatter" was found in the data at a given point in the flow, and collecting a large number of samples was not necessary. Recording data from an oscilloscope was, thus, possible. Thirty to fifty samples at each position were recorded and the Doppler frequencies of all particles were determined. The average of all particle velocities (u) at a point in the flow was used in the correlation of the data. Had the flow been turbulent, this data processing used would, of course, not have been acceptable.

The positioning of the probe volume in the flow field was attained by moving the vertical plate as opposed to the optics. Thus, the plate was moved in the y direction for the traversal of the boundary layer with the LV probe volume. Positioning was performed with micrometer heads. Positioning accuracy is 0.05mm in the y direction and 0.1mm in the x direction.

Triangular Enclosure

The air filled triangular enclosure consisted of two constant temperature tanks and one horizontal adiabatic bottom, as shown in Figure 3.a. Concurrently with these tests, a Wollaston prism schlieren interferometer was used to measure the heat transfer rates in the same enclosure [7]. The face of each tank was polished aluminum (1.27cm thick) and was 10.78cm long (L) and 25.4cm wide (Z). These plates formed the two inclined sides of the triangular enclosure. Six copper-constantan thermocouples were imbedded in the aluminum walls and were within 0.16cm of the faces. One tank was maintained at a constant hot temperature by electrical heater coils placed in the water tank and controlled by variable ac transformers. The temperature of the cold surface was maintained by an ice bath mixture in the second tank. The waters in both baths were continually mixed by electrical stirrers such that the walls remained at uniform temperature.

The bottom surface of the enclosure was fabricated from a 2.54cm thick Bakelite plate and was heavily insulated with urethane foam underneath. Aluminum templates on both ends of the test section were screwed into the two side tanks and bottom Bakelite plate for alignment. The two end plates were made of

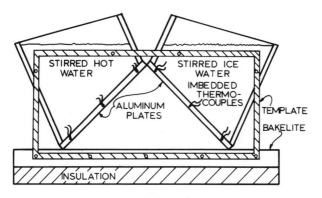

(a) Enclosure Schematic

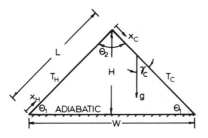

(b) Idealized Enclosure

Figure 3: Triangular Enclosure Assembly

0.64cm optical glass and were sealed onto the ends and held in place with silicon sealant/adhesive. Seams around the corners of the enclosure were also
sealed on the exterior with this sealant.

The idealized enclosure is presented in Fig. 3.b. The values of W, H, θ_1
and θ_2 which were used are 15.24cm, 7.62cm, 45° and 90° respectively. One should
note that the x direction is defined differently for the hot and cold walls due
to the anticipated development of the flow.

The same LV as described above was used to measure the velocities in the
enclosure. The flat plate was replaced by the enclosure and the velocities
between the two glass end plates were measured. Two components of velocity
were measured: those in the x_C and x_H directions. The flow field was traversed
in five approximately equally spaced horizontal planes at distances (h) from
the insulated floor. The test section was moved as opposed to the LV probe
volume as for the vertical plate. Both components of velocity were measured
across the entire test section. Since the flow was laminar, vectorially adding
the averages of the two components to determine the total component (and its
direction) was possible. Positioning again was performed with micrometer heads.
Positioning accuracy is 0.05mm in the horizontal direction and 0.1mm in the
vertical direction.

For the enclosure the angle between the laser beams (θ) was decreased to
0.91 degrees so that the probe volume could be positioned very close to the
surfaces. This was necessary for the measurement of the velocity component
normal to the surfaces, particularly in the corner regions. For this case,

the diameter of the probe volume was approximately 300μm.

Small dust particles were used for the seeding material in the enclosure. Initially, smoke was attempted for this case, but regulating the size distribution of the smoke was not possible. Thus, many more very small particles than scattering particles were introduced with the smoke. As a result, the chamber was very opaque and very noisy signals were obtained. Also, using aluminum powder as seeding material was attempted. The particle sizes were measured with a microscope to be approximately 1 to 3μm. This powder was injected into the chamber but reseeding was often necessary as the particles tended to drop out of the flow.

Thus, to eliminate the very small noise causing particles resulting from the smoke and to eliminate the problem of continual reseeding (and also probably particle dynamics) with aluminum powder, another technique was utilized. Small, finely dispersed chalk dust particles (approximately 1μm) were injected into the chamber via a hyperdermic syringe and needle. The hyperdermic syringe acted as an atomizing jet and only small particles were obtained. The particle sizes were again measured with a microscope and were approximately uniform. For these chalk dust particles, reseeding the flow was only seldom necessary (1 hour of run time or longer), as the particle dropout rate was very low.

EXPERIMENTAL DATA

Vertical Flat Plate

Measurements were taken for several values of x, and approximately ten values of y were used for each x. Experimentally, velocities have been determined for vertical flat plates previously [5,8]. The purpose here is to present data which indicates the LV is capable of making accurate measurements.

The reference temperature rule initiated by Sparrow and Gregg [9] was used in correlating the data. This rule is that a reference temperature (T_r) is used to calculate all fluid properties where

$$T_r = T_w - 0.38 (T_w - T_\infty) \tag{1}$$

and that T_∞ be used to evaluate the coefficient of volumetric expansion ($\beta = 1/T_\infty$).

Velocity and position data have been nondimensionalized as follows such that velocity profiles for vertical and inclined plates approximately collapse to one curve [5,10]:

$$\left. \begin{array}{l} U = (Gr/4X)^{1/2} \hat{u} \\[1mm] Gr = g\beta\cos\gamma(T_w - T_\infty)L^3/\nu_r^2 \\[1mm] X = x/L \\[1mm] \hat{u} = u\nu_r/g\beta(T_w - T_\infty)L^2 \end{array} \right\} \tag{2}$$

$$\left. \begin{array}{l} \eta = Y/(4X)^{1/4} \\[1mm] Y = (Gr)^{1/4}y/L \end{array} \right\} \tag{3}$$

where γ is the angle between the plate surface and the gravity vector (equal to zero for the vertical plate tests).

Typical data is presented in Fig. 4. This data was collected midway between the ends of the plate for $T_w = 86°C$, $T_\infty = 25°C$, and $X = 0.482$. These conditions resulted in a Grashof number (Gr) of 18.6×10^6. The data is compared to the analysis of Yang and Jerger [10] (X = 0.083 and 0.583) and the data of Schmidt and Beckman [8] (X = 0.583). Two values of X from Ref. 10 are presented mainly as a reference demonstrating how much results change with X.

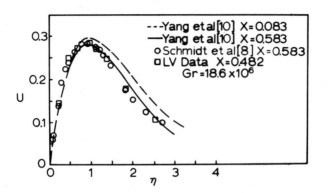

Figure 4: Velocity Profile for Vertical Flat Plate
and Comparison to Previous Results

Present results (X = 0.482) are in excellent agreement with the data of
Ref. 8 (X = 0.583). Differences are generally less than 2%. Although the
data was not taken at exactly the same value of X, Yang's analysis indicates
very little difference should exist between X = 0.482 and 0.583. The present
values of U should be slightly higher than those for X = 0.583. As can be
seen in Fig. 4, this is true. The recorded peak value of U of the present data
is 0.285 (X = 0.482) as compared to 0.280 (X = 0.583).

One problem which arose with making velocity measurements in the open free
convective field was with room currents. With conventional measurement appara-
tuses which use self-induced integration and averaging techniques across a
two-dimensional field, such as Mach-Zehnder interferometers, Wollaston prism
interferometers or streak photography, very small fluctuations are not important.
Very small fluctuations tend to average out in the integration. However, in
making point measurements (LV) these small fluctuations are more pronounced.
Thus, considerable care and time were required to eliminate room currents. The
flow was allowed enough time such that unsteady local air currents dissipated
after each movement of the flat plate assembly. For natural convection in the
enclosure, however, such room currents were not of consequence.

Triangular Enclosure

Measurements were taken midway between the end plates for five values of
h in the chamber and from 20 to 30 measurements were made for each component
at each level. To correlate the velocity data tangent to the nearest surface,
nondimensionalization was once again utlized and Equations (1) to (3) were used.
For the case of the enclosure the free stream temperature (T_∞) was replaced by
the mean temperature ($T_m = (T_H + T_C)/2$), such that comparison to simple flat
plate data was possible. For the enclosure geometry, γ = 45°.

Tangential velocity data near the center of the enclosure is presented in
Fig. 5. The conditions for which this data was collected are T_H = 64°C, T_C =
0°C, and $\overline{H}$ = 0.470. The overall Grashof number based on the total temperature
difference (Gr = $g\beta\cos\gamma(T_H - T_C)L^3/\nu_m^2$) is 6.5 x 10^6 for this case. In Fig.
5, velocity profiles near the two isothermal surfaces are presented and compared
to that of the LV results for the isothermal flat plate (which are also approximately representative of inclined plate results).

On lines across the central portion of this enclosure, one can see that the
velocity profiles exhibit similarity behavior. Little difference is noted between the three sets of data in Fig. 5. Also, in this region the flow was almost

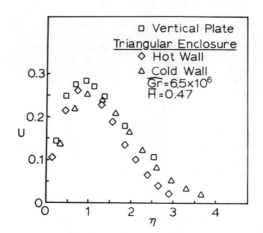

Figure 5:   Velocity Profile for Triangular Enclosure
            for $\overline{H}$ = 0.470 and Comparison to Vertical Flat Plate
            Results

exactly one-dimensional near the surfaces. No velocities normal to either surface were measured for $\overline{H}$ = 0.470. Nondimensionalized velocity profiles at other
values of $\overline{H}$ were also obtained. For brevity and since such profiles would not
supplement the aim of this paper, these figures are not presented here.
       Finally, by vectorially adding the averages of the two components at
selected points in the chamber, a flow map was constructed. This is presented in
Fig. 6 and on this figure is a velocity magnitude scale. One can easily see the
development of the flow and profile shapes on this diagram and observe that in
the central region of the chamber no flow exists. Thus, from this figure, one
can conclude that for this geometry and this Grashof number the air is moving
in the so-called "boundary layer regime." This observation agrees with the
temperature gradient data taken for the same conditions [7].

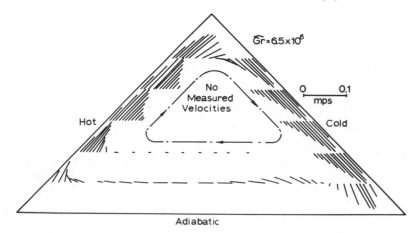

Figure 6:   Flow Patterns in Triangular Enclosure

Particle Dynamics

Two different free convective flow fields were investigated here. The first case was external flow while the second case was internal. For the first case, velocities, which were measured with the LV with smoke and estrained dust, were within 2% of previously measured (streak photography) velocities and within 2% of theoretical results. For this case it is rather obvious that particle dynamics were not influential.

For the second case, atomized chalk dust particles were injected into the test cell. The effects of such particles in free convection were not known previous to this study. Nondimensionalized velocity profiles were presented near both hot and cold surfaces of the enclosure and compared to inclined plate correlation. Near the central horizontal plane of the enclosure corner effects should be minimal and the flow should be essentially one-dimensional. The non-dimensional velocities did approximately collapse onto one curve as expected and compared very well to the simple inclined plate correlation. The small differences can be attributed to the fact that the boundary layer development will be different for simple inclined walls and the enclosure walls because of the flow near the corners. Also, near the walls the velocity vectors were tangent to the surfaces and no normal components were measured. This indicates the vertical gravity vector was not influencing the flow. These observations lead one to conclude that particle dynamics are not influencing the measurements.

Also, for the large Grashof numbers used in the second case the flow was expected to be in the "boundary layer regime" [3]. For this regime, essentially no flow exists in the central portion of the enclosure. By examining Fig. 6 one can see that no velocities were detected in this region. Had particle lag been present, a significant number of signals in the vertical direction would have been measured here due to the particles falling out of the flow near the apex. This observation agrees with the fact that very little reseeding was necessary after each test began.

Thus, since in the central region of the test section the velocity curves approximately collapsed onto a nondimensionalized curve, and no normal components were measured near the walls, and since no vertical velocities were measured in the center of the flow field, the effects of particle dynamics in the enclosure have been concluded to be negligible.

CONCLUSIONS

The aim of this paper was primarily to demonstrate the value of laser velocimeters in both external and internal laminar free convection flows. Overall, accurate results were obtained for free convection over a vertical isothermal flat plate where flows had previously been examined with streak photography and in a triangular enclosure where results were not available. Specific conclusions include: (1) The LV measurements over the flat plate were within 2% of previous measurements and theoretical results. (2) One must be careful in external flows to eliminate room currents or other disturbances. (3) Smoke and entrained dust were the most efficient seeding materials for external flows. (4) Injected atomized chalk dust was the most efficient seeding material for internal flows. (5) For the particular enclosure geometry and conditions examined the flow was in the "boundary layer regime" and velocity profiles approximately collapsed to one curve on a nondimensionalized plot near the center of the geometry.

ACKNOWLEDGEMENTS

The authors gratefully acknowledge the support of the Department of Mechanical and Aerospace Engineering, University of Virginia, for the experimental apparatus and optical system.

REFERENCES

1. Ostrach, S., "Natural Convection in Enclosures," Advances in Heat Transfer,
    Vol. 8, Academic Press, New York, 1972, pp. 161-227.

2. Raithby, G.D. and Hollands, K.G.T., "A General Method for Obtaining Approxi-
    mate Solutions to Laminar and Turbulent Free Convection Problems," Advances
    in Heat Transfer, Vol. 11, Academic Press, New York, 1975, pp. 265-315.

3. Eckert, E.R.G., and Carlson, W.O., "Natural Convection in an Air Layer
    Enclosed Between Two Vertical Plates with Different Temperatures," Int. J.
    Heat Mass Transfer, Vol. 2, Nos.1/2, March 1961, pp. 106-120.

4. Sernas, V., Fletcher, L.S., and Aung, W., "Heat Transfer Measurements with
    a Wollaston Prism Schlieren Interferometer," ASME Paper, 72-HT-9, August,
    1972.

5. Kierkus, W. T., "An Analysis of Laminar Free Convection Flow and Heat Trans-
    fer About an Inclined Isothermal Plate," Int. J. Heat Mass Transfer, Vol. 11,
    No. 2, February 1968, pp. 241-253.

6. Amenitskii, A.N., Rinkevichyus, B.S., and Solovev, G.M., "Measurements Using
    the Doppler Effect of Small Velocities in Flows Occurring in the Free Con-
    vection of Fluids," Soviet Physics-Doklady, Vol. 17, No. 11, May 1973,
    pp. 1078-1079.

7. Flack, R.D., Konopnicki, T.T., and Rooke, J.H., "The Measurement of Natural
    Convective Heat Transfer in Triangular Enclosures," Recently submitted to
    ASME for possible publication.

8. Schmidt, E. and Beckmann, W., "Das Temperatur-und Geschwindigkeitsfeld vor
    einer Wärme abgebenden senkrechten Platte bei natürlicher Konvektion,"
    Tech. Mech. u. Thermodynamic, Bd. 1, Nr. 10, Oct. 1930, pp. 341-349, cont.
    Bd. 1, Nr. 11, Nov. 1930, pp. 391-406.

9. Sparrow, E.M. and Gregg, J.L., "The Variable Fluid Property Problem in Free
    Convection," Trans. ASME, Vol. 80, 1958, pp. 879-886.

10. Yang, K.T. and Jerger, D.W., "First Order Perturbations of Laminar Free Con-
    vection Boundary Layers on a Vertical Plate," Trans. ASME, Journal of Heat
    Transfer, Vol. 86, 1964, pp. 107-115.

Feasibility of Velocity Field Measurement in a Fluidized Bed with a Laser Anemometer*

C. P. WANG, J. M. BERNARD, and R. H. LEE[†]
The Ivan A. Getting Laboratories
The Aerospace Corporation
El Segundo, CA 90045

ABSTRACT

A laser anemometer has the unique capability of measuring velocity field and turbulence parameters without perturbing the flow. Hence, it is possible to apply the laser anemometer to a fluidized bed provided that the refractive indices of all of the materials comprising the fluidized bed are identical. To demonstrate the feasibility of this technique, we have constructed a rotating flow with glass spheres in an index-matched mixture of benzyl and ethyl alcohols. A digital correlator is used with a phase-modulated laser anemometer system to measure the flow field and rms fluctuation velocities. Signal processing techniques, preliminary experimental results, and the criteria for the required index match are discussed.

INTRODUCTION

The fluidized bed reactor[1], although in existence and under study for some time, has become the subject of renewed interest as a result of the role played in various energy-related areas such as coal conversion technology. Various experimental and theoretical studies have been done in these areas[1-3]. However, because of the solid particles in the fluidized bed, it is extremely difficult to measure the local flow velocities and turbulent fluctuation velocities by conventional techniques. Circulation patterns and particle velocities have been obtained using tracer particles in a transparent bed in which glass particles have been fluidized by a liquid of the same refractive index[4]. Other work in this area includes the measurement of fluid velocity inside a fixed bed using a laser anemometer[5].

To measure the local flow velocity and turbulent velocity inside a fluidized bed, a new diagnostic technique is conceived. It is noted that glass beads of different diameters can be fluidized by a liquid at various flow rates, and the local velocities and turbulent velocities in such a system may be measured by means of a laser anemometer[6-8], provided that the refractive index of the liquid is matched to that of the glass beads. These direct measurements in a fluidized bed will provide much needed data for the understanding of the fluidized bed and for model verification.

The main objective of this work is to demonstrate that a laser anemometer can indeed be used to measure the local and turbulent velocity field in a transparent bed, and to determine how closely the refractive index of the glass beads and the liquid must be matched, and to determine the effect of

*Work supported by National Science Foundation, Grant ENG 76-04584

[†]Advanced Program Division, Development Operations is Dr. Lee's Aerospace affiliation.

signal drop-out.

SIGNAL PROCESSING TECHNIQUE

Laser anemometry, a powerful technique for flow velocity measurement, is based on Mie scattering and the Doppler effect[8]. It has been success-fully applied for various flow velocity measurements in wind tunnels, jet, combustion flow, and blood flow[6]. To apply this technique to measure local and rms (root-mean-square) fluctuation velocities in a transparent two-phase flow, signal processing problems may arise. For example, there is signal drop-out due to the fact that no scatterers are present when the sampling volume lies momentarily within a glass bead. Also, how the fluid and bead velocities can be separately determined, and how serious the degradation of the signal is due to refractive index mismatch and non-uniformities in both the glass beads and the fluid. Because of these difficul-ties, we choose to use the most sensitive and flexible signal processing technique, photon correlation[6, 9].

The photon correlation technique, by comparison, has remarkable power. It covers a wide dynamic range, is sensitive to extremely low scattered light intensities, and does not require a continuous signal. The Fourier transform of the correlation function $G(\tau)$ is the power spectrum $P(\omega)$. i.e.

$$P(\omega) = \frac{2}{\pi} \int_0^\infty G(\tau)\cos\omega\tau\, d\tau \tag{1}$$

Hence both mean velocity and rms fluctuation velocity can be obtained[6].

Detailed discussions of the photon correlation technique appear in Refs. [6] and [9]. Briefly, as was shown in Ref. [6], the photon correlation func-tion $G(\tau)$ can be expressed as

$$G(\tau) = b_1 + b_2 \cos b_3\tau \exp(-b_4\tau^2) \tag{2}$$

where b_1 and b_2 are constants, $b_3 = \frac{2\pi\mu}{S}$, $b_4 = 2\pi^2(\frac{\sigma^2}{S^2} - \frac{\sigma_a^2}{S^2})$, μ is the most probable velocity, σ is the standard deviation of the velocity distribu-tion function, S is the fringe spacing, and σ_a is the apparent fraction of turbulence due to the finite number of fringes. The turbulence intensity

$\eta = \sigma/\mu$ can be expressed as

$$\eta \simeq \frac{1}{\pi}\left|\frac{2}{3}\ln\left(\left|\frac{h_1}{h_2}\right|\right)\right|^{1/2} \tag{3}$$

when the turbulence intensity is low ($\eta \leq 20\%$). Here h_1 is the first negative and h_2 is the succeeding positive peak height, about a chosen mean, obtained from the photon correlation function.

To analyze the signal drop-out problem, it can be shown that the effect of signal drop-out is to increase the power spectrum width. Assume an average signal duration of τ, then the measured power spectrum $P_m(\omega)$ is a convolution product of the continuous signal and the gate function with opening duration τ, i.e.

$$P_m(\omega) = P_c(\omega) * P_\tau(\omega) \tag{4}$$

where $*$ means convolution product, $P_c(\omega)$ and $P_\tau(\omega)$ are the power spectrum of the continuous signal and the gate function, respectively. For a Gaussian spectrum, the measured half width is the sum of the true half width and the gate half width. i.e.

$$(\Delta\omega)^2 = (\Delta\omega_c)^2 + (\Delta\omega_\tau)^2 \tag{5}$$

Let us examine a typical case, in which the mean velocity is 15 cm/sec, voidage is 0.73, and rms velocity fluctuation is 4 cm/sec. In our typical experimental setup, this rms velocity corresponds to a spectrum width $\Delta\omega_c$ = 2.7 x 10^3 Hz. For a glass bead diameter of 6 mm and voidage of 0.73, the average clearance is 2.4 mm. The transit time is τ = 2.4 mm/150 mm/sec = 1.6 x 10^{-2} sec. This corresponds to a spectrum width $\Delta\omega_\tau$ = $1/\tau$ = 62.5 Hz. This is much smaller than that due to the rms velocity, $\Delta\omega_c$. Hence $\Delta\omega_\tau$ is negligible.

It is noted that the surface of the glass beads scatter large amounts of light, as they are equivalent to large scattering particles. It is possible to set the clipping level high enough such that only large bursts of signals can contribute to the correlation function. Hence the measured velocity corresponds to the glass beads velocity. On the other hand, we can set the clipping level low enough or use double clipping such that large bursts of signals are saturated and their contribution to the correlation function is minimized. Hence the measured correlation function corresponds to the local flow velocity. More detailed calculations and experimental verification of this discrimination technique are in progress and will be reported later.

EXPERIMENTAL SETUP AND RESULTS

A flow system, shown in Fig. 1, has been constructed for the liquid fluidization of glass spheres. It is designed to provide steady, uniform, and laminar flow at the inlet of the 7.6 cm diameter x 45 cm long test section. It is instrumented to measure the liquid flow rate and the pressure and temperature differences across the bed. In this gravity-fed system (chosen to minimize pump-induced velocity fluctuations), flow rate can be varied from a few cm/sec to a few hundred cm/sec. Preliminary tests of the flow system using water and glass beads have shown fluidization [3].

However, because the index matched fluid contains ethyl alcohol and thus is flammable, an elaborate safety analysis has been done, and explosion-proof pumps, and a vented, completely enclosed facility for the flow system and catch tanks have to be installed before we can run the refractive index matched fluid [3]. Hence we have not yet been able to run the flow system. In order to demonstrate the principle and to investigate the performance characteristics of the laser anemometer, a small rotating flow system with glass beads and refractive index matched fluid was set up. As shown in Fig. 2, the bottom disc, 15 cm in diameter, was rotating at a constant velocity. The side wall was stationary and the top surface was free. The sum of the laser beam path length and the scattered light path length inside the fluid was about 10 cm.

The laser anemometer system using the differential heterodyne arrangement [7] is shown in Fig. 3. The illumination optics provided two equally intense beams which intersected in the test section. The scattered light was then collected by a lens and aperture system and fell onto a photomultiplier. The data processing apparatus was a digital correlator (Malvern K7023), which computed in real time the correlation function of the arrival of photons at the photomultiplier. With correlation delay times adjustable between 50 nsec and 1 sec, this apparatus is ideally suited to measuring mean and rms fluctuating velocity profile in our solid/liquid fluidized bed.

To determine the velocity sense and to aid interpretation of turbulent distributions, a solid-state electro-optic phase modulator [10] (Malvern K9023), which can shift the laser frequency up to 1 MHz, was employed. Using this phase modulator, the apparent turbulence intensity can be increased or decreased at will by applying different phase-shifting frequencies [6,9].

Critical to the minimization of interference from the solid glass spheres on the laser anemometer measurement is the matching of the refractive index of the liquid to that of the glass spheres. Previous workers have used

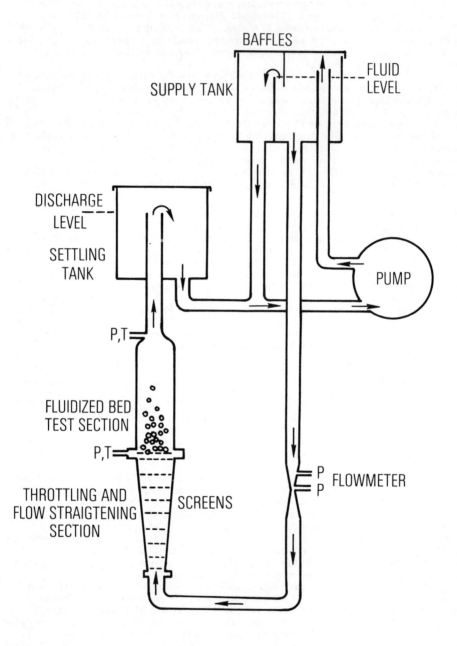

Figure 1. Schematic of fluidized bed.

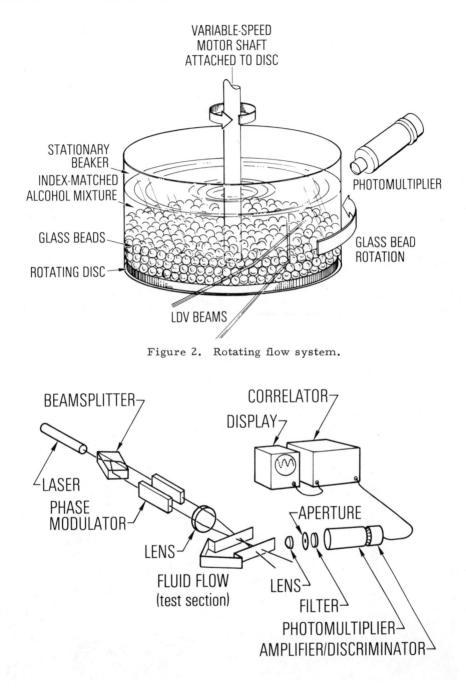

Figure 2.   Rotating flow system.

Figure 3.   Laser anemometer system.

silicone oils [5] and dimethyl phthalate [4]. However, both of these liquids have high viscosities. We found that a mixture of 65% benzyl alcohol and 35% ethyl alcohol, having refractive indices of 1.54 and 1.36 respectively, matched the refractive index of the glass beads very well. A refractive index difference of less than 10^{-4} can easily be achieved. A photograph of this index matching is shown in Fig. 4 along with the appearance of the spheres in pure benzyl alcohol for comparison. The viscosity of the alcohol is relatively low, hence high Reynolds number flow can be achieved.

To demonstrate feasibility, we have performed some preliminary measurements in a transparent two-phase rotating flow using the apparatus shown in Fig. 2. Photon correlation functions with/without flow, and with/without glass beads have been obtained as shown in Fig. 5. Figure 5a shows the correlation function in a stationary fluid without glass beads. The signal frequency corresponds to the frequency shift due to the phase-modulator. Figure 5b shows the correlation function in a rotating flow without glass beads. The signal frequency corresponds to the sum of phase-modulator frequency 48.4 kHz and Doppler frequency 26.8 kHz. This corresponds to a mean velocity of 24.2 cm/sec, agreeing with the velocity expected from theory [11]. Since this is a laminar flow, the rms fluctuation velocity is very small. Figure 5c is the correlation function in a stationary fluid index matched with the glass beads. The signal frequency corresponds to the frequency shift due to the phase-modulator. Although the correlation function is noisier than that in Figure 5a, the mean frequency can easily be determined. In fact, this mean frequency is not too sensitive to the matching of refractive index, the mean frequency can still be determined with a refractive index mismatch up to $\Delta n = 10^{-3}$. Figure 5d is the correlation function in a rotating flow with index matched glass beads. The signal frequency is the same as in Figure 5b, but the correlation function is decaying. This decaying of the correlation function corresponds to an apparent rms velocity of 1.25 cm/sec according to Eq. 3. This large apparent rms velocity is due to the presence of glass beads.

Fig. 4. Glass beads in benzyl alcohol (right) and a liquid of matched refractive index (left).

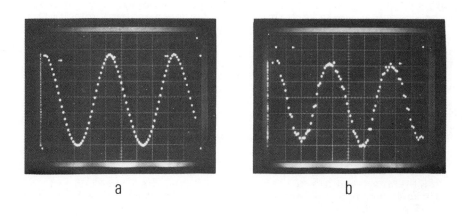

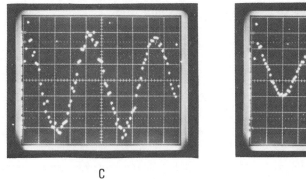

Figure 5.   Photon correlation functions.

   a. Stationary-fluid without glass beads; sampling time 0. 5
       μsec/dot; averaging time 10 min. ; and signal frequency 48. 4
       kHz.
   b. Rotating flow with no glass beads; sampling time 0. 35 μsec/
       dot; averaging time 10 min. ; and signal frequency 75. 2 kHz.
   c. Stationary fluid with glass beads; sampling time 0. 50 μsec/
       dot; averaging time 10 min. ; and signal frequency 48. 4 kHz.
   d. Rotating flow with glass beads; sampling time 0. 35 μsec/dot;
       averaging time 10 min. ; and signal frequency 75. 2 kHz.

CONCLUSION

    In conclusion, our preliminary experimental results indicate that using
a laser anemometer in a transparent two-phase flow, the local and turbulent
velocities can be measured. Critical issues, such as signal drop-out seem
not to be a problem. Also, determination of the mean and rms velocity from
the measured correlation function is not too sensitive to the degradation of
signal due to refractive index mismatch and the non-uniformity of the glass
beads and fluid. For a optical path length of 10 cm in the transparent two-
phase flow, a refractive index mismatch of $\Delta n \leq 10^{-3}$ is tolerable. As for
the separate determination of glass beads and local flow velocity, experimental work is in progress and the result will be reported later.

REFERENCES

1.   D. L. Kearnes, ed. Fluidization Technology , Vol I and II, Hemisphere
      Publishing, N. Y. (1975).

2.   R. J. Bywater, R. H. Lee and P. M. Chung, "A Statistical Flow Model
      for Fluidized-bed Reactors", Presented in the AIAA 8th Fluid and
      Plasma Dynamics Conference, Hartford, Connecticut, June 16-18,
      1975 or AIAA Paper No. 75-874.

3.   R. H. Lee, R. J. Bywater, J. M. Bernard and C. P. Wang, "Turbu-
      lent Structure in Fluidized Beds", unpublished.

4.   B. A. J. Latif and J. F. Richardson, "Circulation Patterns and
      Velocity Distributions for Particles in a Liquid Fluidized Bed", Chem.
      Eng. Sci. 27, 1933-1949 (1972).

5.   W. Johnston, A. Dybbs and R. Edwards, "Measurement of Fluid
      Velocity Inside Porous Media with a Laser Anemometer", Phys. Fluids
      18, 913-914 (1975).

6.   T. S. Durrani and C. A. Greated, Laser Systems in Flow Measurement,
      Plenum Press, N. Y. (1977), Chapter 4.

7.   C. P. Wang, "A Unified Analysis on Laser Doppler Velocimeters",
      J. Phys. E:Sci. Instr. 5, 763-766 (1972).

8.   C. P. Wang and D. Snyder, "Laser Doppler Velocimetry: Experimental
      Study", Appl. Optics 13, 98-103 (1974).

9.   A. D. Birch, D. R. Brown, J. R. Thomas and E. R. Pike, "The
      Application of Photon Correlation Spectroscopy to the Measurement of
      Turbulent Flows", J. Phys D : Appl. Phys. 6, L71-L73 (1973).

10.   R. Foord, A. F. Harvey, R. Jones, E. R. Pike and J. M. Vaughan,
      "A Solid-state Electro-optic Phase Modulator for Laser Doppler
      Anemometry", J. Phys. D : Appl. Phys. 7, L36-L39 (1974).

11.   F. Bien and S. S. Penner, "Velocity Profiles in Steady and Unsteady
      Rotating Flows for a Finite Cylindrical Geometry", Phys. Fluids 13,
      1665-1671 (1970).

Design and Evaluation of a Laser Doppler Velocimeter for the AEDC-PWT 4-Ft Transonic Tunnel[1]

F. L. CROSSWY,[2] **D. B. BRAYTON,**[2] and **D. O. BARNETT**[3]
Sverdrup/ARO, Inc., AEDC Division
Arnold Air Force Station, TN

ABSTRACT

The design details and initial performance evaluation of a two velocity component Laser Doppler Velocimeter (LDV) system for a large transonic wind tunnel are presented. The two color, backscatter type LDV was specialized for free stream, pitch plane flow angle and velocity magnitude measurements using only intrinsic particulate matter as light scatterers. An environmental control system and remote optical alignment techniques were developed since the LDV was located in the hostile environment (vibration, vacuum, inaccessibility, etc.) of the tunnel plenum region. A new type LDV signal processor was developed to process the low level signals produced by the small naturally occurring particles in the tunnel flow. A versatile computer controlled system was developed for high speed data acquisition. Mean LDV velocity and flow angle data are compared to thermodynamic tunnel velocity calculations and aerodynamic flow angle probe data.

INTRODUCTION

Velocity magnitude and flow angle in the vicinity of a test article are two parameters affecting the aerodynamic characteristics of the model. Frequently, such models will be tested and then returned, with modifications, for further testing. For proper aerodynamic evaluation of the modifications it is highly desirable that the original test conditions, including velocity and flow angle characteristics, be duplicated as closely as possible. It is essential, therefore, in obtaining basic aerodynamic model performance data or in assessing the effects of modifications to a model that the local velocity and flow angularity be monitored throughout a test program.

---

[1]The research reported herein was performed by the Arnold Engineering Development Center (AEDC), Air Force Systems Command. Work and analysis for this research was done by personnel of ARO, Inc., a Sverdrup Corporation Company, operating contractor of AEDC. Further reproduction is authorized to satisfy needs of the U. S. Government.

[2]Research Engineer, Instrumentation Branch, Propulsion Wind Tunnel Facility, ARO, Inc.

[3]Formerly Senior Research Engineer Sverdrup/ARO, Inc.; presently Assoc. Professor, School of Engineering, University of Alabama, Birmingham.

The use of mechanical probes to continuously monitor variations in tunnel conditions during model tests in Tunnel 4T is not practical since the probes themselves can induce objectionable aerodynamic disturbances. Inverted model techniques for flow angle compensation are time consuming and therefore costly. Even though tunnel empty (no model) flow angle versus tunnel condition information is available, there is still some uncertainty as to test section flow angle characteristics during model tests. Tunnel calibration data may be used to estimate mean test section velocity but provides no localized velocity information. Experience at AEDC indicated that the Laser Doppler Velocimeter (LDV) might be used to obtain on-line local flow field information.

The goal of the 4T LDV effort was, then, to adapt LDV technology for useful free-stream velocity magnitude and flow angle measurements in a large aerodynamic test facility such as Tunnel 4T. Major objectives of this effort were to (1) specify the 4T environmental constraints and (2) to determine ultimate LDV utility within these constraints, specifically (a) data rate for unseeded flow and (b) measurement uncertainties.

DESIGN CONSTRAINTS

The development of an LDV system for 4T required numerous design innovations and compromises because of the many environmental constraints imposed by the tunnel. The most severe constraints and design consequences were: (1) Inaccessibility - Optical access to the flow required locating the LDV system within the plenum region of 4T, therefore the design provides for vacuum protection and remote optical alignment and operation, (2) Vibration - In this location, objectionable levels of vibration are encountered so that a vibration isolation system and specialized lock-down devices for adjustable optical components were necessary, (3) Restricted Optical Access - The small test section window port (5 in. x 6 in.) in 4T and the limited access to this window constrained the LDV optical design to a coaxial transmitter-receiver configuration as opposed to a more desirable off-axis configuration. The small window also limited the traversing capability of the LDV system to a single direction along a line transverse to the tunnel flow, (4) Intrinsic Particulate Matter - Artificial seeding is not presently used in 4T because of possible contamination of the test article and tunnel instrumentation. In addition, a tortuous path viscous filter is installed in the tunnel circuit to restrict the passage of dustlike particles and H_2O condensate is normally suppressed during tunnel operation. The resultant dearth of light scattering particles necessitated an efficient LDV optical design as well as development of low level signal processing techniques.

APPARATUS

Tunnel 4T - Tunnel 4T is a continuous flow, closed-circuit, variable density, transonic wind tunnel with test section dimensions 4 ft by 4 ft (cross section) by 12.5 ft long. The basic Mach number range is 0.1 to 1.3. Two separate sets of nozzle block inserts can be installed to provide Mach 1.6 or 2.0 operation. Variable porosity walls are employed for reduction of wall interference effects. The tunnel stagnation pressure can be varied from 160 to 3700 psfa and the stagnation temperature can be controlled to a limited extent over the range 80 to 160°F. In addition to the conventional sting-mounted force and pressure tests of aerodynamic models, 4T can be used for highly specialized testing techniques such as captive-trajectory store separation, store drop studies, magnus effects testing and dynamic stability investigations.

Environmental Control System - The location of the LDV system with respect to the 4T test section and plenum region is shown schematically in Fig. 1. The

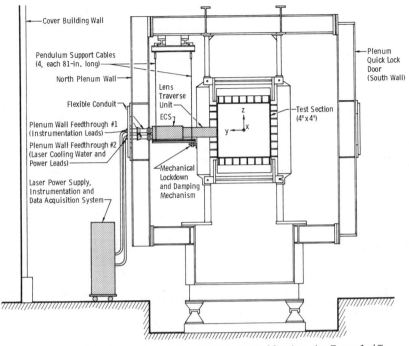

Figure 1.   Laser Doppler Velocimeter Installation in Tunnel 4T
            (Upstream-to-Downstream View)

laser, transmitting optics and receiving optics are mounted inside an Environmental Control System (ECS) which provides a controlled vibration, temperature,
vacuum and dust-free environment. Vibration protection is provided by suspend-
ing the ECS as a pendulum. The natural frequencies in the X, Y, and Z directions are 1, 0.7, and 10.5 Hz, respectively. Temperature sensitive LDV optical
components are mounted on temperature controlled platens inside the ECS enclosure. The laser is housed in a compartment held at atmosphere pressure whereas
the transmit-receive optics compartment assumes plenum pressure.

Two Color Optics - Two color optics were developed for the 4T LDV system
as shown in Fig. 2. Several features are worthy of emphasis. High efficiency
optical interference filters are used to separate the 488.0 nm and 514.5 nm
lines from the two multi-color laser beams. The color separation elements are
mounted on two temperature controlled platens to stabilize their beam steering
characteristics. The parallelism of the two blue and two green beams exiting
the color separation elements can be mechanically adjusted while visually monitoring fringe quality at PV3. Transmitter beam light scattered from the surfaces of lens L5 is prevented from entering the receiver by use of the tele-
scoping baffle tube assembly TBT. Probe volume PV1 is positioned along the
Y-axis by moving lens L5. An invaluable remote optical alignment technique is
implemented by the motor driven mirror M9. By driving the pitch and yaw motors
MTR1 and MTR2 the image of PV1 can be realigned onto aperture AP1 following
slight misalignment of any of the LDV optical elements. The green and blue
fringe planes are inclined at a nominal 45° with respect to the X-axis in order
to eliminate the flow direction "dead zone" problem identified in Ref. 1. The
window W2 was specially prepared since it has been found that surface non-
uniformities, scratches, and adhering particles can shift the spatial orientation of the probe volume fringes by several tenths of a degree.

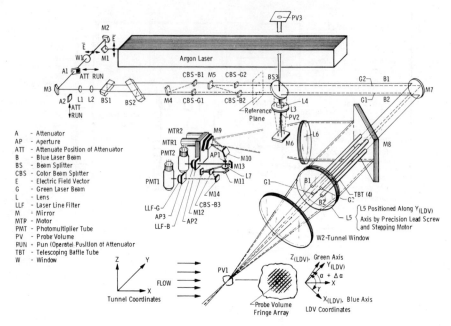

Figure 2.   Two Color LDV Optics

LDV Signal Processors - The LDV signals are processed by AEDC developed counter type instruments termed Doppler Data Processors (DDP). The DDP accommodates signal frequencies from 15 KHz to 100 MHz and has a signal period measurement resolution of 1.0 nsec over the range 15 KHz - 4 MHz and 0.01 nsec over the range 4-100 MHz. The DDP requires 8 relatively well defined signal periods and rejects distorted signal waveforms by requiring the 4 and 5 cycle counts to be 4/8 and 5/8 of the 8 cycle count to within a selectable tolerance of 0.79, 1.5, or 3.0 percent [2].

A new AEDC developed instrument termed the Doppler Data Analyzer (DDA) was evaluated with the 4T LDV system for processing low level, photon limited signals that are normally rejected by the DDP instrument [3]. The latest model of the DDA is a counter type instrument controlled by three high speed microprocessors and is programmed to measure the period of signals containing 1, 2, or 4 signal cycles. The period data is arrayed in a histogram format with the mode of the histogram taken as the best initial estimator for the most probable signal period. A more accurate statistical estimator is obtained by averaging the period data contained in the five histogram cells centered on the mode cell. High speed circuitry is used throughout the DDA so that limited duration (limited by memory capacity) velocity sample rates in excess of 10 MHz and histogram generation rates of 40 KHz are possible (for signal frequencies $\geq$ 50 MHz). The DDA has been developed as a stand alone instrument and is not tied to the data acquisition system discussed in the following paragraph.

Data Acquisition System - A computer based data acquisition system was developed for the 4T LDV effort. The computer has a 32K memory supplemented by a 4.9 megabyte disk memory unit. For data sets consisting of data from two DDP instruments and a digital clock, a data throughput rate of 36,000 sets per second can be achieved. An interactive program with numerous options is used to permit adaptation of the LDV system to particular test requirements. Data are permanently recorded on magnetic tape with on-line data provided by a line printer. The 4T test section free-stream Mach number is computer controlled to within ±0.005. When the Mach number is outside this tolerance LDV data acquisition is automatically inhibited until proper tunnel conditions are reestablished.

MEASUREMENT UNCERTAINTIES

The uncertainty in the value obtained for the velocity or flow angle measured by an LDV is dependent upon many factors. Two of the most difficult to assess are errors resulting from particle-fluid interaction dynamics and the possible statistical bias of the data due to the inherent sampling characteristics of LDV systems. All LDV measurements in 4T to date have been obtained in the free stream and no clear evidence of particle lag or statistical bias effects have been observed.

A widely used expression for the overall uncertainty, U_f, in the measurement of a function $f = f(x_i)$ is given by [4].

$$U_f = \pm(B_f + t_{0.95} S_f) \tag{1}$$

where the x_i are various system parameters, B_f is the overall bias error, S_f is the overall precision error and $t_{0.95}$ is the 95 percentile point of the Student t distribution. An elemental bias error b_{x_i} is associated with each x_i parameter so that the overall bias error is given by

$$B_f = \left[\sum_{i=1}^{N} \left(\frac{\partial f}{\partial x_i} b_{x_i}\right)^2\right]^{1/2} \tag{2}$$

Similarly, the overall precision error is related to the elemental precision errors by

$$S_f = \left[\sum_{i=1}^{N} \left(\frac{\partial f}{\partial x_i} s_{x_i}\right)^2\right]^{1/2} \tag{3}$$

The various elemental errors associated with an LDV measurement are obtained from considerations related to the operations of (1) calibration, (2) signal processing and data acquisition, and (3) data reduction.

Single Velocity Component LDV - The initial LDV measurements in 4T were single velocity component measurements in the x-axis direction (Figs. 1 and 2) using a dual-scatter, backscatter optical configuration. The uncertainty in each V_{x_i} velocity sample was found to be

$$U_{V_{x_i}} = \left[A_1 V_{x_i}^2 + 2\left(A_2 V_{x_i}^2 + A_3 V_{x_i}^4\right)^{1/2}\right] \tag{4}$$

where $A_1 = 7.3 \times 10^{-7}$ sec/ft, $A_2 = 2.6 \times 10^{-5}$, $A_3 = 8.6 \times 10^{-12}$ sec^2/ft^2. The A_1 coefficient stems from the signal processing operation and reflects the signal-to-noise (S/N) ratio dependent bias error of the DDP instrument. The A_2 term represents the precision error associated with the LDV system calibration procedure. The A_3 coefficient represents a signal processing precision error which encompasses the basic time resolution capability of the DDP and the precision error induced in the measurements by S/N ratio effects. The A_3 precision

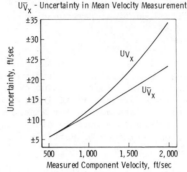

Figure 3.   Uncertainty in the Measurement of a Single
             Velocity Component

error can be significantly reduced by exploiting time interval averaging techniques [5 and 6].   Therefore, the uncertainty expression for mean velocity measurements, $U_{\overline{V}x}$, is the same as Eq. 4 except that the A_3 term approaches zero. Figure 3 is a plot of Eq. (4) versus measured velocity.   It can be seen that the uncertainty in mean velocity measurements is approximately ±1.1 percent of the measured value throughout the velocity range.   The uncertainty in the single velocity sample values starts at ±1.1 percent at 500 ft/sec and increases to ±1.7 percent at 2000 ft/sec.

Two Velocity Component LDV – The two velocity component LDV shown in Fig. 2 was set up to measure velocity magnitude and flow angle in the X–Z plane. The expression for a velocity magnitude sample is given by

$$V_{xz_i} = \left[\left(\frac{V_{Gi} - V_{Bi} \cos \gamma}{\sin \gamma}\right)^2 + V_{Bi}^2\right]^{1/2} \tag{5}$$

where V_{Gi} is the velocity vector projection measured by the green fringe system, V_{Bi} is the projection measured by the blue fringe system and γ is the angle subtended by the blue and green measurement axes (Fig. 2).   The expression for a flow angle sample is given by

$$\phi_{xz_i} = (\alpha + \Delta\alpha_i) - \cos^{-1}\left(\frac{V_{Gi} \sin^2 \gamma}{V_{Gi}^2 - 2V_{Bi}V_{Gi} \cos \gamma + V_{Bi}^2}\right)^{1/2} \tag{6}$$

where α is the static angle subtended by the green measurement axis and the x-axis (Fig. 2) and $\Delta\alpha$ represents small vibration induced variations about α. The angle α is determined during the LDV calibration procedure and $\Delta\alpha$ is measured by an inclinometer which monitors the rigid body inclination of the LDV optical system in the X–Z plane.   The inclusion of the γ factor in Eqs. 5 and 6 is necessitated by the slight non-orthogonality ($\gamma = 89.1°$) of the blue and green axes.   Initial failure to account for axis non-orthogonality resulted in data reduction errors of about one percent in velocity magnitude and up to several tenths of a degree in flow angle.   Mean velocity magnitude, $\overline{V}_{xz}$, and mean flow angle $\overline{\phi}_{xz}$, estimates are determined by replacing V_{Bi}, V_{Gi}, and α and $\Delta\alpha_i$ by their arithmetic means.

Equations (5) and (6) were substituted into Eq. (1) along with specific measured sets of parameter values to compute the uncertainties for the XZ plane velocity and flow angle measurements. These uncertainties were then included in the data plots shown in Figs. 5-6.

TEST RESULTS

Single Velocity Component Measurements – The single velocity component LDV system was used to evaluate (1) LDV performance using only intrinsic tunnel particulate matter as light scatterers and (2) the newly developed DDA instrument.

Under normal tunnel operating conditions (no H_2O condensate) LDV operations were generally unsatisfactory because of the lack of sufficient size light scattering particles. However, particulate size and number density were found to be ideal for LDV operations over the Mach number range 0.6 to 1.3 when low to moderate levels of H_2O condensate were produced in the tunnel test section. Under these conditions LDV velocity sample rates as high as 1000 per second were often observed. However, when H_2O condensate was formed at Mach 1.6 and 2.0 the particulate size was so small and number density so large that the photo-detector signals were completely chaotic with the result that LDV operations were not possible.

Figure 4 is typical of the LDV data recorded in 4T. LDV mean velocity measurements were compared to U_∞, the free-stream velocity derived from tunnel calibration procedures. Data points denoted by the circles were obtained from 1000 DDP velocity samples while the triangles represent data derived from 30,000-75,000 DDA velocity samples. The LDV mean velocity data is seen to compare with U_∞ to within 1.5 percent for the entire Mach number range shown. This data comparison is affected, to some degree, by the fact that the LDV is essentially a point measurement device whereas U_∞ is a gross average quantity.

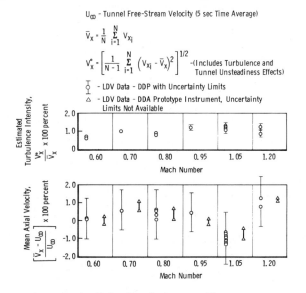

Figure 4.   Single Velocity Component Measurements at the
            Tunnel Centerline

Unpublished pitot-static probe data show similar velocity variations about U_∞. Therefore, some of the differences between $\overline{V}_x$ and U_∞ are probably attributable to local velocity deviations from U_∞. The relative turbulence intensity estimates are seen to vary between 0.6 and 1.2 percent over the Mach number range but, these measurements are possibly affected by tunnel unsteadiness.

Figure 4 shows consistent agreement of the DDA and DDP data. Further observations during this testing period showed that the DDA met its major design goal of being able to acquire data from small particulate matter. This was demonstrated by the fact that the DDA was able to acquire accurate velocity data even after the DDP had ceased to function because of insufficient signal cycles produced by the small particles.

Two Velocity Component Measurements - The two-color LDV system shown in Fig. 2 was used to acquire XZ plane flow angle measurements for comparison with aerodynamic flow angle probe data. Velocity magnitude was compared to the freestream velocity. Figure 5 shows a Y-axis scan from the tunnel centerline toward the north tunnel wall. The discontinuity in the data was caused by a slight change in diffuser flap position during the LDV scan. The LDV data is seen to be in good agreement with the U_∞ and flow angle probe data. These data were acquired under normal tunnel conditions (extremely low levels of H_2O condensate) using the DDP instrument. Consequently, the time required for data acquisition was objectionably long for routine utilization during model tests.

Figure 6 shows a Y-axis scan at Mach 0.4. As at Mach 0.8 the LDV data is seen to be in good agreement with the U_∞ and flow angle probe data. A significant tunnel condition change during the LDV scan produced the noticeable discontinuity in the data. It is noted that in response to the tunnel condition

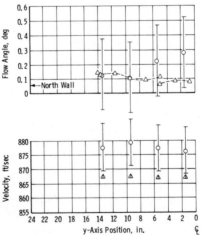

Figure 5.   Y-Axis Scan - X-Z Plane Flow Angle and Velocity Magnitude Measurements at Mach 0.8

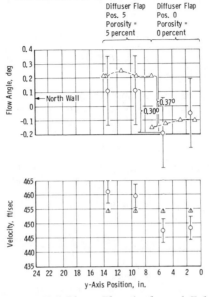

Figure 6.   Y-Axis Scan - X-Z Plane Flow Angle and Velocity Magnitude
            Measurements at Mach 0.4

change the flow angle probe data shows a discontinuity of 0.37 degrees compared
to 0.30 degrees for the LDV data. For the first set of tunnel conditions the
LDV velocity was about 1.4 percent lower than U_∞ whereas for the second set of
conditions the LDV velocity was about 1.4 percent higher than U_∞. As previously
noted, this is probably attributable to the fact that the LDV measures local velocity whereas U_∞ is a gross average quantity.

At the end of the testing period during which the data of Figs. 5 and 6
were acquired, an experiment was conducted at Mach 0.8 to determine if H_2O condensate formation could be controlled and thereby exploited when required for
LDV operations. The only change in tunnel procedure was that tunnel circuit
make-up air was drawn from an atmospheric source rather than through the desic-
cant material normally used. Under dry conditions the LDV data rate was about
one velocity sample per second while, with the atmospheric air source, condensate was formed and the data rate jumped to 200 velocity samples per second.
About 10 minutes were required to change over from the atmospheric source and
return to normal tunnel operating conditions.

CONCLUSIONS

This study has shown that usefully accurate flow angle and velocity magnitude data can be obtained by LDV techniques in a large aerodynamic test
facility like 4T. However, exclusive reliance upon intrinsic dust-like particulate matter for light scatterers imposes a fundamental limitation on the LDV
data productivity. Further development and application of low level signal processing instruments such as the DDA and the photon correlator should provide at

least a moderate increase in data productivity. When thermodynamically possible and aerodynamically permissable, controlled formation of H_2O condensate can provide an ideal natural source of particulate matter for LDV operations at transonic velocities. In contrast, condensate size was too small and number density too large for useful LDV operations in 4T for Mach numbers above 1.3.

An analysis of the 4T LDV system indicates an uncertainty in mean XZ plane velocity magnitude measurements of about ±1.1 percent and an uncertainty in mean flow angle measurements of about ±0.25 degrees. The largest portion of these uncertainties is caused by precision errors associated with the presently used LDV calibration techniques.

Successful operation of the LDV system was preceded by extensive efforts to counter the environmental factors (vibration, vacuum, etc.) encountered in Tunnel 4T. An invaluable remote optical alignment technique was developed to compensate for slight optical misalignments induced by sustained periods of operation in the tunnel vibration environment.

The two-color LDV optics developed for use in 4T represent a technological advance compared to the polarization and dual Bragg cell type two-velocity component optics previously developed and used at AEDC in that (1) 100 percent more of the available laser power is utilized and (2) the channel-to-channel cross talk is less.

The good agreement of LDV data with the free-stream tunnel velocity and aerodynamic flow angle probe data was encouraging. However, it should be emphasized that attainment of accurate LDV data in a large aerodynamic test facility like 4T can require expenditures of design and operations efforts at least an order of magnitude greater than simpler applications where such factors as tightly scheduled tunnel operations, hostile environment (vibration, etc.) and remote and inaccessible location of the LDV system are not encountered.

REFERENCES

1. Brayton, D. B., Kalb, H. T., and Crosswy, F. L. 1973. A Two-Component Dual-Scatter Laser Doppler Velocimeter With Frequency Burst Signal Readout. Applied Optics, Vol. 12, No. 6, pp. 1145-1156.

2. Kalb, H. T., Brayton, D. B., and McClure, J. A. 1973. Laser Velocimetry Data Processing. AEDC-TR-73-116.

3. Kalb, H. T. and Cline, V. A. 1976. New Technique in the Processing and Handling of Laser Velocimeter Burst Data. Review of Scientific Instruments, Vol. 47, No. 6, pp. 708-711.

4. Abernethy, R. B., Thompson, J. W., et al. 1973. Handbook - Uncertainty in Gas Turbine Measurements. AEDC-TR-73-5.

5. Hewlett-Packard Application Note 162-1. "Time Interval Averaging."

6. Barnett, D. O. and Giel, T. V. 1976. Application of a Two-Component Bragg Diffracted Laser Velocimeter to Turbulence Measurements in a Subsonic Jet. AEDC-TR-76-36.

Ocean Laser Velocimetry Systems: Signal Processing Accuracy by Simulation

WILLIAM T. MAYO, JR.
Spectron Development Laboratories, Inc.
3303 Harbor Boulevard, Suite G-3
Costa Mesa, CA 92626

ABSTRACT

This paper summarizes recent sponsored research by SDL concerning the accuracy limitations of multiparticle and photon noise effects on ocean-going laser velocimeter systems. An automatic threshold selection technique has been shown to help reduce multiparticle errors. A simple perturbational theory for predicting the rms errors for a burst counter system has been demonstrated experimentally and by direct photon noise simulation. Calculated Mie scattering coefficients for clear ocean water are also presented.

SDL has developed computer software which simulates the LV system physics and electronic detection including the random realization of specified particle size and index distributions; random particle arrival; signal amplitudes calculated from Mie scattering, laser power, optical geometry, propagation losses, and detector parameters; random realization of the individual photon events which constitute the signal; real photomultiplier tube effects, the effects of bandpass filters; and several models of burst-counter and photon correlation processors.

We have applied the simulation software specifically to the problem of a high-accuracy argon laser velocimeter system for a range of 3 meters in the clear ocean. Relevant particulate size data is used. The results depend severely on particle index of refraction which is not well known.

INTRODUCTION

There are many potential applications of laser velocimeters in the sea. These include the study of ocean wave action, turbulence and material transport, currents, structure generated flow fields, and other applications. Spectron Development Laboratories has conducted a study for ARPA [1] during the period May 1977 - May 1978 concerning the limitations of accuracy of fringe laser velocimeter systems for use in clear ocean water. This study utilized fundamental data collected or computed in a related NAV SEA study [2]. The emphasis for the ARPA work was to determine whether multiple-particle errors and/or photon noise prevent the attainment of precision of approximately $1:10^3$ to $1:10^4$ with averaging techniques at ranges of 1 to 3 meters. The study of refractive fluctuation errors which has been included in the NAV SEA efforts [2] were specifically omitted in this work. Clock count errors which limit higher speed air flow measurements are within the desired accuracy range if a 500 MHz counter detection system is used and were not considered further.

In restricting the problem to the error limitations due to photon noise and multiple particle effects, we address fundamental concerns of laser velocimetry. There is strictly almost no such thing as the mythical "single-particle" LV signal. Most natural particulate distributions have enough of the

473

very small particles [3] that there are always many in the probe volume. When a single particle of significantly larger size is present, the signals due to the smaller particles have only a small effect. Thus, due to the shape of naturally occurring size distributions, it is also difficult to obtain the often referred to many-particle case with Gaussian field limit. When the detection threshold is changed, the detected error level changes. These errors, which may be too small to be rejected by gross error rejection circuits, arise out of the perturbation of the zero crossings in a marked inhomogeneous filtered Poisson process [3]. The behavior of such errors is highly dependent on the scatterer size distribution, the composition (index of refraction distribution), and the size of the probe volume. We have chosen to study such errors by digital computer simulation techniques and have developed an algorithm for automatic threshold adjustment to avoid large multiparticle error effects for many particle distributions.

The statistics of photon noise becomes Gaussian [3] when there are many photoelectrons per filter resolving interval [3]. While this makes the mathematical treatment of errors simpler, one should always keep the light levels (particularly the background light levels and flare light) as low as possible. For this work, we have derived a perturbational model for small errors at high "signal-to-noise ratios" and performed simple experiments which validate the simple theory. We have also completed a much extended version of the simulation software developed earlier for NASA Langley [4] which allows direct study of burst counter processor errors in the presence of filtered inhomogeneous Poisson noise.

The development of our simulation software is documented in detail, along with the results which we summarize only briefly below, in the final report for our ARPA contract [1].

SCATTERING AND EXTINCTION MODELS

The current state-of-the-art for ocean particulate models is briefly reviewed in Reference 2.

A number of individuals have provided size distributions of ocean particulates and have made reasonable assumptions concerning the average index of refraction of these particles [2]. Hyperbolic distributions (commonly called Jung's distributions) are of the form:

$$N(d > y) = ky^{-m} \qquad (1)$$

where N(d > y) = cumulative number of particulates per cubic centimeter
            greater in diameter than y (expressed in micrometers),
       m = characteristic slope of the distribution, and
       k = numbers of particulates per cubic centimeter greater
            than one micrometer diameter.
Typical clear ocean water has values of k between 2000 and 20,000/cc. Typical slopes are m = 2.7 for the 1–10 micrometer diameter range. Conservation of volume (partial volume due to particulates) shows that m must exceed 3 for larger particles [5].

The above type of size data is generally obtained by either Coulter counter measurements of small samples or microscope inspection of small samples. Both procedures are inadequate for determining the distribution of the larger particles (greater than 10 μm in diameter) due to the small sample sizes. This is unfortunate, since calculations show that in some situations, particles in the 10–100 μm size range are known to exist, but little data is available.

Some of the most thorough index of refraction modeling efforts have been made by Gordon and his associates. In a fairly recent paper, Brown and Gordon [6] provide clear coastal water models with a small-fraction index of refraction (with relative occurrence probability) and a large-fraction index of refraction.

The deduction of the effective values of the indices was based on trial and error fits of average Mie scattering computations to measured angular volume scattering functions (VSF). The size distributions used were piecewise linear fits on a log-log plot of Coulter counter size distributions with the small-fraction number densities selected somewhat arbitrarily to satisfy the VSF data. The procedure in such modeling attempts is to use a priori assumptions that organic components have indices in the 1.01 to 1.05 range and mineral components fall in the 1.15 to 1.20 range. The resulting fits of the predicted VSFs agree reasonably well with measured data, except at very small forward-scatter angles (indicating lack of correct large particle number data) and back-scatter angles near 180° (where no scattering data was measured).

The performance of ocean going laser velocimeter systems will depend great-ly on the type of particulates. Figure 1 is a plot of the differential Mie scattering cross section versus size for the two different indices of refrac-tion 1.03 and 1.15, for wavelength 488 nm/1.33. We have also made polar plots of the Mie scattering coefficients and find that the variation with angle is much less than that which occurs for a typical aerosol particle.

The forward scatter due to particles in water is nearly independent on the index of refraction. Typical narrow beam 1/e extinction lengths in deep clear ocean water are 5 to 20 meters for the blue-green wavelengths.

In the simulations that are described below, we have considered a "worst case" of 2000 particles/cc with index of refraction = 1.03+j0 and a "best case" of 20,000 particles/cc with index of refraction = 1.15+j0, both in the range $1 < d < 10$ micrometers. More exact simulations will be possible if the size distribution is simultaneously measured with the backscatter coefficients in ocean experiments.

THREE METER LV DESIGN AND SIMULATION

We have used the simulation software to look at the behavior of several different sets of optical system parameters. Multiple particle problems be-come severe even for clear ocean water unless the probe volume size is kept

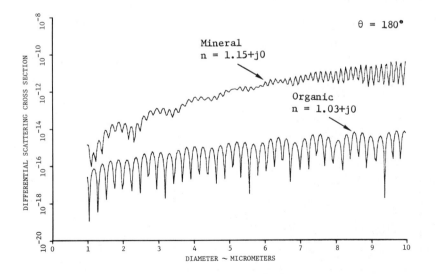

Figure 1. Calculated Mie Scattering Cross Sections for Waterborne Particulates; 180° Backscatter.

small. We do not know whether random refractive effects will allow the diffraction-limited focussing of 0.75 cm diameter beams at a range of 3 meters.
We have assumed such transmitter beam sizes with beam angle of 0.03 radians. A
one watt argon laser operating at free-space wavelength 488 nm is assumed along
with optical efficiency of 0.3 and detector quantum efficiency of 0.2. The
photomultiplier tube (PMT) charge gain is assumed to be only 10^5 with exponential single photoelectron pulse height statistics. A Gaussian bandpass filter
with 300 KHz bandwidth is assumed. The processor model determines zero crossings by linear interpolation between the periodically spaced sample values. A
Macrodyne-type error circuit which requires sequential crossing of thresholds
above and below zero in between the zero crossings is simulated. A histogram
generator which sorts the normalized errors is also included. Figure 2 is a
block diagram of the computer software. The details in the description of the
simulation have been left for the ARPA report [1].

MULTIPARTICLE SIGNALS

Error Versus Threshold Level

Figure 3 is a compressed graph of a section of the signal simulated for
the above-mentioned optical system, particle index of refraction n = 1.15+j0,
for a Jung's distribution with m = 2.65 in the 1 to 10 micrometer range, with
20,000/cc greater than 1 micrometer. This simulation was strictly classical
with no photon noise. All input bursts were given the same known frequency.
The average number of particles in the probe volume greater than 1.0 micrometers was seven, but the signal appears to be a single-particle signal at high
thresholds due to the large signal dynamic range.

The basic error level of the simulation and the machine precision was
established by tests with nonoverlapping bursts as: rms error = 0.26×10^{-4}.

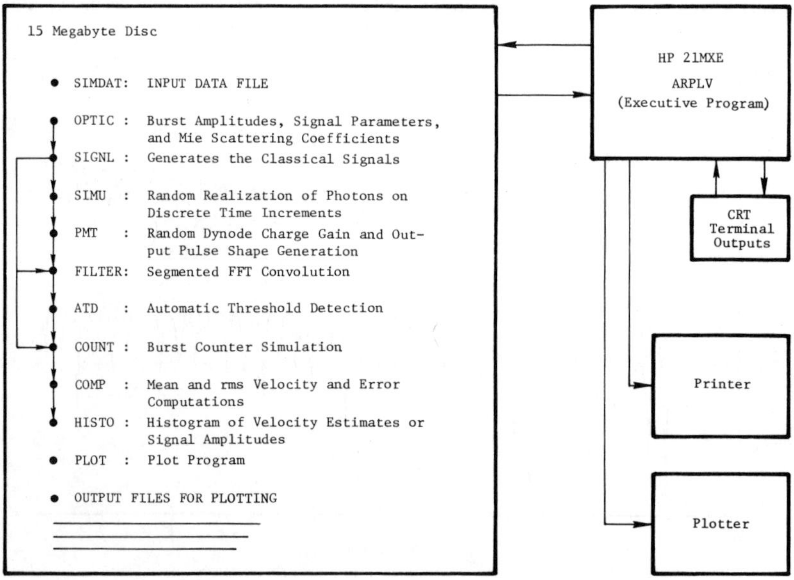

Figure 2.   Block Diagram of Simulation Software.

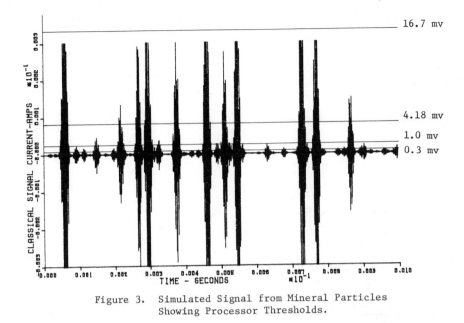

Figure 3. Simulated Signal from Mineral Particles
Showing Processor Thresholds.

(The individual burst frequency errors are normalized by subtracting the known
input frequency and dividing this deviation by the known frequency.)

Figure 3 also shows horizontal lines which indicate four choices of threshold current on the vertical scale. The corresponding threshold voltages (mul-
tiply by 50 ohm load) were used as indicated for the Macrodyne counter simulation threshold setting and the simulated signals were processed. The normalized
error histograms for the lowest and highest of the four thresholds are shown in
Figure 4. The effect of raising the threshold is to simultaneously decrease
the data rate and to decrease the small multiparticle errors.

In order to establish some repeatability to our error and data rate estimates, we ran three statistically independent realizations of 1000 signal
bursts each. The processor simulation was used twelve times, three for each of
the four thresholds. We have plotted the resulting data rate and rms error
curves for each of the three runs on the same graph. This is reproduced as
Figure 5.

Automatic Threshold Effects

After some thought, one realizes that essentially "single-particle" signals
can be obtained from "multiparticle" signals by setting the threshold correctly.
Since conditions often change, this should be done automatically. This is easy
to do in principle: We decide what fraction of the time we wish the signal to
be above threshold (duty factor) to obtain a high probability of high-quality
signals. Next, the desired rate is computed based on the measured velocity and
the burst duration inversely related to the velocity. Finally, the threshold
is lowered or raised according to search logic by comparing the actual threshold crossing rate to the calculated desired rate.

The following example makes this clearer. Assume a desired duty factor
RHO, the ratio of the burst width $BW = KB/V$ and the mean time between triggering the counter, T_B. Using a real time calculation, an automatic threshold
circuit computes BW using the input optical constant KB and the measured velocity, V, it computes T_B from a threshold crossing rate meter and compares the

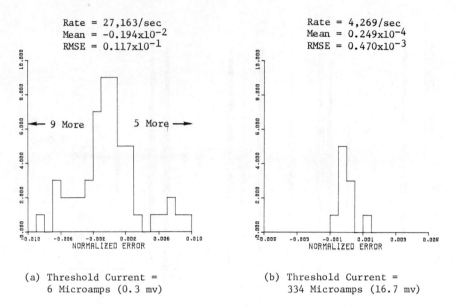

(a) Threshold Current =
    6 Microamps (0.3 mv)

(b) Threshold Current =
    334 Microamps (16.7 mv)

Figure 4.   Multiparticle Error Histograms Versus Threshold Setting.

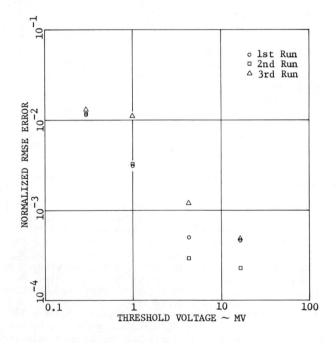

Figure 5.   Error Versus Threshold for Three Independent
            Simulation Runs.

computed duty factor BW/T_B with the desired value RHO. If these do not agree, the threshold is changed according to a converging algorithm. We designed such an algorithm and included it in our ARPLV software. The top two thresholds in Figure 3 were determined automatically by the software. The ratio of the resulting data rates was almost exactly two to one, the ratio of the input values RHO = 0.15 and 0.3. As the error histograms show, these thresholds clearly avoid the large multiparticle errors which result at the lower thresholds.

PHOTON NOISE ERROR THEORY

Figure 6 shows a low-level LV signal with significant photon noise effects. For a higher level signal, the noise distortions would not be so obvious visually, but would still produce small errors in the desired signal period measurement. Catastrophic errors, such as extra or dropped cycles, are rejected by commonly used error rejection circuits. The errors we are concerned with here are small perturbations of the two zero crossings at the start and the stop of a single-particle measurement due to photon noise. The computer simulation directly synthesizes such errors at low and medium "signal-to-noise" ratios, but the approach of synthesizing all the photoelectrons directly becomes inadequate at very high signal levels. In order to examine very small noise-produced errors from large signals, we have devised a perturbational theory. We have verified this theory both by simulation and experiment at low and medium signal-to-noise levels. It should be even more appropriate at very high signal levels.

In general, the noise in LV signals is nonstationary filtered Poisson noise [3] and is not subject to simple discussions. For large signals, the noise becomes nonstationary Gaussian noise [3] with time-varying statistical power proportional to the instantaneous optical power. We will use the assumption (which may not be true due to dispersive time delay in some filters) that the noise power at the zero crossing is determined by the optical power value of the pedestal at the same location. This is reasonable in that with the

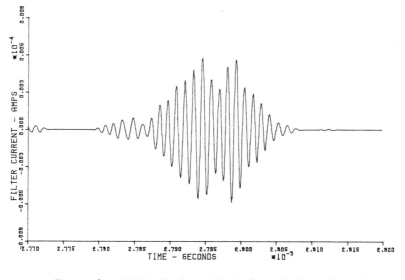

Figure 6.   LV Simulation with Biological Particles --
The Largest Burst.

pedestal removed, the former pedestal crossing becomes the zero crossing. First, however, we treat a constant amplitude sinusoid.

We consider a simple sinusoid,

$$s(t) = a \, \text{Sin} \, (\omega_s t) \qquad\qquad (2)$$

and bandpass filtered zero mean Gaussian noise $n(t)$ with signal-to-noise power ratio,

$$S/N = a^2/2 < n^2(t) > \; >> 1 \qquad\qquad (3)$$

where $< >$ denotes a long time or statistical average. The signal period measurement is given by:

$$\hat{T} = (t_2 - t_1)/N \qquad\qquad (4)$$

where t_2 and t_1 are two zero crossing times separated by N cycles of signal. We have derived [1] an expression for the mean squared error of $\hat{T}$ as:

$$<\hat{\varepsilon}_T^2> = \frac{2}{N^2 a^2 \omega_s^2} [R_{nn}(0) - R_{nn}(t_2 - t_1)] \qquad\qquad (5)$$

where

$$R_{nn}(t_1, \; t_2) = <n(t_1)n(t_2)>$$

is the autocorrelation function of the noise.

Generally, for a white noise source, the noise power spectrum after filtering has the shape of the magnitude square of the filter voltage characteristic. The noise autocorrelation is the inverse Fourier transform of the power spectrum. If the filter passband is wider than the spectral width of an individual signal burst, then the inverse transform is narrower than the time duration of the signal. Thus, the value of $R_{nn}(t_2-t_1)$ would generally be near zero, i.e., the noise perturbations at the two zero crossings t_1 and t_2 will usually be nearly uncorrelated. With the last approximation we obtain:

$$<\hat{\varepsilon}_T^2> \approx \frac{2<n^2(t)>}{N^2 a^2 \omega_s^2} \qquad . \qquad\qquad (6)$$

Recalling that

$$\omega_s = 2\pi f_s = 2\pi/T_s \qquad , \qquad\qquad (7)$$

we may solve for the normalized rms error, σ_T, of the period measurement as a fraction of the true period as

$$\sigma_T = \frac{\sqrt{<\hat{\varepsilon}^2>}}{T_s} = \frac{1}{2\pi N \sqrt{S/N}} \qquad . \qquad\qquad (8)$$

Let us assume that an optical signal is incident on a photomultiplier tube (PMT) with instantaneous power

$$P(t) = P_0(1 + \cos \omega t) \quad . \tag{9}$$

The resulting AC photocurrent after bandpass filtering is

$$i_s(t) = Ge\lambda_0 \cos \omega t + n(t) \tag{10}$$

where G is the dynode charge gain, e is the electronic charge, $\lambda_0 = \eta P_0/h\nu$, where η is the detection quantum efficiency, $h\nu$ is the energy of a single photon, and $n(t)$ is the photon noise. If we consider only the noise at the instant that the $\cos \omega t$ term is zero, i.e., that due to the average incident signal power, then the mean square sinusoidal signal current divided by the mean square noise may be shown (ideal PMT) to be

$$S/N = \lambda_0/4B \tag{11}$$

where B is the electronic bandwidth of the bandpass filter.

If we make the very optimistic assumptions that the photomultiplier is ideal, that the signal visibility is unity, and that the background light level is negligible in comparison with the signal pedestal, then we may relate the above results to single-particle LV signal errors. To do this, we further assume that under these conditions, the detection threshold is less than or equal to the signal pedestal throughout the measurement. Thus, if i_T is the threshold, the mean photoelectron rate λ_0 is greater than or equal to i_T/Ge. By using this condition in Equations (11) and (8) we obtain an approximate photon noise error equation which could be obtained in principle with an excellent optical system.

We have tested Equation (8) experimentally using the signals produced by a light emitting diode modulated by a fixed 100 KHz signal, accurate to better than $1:10^5$, and detected by an RCA 931A photomultiplier tube. The signals were filtered by a Krohn-hite Model 3200 bandpass filter with several different bandpass settings and detected by a Macrodyne Model 2096 counter. The rms noise was measured by switching off the modulation source and measuring the noise produced by the DC value of the detected light using an HP 3400A rms voltmeter. The mean-square signal was obtained by measuring the rms value of signal plus noise, squaring, and subtracting the squared rms noise. The ratio of mean-square signal voltage to mean-square noise voltage was then determined. The variation in S/N was obtained by using a calibrated optical attenuation filter to reduce the incident LED light intensity.

Figure 7 shows a few of the results of the tests. The results are documented in detail in Reference 1. Generally, the agreement between Equation (8) and the experimental results were excellent at signal-to-noise ratios above 10 and less than 500. At the higher values of signal-to-noise ratio, the experiment was limited by the readout precision of the Macrodyne counter to 1:1024 (10 bits). At the lower values of S/N, the perturbational error derivation is not appropriate.

APPLICATION TO THE OCEAN LV SYSTEM

The top threshold current in Figure 3 was 334 microamps. The value of G was 10^5, so the corresponding λ_0 is 2×10^{10} photoelectrons/second. If we assume a filter bandwidth of 300 KHz, the theoretical signal-to-noise ratio is 1.74×10^4 and the predicted normalized rms error per burst is 1.5×10^{-4}. Thus, our simple theory predicts excellent error behavior for the 3 meter, 1 watt LV system previously discussed when the particles are assumed to be mineral (n = 1.15) in composition.

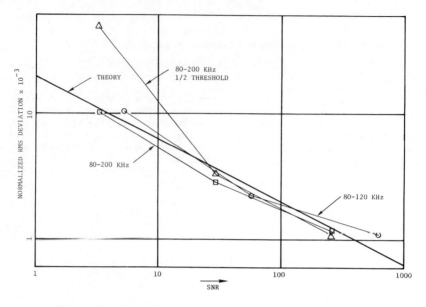

Figure 7.   Normalized RMS Error vs. SNR: Experimental
and Theoretical.

For the worst assumption that the particles are biological in composition
and sparsely distributed (2000/cc with n = 1.03), we see from Figure 1 that the
signal levels will be approximately four orders of magnitude lower.   Ideally,
according to the simple SNR error theory, this would suggest that the rms error
level would be increased to 1.5×10^{-2} corresponding to a threshold current of 33
namps.   (A higher gain PMT could be used; we are keeping the assumption of G =
10^5 for continuity.)   Unfortunately, at such small signal levels, the single
photoelectron filter impulse response level is reached and the theory becomes
invalid.   Also, the assumptions of zero background light may not be valid.

As a final exercise of the simulation software, we used worse case assumptions on particles with 1.03 index of refraction and ran the entire simulation
with a 300 KHz wide Gaussian bandpass filter and no background light.   The signal shown above in Figure 6 is the largest of the signals realized out of approximately 200 signals.   By setting the threshold up to 100 namps and triggering the counter on only the largest particles, we obtained an estimated single
burst rms error of 1.35 percent from the four signal bursts which exceeded the
threshold.   This error level exceeds that predicted by the simple signal-to-
noise theory by a factor of 1.5.   This is not alarming since it is at a S/N
value equal to 5, the number of samples for the simulation is statistically
small, and the simulation includes any multiparticle errors which may have been
present.

CONCLUSIONS

The detailed design of laser velocimeter systems for the ocean is highly
dependent on the index of refraction and backscatter amplitude distribution
and on the maximum useable coherent transmitter beam diameter.   In this study
we have assumed the use of 0.75 cm transmitter beam diameter at a range of 3
meters, with a 1 watt argon laser.   The results use reasonably well accepted
particle size number density distributions.

The results show that if mineral particles with index of refraction 1.15 are assumed, the error is limited by multiple particle effects to less than 0.5 percent if sufficiently high thresholds are used to insure low probability of more than one of the larger particles in the probe volume at once. However, the error level is due to interference by smaller particles also present in the probe volume; the photon noise is negligible by comparison. For the assumption of nearly transparent organic particles with index of refraction 1.03, there are almost no detectable classical signals. With the assumption of no background light, which seems impractical at such low signal levels, a few bursts were observed to produce detectable signals with a 1.35 percent rms error level.

We have developed several computational tools of general interest in LV error prediction. The further application of these tools to detailed design studies for ocean LV systems will require improvements in the data base of single particle backscatter distributions and the useable transmitter beam diameters.

ACKNOWLEDGEMENT

We wish to thank our sponsors, The Advanced Research Projects Agency, with technical direction by P. A. Selwyn, and the Naval Underwater Systems Center, with direction from W. J. Stachnik. We also would like to thank Carolyn Greenwood, SDL, for her contributions to the simulation software.

REFERENCES

1. W. T. Mayo, Jr., "Internal Wave Measurement," Final Report on ARPA Order 3386, Contract N00140-77-C-6670, prepared for the Naval Underwater Systems Center by Spectron Development Laboratories, June 1978.

2. W. J. Stachnik and W. T. Mayo, Jr., "Optical Velocimeters for Use in Seawater," OCEANS '77 Conference Record, Los Angeles, CA., October 1977.

3. W. T. Mayo, Jr., "Modeling Laser Velocimeter Signals as Triply Stochastic Poisson Processes," Proceedings of the Minnesota Symposium on Laser Anemometry, October 1975.

4. W. T. Mayo, Jr., "Study of Photon Correlation Techniques for Processing of Laser Velocimeter Signals," prepared for W. W. Hunter, Jr., NASA Langley Research Center, NASA CR-2780, February 1977.

5. G. Kullenberg in Optical Aspects of Oceanography, edited by N. G. Jerlov and E. S. Nielsen, Academic Press, New York, 1974.

6. O. B. Brown and H. R. Gordon, "Size-Refractive Index Distribution of Clear Coastal Water Particulates from Light Scattering," Applied Optics, 13, p. 2874, December 1974.

PARTICLE DIAGNOSTICS II

Chairman: **W. H. STEVENSON**
Purdue University

Application of SPART Analyzer for Monitoring Real-Time Aerodynamic Size Distribution of Stack Emission

M. K. MAZUMDER, R. E. WARE, J. D. WILSON,
L. T. SHERWOOD, and P. C. MCLEOD
Graduate Institute of Technology
University of Arkansas
Little Rock, AR

ABSTRACT

A single particle aerodynamic relaxation time (SPART) analyzer for measuring the aerodynamic size distribution of particles and droplets suspended in air is described. The analyzer employs a laser Doppler velocimeter to measure the motion of particulates suspended in an acoustic radiation field generated by an acoustic transducer. A microphone is used to detect the acoustic excitation. The amplitude and the phase of the particulate motion depend upon the aerodynamic diameter of the particulate. The SPART analyzer measures the relative phase lag of each particulate with respect to the acoustic excitation and computes the aerodynamic diameter. The instrument can count and size particles and droplets at a maximum rate of 10,000 particulates per minute in the range of 0.1 to 10.0 μm in aerodynamic diameter. This range includes the major mass fraction of respirable aerosols. The instrument has a sensing volume of 1.3×10^{-6} cc and, thus, can be used to measure size distributions at the high particulate concentration generally encountered in stack emissions.

INTRODUCTION

Both climate modification and biological hazards arising from aerosols produced by anthropogenic and natural sources are related to aerosol size distribution, concentration, and composition. There have been extensive attempts to develop and use real-time aerosol size spectrometers to characterize the size distribution of aerosols with high temporal and spatial resolution. One of the most significant size-related classifications of aerosols is the aerodynamic size distribution [1]. For example, the deposition of particles in the respiratory tract is primarily determined by the aerodynamic diameter of the particles. The aerodynamic diameter of a particle is defined as the diameter of a unit density spherical particle having the same terminal settling velocity. The use of aerodynamic size obviates the need for knowing details of the nature of the particles, such as shape, density, composition, and refractive index. All of the relevant physical factors are collectively expressed in their most significant form, which is the aerodynamic behavior of the particle [2].

One simple method of aerodynamic size analysis utilizes the differential settling velocities of particles in an aerosol to separate them according to size. The terminal settling velocity of a particle depends upon the aerodynamic diameter of that particle. In general, these instruments require particle precipitation and subsequent determination of size by microscopic methods. It would be more desirable to measure particulate size distribution in real time and in situ, that is, without removing the particles from their aerosol phase.

Probably one of the most widely used real-time automatic sizing devices, the optical-scattering instrument [3], utilizes the measurement of the intensity of light scattered by individual particles. The intensity of the scattered light varies in a complex manner with the wavelength of optical radiation, the scattering angle, and the size, shape, and index of refraction of the scatterer. Thus, light-scattering properties of the scatterer must be known to correlate the measured optical diameter with the aerodynamic diameter.

A recent development in particle size analysis is the electrical mobility analyzer [4], which measures the electrical mobility of charged aerosol particulates. The cumulative electrical mobility of the particulates can be related to the cumulative aerodynamic diameter. The instrument does not measure the aerodynamic diameter of individual particles. Also, its resolution is poor for particles above 0.6 μm in diameter, and it is inapplicable for particles over 1.0 μm in diameter.

Described here are the development and characteristics of a single particle aerodynamic relaxation time (SPART) analyzer that is capable of measuring the aerodynamic size distribution of airborne particulates in the range of 0.1 to 10.0 μm in real time. The instrument's potential application to monitoring the particulate effluent in stack emissions is discussed. The theory of operation of this instrument has been reported earlier [5,6]. Only a brief description of the operating principles, characteristics of the instrument, and experimental results on fly ash aerosol are reported here.

SPART ANALYZER

The SPART analyzer (Figure 1) consists of three main components: (1) an LDV; (2) a relaxation cell containing an acoustic transducer and a microphone; and (3) a microprocessor-based data-processing, -storage, and -display system. The analyzer has two outputs: a line-printer printout and a punched paper tape. The paper tape is used to generate various size and mass distribution plots using an Interdata minicomputer and a Versatec printer.

Principles

The instrument measures the phase lag of individual particles with respect to ultrasonic excitation of known frequency as each particle transits the LDV sensing volume, which is defined by the intersection of the two laser beams.

As each particle transits the sensing volume, it oscillates in the direction of propagation of the acoustic excitation. However, there is a phase lag between the acoustic excitation and the resultant particle motion.

The motion of a small, free spherical particle in a locally uniform acoustic field can be written from Stokes' fluid resistance law:

$$\tau_p(dv_p/dt) + v_p = u_g, \tag{1}$$

where τ_p is the dynamic relaxation time of the particle, v_p is the particle velocity, and u_g is the velocity of the medium (i.e., acoustic particle velocity). Both v_p and u_g are time-dependent quantities. The dynamic relaxation time of a spherical particle of density ρ_p can be written as

$$\tau_p = \rho_p d_p^2 C_c/18\mu, \tag{2}$$

where μ is the viscosity of the medium and C_c is the Cunningham correction factor for molecular slip. For any particle, irrespective of its shape and density,

$$\tau_p = \rho_o d_a^2 C_{ca}/18\mu, \tag{3}$$

where d_a is the aerodynamic diameter, ρ_o represents unit density, and C_{ca} is the Cunningham correction factor referred to d_a.

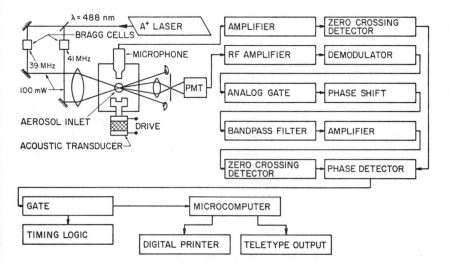

Figure 1. Block diagram of the SPART analyzer.

If ω is the angular frequency of acoustic excitation, u_g can be written as

$$u_g = U_g \sin\omega t, \tag{4}$$

where U_g is the velocity amplitude of the fluid medium. Equation (1) can be solved as

$$v_p = V_p \sin(\omega t - \phi), \tag{5}$$

where V_p is the velocity amplitude of the particle and ϕ is the phase lag. The magnitude of the ratio of the velocity amplitudes (V_p/U_g) or the phase lag ϕ between u_g and v_p is given by

$$V_p/U_g = 1/(1 + \omega^2\tau_p^2)^{1/2} \tag{6}$$

or
$$\phi = \tan^{-1}\omega\tau_p. \tag{7}$$

The aerodynamic diameter of each particle or droplet is determined from the measured value of ϕ. Because the relative phase lag of a particulate is a function of its aerodynamic diameter and the frequency of acoustic excitation, which is known, the aerodynamic diameter can be determined without calibration.

Operation

During operation of the SPART analyzer, the LDV measures the particle motion while a microphone detects the acoustic field. An aerosol sample is drawn through the LDV sensing volume, and the motion of each particle is measured by detecting the light scattered by the particle as it oscillates in the sensing volume. The relative phase lag between the microphone signal and the LDV signal is measured by a data-processing circuit, the output of which is fed to a microprocessor. The microprocessor then computes the aerodynamic diameter from the measured value of phase lag and, as the aerodynamic diameters of other particulates are measured, displays the aerodynamic size distribution of the sampled aerosol. The output of the microprocessor is later processed by a minicomputer that generates and plots the size, mass, and cumulative distributions of the sampled aerosol.

Operational Features

The instrument has the following operational features.

(A) Size Range
     0.1 to 10.0 μm in aerodynamic diameter.

(B) Number of Channels in the Above Range
     128 channels

(C) Theoretical Resolution

Aerodynamic Size Range (μm)	Resolution (μm)
0.1 to 0.2	0.05
0.2 to 0.3	0.04
0.3 to 0.5	0.03
0.5 to 1.5	0.02
1.5 to 1.9	0.03
1.9 to 2.1	0.04
2.1 to 3.0	0.10
3.0 to 4.0	0.20
4.0 to 5.0	0.50
5.0 to 10.0	>0.50

The term resolution in micrometers refers here to the ability of the instrument, based on calculated values of channel cross-sensitivity, to sense two peaks generated by monodisperse aerosols of different mean particle diameter.

(D) Sensing Volume
     (1) Volume $\approx 1.3 \times 10^{-6}$ cc.
     (2) Calculated value of the cross-sectional area
         $(A_s) \approx 2.8 \times 10^{-4}$ cm^2
     (3) LDV fringe spacing $(d_i) \approx 0.88$ μm

(E) Aerosol Sampling
     (1) Flow rate: 75 cc/min
     (2) Inlet tube diameter: 3 mm
     (3) Velocity $(U_s) = 15.8$ cm/sec
     (4) Effective flow rate through the sensing volume:
         1.0×10^{-2} cc/sec

(F) Maximum Count Rate and Coincidence Loss
     (1) 35 counts per second (1% coincidence error) at $U_s = 32$ cm/sec
     (2) 350 counts per second (10% coincidence error) at $U_s = 32$ cm/sec

Actual Particulate Concentration (N_o) (particles/cc)	Coincidence Loss (%)	Indicated Concentration (N) (particles/cc)
5×10^3	1	4.95×10^3
5×10^4	10	4.54×10^4
5×10^5	100	1.50×10^5

Count rate (N_t) is equal to the product $N_o U_s A_s$, where N_o is the number of particles per cubic centimeter, U_s is the sampling velocity in centimeters per second, and A_s is the effective cross-sectional area of the sensing zone in square centimeters.

(G) Maximum Number of Counts in One Channel
$10^6 - 1$

(H) Count Modes
Automatic and manual

(I) Printer Output Format
Total time in seconds
     [Channel 1]
Diameter in micrometers
Total number of counts in channel
$dN/d(\log d_a)$
$dV/d(\log d_a)$
     [Channel 2]
          * * *
     [Channel 128]
Diameter in micrometers
Total number of counts in channel
$dN/d(\log d_a)$
$dV/d(\log d_a)$
     The mass median aerodynamic diameter (MMAD) can be obtained from the volume distribution. The mass median diameter (MMD) for spherical particles of density (ρ_p) is related to MMAD ($\rho_o = 1$) by MMD = $(\rho_p/\rho_o)^{1/2}$ MMAD.

(J) Teletype Paper Tape
     The Teletype output is the number of counts in each channel, listed in order of increasing channel number, which corresponds to increasing particle size. There are ten channels on each line.

(K) Operating Temperature
     The SPART analyzer currently is being operated with the temperature inside its housing being maintained at 37° C by a temperature controller. The

temperature inside the housing--which contains the relaxation cell, laser, and optical components--can be held within ±0.2° C of a chosen value. Operation of the SPART analyzer at body temperature permits the instrument to measure aerodynamic diameter at the temperature most relevant to evaluation of the inhalation hazard. If the aerosol sample also were humidified to 98 percent relative humdity, the measured aerodynamic diameter would be related to the ambient condition that is encountered by an inhaled aerosol inside the human lung.

TEST RESULTS

The SPART analyzer has been tested against several test aerosols that included both monodisperse and polydisperse suspensions of polystyrene latex spheres (PLS). Typical test results for PLS aerosols are shown in Figure 2.

APPLICATION TO MEASURING FLY ASH SIZE DISTRIBUTION

The sensing volume (V_s) of the SPART analyzer can be varied from 10^{-5} to 10^{-7} cc by adjusting the geometry of the transmission and receiving optics and by choosing the proper optical components. A smaller sensing volume will insure measurement of high particulate concentrations, such as those that exist

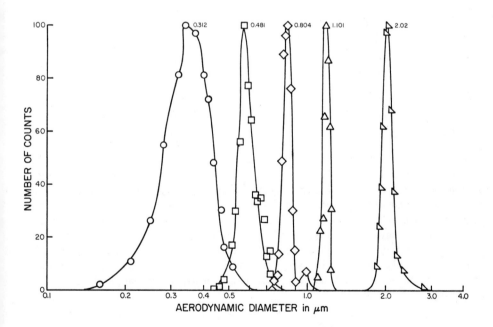

Figure 2.   Typical size resolution of the SPART analyzer when tested against five monodisperse aerosols containing PLS of known diameter. Each peak is labeled with the appropriate known diameter.

inside electrostatic precipitators, without appreciable coincidence losses.
The coincidence losses are caused by the simultaneous appearance of two or more
particulates in the sensing volume. When this occurs, the LDV signal is
distorted, and the analyzer discards the measurement.

Figure 3 shows the normalized size and volume distributions of fly ash
aerosol as measured by the SPART analyzer. The fly ash was redispersed in a
50-liter container from which the aerosol sample was drawn. The count median
diameter was 1.69 μm, and the mass median aerodynamic diameter was 5.04 μm with
a geometric standard deviation of 1.83.

Because the SPART analyzer measures aerodynamic diameter, the effective
geometrical diameter of fly ash that will be measured will extend below 0.1 μm.
Assuming a density (ρ_p) of fly ash of 2.4 g/cc and assuming the particles are
solid and spherical, the relationship between the geometric diameter and the
aerodynamic diameter is given in Table I.

TABLE I
Relationship of Aerodynamic Diameter to Geometric
Diameter of Fly Ash of 2.4 g/cc Density

Aerodynamic Diameter (μm)	Estimated Geometric Diameter (μm)
0.1	0.06
0.2	0.13
0.5	0.32
1.0	0.65
2.0	1.29
5.0	3.23
10.0	6.45
20.0	12.90
25.0	16.13

Fly ash particles are morphologically heterogeneous [7], and the majority
of the particles are nearly transparent spheres, including solid spheres,
cenospheres, and plerospheres. The remaining fraction consists of nonspherical
particles that are amorphous and irregular in shape. Thus, the relationship
between the aerodynamic diameter and the geometric size will depend on the
particulate density and shape. The SPART analyzer measures the aerodynamic
diameter directly, which will permit an assessment of potential lung retention
of the inhaled aerosol without any additional knowledge of particulate
morphology.

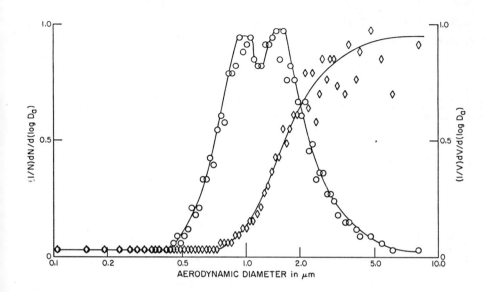

Figure 3. Aerodynamic size (O) and volume (◊) distributions of fly ash
           aerosol.

ACKNOWLEDGEMENTS

The development of the SPART analyzer has been supported in part by EPA
Grant No. R-804429-02-0 and by NIH Grant No. HL20024-02. The authors
gratefully acknowledge the technical assistance of R. G. Renninger and M. K.
Halbert and the helpful discussions with M. K. Testerman, C. W. Lewis, and L.
A. Higgins.

REFERENCES

1.   Mercer, T. T., Aerosol Technology in Hazard Evaluation, Academic Press,
      N.Y., p. 35 (1973).
2.   Morrow, P. E., Am. Rev. Respir. Dis., 110; 88 (1974).
3.   Davis, C. N., Ed., Aerosol Science, Academic Press, N.Y., p. 28 (1966).
4.   Liu, B. Y. H., and Pui, D. Y. H., J. Colloid Interface Sci., 47; 155
      (1974).
5.   Kirsch, K. J., and Mazumder, M. K., Appl. Phys. Lett. 26; 193 (1975).
6.   Mazumder, M. K., and Kirsch, K. J., Rev. Sci. Instrum., 48; 622 (1977).
7.   Fisher, G. L., et al., Environ. Sci. Technol., 12; 447 (1978).

Simultaneous Measurement of Particle Size and Velocity Via the Scattered Light Intensity of a Real Fringe Laser Anemometer

D. M. OGDEN and D. E. STOCK
Department of Mechanical Engineering
Washington State University
Pullman, WA 99164

ABSTRACT

The scattered light from spherical particles irradiated by a real fringe laser anemometer was calculated. The scattering theory of Mie was applied to each beam separately and the resulting fields added. The intensity was then numerically integrated over a finite receiving aperture. The analysis shows that by varying the incident angle of the beams and the size of the receiving aperture, the system can be optimized so that particles from sub-micron to approximately 30 μm can be sized by measuring the scattered light intensity. Thus, together with the Doppler frequency, the size and velocities can be obtained. Both forward and backscattering modes were examined.

INTRODUCTION

In recent years, the electrostatic precipitator has become of paramount importance to the power industry as a tool for the removal of fly ash from coal fired power plants. An efficient design of the precipitators requires an understanding of the fluid and particle dynamics. This requires the measurement of both particle velocity and size. The dominant forces governing the particle motion, aerodynamic and electrostatic, are strong functions of the particle diameter. Thus, the size and velocity must be measured simultaneously.

The ability of a dual beam laser anemometer system to measure particle velocities has been well established in the literature [1,2]. In addition, some work has been done on the simultaneous measurement of particle size and velocity. Durst and Umhauer [3] used an LDA system to measure the particle velocity and a separate white light light source to measure the particle size. They were able to measure particles down to about 1 micron. Ungut, et al [4] have used the scattered light intensity of the dual beam LDA to measure particle velocity and size in the range 30-200 microns. Visibility techniques have been used to measure particles on the order of a micron [5], but, in general, the particle diameter must be less than a fringe spacing. Durst and Elvasson [6] suggested that the dual beam LDA might well be optimized so that particles in the size range 0.1 to 50 microns could be measured.

The purpose of our research was to determine a method of particle sizing with the dual beam LDA that would require the least modification to our present counting type signal processing unit. Because of this we decided to use the scattered intensity of the LDA rather than employing an individual white light source or measuring the visibility parameter.

The size range of interest for electrostatic precipitator work is sub-micron to about 20 microns. In addition, the real index of refraction varies from 1.3 to 1.7 and the imaginary part from 0.0 to .1. If the scattering intensity is to be useful for particle sizing in this range of particle size and index of refraction, it must be, on the average, at least, a monotonic function of particle diameter. It should be relatively insensitive to the refractive index and the signal should be fully visible. In this paper, the dual beam laser anemometer is examined based on the scattering theory of Mie, to determine the geometry of the system that will satisfy the above requirements for particle sizing. Both forward and backscatter systems are considered.

ANEMOMETER SYSTEM AND GEOMETRY

    A schematic of a real fringe laser anemometer is shown in Fig. 1. The beam of a laser operating in the TEM_{oo} mode is split into two equal intensity, linearly polarized beams. The polarization vectors are perpendicular to the XZ plane. The incident half angle of the beams is γ. The receiving optics for

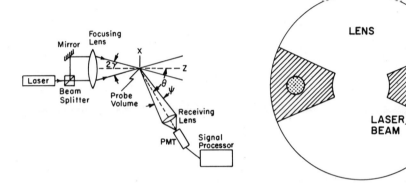

Fig. 1   Dual beam LDA system
         and applicable geometry.

Fig. 2   Schematic of receiving
         lens indicating masked
         region.

collecting the scattered light is centered at angle θ from the positive Z axis. Only forward ($\theta = 0°$) and backscatter ($\theta = 180°$) systems will be considered. The circular receiving lens subtends an angle ψ. Fig. 2 shows a schematic of the receiving lens. In the event that ψ is larger than 2γ and the system is operating in the forward scatter mode, the lens is masked to block the unscattered beams.

CALCULATION SCHEME

    The calculation scheme is based on the assumption that the particles are spherical and centered at the intersection of the two incident beams (probe volume). The beams are assumed to form two coherent plane waves and the distance traveled after splitting adjusted so that they are in phase. The variation of the intensity in the probe volume in the Z direction is Lorenzian while in the XY plane it is Gaussian. Because of these variations, the intensity near the center of the probe volume is nearly uniform. There is, for example, less than 1% variation in intensity over the cross section of a 20 micron particle centered on a beam of 1.5 mm diameter. We will assume then that the particles are irradiated by uniform intensity, plane wave, coherent radiation.

    Using these assumptions, Mie scattering theory was applied. Because the beams are coherent, interference must be considered. Thus,

$$E_{sca} = E_1 + E_2$$

where E_1 and E_2 are the scattered fields from the two incident beams. The scattered intensity is then given by

$$I_{sca} = E_{sca}^2 \ ,$$

where I_{sca} is the scattered light intensity at a point in the receiving lens. From the Mie scattering theory I_{sca} can be written in the form

$$I_{sca} = \frac{I_o}{K_o^2} \, i \ .$$

I_o is the center line intensity of the beams, K_o is the wave number and i contains the amplitude functions of the Mie theory [7]. The details of this procedure are presented by Hong and Jones [8].

The power (P_{sca}) incident upon the receiving optics is just the integral of the scattering intensity.

$$P_{sca} = \frac{I_o}{K_o^2} \int_S i \ ds$$

where S is the area of the receiving lens. A non-dimensional parameter f is defined as

$$f = \int_S i \ ds,$$

and is proportional to the scattered power. The procedure for calculating f is to calculate i from the Mie theory and integrate it over the receiving lens. The algorithm for calculating the amplitude functions, contained in i, is given by Wickramasinghe [9]. The integration was done numerically using a trapezoidal scheme. We found that 200 points in the receiving lens was sufficient for calculating f.

The visibility parameter is given by

$$V = \frac{P_{sca,max} - P_{sca,min}}{P_{sca,max} + P_{sca,min}} \ .$$

$P_{sca,max}$ is the scattered power when the particle is in the center of the probe volume. $P_{sca,min}$ is the scattered power when the particle is centered on the first dark fringe. In terms of the function f

$$V = \frac{f_{max} - f_{min}}{f_{max} + f_{min}} \ .$$

In the next section, a summary of the dependence of f and V on the incident angle γ, the angle subtended by the receiving lens ψ, and the index of refraction is shown. A listing of the program and a complete set of full scale plots are given in reference [10].

RESULTS

In Fig. 3 is shown the results of varying the incident angle. The wavelength of the incident light, λ, is .5145 μm. As γ increases, the intensity

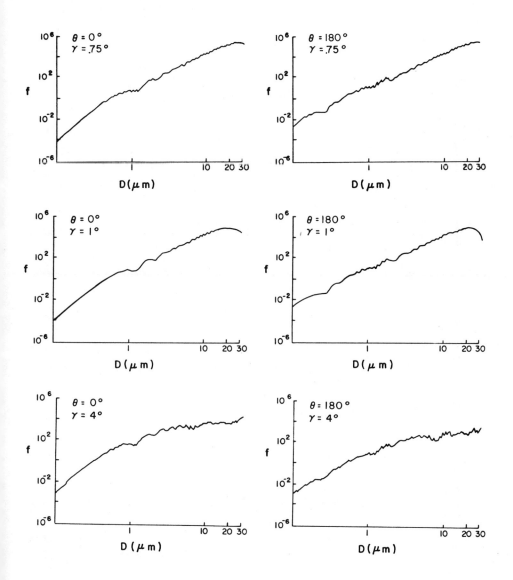

Fig. 3   Plots of the amplitude function f for different values of the
incident angle γ. The fixed parameters are: N= 1.5, M= .001,
ψ= 10°, λ= .5145 μm.

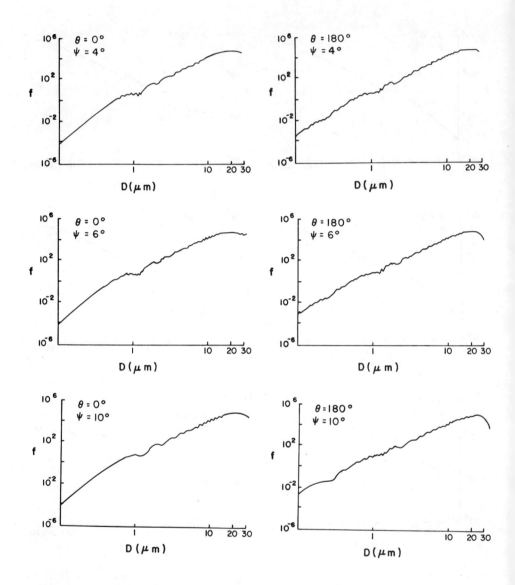

Fig. 4 Plots of the amplitude function f for different values of collection
        angle ψ. The fixed parameters are: $N = 1.5$, $M = .001$, $\gamma = 1°$,
        $\lambda = .5145$ μm.

flattens for larger diameters. This decreases the range over which the intensity is near monotonic. Hence smaller angles are desirable. Notice that for a
Y of 1° the intensity on the average, monotonically increases out past 20 microns,
which is satisfactory for the range of particles of interest.

The results of varying the aperture size are shown in Fig. 4. It can be
seen that for the range of aperture sizes investigated, there is very little
effect on f. In terms of signal strength, especially in back scatter, it is
desirable to have a larger collecting optics. Based on this analysis we decided that an incident angle of 2° (Y = 1°) and a collecting optics of 10° are
optimum for giving a near monoton-ically increasing intensity which is adequate
for particle sizing submicron to 20 micron particles.

The visibility of the signal for the choice of Y and ψ was investigated.
Fig. 5 shows the visibility function for several values of Y. Notice that in

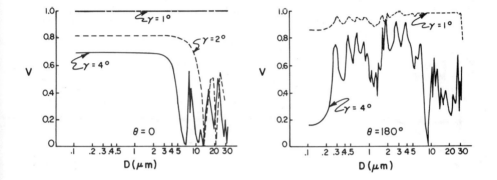

Fig. 5 Visibility plots for various values of the incident
        angle . The fixed parameters are: N = 1.5, M = .001,
        ψ = 10°. λ = .5145 µm.

forward scatter, as the incident angle decreases, the visibility becomes flat-
ter and larger. For a Y of 1° and ψ of 10° the signal is fully visible.
Thus, the geometry which optimizes the scattered power, also optimizes the
visibility. The same trends are true for the backscatter mode. Although the
visibility oscillates much more, for the optimum geometry it is quite large.
Further investigations showed that for the optimum geometry, the visibility was
nearly insensitive to the refractive index in the range of interest to the
electrostatic precipitator. Finally we must determine how sensitive the intensity function is to the index of refraction. Fig. 6 shows the dependence of
the intensity function on the imaginary index of refraction, M. Fig. 7 shows
the dependence on the real index of refraction, N. Oeseburg [11] defines a
parameter A as follows:

$$A = 2(D_L - D_S)/(D_L + D_S)$$

where D_L is the largest value of the diameter where the intensity function is
multi-valued, and D_S is the smallest value of the diameter. The parameter A is
a measure of smallest interval in particle diameter that can be distinguished.
Most commercial counters use an A value of 0.6 for their design criteria. The
A values for the range of index of refraction examined are all less than this
value. Thus, for any given index of refraction, the intensity function is
sufficiently well-behaved for particle sizing.

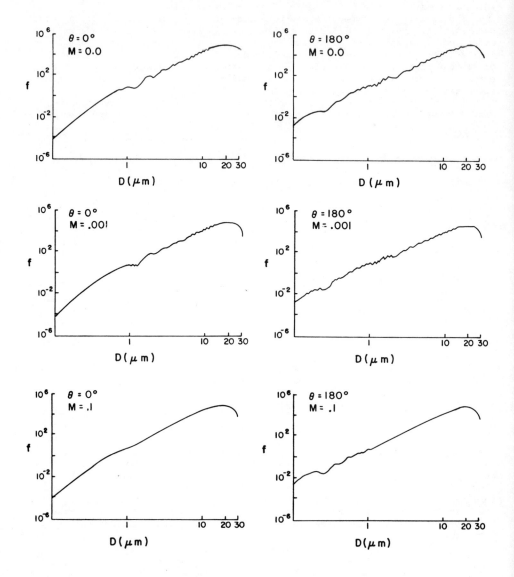

Fig. 6   Plots of the amplitude function f for various values of the imaginary index of refraction M. The fixed parameters are: N = 1.5,
γ = 1°, ψ = 10°, λ = .5145 µm.

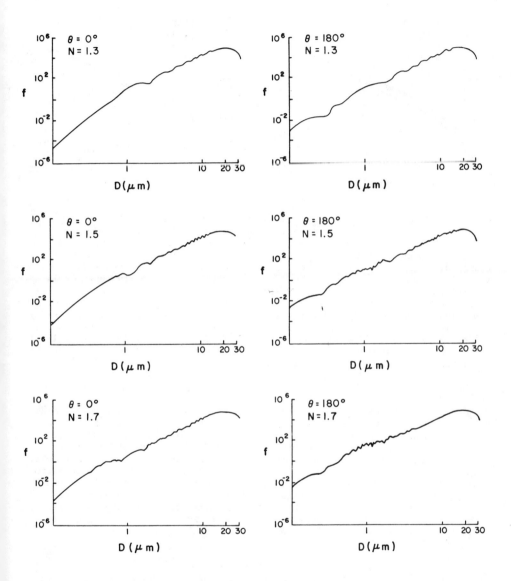

Fig. 7   Plots of the amplitude function f for various values of the real
index of refraction N. The fixed parameters are: M = 0.0, γ = 1°,
ψ = 10°, λ = .5145 μm.

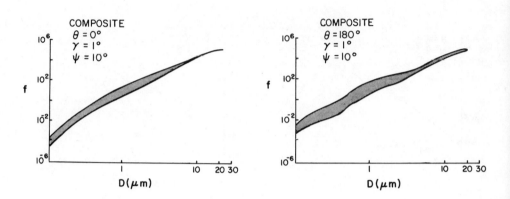

Fig. 8   Composites of the amplitude function f for the
range of refractive indices seen in Figs. 6 and 7.

Fig. 8 shows a composite for the full range of indices of refraction
examined. For forward scattering, the maximum A values remain below 0.6.
Thus, in forward scatter for the optimum geometry, the intensity function is
sufficiently insensitive to the refractive index to be used for sizing particles. Fig. 8 reveals that in backscatter the intensity function is more
sensitive to the refractive index. The largest A value is about 0.8. This may
be somewhat large, but it should be noted that these large values occur at
diameters less than a few microns. Thus, while for small particles, f is insensitive to the diameter, for most of the particle sizes out to 20 microns,
the intensity is sufficiently well-behaved.

DISCUSSION

We have shown that the incident angle and the optics size can be optimized
to give a function f of the particle diameter which behaves well enough to be
used for particle sizing, the optimum values being $\gamma = 1°$ and $\psi = 10°$. The
signal is fully visible both in forward and backscatter. In forward scatter, f
was found to be relatively insensitive to the refractive index of the particle.
The backscatter mode, however, was rather sensitive to the refractive index for
smaller diameter particles, though less sensitive for larger particles. Finally, in terms of the scattering characteristics, it appears as though a dual
beam laser anemometer system can be used to simultaneously measure particle
velocities and size, in a range from submicron to 20 microns and a range of
refractive indices applicable to electrostatic precipitators.

REFERENCES

[1] F. Durst and J. H. Whitelaw, "Optimization of Optical Anemometers," Proc.
    Roy. Soc. Lond., A. 324, 1971.

[2] J. T. Jurewicz, D. E. Stock, and C. T. Crowe, "Particle Velocity Measure-
    ments in an Electrostatic Precipitator with a Laser Velocimeter," AICHE
    Symposium, Vol. 73, p. 131, 1977.

[3] F. Durst and H. Umhauer, "Local Measurements of Particle Velocity, Size

Distribution and Concentration with a Combined Laser Doppler Particle Sizing System," Proceedings of the LDA Symposium, Copenhagen, 1975.

[4] A. Ungut, A. J. Yule, D. S. Taylor and N. A. Chigier, "Simultaneous Velocity and Particle Size Measurements in Two Phase Flows by Laser Anemometry," AIAA 16th Aerospace Sciences Meeting, Huntsville, Alabama, January 1978.

[5] W. M. Farmer, "Measurement of Particle Size, Number Density and Velocity Using a Laser Interferometer," Applied Optics, Vol. 11, No. 11, Nov. 1972.

[6] F. Durst and B. Eliasson, "Properties of Laser-Doppler Signals and Their Exploitation for Particle Size Measurements," Proceedings of the LDA Symposium, Copenhagen, 1975.

[7] Milton Kerker, "The Scattering of Light and Other Electromagnetic Radiation," Academic Press, 1969.

[8] N. S. Hong and A. R. Jones, "A Light Scattering Technique for Particle Sizing Based on Laser Fringe Anemometry," J. Phys. D: Appl. Phys., Vol. 9, 1976.

[9] N. C. Wickramasinghe, "Light Scattering Functions for Small Particles," Wiley, 1973.

[10] D. M. Ogden, "Scattering Amplitude Functions for a Dual Beam LDA," TEL Report No. 78-21, Thermal Energy Laboratory, Dept. of Mechanical Engineering, Washington State University, 1978.

[11] F. Oeseburg, "The Influence of the Aperture of the Optical System of Aerosol Particle Counters on the Response Curve," Aerosol Science, Vol. 3, 1972.

On-Line Particle Diagnostics Systems for Application in Hostile Environments

W. D. BACHALO
Spectron Development Laboratories, Inc.
3303 Harbor Boulevard, Suite G-3
Costa Mesa, CA 92626

ABSTRACT

A particle sizing instrument has been designed to make in situ real time measurements in the range of 0.5 to 25 µm and the flow velocity in flue gas pipes. The two single particle scatter detection techniques used, scattering intensity ratioing and particle sizing interferometry, are discussed. A description of the optical arrangement is given.

INTRODUCTION

The development of reliable particle diagnostics instrumentation to be used in a broad range of applications is of vital importance. One of the immediate needs is in the area of coal combustion. Whether the coal is burned in a conventional way or in a fluidized bed, monitoring of the particulate formation and cleanup systems is imperative. Pressurized fluidized bed combustion systems have additional requirements on particle control because of the threat of erosion and corrosion to the gas turbine blades. As well as coal combustion applications, the use of low grade fuels in power plants requires research on the fuel spray droplet formation and on soot formation. The control of particle emissions from all of the combustion systems stands as a high priority because of the threat to our environment.

Instrumentation that will be operated in combustion facilities face difficult environments. The high temperatures and pressures involved render many existing techniques unusable. High particulate concentrations and the need for in situ real time measurements in relatively large facilities eliminate most of the remaining methods. A broad range of particle sizes needs to be detected (~ 0.05 µm to 50 µm) and measurement of such small sizes requires that the techniques used have the sensitivity necessary to work in the dirty environments. Furthermore, the measurements must be relatively independent of the transient levels of contamination on the access ports and source-beam extinction. No single technique is available that will satisfy all of the measurement criteria.

The sizing techniques that exist fall into categories based on sampling and on optical methods. Sampling requires the extraction of a sample from the flow and the analysis is then performed externally with instrumentation including microscopy, scanning electron microscopy, weighing, or a combination of these. Situations exist in which sampling techniques are adequate, economical, and even superior. However, the sampling can disturb the flow field and the particle distribution even when the sample is withdrawn isokinetically. More serious is the fact that sample preparation and analysis frequently requires excessive amounts of time.

There are several advantages to applying optical methods. With optical
methods, in situ particle size and velocity measurements in real time are possible.
Laser light scatter detection sizing techniques that measure single particles at
a time offer high information content, good spatial resolution, and signal-to-
noise characteristics. Sophisticated electronics circuitry is available for use
with these techniques to handle large quantities of data efficiently while the
measurements are made non-intrusively in environments wherein material sampling
probes would be rendered inoperable. Two optical techniques that perform size
measurements independent of the absolute scattering intensity, particle sizing
interferometry, and scattering intensity ratioing have been incorporated into the
particle diagnostics system described in this paper. This system can measure particles in the size range of 0.5 µm to 25 µm in diameter as well as the particle
velocity, thus allowing the determination of particle concentration. Relevant
features of these systems produced by Spectron are discussed.

PARTICLE SIZING INTERFEROMETRY

The realization that the laser velocimeter signal amplitude modulation was
dependent on the particle size and properties provoked several attempts at deriving the relationships leading to particle size determinations. The first such
attempt was the work of Farmer[1]. Under the assumption of paraxial scattering, the
light waves scattered from each beam were taken as identical over the collection
aperture, thus eliminating the effects of the scattering properties of the particles from the analysis. The scattered light intensity was evaluated as the integral of the fringe system intensity over the particle taken as a disk of diameter
d. The conclusion was drawn that the visibility in forward or backscatter was
given by:

$$\overline{V} = \frac{|J_1(\pi d/\delta)|}{(\pi d/\delta)} \tag{1}$$

where J_1 is the first order Bessel function of the first kind, d is the particle
diameter, and δ is the fringe spacing.
Robinson and Chu[2] derived relationships from the more rigorous scalar diffraction theory. They showed that for a large collecting aperture, the scattered
light fields would produce visibilities as given by expression 1. Considering
the realistic situation of a finite aperture, the scattered field from each beam
approximated by the Fraunhofer diffraction theory was integrated over the collection lens. The concept is shown schematically in Figure 1 where the Mie scattering diagrams have been traced for the scatter from each beam. The overlapped region contributes to the high frequency component of the signal while the remaining portion of the scatter at the lens produces the low frequency pedestal. In
the treatment of Robinson and Chu, the effects of the collecting aperture were
recognized. For example, with small particles having broad forward scatter lobes,
an insufficient collecting aperture will result in a higher measured visibility.
In a more recent paper, Chu and Robinson[3] showed how the relative index of
refraction of the particles affects the visibility by using the exact Mie scattering theory in the calculations. For small fringe spacings, the visibility curve
showed oscillations when computed for particles with no absorption (zero complex
component of index of refraction). For larger fringe spacings (e.g., 13 µm) the
oscillations were reduced to an error bandwidth of ~ 7 percent and with a small
amount of absorption, the curve was smoothed considerably. Comparisons of the
visibility relationships computed from the exact Mie theory for spheres to those
computed from the Fraunhofer approximations showed reasonably good agreement.
Robards[4], using a similar approach and the diffraction theory to describe
the scattering, demonstrated some additional effects of the collection aperture.
The effect of the beamstops, neglected by the previous researchers, was calculated

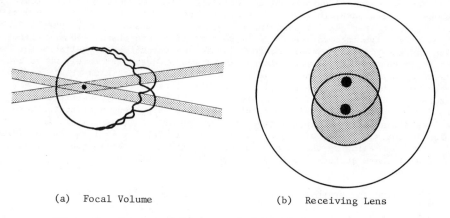

(a)   Focal Volume   (b)   Receiving Lens

Figure 1.   Conceptual Diagram of the Dual Beam Scattering.

by introducing the appropriate limits of integration over the collection aperture.
A single beamstop was shown to produce a significant shift in the visibility
curve and an increase of the secondary peaks.   The overall effect was to reduce
the useful range of the curve to the point where unambiguous measurement could be
made over only a factor of four in size.   However, the use of an individual beamstop for each beam resulted in a curve similar to that of Equation 1.   Furthermore, the secondary peak was suppressed with the proper selection of stop diameter.   The experimental results agreed extremely well with the theory for the
corresponding single and twin beam stop cases.

Because of these findings, it is important to compute the visibility curve
for the optical geometry and particle size range used.   The assumptions of equal
beam intensities, a small beam intersection angle with the beams focused at the
crossover point, and a waist of the focused beam much larger than the particle
must be satisfied by the optics.   A high quality beam splitter and focusing optics
can easily satisfy these criteria.   The fields scattered from each beam can be
expressed as:

$$E_1 = E_o \exp \left[-(x_1^2 + y_1^2)/b_o^2\right] \exp (-i\omega_o t + ikz_1)$$

$$E_2 = E_o \exp \left[-(x_2^2 + y_2^2)/b_o^2\right] \exp (-i\omega_o t + ikz_2)$$

(2)

where E_o is the magnitude of the field on the centerline of each beam, b_o is the
$1/e^2$ radius of the beam, x_i and y_i are the beam coordinates normal to the beam,
and z_i is the direction of the ith beam, $k = 2\pi/\lambda$.   The photodetector sees the
collected intensity:

$$i_p \sim \iint |E_1|^2 + |E_2|^2 + |E_1||E_2| \cos\gamma \, da$$

(3)

where γ is the phase angle.   The visibility can be written as:

$$V = \frac{2 \iint_A |E_1||E_2| \, dA}{\iint_A |E_1|^2 + |E_2|^2 \, dA}$$

(4)

For particles large enough for diffraction theory to be valid, the particle can be replaced by a black disk in the form and size of its geometric shadow area. It is important to note that the prediction of the absolute scattered intensity is irrelevant since the visibility measurement depends on the shape of the scattering distributions. The principle of Babinet[5], which states that the diffracted field from an aperture is equal and opposite in sign to that of a disk of similar geometry, is implied. Thus, for a spherical particle the Fraunhofer diffraction, found by taking the Fourier-Bessel transform of the aperture field distribution[6], can be used to describe the scattering. Substitution of the expressions for the scattered fields with the assumption of equal illuminating beams leads to the expression for visibility (Reference 4) given by:

$$
V = \frac{\iint_A J_1 \frac{(\beta \ell_1)}{\beta \ell_1} J_1 \frac{(\beta \ell_2)}{\beta \ell_2} \, dA}{\iint_A \left[\frac{J_1(\beta \ell_1)}{(\beta \ell_1)} \right]^2 \, dA} \tag{5}
$$

In this expression, ℓ_1 and ℓ_2 are the distance from beams 1 and 2 at the lens to any point on the plane of the receiving lens and $\beta = \pi(d/\delta)s$ where s is the distance between the beams at the collecting lens.

The expression must be integrated numerically over the receiving aperture A excluding the beamstops. Calculations have been carried out for a number of receiving lens geometries and some of these results are shown in Figure 2. The curves were calculated for the optical geometry of the instrument in which the laser wavelength is 0.488 μm, fringe spacing, δ = 25 μm, collecting lens diameter, L = 100 mm, and focal length, R = 250 mm. Since the radius of the first zero of the Airy pattern which is given by

$$
r_o = 1.22 \frac{\lambda R}{d} \tag{6}
$$

is smaller than the lens diameter, no aperturing effects will occur for the size range measured (3 to 25 μm). Dramatic variations in the visibility curves are realized with changes in the beamstops. The elliptical beamstop was used on the present system because there are three beams impinging on the lens. With that geometry, the curve is not significantly different from the results of Reference 1. However, the visibility is lower than the curve of Reference 1 by as much as 10 percent and the secondary peak is higher. A pie-shaped mask blocking the two 90° segments centered on each of the beams, suggested by Roberds[7], can be used effectively to reduce the probe volume if necessary by allowing an increased beam intersection angle for the same particle size range.

When making measurements of irregular-shaped particles, some consideration is required regarding the instrument's response. The signal visibility is affected primarily by the particle dimension normal to the fringes. If the particles have dimensions that are not significantly different in one direction than another (i.e., are approximately isometric) and are randomly orientated, then the average diameter measured from a large number of samples should approach that of an equivalent sphere.

SCATTERING INTENSITY RATIOING

The method of sizing particles by measuring the light they scatter at two or more discrete angles in the near forward direction was first suggested by Hodkinson[8]. Gravatt[9] produced an instrument based on this concept recognizing that the technique minimizes the effect of particle index of refraction and

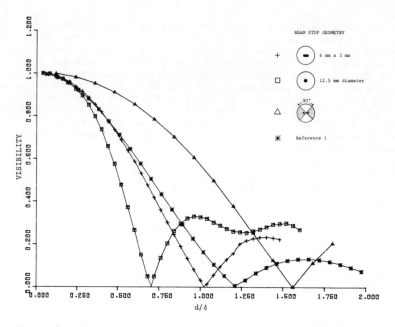

Figure 2.   Visibility Curves for Various Aperture Configurations.

eliminates the effect of source and scatter intensity variations, the effect of
random particle trajectory through the Gaussian beam and non-uniform particle
velocities. Hirleman[10] has made several important contributions in the development of the technique including an analysis of the scattering cross section versus
probe volume effects and the use of multiple collection angles. The concept is
shown schematically in Figure 3. With this technique, it is the shape of the forward scatter lobe that is used as a measure of the particle size. The ratio of
the light scatter by a particle of diameter d and measured at two angles, θ_1 and
θ_2, is given by:

$$R(\theta_{1,2}) = \frac{[i_1(\theta_1,\alpha,n) + i_2(\theta_1,\alpha,n)] \sin\theta_1 \Delta\theta_1}{[i_1(\theta_2,\alpha,n) + i_2(\theta_2,\alpha,n)] \sin\theta_2 \Delta\theta_2} \qquad (7)$$

where i_1 and i_2 are the Mie coefficients for a particle of size parameter, $\alpha =
\pi d/\lambda$, and index of refraction n is independent of the incident intensity.
Representative Mie scattering diagrams have been plotted in Figure 4 for particles of index of refraction n = 1.57 - i0.56.
     Using the calculated Mie coefficients in Equation 7, the relationships be-
tween the intensity ratio and the particle diameter were calculated and are shown
in Figure 5. In Figure 5(a) the ratio for non-absorptive particles shows some
oscillation about the curve for absorptive particles which produces uncertainty
in the measurement. Fortunately, the present instrument is being applied to the
measurement of combustion-related particulates consisting of fly-ash, unburned
carbon, sorbent, and other materials combined with the coal, all having a relatively high absorption. Figure 5(b) is the curve computed for the angle pair
used in the instrument. The angles were selected to work over the size range of
0.5 to 3 μm in diameter with reasonable sensitivity to the small size range.
     Although the measurements are relatively insensitive to the width of the
collecting annular aperture (defining $\Delta\theta$), care is required in the collection
optics design. Since the large angle has a shorter depth of field than the small

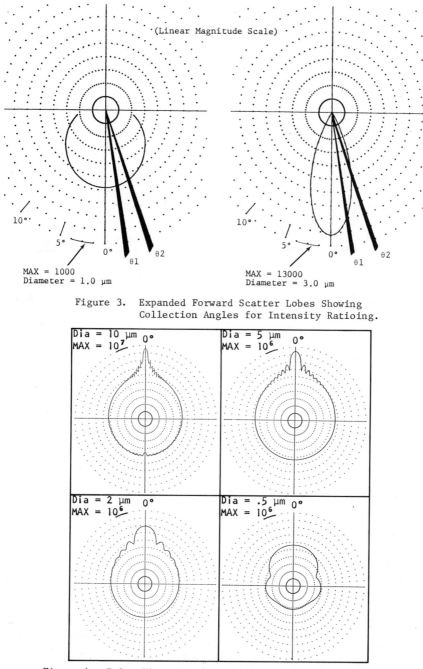

Figure 3. Expanded Forward Scatter Lobes Showing
           Collection Angles for Intensity Ratioing.

Figure 4. Polar Plots Showing the Magnitudes of the Mie
           Scattering Coefficients for Particles of
           Refractive Index n = 1.57-i0.56 and λ = 0.5145 μm.

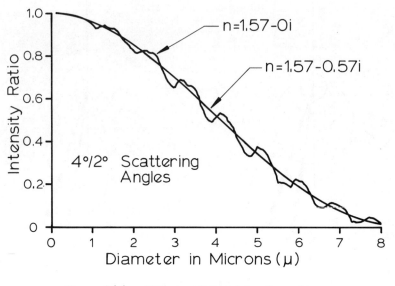

Figure 5(a).   Effects of Particle Absorption on
               Scattering Intensity Ratio.

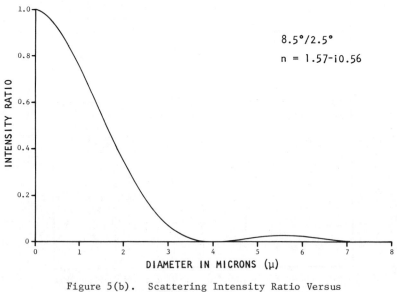

Figure 5(b).   Scattering Intensity Ratio Versus
               Particle Size.

angle, partially masked signals on the large angle may be combined with full amplitude signals of the small angle. This can be minimized by using a smaller annular aperture and photomultiplier tube aperture on the smaller angle collection system. The number of coincident signal bursts should not be significantly less than the bursts from the small angle photomultiplier tube.

Because the particulates measured cannot be expected to be spherical, consideration of the instrument response to non-spherical particles must be made. Hodkinson[11] has showed that the forward scatter for a polydisperse system of irregular particles was similar to that of spherical particles of equivalent projected cross-sectional area. If the Fraunhofer diffraction theory is considered, it can be seen that the extent of the angular scatter by an irregularly-shaped particle depends on its dimensions in the respective directions (e.g., consider the scatter of a rectangle). When the scatter is detected with a circular aperture centered in the forward direction, the scatter from individual particles is averaged over the aperture and produces a measurement approaching that of an equivalent sphere. Hirleman[10] investigated the scatter from irregularly-shaped particles (i.e., agglomerated polystyrene spheres) and found that the scattering patterns had forward lobes that appeared similar to that of spheres of equivalent diameter.

Since this technique measures a single particle at a time, an upper limit on the particle concentration is set by the requirement of having a low probability of having more than one particle in the probe volume at a time. Assuming that the particles are Poisson distributed, the probability of having n particles in the prove volume, V, at one time is given by:

$$P(n) = \frac{\exp(-\bar{N}) \, \bar{N}^n}{n!} \tag{8}$$

where $\bar{N}$ is the average number of particles in the probe volume at any time. For a practical probe volume of 2×10^{-6} cm^3, a concentration of 10^5 particles/cm^3 would produce multiple particle signals only 1.7 percent of the time. Ho, et al[12], showed that if more than one particle was in the probe volume at one time but the peaks were separable, the two pulses could be processed independently using peak detection. With this method of signal processing and a possible probe volume as small as 2×10^{-7} cm^3, the possibility of making measurements at concentration as high as 10^7 particles/cm^3 was stated.

INTEGRATED PARTICLE SIZING SYSTEM

It is clear that an in situ particle sizing instrument will be required to operate over a size range greater than one decade without manipulations of the optics. The estimated size range of interest for the proposed application was 0.5 to 20 μm in diameter. Particle sizing interferometry could not be used in the small size range (0.5 to 3 μm) because the width of the scattering lobe is approximately ±50° (see Figure 4). For this reason, scattering intensity ratioing was incorporated into the system.

An argon-ion laser was used to produce both a blue (0.4880 μm) and a green (0.5145 μm) wavelength simultaneously. Figure 6 is a schematic drawing of the optics system. The wavelengths are separated with a dispersion prism and directed through the respective optical elements. Since the ratioing technique must be sensitive to the smaller size range in which the particles are, in general, more plentiful, the beam is expanded to achieve a focused spot diameter of approximately 80 μm to e^{-2} of the peak intensity level. The blue beam is split into two equal intensity beams and focused to a crossover of approximately 250 μm in diameter. Particles passing through the focal volume scatter light at both wavelengths. A dichroic mirror separates the signal by wavelength and

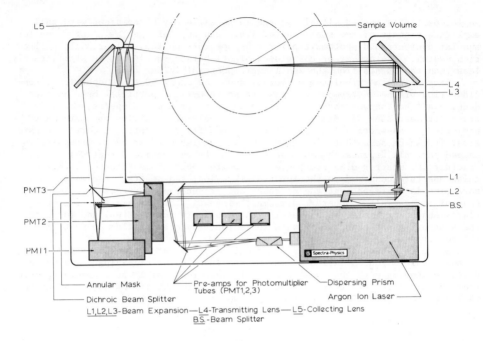

Figure 6.   Integrated Particle Sizing System Combining
            Particle Sizing Interferometry and Scattering
            Intensity Ratioing.

directs the light to the appropriate photomultiplier tubes.   The ratioing signal
is passed through an annular mask to separate the scatter at the two set angles.
      In the electronics, the signals are low pass filtered before being processed.
The scattering intensity ratioing signals are integrated over the burst to eliminate the effects of any remaining noise on the signal.   Ratios of the integrated
values are taken and a 10-bit binary number proportional to the ratio is output.
Particle sizing interferometry is more sensitive to noise on the signal so more
sophisticated electronics processing is required.   The usual 5/8 (or other count
ratio) periodicity check is used primarily to reject multiple particle signal
occurrences.   To discriminate against noise, the multiple level signal test developed by Macrodyne was used.   This technique has been proven to be exception-
ally effective in the rejection of noise and can decrease the number of signals
otherwise rejected by the periodicity check.   Signals that pass the test are
separated into the high frequency (Doppler) and the low frequency (pedestal) components, integrated and the ratio is taken.   The output is a 10-bit binary number
proportional to the visibility.
      A microprocessor-based histogram generator that can store two histograms
simultaneously (one from each technique) is used to manage the data.   Samples
are plotted on the video monitor as they are accumulated.   When a preset number
of samples have been accumulated, the histograms are plotted and the data is reduced to particle size and tabulated on hard copy.   The system can operate at
data rates as high as 10^4/sec.   Photographs of the system are shown in Figure 7.

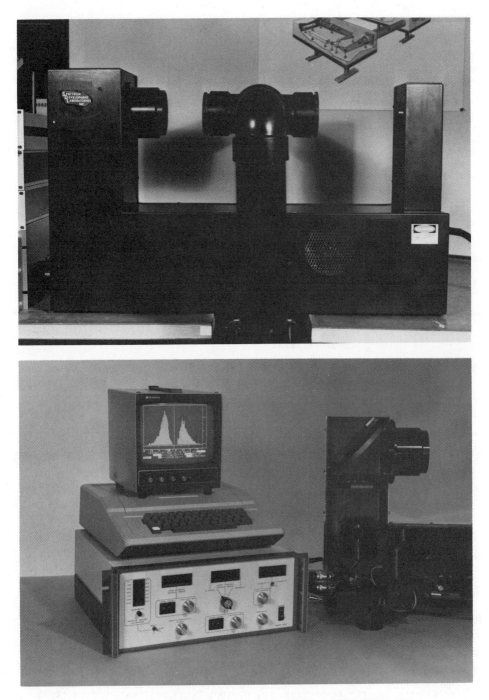

Figure 7.   Particle Sizing Instrument Shown with
            the Test Facility.

CONCLUSIONS

The concepts incorporated into the particle sizing instrument have been shown to offer a reliable means of producing real time in situ particle size measurements. Because of the relative complexities involved in these systems, care is required in their design and application. With the integrated optics system, measurements can be made over a realistic particle size range without any manipulations by the operator. In the past, such manipulations have led to instrument failure or unreliable data.

Related signal processing systems are currently being designed incorporating the latest 16-bit microprocessor technology to improve the reliability of the processor and to automate many of the instrument functions. Because setting up the electronics to the prevailing signal conditions is critical, the automation of these functions is essential.

The first instrument of this type was produced for the Argonne National Laboratory and is being tested in their Solids/Gas Flow Test Facility. A second instrument is being developed for installation on the Curtiss-Wright Small Gas Turbine Pilot Fluidized Bed Facility.

REFERENCES

1. W. M. Farmer, "The Interferometric Observation of Dynamic Particle Size, Velocity, and Number Density," Ph.D. Thesis, University of Tennessee, 1973.

2. D. M. Robinson and W. P. Chu, "Diffraction Analysis of Doppler Signal Characteristics for a Cross-Beam Laser Doppler Velocimeter," Applied Optics, Vol. 14, No. 9, September 1975.

3. W. P. Chu and D. M. Robinson, "Scattering from a Moving Spherical Particle by Two Crossed Coherent Plane Waves," Applied Optics, Vol. 16, No. 3, March 1977.

4. D. W. Roberds, "Particle Sizing Using Laser Interferometry," Applied Optics, Vol. 16, No. 7, July 1977.

5. H. van deHulst, The Scattering of Light and Other Electromagnetic Radiation, Academic Press, New York, 1967.

6. J. W. Goodman, Introduction to Fourier Optics, McGraw-Hill, New York, 1968.

7. D. W. Roberds, Private Communication.

8. J. R. Hodkinson, "Particle Sizing by Means of the Forward Scattering Lobe," Applied Optics, Vol. 5, No. 5, p 839, May 1966.

9. C. C. Gravatt, "A New Light Scattering Method for the Determination of the Size Distribution of Particulate Matter in Air," Proceedings of the 3rd International Clean Air Congress, Düsseldorf, 1973.

10. E. D. Hirleman, "Optical Technique for Particulate Characterization in Combustion Environments: The Multiple Ratio Single Particle Counter," Ph.D. Thesis, Purdue University, 1977.

11. J. R. Hodkinson, "Light Scattering and Extinction by Irregular Particles Larger than the Wavelength," Proceedings of Interdisciplinary Conference on Electromagnetic Scattering, pp. 87-100, Permagon Press, Oxford, 1963.

12. C. W. Ho, P. W. Tveten, P. W. Chan, and C. Y. She, "Particle Size Measuring
     Device in Real Time for Dense Particulate Systems," Applied Optics, Vol. 17,
     No. 4, February 1978.

Particle Sizing Interferometer Measurements in Rocket Exhausts

W. M. FARMER, K. E. HARWELL, J. O. HORNKOHL, and F. A. SCHWARTZ
Gas Diagnostics Research Division
The University of Tennessee Space Institute
Tullahoma, TN

ABSTRACT

Particle sizing interferometer measurement of particle size distributions, velocity and number density in samples of rocket propellant exhausts is described. Signal visibility and scattered intensity for individual particles was measured for a nominal size range of 0.3 to 6 micrometers. Calibration tests of the instrument were obtained by firing small propellant samples into a 20-m^3 chamber and comparing the measured particle size distributions with those obtained from estimates using multiple wavelength transmissometers and a commercial optical particle counter. Particle sizing interferometer estimates of particle size distributions and number densities required the development of models which could account for sample space dependence on particle size and illuminating fringe contrast variation. Results obtained by using these models on particle size data from calibration and rocket motor test firings shows the need for well-controlled, insitu-perturbationless particle size measuring techniques for hostile environments.

INTRODUCTION

A knowledge of particle size distributions in the primary and secondary smokes generated by solid rocket motor combustion processes is of fundamental importance for accurately modeling both the combustion process and subsequent smoke formation. Such models serve the two fold purpose of characterizing the optical signature of propellants and of predicting general propellant combustion properties from the knowledge of the propellant chemistry.

For all practical purposes it is impossible to mechanically sample real rocket exhausts during an actual firing. The environment is fluctuating too rapidly and a probe perturbs the flow. Optical techniques are therefore the only truly suitable means of measuring particle size. However, there are extreme conditions which must be overcome if particle size measurements are to be obtained with any optical instrument. One of the most significant problems in this regard is that associated with background light in the exhaust. This light results from radiant gas emission and the black body radiation of hot particle located in the exhaust. The difficulty in measuring such particles is further compounded by the facts, that 1) they may have extremely high velocity (on the order of kilometers per second), 2) the number density may often be very high (at times exceeding 10^7/cc) and 3) the particle chemistry is probably not known leading to large uncertainties in particle indices of refraction. Furthermore, the particles can be expected to be small (of the order of 1-10 micrometers or less) which precludes numerous photographic and holographic techniques which might be used otherwise. During the past year the Gas Diagnostics

518

Division at The University of Tennessee Space Institute has been involved in preliminary studies to determine the feasibility of applying a particle sizing interferometer (PSI) to the measurement of such particles. To this end both analytical feasibility studies and experimental research has been conducted to determine the magnitude of the problems which may be encountered in obtaining these kinds of particle size measurements. The experimental studies which will be reported in this paper have consisted of PSI measurements of rocket exhaust particles obtained from the combustion of small samples of rocket propellant fired into a large (20-m^3) mixing chamber where the exhaust is mixed mechanically to maintain particle suspension. These PSI measurements have been compared with those obtained with a commercial optical counter and with a multiple wavelength transmissometer. Additionally some preliminary measurements have been made of particles in the edge of a small rocket motor plume. The purpose of this paper is to describe the results of the preliminary experimental measurements in the mixing chamber, how the measurements from the various instruments compare, and some of the features of the measurements in rocket plumes.

OPTICAL SYSTEM DESCRIPTION

Two slightly different forward scatter optical system geometries have been used for these measurements. The first geometry, shown in Fig. 1, is that which was used for the measurements in the mixing chamber. It is a forward scatter dual-scatter or differential Doppler system. The distinguishing feature about this system is the collection of forward scattered light with a variable focus receiver system which reflects the light back outside the mixing chamber where it is then relayed to a photomultiplier tube. The purpose for this particular geometry results from the fact that the chamber may be cooled to less than -30°C or heated to greater than +30°C and the relative humidity may be controlled between 20% and 100%. Furthermore, since the rocket propellants tested can be quite corrosive (particularly those which produce hydrogen chloride exhaust products) it was desirable to minimize environmental exposure. The laser used in this system is a Spectra Physics model 124A which produces approximately 30 milliwatts of light power. The fringe period generated by the transmitting

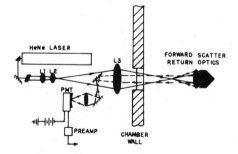

Fig. 1 PSI optical geometry for
mixing chamber measurements.

optics was typically set to about 6 microns. Because the lenses used in the transmitter portion of the system have relatively low F numbers (~2-3) with spherical aberation and astigmatism, the sample space or probe volume had an elliptical cross-section.

The cross-section was approximately 50 interference fringes wide in a

direction normal to the fringe planes and approximately 20 fringe periods wide
in a direction parallel to the fringe planes. An adjustable slit was used in
the receiver geometry immediately in front of the photomultiplier tube (EMI-
9781R). It was set at a sufficient width to allow the return image of the
geometric probe volume to pass with little or no diffraction. The photomultiplier tube housing was connected to a dual channel preamp which was used to
transmit the signal to the visibility and signal time period processing electronics. The preamp separately amplified the mean and the Doppler components
of the signals.

   The second optical system which has been used in the rocket plume measurement is shown in Fig. 2. Because there was no need to separate the transmitter
and receiver system, a standard forward scatter optical system geometry was
used. The laser, preamp, and electronics systems were identical to those in the
first system. This system was mounted in a box for environmental protection
during motor firings. The box was built of 3/4" plywood and painted black to

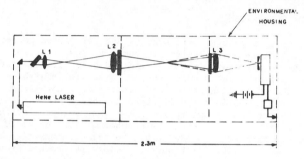

Fig. 2   PSI optical geometry for rocket
         plume measurements.

minimize interference with transmissometer measurements which were made through
the plume. The entire inner portion of the box was lined with foam rubber 2"
thick. This was found to be more than sufficient both for temperature insula-
tion (measurements were made in West Virginia at the end of the previous winter)
and for protection from the shock waves present during motor ignition. Both
optical systems were mounted on a 1.3-cm thick, 0.6 x 1.2-m aluminum base plate.
The fringe period in the second system was set to be approximately 10 microns
and the limiting aperture at the photomultiplier tube was set such that the
depth of field determined by the receiver lens aperture combination was approximately 1/4 that of the geometric probe volume.

ELECTRONIC SYSTEM DESCRIPTION

   A block diagram of the electronic system used for processing the scattered
light signal and for acquiring and manipulating the data is shown in Fig. 3.
In this configuration the signal is divided into two portions, one portion
enters the visibility processor and one portion enters the velocity processor
and data acquisition system. The velocity processor provides signal logic tests
for multiple particles signals, signal dropouts, and acceptable particle trajectory. These tests are used as the control logic for accepting or rejecting
signals measured by the visibility processor. The lower portion of the figure
shows the approach we have taken in obtaining a visibility measurement. One
portion of the signal is integrated directly. This integral yields a mean value,
or pedestal portion of the signal. The second portion is filtered to remove

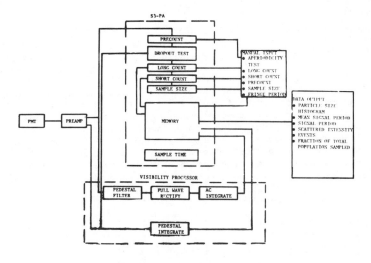

Fig.   3   Block diagram of signal processing
and data management system.

the pedestal, and then full wave rectified and integrated. This measurement
yields a value which is proportional to the energy in the sinusoidal portion of
the signal. In each leg of the visibility processor four multiplexed integra-
tors are used in order to obtain a signal dynamic range of approximately $1:10^4$
with three digit accuracy. After the integrations are complete the values for
the AC (sinusoidal) and the DC (mean value) components of the signal are sent
to a small microprocessor memory where they are recorded along with the signal
time period measurement. The velocity processor can be programmed to perform
signal time period averages over a large and small number of signal periods.
This number can range between 2 and 256 on both the large and small averages.
The acceptable aperiodicity in the signal is determined by the difference in
the large and small number of cycles time averages. For brevity these averages
are called long and short counts. The acceptable aperiodicity is also programmable from 0.1 to 50%. The velocity processor has a so-called dropout
detector which allows the processor to immediately recycle the instrument if
particle trajectories or noise are such that the signal has an insufficient
number of cycles to meet the criteria imposed upon the signal processor logic.

After a sufficient data set has been obtained (the number of measurements
which can be entered into memory is also programmable), a set of prechosen
software programs are used for data manipulation and computation. One program
computes the statistics associated with particle velocity. The parameters computed are 1) mean particle velocity, 2) standard deviation of the velocity distribution expressed as a percent of mean velocity, and 3) the Kurtosis of the
velocity distribution. The particle size analysis includes histograms of particle
size based on a spherical particle visibility function which is permanently in
memory and the correlation of two measurements of particle size. The measurements
are determined from a visibility and from the scattered intensity magnitude of the
signal. Finally the entire data set which is in memory, i.e., the signal time
period, the magnitude of the signal mean value and the magnitude of the AC
integral can be displayed along with the equivalent particle sizes from signal

visibility and relative scattered magnitude.

EXPERIMENTAL RESULTS

The optical system shown in Fig. 1 was used in the mixing chamber tests. Calibration tests were performed in the mixing chamber using urea-formaldehyde resin particles which are spherical, translucent, and remarkably monodisperse for a non-liquid particle. The PSI system was focused approximately 15" inside the mixing chamber and approximately 30" from the floor of the chamber. The optical commercial optical counter sampling tube was located near the top of the chamber. The urea-formaldehyde was injected into the chamber and main-tained in suspension by a set of small electric fans which maintained a counter-clockwise circulatory pattern near the top of the chamber. Figure 4 shows a set of histograms obtained with the PSI for the urea-formaldehyde measurements.

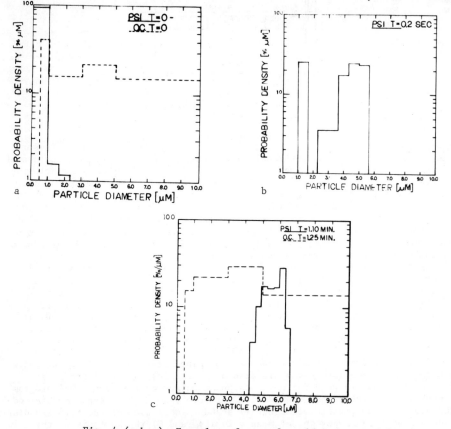

Fig. 4 (a,b,c)   Examples of urea-formaldehyde particle size
                  histograms obtained with a PSI and a commercial
                  optical counter.

The histograms from the PSI are plotted as solid lines while those obtained with the commercial optical counter are plotted as dashed lines. There exists ob-vious discrepencies between the two sets of data. The commercial counter con-sistently counts a large number of particles less than 1 micron. The mode sizes for the two distributions in the urea-formaldehyde range roughly agree. It is interesting to observe the time history dependent size distribution as observed with the PSI system. We note that before particle injection the background particle sizes were predominately less than about 1.5 microns and the particle number density is surprisingly low (see Fig. 5). The PSI sample space is located well below the plane circulation fan and air movement is of relatively low volume in this area. The commercial counter on the otherhand has its sampling tube located near the top of the chamber where there is considerable air circulation and it mechanically withdraws an air sample at a fairly high rate. We notice that after the urea-formaldehyde particles are injected the background particle count immediately becomes over-shadowed by the urea-formaldehyde particles until after approximately one minute after dust injection, the background count is virtually non-existent in the particle size histogram. We note also that the size distribution parameters remains roughly constant throughout the sample time of the experiment although there are minor changes in the standard deviation of the measurements. This variation is to be expected due to the different number of particles which were sampled during that time. Figure 5 plots the number density of the urea-formaldehyde particles as estimated from the PSI measurements using some of the simple models previously described.[2] A least squares linear logarithmetic analysis was found to fit the data. It should be expected that the particles will settle out of the chamber and be driven to the walls by the circulation fans. Thus, the number density will decrease with time and theoretical predictions suggest that the decrease should be exponential.[4] It is interesting to note that near the beginning of the set of measurements, the optical counter and the PSI were in close agreement on the particle number density and that as time went on the number density observed with the PSI system fell off at a higher rate than that obtained with the commercial counter which sampled the aerosol at a constant rate. This data thus hints at the possibility of data biasing errors in terms of number density measurements due to the location of mechanical sampling probes and in mechanically withdrawing a sample at a constant rate.

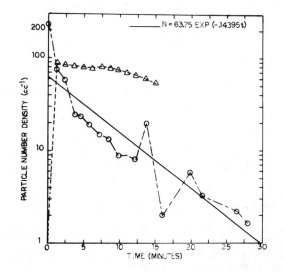

Fig. 5   Urea-formaldehyde particle number density
in the mixing chamber as a function of time.

After the urea-formaldehyde tests, the PSI system was applied to measurement of rocket exhaust particles obtained from small propellant samples fired into the mixing chamber. For the measurements reported here the mixing chamber was kept at a temperature of approximately 20.2°C and a relative humidity of 20%. Figure 6 shows an example of the estimated number density as a function of time for these measurements. The number density as determined by the commercial optical counter is plotted in the same figure and again shows that while initially the particle number densities as determined by the two instruments are in reasonable agreement, the optical counter which is mechanically withdrawing an aerosol sample yields a much slower particle fallout rate than that determined by the PSI. Figure 7a shows the size distribution approximately 1.33 minutes after propellant ignition and represents approximately 440 particle measurements. It reveals a distinctly bimodal size distribution with one mode occurring at less than 0.5 micrometers. We observe that there is good agreement in the commercial optical counter and PSI measurements for the values of the small mode size and near the edge of the large mode size during the time interval between 1.33 and 3.96 minutes (Fig. 7b). The small size mode peaked at about 72% probability of occurrance and then decreased to zero. The large mode on the otherhand fell to about 4% and then rose to a value of about 15-20%. The small size mode continued to decrease and the large size mode increased until at approximately 8.6 minutes (Fig. 7c) the small size mode had virtually disappeared leaving only the large size mode distribution which was

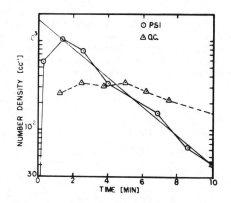

Fig. 6   Particle number density as a function of time in the mixing chamber for a rocket propellant exhaust sample.

close to the background size distribution before ignition.

Figures 7a-c were chosen to make the PSI measurements correspond as closely in data recording time with the commercial optical counter as possible and still be representative of the overall time history of the measurements. The PSI histograms are also plotted in terms of 0.55 micrometer intervals in order to more closely correspond to the commercial counter intervals for the smaller sizes although the data was acquired in 0.275 micrometer intervals.

It is difficult to directly compare the results between the PSI and

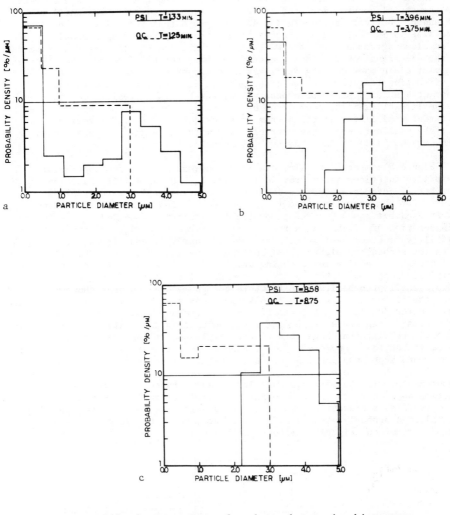

Fig. 7 (a-c)   Comparison of rocket exhaust size histograms
             obtained with a PSI and commercial optical
             counter in the mixing chamber.

commercial optical counter because the commercial counter sample tubing is outside the test chamber. The aerosol sample withdrawn from the chamber will be
cooled when the chamber is above ambient temperature (as was this case) and
warmed when it is below. Furthermore, the commercial counter size increments
were much coarser than those of the PSI and because the index-of-refraction of
the particles is not certain, its response function is suspect. Finally, the
commercial counter sampled approximately 200 times more volume per unit time
than the PSI.

There are some interesting discrepencies in the two sets of measurements. The commercial counter does not reflect the bimodal distribution until very near the end of the test sequence. Its data suggests that initially the predominent mode size is less than 0.5 micrometers and that its resulting probability density is in close agreement with the PSI. It also shows a significant size fraction in the size range of 0.5 to 1 micrometers which the PSI does not and it shows the particle number density decreases at a slower rate than that of the PSI. In attempting to resolve these discrepencies it might be argued that the apparent bimodal distribution observed by the PSI is in reality due to a decrease in particle size resolution predicted by other workers who have used different visibility functions from those used here.[5-7] Their work suggests that the smallest resolvable particle size with the signal processor used in this work would be of the order of 1.6 micrometers.

Figure 8 compares the size distributions estimated using appropriate visibility functions taken from Refs. 3 and 5. The visibility function chosen from Ref. 5 assumes a particle index-of-refraction of 1.5 with a zero absorption index. For the experiment the fringe period was approximately 6.6-μm and receiver F number was 4.8. The theoretical calculation from Ref. 5 was for a fringe period of 6.3-μm and a receiver f number of 5. Figure 8 shows that the visibility function from Ref. 5 does not adequately reconcile the differences in the data obtained with the PSI and the commercial counter. This visibility function shows modal sizes existing which are well beyond any estimated by the visibility function in Ref. 3 or the commercial counter. Furthermore, it also shows a small probability density where the commercial counter shows it is large. Thus, the differences in the size distributions measured by the PSI and commercial counter do not seem to stem from an uncertain visibility function. Numerous Hueristic arguments may be put forth to explain these discrepencies. For example, the different locations of the sample spaces for the two instruments, the fact that the commercial optical counter is calibrated in terms of latex spheres and hence the particle sizes determined by the optical counter are equivalent latex sphere sizes could all contribute to the discrepencies in the measurements. If the particles are irregularly shaped then the visibility function applied to the size determination will obviously be erroneous or can be interpreted as representing some kind of average size for the particle shape. A final, and probably most significant factor, could be large errors in the size distribution weighting factors which are used.

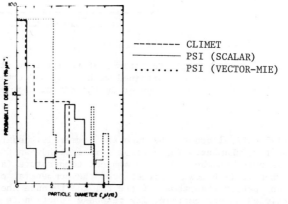

Fig. 8   Comparison of particle size histograms obtained using a visibility function derived from 1) scalar scattering theory, 2) Mie scattering theory, and 3) a commercial optical counter.

When the results from the PSI or the commercial counter are compared to those of the multiwavelength transmissometer there is considerable disagreement. Typical transmission values for visible wavelengths were between 60 and 80%. If it is assumed that all particles measured by the optical counter and the PSI are about 3 micrometers in diameter (the case for minimum transmission), then the estimated transmission would be greater than 93%. The difference between the measured transmission and that predicted from single particle counter measurements may be attributed to at least 2 factors. First, transmission estimates based on size distributions obtained with optical counter measurements have not accounted for the absorption cross-section of the particles which is an unknown factor for all three sets of measurements. If the absorption cross-section of the measured particles is significant then it follows that 1) the commercial counter would become less sensitive (its response function decreases) for all particle sizes greater than 0.5 micrometers,[8] 2) then if the PSI is resolution limited as one theory suggests, the large mode size measured by the PSI would shift form about 2.9 micrometers to about 4.3 micrometers and 3) the optical counter would show fewer measurements of particle sizes larger than about 5 micrometers. In view of the evidence a reasonable conclusion would seem to be that the absorption cross-section for the particles is large and that the PSI is not resolution limited for sizes less than 1.5 micrometers. A large absorption cross-section suggests particles which have a high metallic content which is consistent with sampling done for chemical analysis of the exhaust products. A second possible discrepency may result from high number densities (for particles less than 0.3 micrometers) which probably cannot be detected by the PSI or commercial optical counter but which can yield significantly lower transmissivities. These particles can result from condensation nuclei which may grow or decay according to a thermodynamic characteristic of the sample chamber (see Ref. 9). However, if the parameters for a log-normal size distribution are estimated from the multiple wavelength transmissometer measurements, the number densities which could be measured by the individual particle instruments still do not agree with those predicted by the distribution function. Furthermore, for the number densities which are estimated (10^5/cc for a diameter of about 0.04 micrometers) the signals generated by the PSI would probably show an incoherent scattered light background resulting from multiple particles simultaneously in the sample space. Real time observation of the PMT current with an oscilloscope showed no such background.

PSI measurements made in the edge of a small rocket plume were made during the optical system shown in Fig. 2. The size range covered by the PSI was roughly 0.5 to 10 micrometers. The sample space of the PSI system was focused at the edge of the plume such that the particle velocity measured was only 3 or 4 m/sec. Several different solid propellant formulations were examined. Each propellant was found to have a unique particle size histogram. The particle modal sizes ranged between 0.5 and 10 microns. During operation only approximately 0.3 seconds of the total rocket motor firing could be examined because of the limited data memory possible with the current system (900 measurements.)

SUMMARY AND CONCLUSIONS

The feasibility of using a PSI system to obtain insitu particle size distributions from rocket exhaust propellant has been demonstrated. Measurements have been made of propellant samples fired into a small mixing chamber and in the edge of small solid motor plumes. The PSI data show that the size distributions generated by these devices are strong functions of time indicating the need to be able to sample such distributions rapidly and to obtain the data in large quantities. The PSI measurements show a bimodal size distribution with mode sizes less than 0.5 and at about 3.0 micrometers for a rocket propellant

sample measured in the mixing chamber. Comparison of the PSI and commercial
optical counter data shows reasonable agreement in the probability density
function of sizes less than 0.5 micrometers and greater than 1 micrometer.
There is a discrepency between the measurements in the 0.5 to 1 micrometer
range. When the particle data from the PSI and commercial optical counter is
compared with that obtained with multiple wavelength transmissometers there is
considerable disagreement. Transmissometer data suggest that either the particle sizes are much smaller or number densities are much higher than detectable
with the single particle instruments for the measurable size ranges. The results show that no one particle size instrument can yet give convincing and un-
questionable measurements. The PSI data suggests that insitu measurements may
vary significantly from instruments which must withdraw a sample from the
original environment. Extreme caution must therefore be exercised in interpreting experimental results for theoretical predictions. These measurements have
clearly demonstrated that 1) secondary smoke measurements are possible with a
PSI system, that 2) number densities are not prohibitively high for this size
range and that 3) the rocket plume measurements show that the background light
is not significant for points of observation near the plume edge.

REFERENCES AND FOOTNOTES

1. See for example, M. Born, and E. Wolf, Principles of Optics, 3rd Revised
    Edition, Pergamon Press, New York, 1964, pp. 395-398.

2. W. M. Farmer, "Sample Space for Particle Size and Velocity Measuring
    Interferometers," Applied Optics, 15, 1984 (1976).

3. W. M. Farmer, "Measurement of Particle Size Number Density, and Velocity
    Using a Laser Interferometer," Applied Optics 11, 2603 (1972).

4. N. A. Fuchs, The Mechanics of Aerosols, Pergamon Press, New York, 1964,
    pp. 250-257.

5. W. P. Chu and D. M. Robinson, "Scattering From a Moving Spherical Particle
    by Two Crossed Coherent Plane Waves," Applied Optics, 16, 619, (1977).

6. D. W. Roberds, "Particle Sizing Interferometry," Applied Optics, 16, 1861
    (1977).

7. R. J. Adrian and K. L. Orloff, "Laser Anemometer Signals: Their Visibility
    Characteristics and Application to Particle Sizing," Applied Optics, 16,
    677, (1977).

8. D. D. Cooke and M. Kerker, "Response Calculations for Light-Scattering
    Aerosol Particle Counters," Applied Optics, 14, 734, (1975).

9. E. Miller, "The Dynamics of Secondary Smoke Generation in Smokeless Solid
    Propellant Plumes," 1975 JANNAF Propulsion Meeting, CP1A Publication 266,
    Paper Unclassified.

ACKNOWLEDGEMENTS

    Portions of the research reported here were supported under U.S. Army
Contract DAAK40-77-C-0123.

WILD CARD SESSION

Chairman: **F. DURST**
University of Karlsruhe

LDV Measurements on Propellers

J. P. SULLIVAN
Purdue University
West Lafayette, IN

ABSTRACT

LDV measurements of the fluid velocity around propeller blades provides sufficient information to calculate the overall performance parameters and the details of the blade loading distribution.

INTRODUCTION

A recent comparison (Ref. 1) of advanced propulsion systems for future subsonic transport aircraft shows that the use of high disk loading turboprop propulsion systems has the potential for a large decrease in fuel usage and reduction of DOC when compared to conventional turbofans. The potential gains are dependent on the successful development of high efficiency propellers at M = .8 cruise.

Initial analysis and experiments on a new series of propellers, Ref. 2 and 3, incorporated advanced aerodynamic concepts in an effort to improve propeller effeciency. The nacelle flow field was tailored and area ruled and the blades were thin and swept to suppress compressibility losses.

The current propeller research at Purdue is aimed at predicting and measuring the induced velocities and induced losses that occur on these heavily loaded swept propeller blades.

EXPERIMENTAL SET UP

The experimental program involves wind tunnel tests of a variety of propeller models (.3m diameter) with variations in sweep, loading, etc. A two-component LDV system is used to measure velocities around the propeller from which the induced velocities and power and thrust distribution on straight and swept blades can be found.

The tests were performed in the 1.07m by 1.37m test section of the Purdue subsonic wind tunnel. Oil seed particles (1 - 2 Micron diameter) generated by a standard mist lubrication unit were introduced at a point in the stilling section.

The LDV system employed is a conventional two-color dual scatter system. A computer controlled zoom lens allows remote positioning of the probe volume.

531

Figure 1. Swept Blade Propeller in Wind Tunnel

A TSI counter processor (MODEL 1990) is used to convert the LDV signal from the photomultiplier tube to an analog signal proportional to velocity. The analog signal is displayed on an oscilloscope and is input to a DEC 11/03 computer for digital processing.

Two sets of propeller blades have been constructed and tested in an eight bladed configuration. Both blades were designed with the same twist and chord distribution but one was straight and the other swept. The propeller is driven by a hydraulic motor mounted inside the nacelle. A photograph of the swept bladed propeller in the wind tunnel is shown in Figure 1.

RESULTS

Some examples of the axial velocity measured with the LDV system are shown in Figures 2,3 and 4. These oscilloscope traces of the axial velocity versus time as the propeller blades go by the probe volume are generated by the analog output of the LDV signal processor. Figure 2 is a measurement at

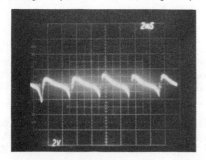 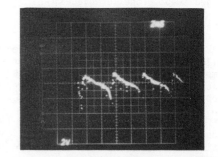

MULTIPLE SWEEP   SINGLE SWEEP

Figure 2. Axial Velocity Downstream of Propeller
Velocity Scale - 10.4m/sec/div

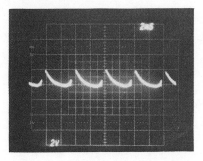
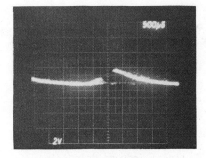

Figure 3. Axial Velocity Between Blades
Velocity Scale - 10.4m/sec/div

the 3/4 radius point and a small distance in front of the propeller blade.
Then sudden decrease in velocity occurs as the stagnation point of each blade
passes near the LDV probe volume. Figure 2a is an overlap of multiple oscilloscope traces and Figure 2b is a single trace at the same test conditions and
mean data rate of approximately 10KHZ. There is no data at the beginning of
the trace of Figure 2b because the stream tube containing the seed particles
has meandered out of the probe volume.

When the LDV probe volume is moved in between the blade row, the axial
velocity versus time contains a gap in the data when the blades hits the probe
volume as shown in Figure 3. It is evident that blade is lifting, since the
velocity on the upper surface is greater than on the lower surface. Downstream of the propeller, the viscous wake of the blade gives rise to a dip in
the velocity versus time curve as shown in Figure 4. It appears that with an
expansion of the time scale and appropriate averaging, the section drag coefficient can be obtained using a momentum deficit method. Methods for computer
processing the LDV signal are currently in progress.

The average velocities downstream of the propeller are shown in Figure
5. The axial velocity and the velocity components at ± 45° to the axial
directions were measured. From these measurements the swirl velocity, swirl
angle and power coefficient can be calculated.

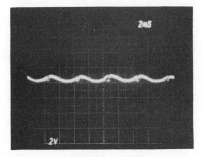
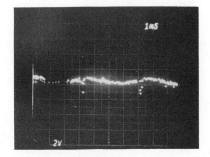

MULTIPLE SWEEP   SINGLE SWEEP
Figure 4. Axial Velocity Downstream of Propeller
Velocity Scale - 10.4m/sec/div

CONCLUSION

The initial LDV measurements at Purdue indicate that the following information is available from the LDV system:

1. Overall Performance - C_T, C_P
2. Swirl Angle Distribution
3. Blade Loading Distribution - dC_T/dx, dC_P/dx
4. Local C_ℓ and C_d
5. Interferance Effects-Spinner, Nacelle, Wing, Body
6. Shock Location
7. Local Induced Velocities
8. Wake Characteristics - Strength and Position of Vortex Sheet
9. Stall Pattern

ACKNOWLEDGEMENT

This work was supported by NASA Lewis Research Center.

REFERENCES

1. Neitzel, R. E., "Future Subsonic Transport Engine Technology Improvements and Resultant Propulsion Alternatives", Journal of Energy, Vol. 1, No. 3, May-June 1977.

2. Rohrbach, C. and Metzger, F. G., "The Prop-Fan - A New Look in Propulsors", AIAA Paper No. 75-1208.

3. Mikkelson, D. C., Blaha, B. J., Mitchell, G. A., Wikete, J. E., "Design and Performance of Energy Efficient Propellers for Mach 0.8 Cruise", NASA TMX-73612.

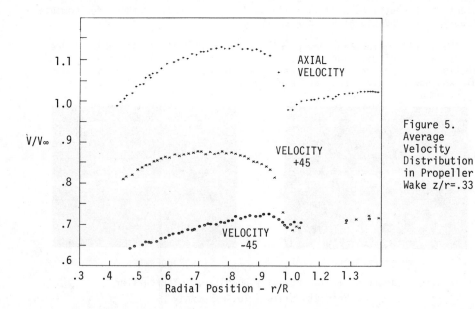

Figure 5. Average Velocity Distribution in Propeller Wake z/r=.33

Wild Card Discussion Session

Edited by
W. H. STEVENSON

The Wild Card Session, chaired by Franz Durst, was a wide open discussion on the general topic of three-dimensional measurements. John Sullivan began the session with his paper on "LDV Measurements on Propellors." This illustrated an area where the ability to make 3-D velocity measurements would be very useful. The session was then opened to comments from the floor relating to the participants' successes and failures in multi-component LDV measurements. We have not attempted to provide a complete transcript of the remarks made, but only to highlight important points and draw conclusions where appropriate.

W. Yanta discussed problems he was having with 2-D measurements of the supersonic flow around a projectile at a small angle of attack. A rather significant transverse (to the projectile) velocity component was being measured when the projectile was not spinning, even though this velocity should be zero for the flow conditions involved. No conclusions as to the source of this effect were reached, but if it is a real effect then that is useful information which would not be provided by a simple 1-D measurement. On the other hand, if it is an artifact of the measurement technique this would be worth knowing also.

Several individuals reported on successful 3-D measurements using various optical configurations including conventional two-color 2-D dual beam systems with a third dual beam system set up at right angles. This, of course, is extremely difficult to implement in a scanning system or where optical access is limited. Acquisition of the third (axial) component by a direct backscatter reference beam arrangement has been done successfully by R. Pike and others. This requires very careful optical design and the Doppler frequencies can become very high for axial velocities above a few meters per second.

Cross polarization rather than two colors has been used to separate the signals from perpendicular velocity components by some workers. This seems to work satisfactorily in most cases. However, J. Meyers noted that he had been unable to use such a system when Bragg cells were inserted for frequency shifting. In spite of all attempts to eliminate the problem serious crosstalk between the two channels were present. No conclusions as to the source of this difficulty could be reached.

Another approach to signal separation is to use a single color with a dual Bragg cell which puts a different frequency shift on each perpendicular component. This has the advantage of requiring only a single detector with suitable electronic filtering of the output to separate the signals. J. Meyers and others have used this method successfully. Meyers pointed out, however, that there are dynamic range problems if large velocity fluctuations exist, since the passbands for the two signals must be well separated to avoid crosstalk. This

535

problem would be accentuated in a 3-D system of this type.

W. Stevenson raised the question of the poor signal quality often obtained with the blue line (0.488 µm) of the Argon laser. This initiated a spirited discussion on the subject. The problem appears to arise from unwanted higher order transverse modes. J. Meyers indicated that he had cured the problem by careful optical alignment of the plasma tube bore--a procedure which was repeated at monthly intervals. P. Dimotakis stated that imperfections in the Brewester windows and dust on the windows seemed to be the sources of the difficulty. Very careful cleaning of the windows was recommended. (This also prevents permanently "burning in" surface contamination.) The issue was complicated further when A. Boutier noted that he had solved the blue line problem by replacing the laser mirrors.

Apparently there are several possible effects which will induce the higher order modes and degrade signal quality at 0.488 µm. This line is more sensitive to small defects in the optical cavity than the 0.5145 µm line and therefore great care must be taken to make the cavity as nearly perfect as possible. (Several colorful comments were made about the relative quality of various lasers. We will avoid the rather strong temptation to put them in writing.)

Some general discussion on various optical arrangements for obtaining 3-D measurements was held. There is obviously no completely satisfactory approach at present. The axial reference beam system, in spite of its problems, seems to offer as much hope as any. Some careful experimental studies of such a system under different conditions should be carried out to provide a base of information for design.

A final topic was the subject of multiplexing signals from the individual channels in a 3-D system to avoid the cost associated with three separate Doppler signal processors. The basic requirement is introduction of time delays on two of the three signals to separate them so that they can be fed sequentially to a single processor. Conventional delay elements will not provide a variable delay of sufficient magnitude. The idea of using surface acoustic wave (SAW) devices was suggested. These may have the necessary characteristics.

It is apparent that very little attention has been given to the problem of 3-D measurements. Even 2-D measurements are not routine, although progress is being made. There are areas where 3-D capability would be very desirable and hopefully some research on this topic will be conducted in the near future.

SESSION XI
PANEL DISCUSSION

Moderator: **H. D. THOMPSON**
Purdue University

Panel Discussion

Moderator:

H. DOYLE THOMPSON

Panel Members:

FRANZ DURST
HANS PFEIFER
WARREN STEVENSON

THOMPSON: We'll just go right ahead here. We'd like to continue on with some of our previous discussion and then get into some other problem areas. We'd like to make it so you can participate as much as possible. We're recording this session downstairs and are going to try to transcribe it, so please speak up or even come up and use this microphone for your comments. We've asked Franz Durst, Warren Stevenson and Hans Pfeifer to sit up front. It makes the panel official by having them up here. We'll just continue with the question we ended up with in terms of "Anything New in Signal Processing?" Do you have something you'd like to say Dr. Pfeifer?

PFEIFER: Warren asked me to stimulate discussion on signal processing and I want to mention four different points. Two of them are concerned with hardware and two with software. Many people, especially newcomers in the field, ask what is the best data processing system for their special application. I think there is no best data processing system. Basically we have frequency tracking, we have counting, and we have photon correlation. In some cases, of course, there is nearly no choice. For example, if we consider measurements in wind tunnels at very high speeds, if we have access to both sides of the wind tunnels, and if we have money enough to buy an argon laser, then, there is no choice--counters are the most suitable system. In other cases, for example, if you consider the backscatter mode over long distances, then probably photon correlation is the best way to get the information. There are some special data processing systems like transient recorders and digital processing which may be helpful if the signal-to-noise ratio is bad but not so bad that photon correlation must be used. All these types of data processing can be reliable and I, personally, don't expect very significant improvement in them. We heard that the photo correlator now has 10 nanosecond resolution. Dual counter systems are available with clocks up to 1000 megahertz. There is no improvement to be expected in frequency trackers and so on. Therefore, I don't expect any major innovations in this field.

On the second point, that is data reduction, I think we can expect some improvement by means of micro processors. You know, for example, photon correlators and counters are normally not stand alone instruments. In most cases, we must use a mini computer. This is not a problem in wind tunnel applications, for example, because there is nearly always a mini computer available. But in many cases we have just some raw data in the form of some bits and so I hope we can get very essential improvements in data reduction hardware in the near future.

On the side of the software, I feel that the bias problem is not yet solved. The bias problem, of course, is not only a problem of data processing, it is also a problem of data reduction. I think work must still continue in this field and

539

my personal feeling is that there will not be a complete solution which is applicable to all the different types of flows.

As a final point, which also concerns software, we can still expect some improvement here in data reduction, in the areas of auto-correlation, power spectra, and cross-correlation. I'm a little surprised that no one talked about cross-correlation, for example, between velocities and other physical quantities such as the velocities in a jet and the acoustic sound waves outside. Certainly, we can look for innovations here. I think this should be enough to stimulate discussion and I will stop here.

THOMPSON: Comments?

YULE: On the subject of cross-correlation, I might mention that at Sheffield we're just starting correlations of temperature and velocity - the spatial correlations and also there has been work on pressure and velocity correlations in jets and flames. There are no special problems involved, it's just the matter of data reduction.

THOMPSON: How do you make your pressure measurements?

YULE: Streamline microphones. We simply take all the data through a multiplexer, put it on disc and process it. This is a counter system. Each time we have a velocity signal, we correlate relative to that signal in time.

THOMPSON: What is the spatial resolution of your microphone?

YULE: It's not very good. That is a problem. It's about 0.7 cm.

KUMASAKA: How do you know that you are not just correlating noise? We find that wind blowing over the microphone generates its own self noise.

YULE: The experiment was undertaken in a cold jet which is being forced by a large speaker, so we have quite a strong pressure field.

PIKE: On the subject of cross-correlating variables, there has been some nice work at the Gas Board in Birmingham in England where they correlated the Raman scattered light from a methane air jet with local velocity. In this case they are looking at concentration velocity correlations. I would like to say that Hans Pfeifer is very provocative by saying that there's no choice but a counter in forward scattering. (Laughter). After all the horror stories we have heard about counters, I'm surprised he can say that. I would say there was no choice either, but not that direction.

DIMOTAKIS: In your comment about the work at Birmingham, do you mean time resolved Raman scattering and velocity correlation? You can't mean that surely. What do you mean by correlation?

PIKE: What they measure is the intensity of Raman scattered light as a function of time to give a concentration versus time history of methane in the methane-air mixture.

DIMOTAKIS: What laser are they using?

PIKE: They are using something like a 10 watt argon. It's true. It's published in Jour. of Phys. D about six months or a year ago.

SELF: I just mention that this calls to mind that Charles Kruger and his student have a nice way of measuring point temperature in high temperature flames by laser fluorescence using a dye laser, exciting one sodium line and looking at the light emitted on the other. This looks like a very nice way of getting temperature fluctuations. They haven't done any correlation with anything else yet, but that obviously is much easier than Raman. You do have to have sodium or something in the flow.

PIKE: I'd express the opinion that this is a forward looking area, i.e., a cross-correlation of various variables with velocity, and it certainly has to grow as the field matures.

BOUTIER: I want to mention that there is a team at ONERA working in the area of jet noise studies. They are using several methods - laser velocimetry as well as infrared measurement of temperature and concentration. They have correlated these two measurements very recently and have obtained good correlation between velocities, temperatures, and densities through the jet. Also they made fluorescence measurements through the jets from combustion products and obtained good correlation between velocity and the frozen properties.

PIKE: Just a word about this, there's a whole field of activity which we now call fluorescence correlation spectroscopy. We have meetings on it.

DURST: I would like to briefly comment on the counter-correlator question. I would also have not made the distinction as strongly as Hans made it, but there's one thing to say, fast time resolved measurements have not been possible so far with correlators. There you need a counter.

PIKE: That's indeed why we have developed this new system.

WARSHAW: We haven't correlated anything yet, but we've used pulse dye laser operating in the blue - one joule, one microsecond with about an angstrom line-width - to give us a 5% $\Delta T/T$ measurement in a hydrogen-air flame. That's with a single shot, so that's essentially a point measurement in time and space. The beam is about 1 cm in cross section and about 4 or 5 times diffraction limited.

DURST: May I possibly switch you over to particles? I think equally important are combined particle-velocity measurements. I know that fluid mechanics people have advanced the field of laser Doppler anemometry tremendously by measuring particle velocities and using that information as fluid velocity, if the particles have been chosen to be small enough. There is, however, an entire field of people interested in just the particle velocities. I think that combined fluid velocity - particle velocity measurements are significant and should be made.

THOMPSON: I think another area that needs to be looked at is particle density distribution. When you start to look at the biases present in a turbulent flow, if you don't know the size and distribution of the particles, you have a very difficult time sorting out the data. Many of the analytical studies made to determine biases assume uniform particle distribution and uniform size. Even if you start out with a uniform seeding its very likely that in a turbulent flow you will end up with a non-uniform distribution. I'm not sure that count rate actually reflects actual particle density distribution.

DIMOTAKIS: There are two things I'd like to address. One is in response to Dr. Pfeifer's comment on the bias. I really think the sampling bias that results from the velocity can be handled. The one sampling bias we cannot handle is the one, I suspect, that gave me trouble in the Reynolds stress measurements where there are correlations between number densities and some measured quantities. That is much more insidious and much harder to handle. The second point I would like to made is that I believe a grid system for velocity measurement is very important in turbulence. Our outlook on turbulence is changing and the view we have at Cal-Tech is that turbulence is really composed of fairly large flow patterns, large structures if you will, and the flow appears to be described fairly well by the kinematic interaction of these large vorticies. The small fluctuations represent just the hair on the animal as Don Coles says and they really amount to very little of what's going on. If you remove this hair and compute stresses, for example, you'll find that you have accounted for something like 80 or 85 percent of what's happening. Now if you consider a point measurement, it's easy to see why these animals have not been observed in the past. What you need is to know what the whole velocity field is like at any one time. If you are measuring at any one point, you will be blind to these structures unless the phenomenon is completely periodic or quasiperiodic as it is in the wake of a cylinder. We have always considered this a singular case until now, but I think it is not. It's just one end of the spectrum. The next case is the two dimensional shear layer which is a little bit less organized but still quite organized. I think research on turbulent boundary layers is identifying the structures. I have begun some work on a round jet in the far field, not the near region, and I'm beginning to identify flow structures again. In all those cases, it has been necessary to make a field measurement. The point measurement is quite inadequate to reveal the structure and I think we are fooling ourselves if we continue pressing the point measurement. All the spectra and auto correlations and moments and skewness and kurtosis factors are not going to give you a thing. Now I know what it means to design a multi-point system when we all have troubles with one point, but I really think that until we come to that progress in turbulence, we will have to wait.

PFEIFER: I agree with that statement. We found nearly the same thing during the study of free jets. There's not a Gaussian turbulence but there are very large coherent structures in such a flow. These structures, of course, can only be detected if there is more than one measurement point.

PIKE: I would agree that, of course, the one feature of turbulent flow fields is that they are multi-scale phenomena. One of the prime features of a turbulent flow field is the number of scales of length involved. There is a whole field of activity which is currently very popular and has been for many years of propagating laser beams through turbulence. We are not looking at point velocities in this case, but at the scintillation of the beam as it gets through a turbulent field. Recently, quite a lot of new discoveries about turbulence have been made this way in our laboratories. I don't have time to discuss anything in detail here, but perhaps I could refer you to a recent Physical Review Letter by Dr. Jakeman of our laboratories on the significance of K - distributions in turbulence which I think some of you may find quite interesting.

PFEIFER: May I go back to the particles. I want to ask a question. We made some theoretical calculation on backscatter light intensity from small particles and found, for example, that for particles with a diameter of 10 microns (I don't remember the refractive index) that a change in diameter of 10^{-3} may change that backscatter intensity by one order of magnitude. I would like to know how it is possible to determine particle diameter by means of backscatter light in this case as has been inferred in some of the papers we have heard.

SELF: That is the nature of Mie scattering. In backscatter, you get enormous excursions in the amplitude with very small changes of particle size and refractive index. That's why I think the best hope for particle scattering is to stay in the forward diffraction lobe which is the only part of the Mie scattering that can be easily characterized.

STEVENSON: I agree with that statement. Unfortunately none of the authors are here at this point, but there was some hint in papers presented today that one could use the fringe visibility method successfully in backscatter which we've all assumed was impossible or at least ridiculous. Of course visibility is not intensity dependent directly which may offer some hope. I'm not convinced yet, but at least some people are thinking there may be a solution.

DURST: I would not go so far as to say one should only work with the forward lobe. If you do your Mie calculations and this strong dependence shows up in your case then you obviously wouldn't go off axis. However, if your calculation shows that your instrument behaves sensibly in a certain direction then I think you should take advantage of it.

WITTIG: You have a chance to eliminate this problem, at least to a certain degree, if you take the ratio. You are then independent of the intensity.

STEVENSON: That really doesn't work in backscatter.

WITTIG: It does under certain conditions for certain indices of refraction. But it does depend on the particles obviously, so you have to know what you have.

PIKE: A feature of Mie scattering which I don't know if everyone knows is that it still goes up as r^6 even though you're in the Mie region when you go to the forward direction. I'm not saying it's exactly r^6. I'm saying you can go through the Rayleigh region up to, say, 0.2 micron and it then carries on as a 6th power more or less well into the Mie region. Therefore, if you want a rough estimate of what the scattering is in the forward direction you can still use the Rayleigh theory. You don't have to do a Mie calculation.

STEVENSON: I would like to change the subject if I could. At the previous Purdue meetings there were always one or two things that generated controversy and did not really get resolved. This seems to have occurred again. Obviously there are still some questions regarding biasing. Not only velocity biasing but the others need to be looked into so that we can reach a definitive understanding of the problem and not just leave it floating around. Also there are still many questions in regard to the two spot system. Is it or is it not better in given situations. A lot of evidence was presented here to indicate that it is better. For example, you can get closer to walls than with a conventional LDV system. But I haven't seen enough measurements yet to convince me. If I had to make a choice to use one of the two instruments, particularly in a highly turbulent flow, I don't think I know enough yet to make that choice. Finally, I think particle sizing is still an area where there is, if not controversy, then certainly an opportunity for doing more work to establish the true merit of the many competing techniques. We should be directing our efforts toward resolving such issues.

THOMPSON: Joe Humphrey couldn't stay for this session, but he wanted me to make some comments on engine flows. He feels that research on "real" engines is necessary and commendable, but integrated parallel efforts need to be taken on simpler but more controllable and understandable configurations. The second point is the question of whether or not particles correctly track the flow and how to handle any errors in the calculations. Of course, his interest is in engine measurements, but the question of how far we go toward coordinating our efforts in difficult environments is more general.

YEOMAN: In the U.K., we do have something approaching an integrated effort on engine research and we have an "engine club" comprising all the major automotive manufacturers. They have contributed simple engines which they've worked on for many years using different methods. These are research engines, not production engines, and have been looked at for the last 18 months with LDV techniques. They are pushing us always to work more on production engines, but we do have this parallel program going where we look at these simpler experimental engines and at the same time try to satisfy their requirements by doing contract work on production engines.

ADRIAN: I just wanted to add one thing to Warren's list that I am surprised hasn't been brought up in a conference on measurements in hostile environments. That is the effect of refractive index fluctuations in non-isothermal flows. Certainly in internal combustion engines you have large temperature gradients and consequently large refractive index gradients. These are serious effects that are not easily diagnosed because one can actually see signals that look fine, but give completely wrong velocities.

STEVENSON: Ron, do you have some experimental evidence in regard to that?

ADRIAN: Yes, I can cite two examples. I can't put firm numbers on it, but one case is our experience in measuring thermal convection in water where the temperature fluctuations are only about 0.1 to 0.2 °C. We can see errors that are perhaps 10 to 20 percent of the turbulence intensity.

STEVENSON: That's the error in the measured turbulence intensity, not the measured mean velocity?

ADRIAN: That's right. For the mean velocity, if your beams happen to be fluctuating in all directions equally you can possibly come out with the right result. But that would again depend on the flow. You can also get the wrong mean velocity. I'll quote a study on measurements in flame tubes done by Dave Rice in our Aero Department where his mean velocity measurements always showed a serious defect just behind the propagating flame-front. He could correlate that very strongly with the deflection of the beams due to the refractive index gradient. He's just writing up the thesis now.

PIKE: We've had experience with trouble of this sort in supersonic tunnels looking behind shock fronts. There you don't get the wrong velocity but a very lousy signal because of the beam shaking and wandering about. In our work on the normal shock boundary layer intereaction there are holes in our grid of data points near the shock wave due to this effect. The only way we find to help the situation is to attack the point we want from another angle so that the beams will be going through a different part of the shock wave structure.

DURST: There have recently been some critical comments by Bradshaw on data obtained by laser anemometry. These are correlation data obtained at Cal-Tech. It is also claimed that the turbulence intensities are too high because of the beam fluctuations. We are now taking data in a similar water flow and we controlled the two beams by photodiodes. Only when the photodiodes are at a high level do we take data. The beams might still do something funny, but I think that the probability of bad data has been reduced.

DIMOTAKIS: I'd like to comment on the index of refraction fluctuation problem that has been brought up. We are constructing a blowdown combustion facility that will operate up to sonic speeds where we plan to burn hydrogen and fluorine initially because they are hypergolic chemicals and we don't have flame stability problems that way. The temperatures will be up to 2000° K in the reaction zone with the concentrations we plan to use. You can improve things somewhat by using helium as a diluent because it has an index of refraction that's much closer to vacuum to start with. Also, we will keep path lengths as short as possible by coming in vertically insteady of perpendicularly. This will give some loss in spatial resolution because the long axis of the focal volume is in the wrong direction, but if you have 2000° K, you take what you get. I've estimated that the combination of the short path lengths and the helium diluent is going to solve this problem. We've made measurements in heated water flows and there we found, with the processor I have, that the data rate drops by two or three orders or magnitude. It just will not produce acceptable data when the beams wander. We don't know what to do in water, but in the high temperature gas flows the helium idea really works very well, at least on paper.

SELF: Is the drop out due to the beams not crossing or is it due to not looking at the volume your're illuminating?

DIMOTAKIS: I think it's both.

SELF: Backscatter helps.

DIMOTAKIS: That's right. The other thing I'd like to comment on is partly suggested by what Warren said. I think it's really important for the LDV community to police itself sufficiently. When somebody makes a measurement with an LDV that disagrees by a factor of two with a pitot tube when it should agree, there is nobody in the world who is in a position to check what the problem is besides the person who did that measurement. If that result gets in print, unpoliced, we are compounding a very serious problem that exists regarding the credibility of laser Doppler measurements. In the outside world, the people who do fluid mechanics and aerodynamics research are looking for results. They are not interested in bias or sampling or counters and trackers. They see the shear stress distribution and it's half of what it should be. They don't care why it is, they say that's just garbage. I happened to pick one of those problems and it cost me the better part of three years of my life to find out what was going on. I really feel we have a responsibility here. You shouldn't just walk along and leave such measurements in the literature, but this is happening. You speak to many people, especially those who do not have someone they trust in their own laboratory doing laser velocimetry, and they just won't look at the data at all. I think it's a shame, but unfortunately it's almost justified and I think it is a serious problem. I share Warren's frustration that so many years after these discussions start, issues such as the biasing problem have not been resolved. You see at this conference a spectrum of opinions - people at one end who say it's a big problem and at the other end those who say it doesn't even exist. You would think that after so many years we would have converged to some consensus.

WITTIG: I am one of those people who just wants to make measurements in engines. But if you look at comparable techniques that have been around for a long time, take a pitot tube or whatever, they still have their problems too. I'm completely in agreement with what you say, but on the other hand, we shouldn't condemn the LDV as being unique in this respect.

DURST: This was actually the point I was trying to make in my talk, that the instrument people should now go and work together with the fluids people to avoid these problems you have just indicated. A fluid mechanics expert will make certain checks automatically, such as checking mass conservation in a flow, whereas an instrumentation man will have faith in his instrument, which he should have, and will not cross check the results. I think that such cross checking is necessary.

WITTIG: Coming back to the two-spot system, do you think this has an advantage and if so, when?

STEVENSON: This is one thing I am interested in because it appears to me that a dual focus system has more problems in, say, a combusting flow than a laser velocimeter, because the beam will be defocused plus the spacing may change considerably. You can't really control the data you're getting as well as you can with a laser velocimeter. That's just a gut feeling, however.

FRANKE: But you have just as many problems in the LDV case because of gradients in the fringe pattern. We took an LDV up on a hot roof and looked at the fringe pattern and you could see it move erratically.

STEVENSON: I'm not sure this is quite as serious because the particles move through the probe volume much faster than you can visually observe fringe motion.

PIKE: Isn't the point that you will get an answer from the two spot which will be wrong, whereas with fringes the data will disappear?

FRANKE: No, you'll get an answer, but the beam wander gives you a lot of apparent turbulence on your signal.

STEVENSON: What was the path length?

FRANKE: Thirty meters.

STEVENSON: Okay, I would expect even worse problems with a two-spot system over that distance.

PIKE: Could I make another comment on this particular shopping list of Warren's? I think with this new combined optics instrument that I described earlier in the conference where you can switch from fringes to dual focus if you want, we're beginning to get a feel for this problem. I can certainly stick my neck out a little bit and say that yes, you can measure closer to surfaces with a two spot system. I think there's no question about that. But I also think there's no question that you need more seeding and there's very little question that when you get into degrees of turbulence above 15% you lose your signal.

MOORE: We have found that the two-spot system has a significant advantage in that it gives you the flow angle accurately and can be used at higher velocities. So in fast flows and in things like turbomachinery it has an advantage. In turbulence I would agree that it doesn't have an advantage.

DURST: I think the advantage in turbomachinery still has to be proved. Let me give you the reasons why that instrument has been so successful in turbomachinery. The development work was done at DFVLR by Schodl who had the problem of making measurements in turbomachinery. Roy Pike works in wind velocity measurements and can easily get wind velocity measurements with his instrumentation. If you give your instrument to somebody else, he may not be able to make those measurements. I'm not saying the two-spot system is not best for turbomachinery work, but to have it used in just one situation by one person is not really a general proof of its superiority. May I add that we have tried the dual-focus method in separated flows in water where we had 100% turbulence in some regions and it did not work. Just as Roy indicated, at high turbulence intensity you get no data. Also, the data we did obtain took something like ten times the time it took with a normal laser anemometer system. In regard to the point of making measurements close to walls we usually track in a water flow straight into our walls with an LDV system. The tracker goes quickly bezerk just before we reach the wall, but you can see the profile going into the Bragg shift frequency at the wall, so we do not see the advantage which some people claim.

MOORE: I would like to show some measurments we made in a compressor with a two-spot system. We made point measurements and from the array of points established flow contours. The flow angles are accurate to about half a degree and the velocities to about one percent. We are looking initially with holography at this region and then going back to fill in the details. There are

25 contours here which required about seven hours of running time. Another application is in jet engine studies. This is the system Tony Smart mentioned yesterday. We made measurements downstream of the nozzle to get velocity contours. You can see that the velocity is around 900 meters/second. The measurements were made with a 20 milliwatt helium-neon laser. So in these two situations the two-spot system has performed very well indeed.

THOMPSON: I think we must adjourn. I feel this workshop, and particularly this panel discussion, have been very effective in defining the state-of-the-art and the remaining problem areas. We thank you all for your contributions.

Workshop Participants

ADRIAN, Ronald J.
Department of Theoretical and
 Applied Mechanics
University of Illinois
Urbana, IL 61801

ADAMS, Tim G.
Ford Motor Company
Room E2166, Science Research
 Laboratory
P.O. Box 2053
Dearborn, MI 48121

AGARWAL, Jugal K.
TSI Inc.
P.O. Box 3394
St. Paul, MN 55165

ALLAN, Joseph
AiResearch Mfg. Company
111 So. 34th St.
Phoenix, AZ 85010

ARTMAN, Joseph O.
Carnegie-Mellon University
Electrical Engineering Department
Pittsburgh, PA 15213

BACHALO, William D.
Spectron Dev. Labs
3303 Harbor Blvd., G-3
Costa Mesa, CA 92626

BAJURA, Richard A.
Mechanical Engineering Department
West Virginia University
Morgantown, WV 26506

BALLAL, Dilip R.
School of Mechanical Engineering
Purdue University
West Lafayette, IN 47907

BECHTEL, James H.
General Motors Research Labs
Warren, MI 48090

BECKER, Roger J.
University of Dayton
Dayton, OH 45469

BENNETT, Don
NASA - DFRC
Box 273
Edwards, CA 93523

BENNETT, John C.
United Technologies Research Center
M.S. 16
East Hartford, CT 06238

BOGARD, David G.
Oklahoma State University
Stillwater, OK 74074

BOUTIER, A. R.
ONERA
29 Avenue de la Division
Leclerc 92320 Chatillon, France

BREMMER, Robin J.
TSPC
Purdue University
West Lafayette, IN 47907

BROCKMANN, Eugen
Lehrst. f. Techn. Thermodynamik
RWTH AACHEN
51 Aachen, West Germany

BROWN, Gordon M.
Ford Motor Company
Room S1023, SRL
P.O. Box 2053
Dearborn, MI 48121

BUCHHAVE, Preben
DISA Electronics
555 Charlesgate Circle
East Amherst, NY 14051

CHENG, Robert K.
Lawrence Berkeley Lab, 70-158
University of California, Berkeley
Berkeley, CA 97704

CHEN, Chi
School of Mechanical Engineering
Purdue University
West Lafayette, IN 47907

COLLINS, Daniel J.
Naval Postgraduate School
22479 Ferdinand Dr.
Salinas, CA 93908

CRAIG, Roger R.
AFAPL/RJT
Wright-Patterson Air Force Base
OH 45433

CROSSWY, Frank L.
ARO Inc.
Arnold Air Force Station, TN 37389

DEVLIN, J. Frank
Boeing Vertol
P.O. Box 16858
Philadelphia, PA 19142

DIMOTAKIS, Paul E.
CALTECH 301-46
Pasadena, CA 91125

DRISCOLL, James F.
209 Aerospace Engineering Building
University of Michigan
Ann Arbor, MI 48109

DURST, Franz
Universitaet Karlsruhe
12 Kaiserstrasse
Karlsruhe 1, Germany 7500

EDWARDS, Robert V.
Chemical Engineering Department
Case Western Reserve University
Cleveland, OH 44106

FARMER, W. Michael
University of Tennessee Space
   Institute
Tullahoma, TN 37388

FIGLIOLA, Richard S.
Aerospace Lab
University of Notre Dame
Notre Dame, IN 46556

FINGERSON, Leroy M.
TSI, Inc.
500 Cardigan Rd.
St. Paul, MN 55165

FLACK, Ronald
Department of Mechanical Engineering
University of Virginia
Charlottesville, VA 22901

FRANKE, John M.
NASA - Langley
MS 235A
Hampton, VA 23665

GIEL, Thomas V., Jr.
ARO, Inc., ETF/TAB
Arnold Air Force Station, TN 37389

GOLDSCHMIDT, Victor
School of Mechanical Engineering
Purdue University
West Lafayette, IN 47907

GOUESBET, G.
Faculte des Sciences de Rouen
Labo. Thermodynamique
76130 Mt-St-Aignan, France

GRAY, Donald D.
School of Civil Engineering
Purdue University
West Lafayette, IN 47907

GUPTA, Ashwani K.
Massachusetts Institute of Technology
MIT 66-038
Cambridge, MA 02139

HARWELL, Kenneth E.
University of Tennessee
Tullahoma, TN 37388

HAVIR, Darrell D.
TSI Inc.
P.O. Box 3394
St. Paul, MN 55165

HUANG, Thomas T.
David Taylor Naval Ship R & D Center
Bethesda, MD 20084

HUMPHREY, Joseph A.C.
University of California, Berkeley
Berkeley, CA 94720

JEFFERS, Larry A.
Babcock and Wilcox Company
1562 Beeson St.
Alliance, OH 44601

KENT, J. Christopher
Ford Motor Company
5561 Lakeview Dr.
Bloomfield Hills, MI 48013

KILAND, Ralph
TSI, Inc.
P.O. Box 3394
St. Paul, MN 55164

KOMAR, James J.
University of Illinois
4260 Dixon Dr.
Hoffman Estates, IL 60195

KUMASAKA, Henry A.
Boeing
P.O. Box 3707, M.S. 1W-03
Seattle, WA 98124

LABBE, Jean
ONERA
29 Avenue de la Division
Leclerc 92320 Chatillon, France

LADING, Lars
Riso National Laboratories
DK-4000 Roskilde, Demark

LECONG, Phung
Electrical Engineering Building
Purdue University
West Lafayette, IN 47907

LEFEBVRE, Arthur H.
Head, School of Mechanical Engineering
Purdue University
West Lafayette, IN 47907

LIGHTMAN, Allan J.
University of Dayton
300 College Park
Dayton, OH 45469

LOMAS, Charles G.
DISA Electronics
779 Susquehanna Avenue
Franklin Lakes, NJ 07417

LONG, Marshall B.
Yale University
15 Prospect St.
New Haven, CT 06520

MATEKUNAS, Frederic A.
General Motors Research
12 Mile & Mound Rd.
Troy, MI 48084

MAYO, William T., Jr.
Spectron Dev. Labs.
3303 Harbor Blvd., G-3
Costa Mesa, CA 92626

MAZUMDER, Malay K.
University of Arkansas
1201 McAlmont
P.O. Box 3017
Little Rock, AR 72207

McVEY Ray E.
TSPC
Purdue University
West Lafayette, IN 47907

MELLOR, A. M.
School of Mechanical Engineering
Purdue University
West Lafayette, IN 47907

MEYERS, James F.
NASA Langley
M.S. 235A
Hampton, VA 23665

MOORE, Christopher J.
Rolls Royce Ltd.
P.O. Box 31
Derby, England

MURTHY, S.N.B.
School of Mechanical Engineering
Purdue University
West Lafayette, IN 47907

NIEMOELLER, Donald E.
Arvin Automotive
2505 N. Salisbury
West Lafayette, IN 47906

NORSWORTHY, Keith H.
Boeing
10906 NE 17th
Bellevue, WA 98006

OGDEN, Donald M.
Department of Mechanical Engineering
Washington State University
Pullman, WA 99163

OWEN, F. Kevin
Consultant
P.O. Box 1697
Palo Alto, CA 94302

PIERCE, Felix J.
Virginia Polytechnic Institute
  and State University
Blacksburg, VA 24060

PIKE, Edward Roy
RSRE
St. Andrews Rd.
Malvern, Worcs. U.K.

PFEIFER, Hans J.
German-French Research Institute
Rue de l'Industrie
F68 St. Louis, France

POWELL, J. Anthony
NASA Lewis Research Center
21000 Brookpark Road
Cleveland, OH 44135

RAMBACH, Glenn D.
Sandia Laboratories
Division 8353
Livermore, CA 94550

RASK, Rodney B.
General Motors Research Center
12 Mile & Mound Rd
Warren, MI 48090

RAZINSKY, Eli H.
General Motors Research Center
12 Mile & Mound Rds.
Warren, MI 48090

REGAN, J.
TSI, Inc.
P.O. Box 3394
St. Paul, MN 55165

ROESLER, Timothy
TSPC
Purdue University
West Lafayette, IN 47907

SALMON, Joseph T., Jr.
McDonnell Douglas
377-F Chapel Ridge
Hazelwood, MO 63042

SANTAVICCA, Domenic A.
D234 Engineering Quad
Princeton University
Princeton, NJ 08540

SANTELLI, Nicholas
Naval Ship R & D Center
Code 1552
Bethesda, MD 20084

SANTORO, Robert J.
D315 Engineering Quad
Princeton University
Princeton, NJ 08540

SCHMITT, Randal L.
TSPC
Purdue University
West Lafayette, IN 47907

SCHWARTZ, Fred A.
University of Tennessee Space Institute
UTSI Access Rd.
Tullahoma, TN 37388

SEASHOLTZ, Richard G.
NASA Lewis Research Center
21000 Brookpark Road
Cleveland, OH 44135

SEFFERN, John J.
Purdue University
West Lafayette, IN 47907

SELF, Sidney A.
Mechanical Engineering Department
Stanford University
Stanford, CA 94305

SERAFINI, John S.
NASA Lewis Research Center
21000 Brookpart Road
Cleveland, OH 44135

SIMPSON, Roger L.
Southern Methodist University
Dallas, TX 75275

SKIFSTAD, J. G.
School of Mechanical Engineering
Purdue University
West Lafayette, IN 47907

SMART, Anthony E.
Spectron Development Labs., Inc.
3303 Harbor Blvd., G-3
Costa Mesa, CA 92626

SOREIDE, David C.
Boeing
P.O. Box 3707
M/SIW-82
Seattle, WA 98124

STEVENSON, Warren H.
School of Mechanical Engineering
Purdue University
West Lafayette, IN 47907

STOCK, David E.
Mechanical Engineering Department
Washington State University
Pullman, WA 99164

STONE, Jack P.
Code 6180
Naval Research Laboratory
4555 Overlook Avenue
Washington, D.C. 20375

SUHOKE, Robert
DISA Electronics
779 Susquehanna Avenue
Franklin Lakes, NJ 07417

SULLIVAN, John P.
School of Aeronautics and Astronautics
Purdue University
West Lafayette, IN 47907

SWOPE, Richard D.
Widenen College
17th and Walnut
Chester, PA 19013

THOMPSON, H. Doyle
School of Mechanical Engineering
Purdue University
West Lafayette, IN 47907

TIEDERMAN, William G.
School of MAE: EN218
Oklahoma State University
Stillwater, OK 74074

VELJI, Amin H.
Technical University, Aachen
51 Aachen, West Germany

VISKANTA, Raymond
School of Mechanical Engineering
Purdue University
West Lafayette, IN 47907

WALKER, Curtis L.
USARTL (AVRADCOM)
Propulsion Laboratory
21000 Brookpark Road
Cleveland, OH 44135

WANG, Charles P.
Aerospace Corporation
2350 El Segundo Blvd.
El Segundo, CA 90045

WARSHAW, Sandy
General Electric R & D Center
4B34/K1
P.O. Box 8
Schenectady, NY 12345

WEETMAN, Ronald J.
Mixing Equipment Company
135 Mt. Read Blvd.
Rochester, NY 14611

WEINER, Bruce B.
Malvern Scientific Corporation
7 Holiday Park Drive
Centereach, NY 11720

WHIFFEN, M. Clay
Lockheed Georgia Company
D72-97 2403
Marietta, GA 30063

WILLIAMS, M. Carlson
Pratt & Whitney Aircraft
400 Main St.
East Hartford, CT 06108

WITTIG, Sigmar L. K.
Institut fur Thermische,
Stromungsmaschinen
Universitat Karlsruhe
D 7500 Karlsruhe, West Germany

WITZE, Peter O.
Sandia Labs.
P.O. Box 969
Livermore, CA 94550

WOODWARD, Roger P.
Polytec Optronics Inc.
7118 Fulton St.
San Francisco, CA 94121

WYATT, William R.
TSPC
Purdue University
West Lafayette, IN 47907

YANTA, William J.
NSWC
White Oak
Silver Spring, MD   20910

YEOMAN, Michael L.
UKAEA
AERE Harwell B525
Didcot, Oxon, England

YULE, Andrew J.
Department of Chemical Engineering
  and Fuel Technology
University of Sheffield
Mappin Street
Sheffield, S13 JD, England

ZIMMERMAN, Donald R.
Detroit Diesel Allison
P.O. Box 894, W-16
Indianapolis, IN   46206

Index